Molecular Host Plant Resistance to Pests

BOOKS IN SOILS, PLANTS, AND THE ENVIRONMENT

Editorial Board

Soil Biochemistry, Volume 1, edited by A. D. McLaren and G. H. Peterson
Soil Biochemistry, Volume 2, edited by A. D. McLaren and J. Skujiņš
Soil Biochemistry, Volume 3, edited by E. A. Paul and A. D. McLaren
Soil Biochemistry, Volume 4, edited by E. A. Paul and A. D. McLaren
Soil Biochemistry, Volume 5, edited by E. A. Paul and J. N. Ladd
Soil Biochemistry, Volume 6, edited by Jean-Marc Bollag and G. Stotzky
Soil Biochemistry, Volume 7, edited by G. Stotzky and Jean-Marc Bollag
Soil Biochemistry, Volume 8, edited by Jean-Marc Bollag and G. Stotzky
Soil Biochemistry, Volume 9, edited by G. Stotzky and Jean-Marc Bollag
Soil Biochemistry, Volume 10, edited by Jean-Marc Bollag and G. Stotzky

Organic Chemicals in the Soil Environment, Volumes 1 and 2, edited by C. A. I. Goring and J. W. Hamaker
Humic Substances in the Environment, M. Schnitzer and S. U. Khan
Microbial Life in the Soil: An Introduction, T. Hattori
Principles of Soil Chemistry, Kim H. Tan
Soil Analysis: Instrumental Techniques and Related Procedures, edited by Keith A. Smith
Soil Reclamation Processes: Microbiological Analyses and Applications, edited by Robert L. Tate III and Donald A. Klein

Molecular Host Plant Resistance to Pests

S. Sadasivam
B. Thayumanavan
Tamil Nadu Agricultural University
Coimbatore, India

CRC Press
Taylor & Francis Group
Boca Raton London New York

CRC Press is an imprint of the
Taylor & Francis Group, an **informa** business

First published 2003 by Marcel Dekker, Inc.

Published 2019 by CRC Press
Taylor & Francis Group
6000 Broken Sound Parkway NW, Suite 300
Boca Raton, FL 33487-2742

© 2003 by Taylor & Francis Group, LLC
CRC Press is an imprint of Taylor & Francis Group, an Informa business

First issued in paperback 2019

No claim to original U.S. Government works

ISBN 13: 978-0-367-44671-0 (pbk)
ISBN 13: 978-0-8247-0990-7 (hbk)

**Visit the Taylor & Francis Web site at
http://www.taylorandfrancis.com**

**and the CRC Press Web site at
http://www.crcpress.com**

Library of Congress Cataloging-in-Publication Data
A catalog record for this book is available from the Library of Congress.

Preface

The ever-increasing global population necessitates continued effort to increase food production. Crop yield has to be increased and the loss due to pests and diseases must be minimized. The insect pests, apart from damaging the crops directly, cause various diseases indirectly since they act as carriers of disease. So it becomes imperative to minimize the loss caused by insects. To overcome the loss, farmers are advised to use pesticides which in turn increases the cost of cultivation and further causes environmental problems. Alternatively, scientists are looking for eco-friendly methods to combat the insect pest population in the field.

With the advent of genetic transformation techniques it is now possible to grow crops with alien genes that confer resistance to insects. Transgenic crops can be produced by transferring genes of nonplant or plant origin. The world has already seen transgenic crops with genes from a nonplant origin, (e.g., *Bl* cotton). In the dynamic ecosystem of a farm there is always a pressing need to rotate crops, and in the context of biotechnology-driven agriculture there is a need to rotate genes. The genes already present in plants will naturally be preferred if found more potent than the nonplant genes. These genes will contribute to eco-friendly practices in crop protection.

Although there are varieties of plant chemicals reported to be associated with insect resistance there is no proper cataloguing of these compounds. In this book we attempt to give the available information on these secondary metabolites. The chapters contain information on chemical structure, biosynthesis, bioactivity, mechanism of action, molecular biology, and perspective in order to give the reader a comprehensive review of these chemicals. This book will help the researcher in this exciting field of molecular approaches to insect resistance to get comprehensive information in one volume.

Although we began the book with the intention to finish it in a year, we could not do so for various reasons. As we proceeded, the information was pouring in and we included the most recent information possible.

Many scientists contributed information, advice, and valuable time in reviewing this book. Dr. P. Balasubramanian was instrumental in formulating the title of the book. Dr. S. Mohankumar, Dr. M. Bharathi, Dr. M. Maheswaran, Dr. M. Mohan, Dr. Selvi, Dr. Raja, Mr. Sivakumar, and Dr. Suseela Gomathi were kind enough to go through the manuscript and make corrections. Dr. D. Sudhakar (Tamil Nadu Agricultural University, India), Dr. R. Croteau (Washington State University, U.S.A), Dr. B. A. Halkier (Centre of Molecular Plant Physiology, Denmark), Dr. P. J. Facchini (University of Calgary, Canada), Dr. T. H. Schuler (IACR Rothamsted, UK), Dr. B. L. Moller (Royal Veterinary and Agricultural University, Denmark), Dr. I. T. Baldwin (Max Planck Institute for Chemical Ecology, Germany), Dr. R. Barkovich (University of California, U.S.A.), Dr. I. Feussner (Institute of Plant Genetics and Crop Plant Research, Germany), Dr. R. Verpoorte (Centre for Drug Research, The Netherlands), and Dr. D. J. Chitwood (USDA-ARS, U.S.A.), provided us with reprint material. We owe our thanks to them and all the others in our center who have helped us in various ways; without them the dream would not have become a reality. Our thanks to Marcel Dekker, Inc. for their patience and sustained interest in this area.

S. Sadasivam
B. Thayumanavan

Contents

Contents

1

Introduction

In 1999, the world population crossed the 6 billion mark. Presently, 80% of the world's population live in what are considered to be the lesser developed countries (LDCs). Demographers predict that the world population will continue rising, reaching 10 billion by the year 2050 despite declining birth rates. The majority of this increase will occur in the developing countries, thereby adding another 2–4 billion people to the nations of the LDCs. It is foreseen that the population density in the developing countries will increase from approximately 55 persons/km^2 at present to 90–100 people/km^2 by 2050 (1). Many believe that the single most important challenge facing mankind for the coming decades is increasing population. Thus, there is a continuing need to increase food production, particularly in the developing countries of Asia, Africa, and Latin America. This increase has to come from increased yields from major crops grown on existing cultivatable lands, perhaps on shrinking land area. Though increasing the yield is the goal, minimizing the pest-associated losses will save the harvested yield. The pest-associated loss is estimated at 14% of the total agricultural production (2). Insects not only cause direct loss to the agricultural produce but also have an indirect effect due to their role as vectors of various plant pathogens. Moreover, application of pesticides to control the insect population results in adverse effects on the beneficial organisms, leaves pesticide residues in the food, and finally, poses pollution problems. Thus, there is a greater need to develop alternative or additional technologies that would allow a judicious use of pesticides and provide adequate crop protection for sustainable food production.

Insects and plants are coevolved organisms. Plants when established from an aquatic to a terrestrial environment became involved in competition with a number of predators and subjected to multiple pressures, mainly from insects. Similarly and simultaneously, development occurred in insects compelling them to feed on a variety of plants. In this competition, plants evolved resistance mechanisms by trying to utilize novel metabolic pathways producing compounds toxic to their predators. In modern agriculture, the host plant resistance mechanism is utilized in the eco-friendly techniques of integrated pest management (IPM). It becomes more essential now to understand in detail the mechanism of host plant resistance to further strengthen insect management programs.

1.1   PLANT RESISTANCE TO INSECT PESTS

The types of plant resistance to insect pests are classified as nonpreference, antibiosis and tolerance (3). Kogan and Ortman (4) proposed the term *antixenosis* to more accurately describe nonpreference. *Antixenosis* describes the resistance of a plant to serve a host to an insect herbivore. As a result, insect pests are forced to select an alternative host plant. Antixenosis denotes the presence of morphological or chemical plant factors that adversely alter insect behavior, resulting in selection of an alternative host plant (5). Antibiosis is the category of resistance that includes those adverse effects on the insect life history that result after a resistant host plant is used for food (3). Both chemical and morphological plant defenses mediate antibiosis, and the effect of these in resistant plants ranges from mild to lethal. The death of early instars, reduced size or low weight, prolonged periods of development of the immature stages, reduced adult longevity and fecundity, death in the prepupal or pupal stage, and abnormal behavior are considered to be parameters for measuring the effects of antibiosis. Host plant tolerance is the ability to grow and reproduce in spite of supporting a population approximately equal to that damaging a susceptible host. The existence of tolerance is determined by comparing damage, plant stand, and the production of plant biomass or yield in infested and noninfested plants of the same cultivars. The expression of tolerance is determined by the inherent genetic ability of a plant.

1.2   DEVELOPMENT OF INSECT RESISTANT CROP VARIETIES

Varieties with higher levels of resistance to insect pests have evolved in several crop plants. The strategy for producing insect resistance varieties is to identify resistance gene sources for a given pest and incorporate them into improved plant type varieties through pedigree/back cross-breeding. The insect-resistant crop varieties helped to achieve a major increase in food production in many parts of world. Such varieties have played an important role in the 'green revolution' in South and Southeast Asia. However, that strategy hardly helped to stabilize the yield, as varieties bred for resistance to one pest succumbed to another. So breeding emphasis was shifted to evolve varieties with multiple resistances for which many breeding strategies were employed, such as sequential release, gene pyramiding, gene deployment, and multilines. Wide hybridization, tissue culture techniques, and molecular marker–aided selection are the innovative methods advocated now. Marker-assisted selection (MAS) for a single gene-controlled trait has been successful. However, MAS has not been utilized for many of the complexly inherited traits because tightly linked markers to the quantitative trait loci (QTL) have not been identified. The potentials of MAS could be exploited fully only when the phenotypes are better understood.

1.3   GENETIC ENGINEERING

The introduction of genetic transformation techniques has opened up new vistas for the insertion of genes that confer resistance to insects into the plant genome (6). Genes from very divergent organisms, such as bacteria to spider, have been introduced to develop insect resistant crop plants. Results on the development of transgenic insect-resistant plants were first published in 1987. This was followed with several reports in the last decade. At least 40 different genes conferring insect resistance have been incorporated into crop plants. This new technology is considered as an additional tool for the control of crop pests. Transgenic crops offer many advantages over conventional insecticides. They are more effective in targeting of insects, have greater resilience to weather conditions, are eco-friendly, have reduced operator exposure to insecticides, and give financial savings to farmers. In addition to widening the pool of useful genes, genetic engineering allows the use of several desirable genes in a single event and reduces the time to introgress novel genes into elite backgrounds.

The ideal transgenic technology should be commercially feasible, environmentally benign (biodegradable), and easy to use in diverse agro-ecosystems as well as show a widespectrum of activity against crop pests.

It should also be harmless to the natural enemies, target the sites in insects that have developed resistance to the conventional pesticides, and be flexible enough to allow ready deployment of alternatives (if and when resistance is developed by the pest).

The genes introduced could be classified as being of plant or nonplant origin. Genes from bacteria such as *Bacillus thuringiensis* (Bt) and *Bacillus sphaericus* have been the most successful group of organisms identified for use in genetic transformation of crops for pest control on a commercial scale (7,8). Bt toxins have been transferred and expressed in at least 26 plant species. However, the level of resistance they confer depends, in most cases, on whether native bacterial or truncated, codon-optimized genes have been used. To date, codon-optimized genes have only been transferred into some of these crops, including cotton, maize, potato, broccoli, cabbage, and alfalfa. Generally, these plants express Bt toxins at levels sufficient to cause high mortality of target pests in the field. An alternative strategy has been to integrate the native Bt gene into the chloroplast genome of tobacco plants, resulting in extremely high expression levels (9). Though Bt genes have been found to be successful and hence commercialized, there is a fear of insects developing resistance to Bt gene products (10). To overcome this difficulty, gene pyramiding (more than one toxin) is advocated. Alternatively, the search is on for the phytochemicals conferring host plant resistance.

Many classes of plant proteins and secondary plant substances have been shown to have a toxic or an antimetabolic effect on insects and have been proposed as possible candidates for genetic engineering. Transgenic plants rarely result in 100% control, but tend to retard insect development and growth (11). Plant-derived genes attack different sites in insects than do the synthetic chemicals and may be deployed in combination with exotic genes. The need for genetic transformation of crops to improve crop production in the developing world has been discussed by Ortiz (12), Sharma and Ortiz (13), and Ortiz et al. (14).

Although there are reports on many secondary metabolites as potential candidates for host plant resistance, only two major groups of plant-derived genes have been used to confer insect resistance on crops. They are inhibitors of digestive enzymes (proteinase and amylase inhibitors) and lectins. These protein-coding genes have been transferred into crop plants without major alteration, and expression has been at a similar level to codon-optimized Bt toxins.

At least 14 different plant proteinase inhibitor genes have been introduced into crop plants. Most effort has concentrated on serine-proteinase inhibitors from the plant families Fabaceae, Solanaceae, and Poaceae, which are targeted mainly against lepidopteran species but also against some coleopteran and orthopteran pests. The most active inhibitor identified to date is cowpea trypsin inhibitor (CpTI), which has been transferred into at least 10 other plant species. Inhibitors of α-amylases are the second type of enzyme inhibitor used to modify crop plants. Genes for three α-amylase inhibitors have been expressed in tobacco (15).

Lectins are proteins with an affinity for carbohydrates. Various lectins have been shown to have some toxic activity against species of the insect orders Homoptera, Coleoptera, Lepidoptera, and Diptera. The lectin from snowdrop (GNA) has been shown to have toxicity against aphids and the rice brown plant hopper (*Nilaparvata lugens*). GNA has been expressed in different crops, including potato (16), wheat (17), and rice (18).

Chitinase and tobacco anionic peroxidase genes have been transferred to tomato. The transgenic plants showed different resistance to the peach potato aphid (19). Tryptophan decarboxylase (TDC) from periwinkle expressed in tobacco reduced the production of the whitefly *Bemisia tabaci* by up to 97% (20).

Resistance genes from animals have also been attempted. So far, work in this area has involved primarily serine-proteinase inhibitor genes from mammals and the tobacco hornworm (*Manduca sexta*) (21). A spider insecticidal gene when transferred to rice was found to cause mortality of leaf folder (22).

Genes transferred with constitutive promoters like CaMV 35S, expressed in all plant tissues, may lead to development of resistance by the insects. Moreover, these proteins will be produced even when there is no insect prevalence and thus will drain the energy, resulting in reduced crop yield. It is time to look for tissue-specific promoters like phloem-specific promoter for sucking pests. Wound-induced promoter will be helpful for judiciously channelizing plant energy to fight insect pests.

1.4 PLANT SECONDARY METABOLITES FOR INSECT RESISTANCE

Though some secondary metabolites have been exploited for genetic engineering toward enhancement of insect resistance, many other compounds are yet untapped. The goal need not always be introduction of a new gene; it may be switching on of a silent gene. Although all plants have most of the biosynthetic apparatus for production of secondary metabolites, all do not produce all known metabolites in the quantities required to defend them against insect attack. Thus, a thorough knowledge of the cataloging of secondary metabolites, their bioactivity against insects, biosynthetic pathways and enzymes, mechanisms of action, and gene regulation are required before start engineering the metabolic pathways for better plant protection. It is with this aim in mind that the present book attempts to relay information on the status of the secondary metabolites as related to insect pest resistance.

The morphological and phenological bases for plant resistance are discussed in detail in Chapter 2. The chemistry, classification, occurrence, bioactivity, mechanism of action, molecular studies, and perspectives of various chemical constituents are discussed in the subsequent chapters. Chapters 3–10 deal with nonprotein amino acids, lectins, enzyme inhibitors, cyanogenic glycosides, glucosinolates, alkaloids, phenolics, and terpenoids, respectively. In Chapter 11, the information available on compounds other than the above listed has been included, and Chapter 12 covers biochemicals induced in the plants upon insect attack. In the concluding chapter, the application of the chemical basis of plant resistance to insect pests for sustainable agriculture is discussed.

REFERENCES

1. Fedoroff, N.V.; Cohen, J.E. Plants and population: is there time? Proc. Natl. Acad. Sci. USA. **1999**, *96*, 5903–5907.
2. Oerke, E.C.; Dehne, H.W.; Schonbeck, F.; Weber, A. Crop production and crop protection: estimated losses in major food and cash crops. Elsevier: Amsterdam, 1994.
3. Painter, R.H. Insect resistance in crop plants. Macmillan: New York, 1951.
4. Kogan, M.; Ortman, E.E. Antixenosis—a new term proposed to replace Painter's "nonpreference" modality of resistance. Bull. Entomol. Soc. Am. **1978**, *24*, 175–176.
5. Smith, C.M.; Khan, Z.R.; Pathak, M.D. Techniques for evaluating insect resistance in crop plants. Lewis Publishers: Boca Raton, FL, 1994.
6. Hilder, V.A.; Boulter, D. Genetic engineering of crop plants for insect resistance—a critical review. Crop Prot. **1999**, *18*, 177–191.
7. Gill, S.S.; Cowles, E.A.; Pietrantonio, P.V. The mode of action of *Bacillus thuringiensis* endotoxins. Annu. Rev. Entomol. **1992**, *37*, 615–636.
8. Schuler, T.H.; Poppy, G.M.; Kerry, B.R.; Denholm, I. Insect-resistant transgenic plants. Trends. Biotechnol. **1998**, *16*, 168–175.
9. McBride, K.E.; Maliga, P.; Stalker, D.M.; Hogan, P.S.; Schaaf, D.J.; Svav, Z. Amplification of a chimeric *Bacillus* gene in chloroplasts leads to an extraordinary level of an insecticidal protein in tobacco. Biotechnology **1995**, *13*, 362–365.
10. Hyde, J.; Martin, M.A.; Preckel, P.V.; Dubbins, C.L.; Edwards, C.R. An economic analysis of non-Bt corn refuges. Crop Prot. **2001**, *20*, 167–171.
11. Estruch, J.J.; Carozzi, N.B.; Desai, N.; Duck, N.B.; Warren, G.W.; Koziel, M.G. Transgenic plants: an emerging approach to pest control. Nat. Biotechnol. **1997**, *15*, 137–141.

12. Ortiz, R. Critical role of plant biotechnology for the genetic improvement of food crops: perspective for the next millennium. Electronic J. Biotechnol. (online). 15 December, **1998**, *1* (3).

13. Sharma, K.K.; Ortiz, R. Program for the application of the genetic engineering for crop improvement in the semi-arid tropics. In Vitro Cell. Dev. Biol. (Plant). **2000**, *36*, 83–92.

14. Ortiz, R.; Bramel-Cox, P.J.; Hash, C.T.; Mallikarjuna, N.; Reddy, D.V.R.; Seetharama, N.; Sharma, H.C.; Sharma, K.K.; Sivaramakrishna, S.; Thakur R.P.; Winslow, M.D. Potential for improving agricultural production through biotechnology in the semi-arid tropics. In: *World Commission on Dams Thematic Reviews*. Environmental Issues Series. World Commission on Dams, Vlaeberg, Cape Town, South Africa, 2000.

15. Masoud, S.A.; Ding, X.; Johnson, L.B.; White, F.F.; Reeck, G.R. Expression of corn bifunctional inhibitor of serine proteinases and insect α-amylase in transgenic tobacco plants. Plant Sci. **1996**, *115*, 59–69.

16. Gatehouse, A.M.R.; Down, R.E.; Powell, K.S.; Sauvion, N.; Rahbe, Y.; Newell, C.A.; Merryweather, A.; Hamilton, W.D.O.; Gatehouse, J.A. Transgenic potato plants with enhanced resistance to the peach-potato aphid *Myzus persicae*. Entomol. Exp. Appl. **1996**, *79*, 295–307.

17. Stoger, E.; Williams, S.; Christou, P.; Down, R.E.; Gatehouse, J.A. Expression of the insecticidal lectin from snowdrop (*Galanthus nivalis* agglutinin, GNA) in transgenic wheat plants: effects on predation by the grain aphid *Sitobion avenae*. Mol. Breed. **1999**, *5*, 65–73.

18. Sudhakar, D.; Fu, X.; Stoger, E.; Williams, S.; Spence, J.; Brown, D.P.; Bharathi, M.; Gatehouse, J.A.; Christou, P. Expression and immunolocalization of the snowdrop lectin, GNA, in transgenic rice plants. Transgenic Res. **1998**, *5*, 371–378.

19. Dowd, P.F.; Lagrimini, L.M. In *Advances in Insect Control: The Role of Transgenic Plants*; Carozzi, N., Koziel, M., Eds.; Taylor and Francis: New York, 1997; pp. 195–223.

20. Thomas, J.C.; Adams, D.G.; Nessler, C.L.; Brown, J.K.; Bohnert, H.J. Tryptophan decarboxylase, tryptamine and reproduction of the whitefly. Plant Physiol. **1995**, *109*, 717–720.

21. Thomas, J.C.; Adams, D.G.; Keppenne, V.D.; Wasmann, C.C.; Brown, J.K.; Kanosh, M.R.; Bohnert, H.J. Proteinase inhibitors of *Manduca sexta* expressed in transgenic cotton. Plant Cell. Rep. **1995**, *14*, 758–762.

22. Huang, J.Q.; Wel, Z.M.; An, H.L.; Zhu, Y.X. *Agrobacterium tumefaciens*–mediated transformation of rice with the spider insecticidal gene conferring resistance to leaffolder and striped stem borer. Cell Res. **2001**, *11*, 149–155.

2

Morphological and Phenological Features

2.1 INTRODUCTION

Plants and insects have had coexisting relationships for a long time. Insects were suppressed either by other insects or toxins or by plant defense mechanisms in order to create a balance between the insect pest population and host. Each plant species has a unique set of defense traits ranging from morphological to phytochemical parameters that have behavioral and physiological ramifications for a potential herbivore consumer.

There are three well-known mechanisms of plant defense to insect damage: antixenosis, antibiosis, and tolerance. Antixenosis is a host plant mechanism that includes morphological, physical, or structural qualities that interfere with insect behavior, such as mating, oviposition, and feeding. Disruption of the normal metabolic process affecting the biology of an insect is the basic manifestation of antibiosis. Tolerance represents the plant's ability to suffer a smaller yield loss than that manifested by susceptible cultivars by virtue of its innate capacity to grow and reproduce itself or to repair injury to a marked degree. Antixenosis and antibiosis are expressed in the insect reaction to host plant, whereas tolerance is due to the plant's response to insect attack.

As a resistance mechanism, antixenosis acts as a structural barrier that affects the insect's behavior in selecting its host. The first plant organs

8

contacted during the preliminary stages of host acceptance are surface hairs or trichomes. Antixenosis describes the inability of a plant to serve as host to an insect herbivore. It modifies insect behavior once in contact with the plant host, without effect on metabolism of both plant and insect. There are several modes of resistance used by plants to deter insects. These modes were categorized as chemical and morphological plant defenses. This chapter will discuss the morphological, phenological, and plant surface characteristics associated with plant resistance. As examples of these characteristics, plants present special epidermal characters (glandular trichomes, hairs, etc.), as well as layers of surface waxes that prevent insects from feeding on them. Resistance to oviposition may come from plant characteristics that fails to provide appropriate oviposition-inducing stimuli or oviposition-inhibiting stimuli. Plant characteristics have been recognized for years as important resistance factors. The chemical inhibition of feeding can be through plant emission of allelochemicals such as repellents, deterrents, or inhibitors. Each type of chemical defense engages the plant in a different mechanism to disturb insect behavior.

2.2 COLOR AND SHAPE (REMOTE FACTORS)

Host selection behavior of phytophagous insects is associated with the color and shape of plants. Plant size, shape, and relative transience are correlated with some resistance and seem to be associated with some strategies of chemical defenses. Color and shape of plants remotely affect host selection behavior of phytophagous insects and thus are associated with some resistance (1).

In many instances, healthy, dark green plants are more attractive to insects than yellowing plants under stress. However, yellow-green plants were preferred to green plants by the pea aphid, *Acyrthosiphon pisum* (2). Most of the aphids and other *Sternorrhyncha* insects are attracted to leaves reflecting colors in the 500- to 600-nm range (yellow-green). Alate aphids are attracted to leaves reflecting about 500 nm, regardless of the species of plant, because they seem to be attracted to plants at a physiologically suitable stage of growth (3).

Color and intensity of light reflected from the surface of cabbage leaves affect host selection by *Brevicoryne brassicae* (L.). Red cabbage is least selected by alate aphids but presents most favorable conditions for aphid increase after infestation (4,5). Red-leaved cotton varieties are less attractive to the boll weevil, *Anthonomus grandis* (6). Similarly, red-leaved cabbage varieties are less susceptible to the cabbageworm, *Pieris rapae* (L.) (7) and *P. brassicae* (L.) (8). Purple-pigmented sorghum lines exhibit a high level of resistance to the shootfly, *Atherigona soccata*, whereas they are more susceptible to the red mite, *Oligonychus indicus* Hirst (9). The imported cabbageworm, *Pieris rapae*, is less attracted to the red foliage of the 'Rubine' variety of Brussels sprouts than to the green ones. Oat cultivars with red tiller bases and pubescent shoot bases were less susceptible than other cultivars to attack by the fruit fly, *Oscinella frit* (10). Stemmed varieties of soybean have proved resistant to the bean fly *Ophiomyia centrosematris* (de Meij) (11).

Form perception probably elicits certain generalized behavioral patterns, but no resistance mechanisms among crop plants have been associated directly with plant shape. Certain morphological characteristics may, however, be linked to other resistance factors. For example, okra-leaved cotton suffers less damage for majority of insect pests (12). The above-ground shape of turnip plants was not important to *Hylemya florales*, the turnip maggot, but the varieties, that generally have strong, round, long roots were more tolerant than the varieties with thin roots. The narrow-leaflet isolines apparently tolerated defoliation better than the wide-leaflet isolines because the narrow-leaflet isolines had greater light interception and maintained equal or better yields, despite similar defoliation. In addition to leaf area index, both canopy light interception and light extinction coefficient are important criteria for selecting cultivars tolerant to defoliation by pests in soybean. In sorghum, the varieties with greater height, greater distance between two leaves, and smaller leaf angle were less susceptible to the aphid (13). The thrips population was related to the leaf insertion angle in onion; the greater the insertion angle, the lower the thrips population densities on the plant. It has been suggested that the smaller angles may offer greater protection to the thrips against bad weather conditions (14). Tests with rice leaf folder indicated that longer and tougher leaf blades contribute to resistance. It appeared that rice leaf blade morphology, i.e., width, length, and toughness, may play a vital role in resistance against rice leaf folder. Leaf folder resistance in TKM6 is suspected to be a combination of antixenosis and antibiosis (15).

2.3   CLOSE-RANGE OR CONTACT FACTORS

The majority of recognized physical defense factors of plants operate on contact with the herbivore. The contact factors most commonly found among resistant crop plants are discussed in the sections that follow.

2.3.1 Silica Content

Many plant species, primarily among the Gramineae, Cyperaceae, and Palmae, have deposits of silica on the epidermal walls, which may play a

TABLE 2.1 Examples for Insect Resistance Related to the Presence of Silica in Plants

Plant	Insect	Effect	Ref.
Rice	*Chilo suppressalis*	Mandibles worn out in high-silica lines	16
Rice	*Chilo zacconius*	Larval survival affected	17, 18
Sorghum	*Atherigona varia soccata*	Low incidence in high-silica varieties	19, 20
	Stem borer	Low incidence in high-silica varieties	
Rice	*Scirpophaga incertulas*	Low incidence in high-silica varieties	21–23
Cotton	Leaf hoppers	Low incidence in high-silica varieties	24
Cucumis callosus	Fruit fly	Low incidence in high-silica varieties	25
Okra	Leaf hoppers	Low incidence in high-silica varieties	26

role in resistance mechanisms. There are several examples for resistance mechanisms related to the presence of silicate encrustation (Table 2.1).

2.3.2 Surface Waxes

Plant leaves are protected against desiccation, insect predation, and disease by a layer of surface wax over the epicuticle. Wax chemically refers to an ester form of a long-chain fatty acid and a high molecular weight aliphatic alcohol. Plant waxes vary from a fraction of a percent to several percent of the dry weight of a plant. Cuticular waxes are rather complex mixtures including mainly *n*-alkanes (from C7 to about C62), branched or unbranched, saturated or unsaturated, as well as alcohols and acids. Leaf surface chemicals undoubtedly affect insect behavior. Insects possess sensory apparatus to detect these chemicals by contact or olfaction. There are primary and secondary plant compounds as the plant waxes both can be detectable by the insect (27). Epicuticular waxes affect the feeding behavior of insects, particularly the settling of probing insects, acting as phagostimulants or feeding deterrents. There have been a number of reports about insects' response to chemicals on the leaf surface as obtained from experiments using surface extracts of leaves or pure chemicals that are known to occur on leaf surfaces. Some of the examples are given in Table 2.2.

In "bloom" cultivar of mustard, the culm is heavily waxed. The neonate larvae experience considerable difficulty in climbing because their

TABLE 2.2   Effect of Surface Waxes in Crop Plants to Confer Insect Resistance

Plant species	Insects	Effect	Ref.
Broccoli	*Phyllotreta albionica*	Waxy leaves are more resistant than aglossy leaves	28
Brassica oleracea var. *acephala*	*Brevicoryne brassicae*	Did not colonize on nonwaxy plants	29
Brassica oleracea var. *acephala*	*Aleyrodes brassicae*	Did not colonize on nonwaxy plants	29
Rubus phoenicolasius	*Byturus tomentosus*	Resistance in waxy plants	30
Rubus phoenicolasius	*Amphorophora rubi*	Resistance in waxy plants	30
Castor	*Empoasca flavescens*	Susceptible with waxy bloom coating	31
Castor	*Achaea janata*	Low population in no-bloom and low-bloom varieties	31
Sorghum	All insects	Wax flakes disappear in resistant hybrids	32
Sorghum	Aphids	Waxy leaves with resistance	33
Sorghum	*Tetranychus telarius*	Leaf bloom with resistance	33
Wheat	*Macrosiphon avenae*	Nonglaborous wheats are resistant	34
Sorghum	*Schizaphis graminum*	Bloomlessness with resistance	35
Sorghum	Stem borer	Surface wax as feeding deterrents	36
Sugarcane	*Eldana saccharina*	Stalk surface wax components contributed to resistance	37
Onion	*Thrips tabaci*	Glossy foliage and glossy scapes are more resistant	38

prolegs get stuck in the wax and the larvae never reach the feeding site (39). The behavior of neonate diamondback moth larvae on resistant cabbage leaves was affected by leaf waxes (40). They concluded that chemical and physical attributes leaf waxes act together with some plant characteristics to affect neonate diamondback moth larvae behavior on cabbage plants. They showed that the larvae spent significantly more time walking, walked more frequently, and walked significantly faster on leaves of glossy resistant cabbage (NY 8329) than on the susceptible (Round-up). Consequently, the average speed and walking speed were both significantly greater on the glossy cabbage leaves than on normal leaves. Alkanes are among the most common constituents of all plant waxes. Specific C32 alkane common in the wax of *Vicia faba* caused the insect to probe for longer periods into parafilm sachets whereas the alkane fraction of nonhost deterred feeding (41). Wax may physically prevent the movement of an insect across a leaf surface. Chapman and Bernays (42) suggested that the nature of the surface wax may be "recognized" by insect as indicative of internal constituents of a plant.

2.3.3   Anatomical Adaptations of Organs

Slight variations in the morphological structure of plants may result in altered fitness to herbivores. Quite often such variations alter the effectiveness of the factors causing mortality. Table 2.3 summarizes the available information. Although toughness of leaf tissues is an efficient defense mechanism, breeding of improved varieties often leads to elimination of such characters, especially in crop plants consumed as leaves or fruits due to inverse relationship with quality of the produce.

2.3.4   Solidness and Other Stem Characteristics

There are numerous examples of slight or profound changes in stem features that reduce fitness of a plant for associated herbivores. Resistance to certain stem borers is related to the nature of the stem tissues. Some examples are presented in Table 2.4.

2.4   TRICHOMES

Trichomes are unicellular or pluricellular outgrowths from the epidermis of leaves, shoots, and roots (88). The collective trichome cover of a plant surface is called pubescence. Several authors have attempted to classify the variety of plant trichomes; the reviews by Uphof (88), Hummel and Staesche (89), and Johnson (90) present some of the most widely accepted classifications. Trichomes serve many critical physiological and ecological functions, particularly those associated with water conservation. Johnson (90) discussed the ecological functions of trichomes as defense against herbivores. These latter discussions relate to our interest in trichomes as resistance factors in crop plants.

Insect species respond differently to the presence of plant hairs. Pubescence as a resistance factor interferes with insect oviposition, attachment to the plant, feeding, colonization, and ingestion. However, glabrous forms of plants may be more resistant to some species. In general, the purely mechanical effects of the pubescence depend on four main characteristics of the trichomes: density, erectness, length, and shape. In some cases trichomes

TABLE **2.3** Interactions of Anatomical Adaptations and Insect Attack in Crop Plants

Crop species	Insect	Anatomical nature	Effect	Ref.
Sugarcane	*Diatraea saccharalis*	Intact leaf sheath	Resistance observed	43
Sugarcane	*Chilo sacchariphagus indicus*	Intact leaf sheath	Resistance observed	44, 45
Sugarcane	Mealybugs	Loose-fitting leaf sheath	Resistance observed	46, 47
Sugarcane	Scale insects and eriophid mites	Loose-fitting leaf sheath	Resistance observed	44
Cotton	*Pectinophora gossypiella, Helicoverpa zea, Heliothis virescens*	Absence of nectaries	Reduced oviposition	48, 49
Cotton	Tarnished bug	Absence of nectaries	Resistance observed	50
Cotton	*Amrasca devastans* and *Bemisia tabaci*	Absence of nectaries	Susceptible	51, 52
	Helicoverpa armigera and *Earias spp*	Absence of nectaries	Resistance observed	53
Cotton	*Helicoverpa zea*	Frego bract	Resistance observed	54
	Anthonomus grandis	Frego bract	Resistance observed	55
	Earias spp	Frego bract	Least susceptible	56
	Lygus hespenus	Frego bract	Susceptible	57
Corn	*Helicoverpa zea*	Long tight husk	Resistance observed	58–61
Sorghum	*Peregrinus maidis*	Tightly wrapped leaves	Resistance observed	62
Wheat	*Hydrellia griseola*	Leaf sheath close to stem	Resistance observed	63
Alfalfa	*Bruchophagus roddi*	Spiralling of pod	Resistance observed	64, 65
Brinjal	Shoot and fruit borer	Thin fruits with short calyx and less calyx diameter	Tolerance observed	66

TABLE 2.4 Studies on Relationship Between Solidness and Stem Characteristics and Insect Damage in Crop Plants

Crop species	Anatomical	Insect	Resistance or susceptibility	Ref.
Wheat	Solid stem	Wheat stem sawfly *Cephus cinctus*	Desiccated eggs and impaired larval movements	67–69
Sugarcane	Hardness of the rind and fiber content of stalks	Larvae of *Diatrea saccharalis*	Resistance observed	44, 70
Sugarcane	Denticles on the leaf midribs, lignification of cell walls; number of layers of celerenchymatous cells; number of vascular bundles	Young larvae of *Diatrea saccharalis*	Resistance observed	44, 70
Sugarcane	Hard stem	*Chilo sacchariphagus indicus*	Less feeding	44, 45, 71–74, 75, 76
Sugarcane	Rind hardness	*Chilo infuscatellus*	Less feeding	77
Cucurbita spp.	Closely packed tough vascular bundles	Squash vine borer, *Melittia cucurbitae*	Resistance to penetration and feeding	
Wild Tomato	Thick cortex in the stem	Aphid, *Macrosiphum euphorbiae*	Prevent the stylet reaching the vascular tissue	78
Alfalfa	Lignified vascular bundles	*Empoasca fabae*	Reduced oviposition	79
Turnip, brussels sprout	Leaf toughness	Mustard beetle, *Phaedon cochleariae*	Feeding rates and larval growth were retarded	80
Cowpea	Thickness of pod wall	Cowpea weevil, *Chalcodermus aeneus*	Penetration into pods interrupted.	81
Rice	Thicker hypodermal layers	*Chilo suppressalis*	Resistance observed	82
Rice	Thick layers of sclerenchymatous hypodermis	Stem borer	Resistance observed	83
Rice	Thicker sclerenchymatous layer than parenchyma	Stem borer	Resistance observed	84
Sorghum	Distinct lignification. Thicker walls enclosing the vascular bundle sheaths within the central whorl of young leaves	*Atherigona varia soccata*	Resistance observed	85
Brinjal	Thick layered vascular bundles with lignified cells	*Leuinodes orbonalis*	Penetration into apical shoot retarded	86
Cotton	Twice stiffness of stem tips in resistant varieties	Aphids	Resistance observed	87

possess associated glands that exude secondary plant metabolites. The effect of glandular trichomes may depend on the nature of the exudates, which may be composed of allelochemics such as alkaloids or terpenes (90). Such toxic substances may kill insects on contact or act as repellents. In some plants, sticky exudates glue the insects' legs and impede locomotion. In addition, glandular trichomes exude secondary plant metabolites, which may interfere with locomotion. They are also toxic and exert disruptive effects by the production of allelochemics. The role of pubescence in host plant resistance has been reviewed by Webster (91) and Norris et al. (1).

2.4.1 Trichomes in Relation to Locomotion, Attachment, and Shelter

Nonglandular and/or glandular trichomes entrap, immobilize or impale many insect pests. Johnson (92) describes how the hooked trichomes present on certain varieties of French bean (*Phaseolus vulgaris*) impale the aphid *Aphis craccivora* Koch. The pea aphid, *Acyrthosiphon pisum* (Harris) readily settles on sparsely haired varieties of French bean, such as Zuckerperl, Perfektion, and the densely haired variety Monila, and exhibits probing behavior. However, a considerable percentage of settled aphids leave plants that have hooked epidermal hairs. The remaining aphids become impaled on the plant hairs and die within a week at 20°C, although it has been shown that French beans are physiologically suitable as a food plant (93). Trichomes on seedling stem offer resistance to the spotted alfalfa aphid *Therioaphis maculata* (Buckt.) through nonpreference. However, once aphids reach feeding sites, either by traveling over plant hairs or by being placed on leaves, there is no adverse effect on aphid survival or reproduction (94). In greenhouse trials, pubescent-leaved wheat cultivars (Vel and Downy) attracted a significantly lower number of bird cherry oat aphid *Rhopalosiphum padi* (L.) than glabrous-leaved cultivars (Arthur and Abe). On 'Arthur,' aphids move rapidly and usually begin probing within 5–10 s whereas on 'Vel' they are slow and often leave the plant without probing (95). Certain varieties of bean plants with hooked trichomes resist the leafhopper *Empoasca fabae* (Harris) (96). Trichomes on pubescent soybean cultivars impede normal attachment of *E. fabae* during oviposition and feeding (97). The classic example of hooked trichomes in plant defense is that of the neotropical passion flower veins against larvae of flashy heliconiine butterflies (*Heliconius* spp) (98). Scanning electron microscopic studies have shown the defensive function of these trichomes, which make numerous puncture wounds in the larval integument, immobilizing the larvae, leading to their death by starvation (99). The first-instar larvae of the cotton pink bollworm *Pectinophora gossypiella* Saunders, hatching from eggs laid on the vegetative parts of cotton plants, are disoriented by trichomes; consequently, the number of larvae reaching the bolls is reduced (100,101). Laboratory observations of the movements of newly hatched larvae of *Heliothis virescens* (F.) on the upper and lower leaf surfaces and petioles of four cotton strains indicate that pubescence provides a mechanism of resistance to such movement (102).

The exudates of glandular trichomes cause entrapment and immobilization of some phytophagous insects, apart from their toxic effects. Dense glandular hairs are present on several species of plants, especially those belonging to Solanaceae. When cell walls of these trichomes are ruptured by contact with insects, a clear water-soluble liquid oozes out which, on reaction with atmospheric oxygen, changes into an insoluble black substance that hardens around the appendages of insects. The immobilized aphids *Myzus persicae* (Sulzer) and *Macrosiphum euphorbiae* (Thomas) die within a short time (103). Lapointe and Tingey (104) noted that aphids whose tarsi are coated with type B exudate are far more likely to discharge type A trichomes of *Solanum berthaulti* and become immobilized than those whose tarsi are free of type B exudates (105). Type B exudate is viscous and may physically limit aphid mobility, resulting in abnormal feeding behavior. The type A trichome gland appears to be the major source of the phenolic substrates and phenolases involved in the phenomenon of aphid entrapment and mortality. *Solanum neocardenasii* also bears type A and type B glandular trichomes that entrap aphids (104). The trichome exudates of *S. neocardenasii* darken and harden on aphid tarsi, labia, and antennae and kill the insects. Mortality of fourth-instar nymphs of *M. persicae* and encasement of tarsi and labia by type A exudates increased with a rise in density and volume of type A trichomes (106). Immature stages of the jassid *Empoasca fabae* also become entrapped and are subjected to greater mortality and encasement of body parts by trichome exudates than adults (106). Adults of the whitefly *Bemisia tabaci* (G.) are trapped by the glandular hairs of tomato leaves (107). The glandular trichomes of *Solanum polyadenium* discharge a sticky substance on contact with larvae of the Colorado potato beetle *Leptinotarsa decemlineata* (Say), which accumulates on the tarsi, causing immobilization of a few larvae, while others fall off the plants (108).

In spider mites, the intensity of webbing is influenced by the leaf surface texture. Strong webbers, such as *Tetranychus urticae* Koch and *T. pacificus* McG., prefer rough spaces to glabrous ones (109,110). Leaf trichomes could very well serve as substrates for attachment of webbing. Khanna and Ramanathan (111) have reported that sugarcane varieties with spinous outgrowths on the leaves are susceptible to *Oligonychus indicus* (Hirst) but those without hairs are immune. Pallard and Saunders (112) found that the hairy cotton variety 'Sakel' is highly susceptible to aphids and spider mites. The number of hairs per square centimeter and the number of branches per trichome correlate negatively, while the length of the hair correlates positively with the *T. cinnabarinus* (Boisduval) population in brinjal (113). However, Goonewardene and coauthors (114) pointed out that different mite species behave differently in response to pubescence. Hair density is reported to have a negative correlation with the population of *Paratetranychus ulmi* (Koch) on apple (115), *Tetranychus* spp on cotton (116), and *T. cinnabarinus* on tomato (117). Resistance to *T. cinnabarinus* in tomato varieties (117) and in *Solanum pennelli* and *Lycopersicum hirsutum* (118) is attributed to glandular hairs. Nonpreference of some tomato lines (119,120) and some *Nicotiana* spp (121). by *T. urticae* is also attributed to glandular hair secretions. Glandular hairs of *Pelargonium* (122), strawberries (122–124), and hop varieties (125) are found to possess chemical substances that act as attractants or repellents to *T. urticae*.

2.4.2 Trichomes in Relation to Feeding and Digestion

Pubescent plants are not generally preferred by small herbivores as it is difficult for their feeding organs to reach the site of feeding. Hairiness in cotton is associated with its resistance to the leafhoppers *Amrasca devastans* (Dist.) (126–130), *Empoasca fascialis* Jacobi (131), and *E. maculata* (132). Similar observations have been made for brinjal, *Solanum melongena* L. varieties, resistant to the cotton leafhopper, *Chlorita biguttula* Ishida (133). Trichomes interfere more with the feeding of nymphs than with adults of *A. devastans* (134,135). Resistant varieties have a greater number of long trichomes on the midrib and lamina versus fewer and shorter trichomes on susceptible varieties. Laminar trichomes without adequate length or long trichomes without adequate density are not effective in providing resistance in okra and cotton (136,137). The high degree of resistance in *Solanum mammosum* L. to *Aphis gossypi* G. and *Henosepilachna vigintioctopunctata* (F.) is attributed to the high density, length, and erect disposition of hairs on the leaves (138). According to Webster (91), pubescence also interferes with feeding of the whitefly *Bemisia tabaci* (G.), western cotton rootworm *Diabrotica virgifera* L., sorghum shootfly *Atherigona varia soccata* Rond., cotton leaf worm *Spodoptera littoralis* (Boisduval), cereal leaf beetle *Oulema melanopus* (L.), and the two-spotted spider mite *Tetranychus urticae* (Koch). The role of trichome density and length in the resistance of soyabean to *Trichopulsia ni* (Hubner) has been evaluated on different portions of the leaves of a relatively resistant cultivar (PI 227687) and a relatively susceptible cultivar (Davis). The youngest leaves of PI 227687, which are densely covered with trichomes, are the most resistant to larval feeding. When these trichomes are shaved off, such leaves become as susceptible to attack as sparsely pubescent leaves (139). Trichomes may also interfere with ingestion of food. High mortality of the cereal leaf beetle *O. melanopus* has been observed when it feeds on pubescent wheat varieties. According to Schillinger and Gallun (140), death of the young larvae results from an unbalanced diet because the larvae feed on hairs rich in cellulose, lignin, and fiber. Kogan (141) also found that pubescence reduces the quality of food of the Mexican bean beetle *Epilachna varivestris* Mulsant and causes greater larval mortality. Consumption and utilization parameters were higher in 'Clark' "glabrous" and "curly" than in "normal" and "dense pubescent" types. Survival was 10–25% higher on the glabrous and curly pubescent types. These results suggest that pubescence reduced the quality of the food ingested and caused greater larval mortality. Much of the reported resistance of hairy plants to Cicadellidae, Aphididae, and other small sap-sucking insects is based on estimates of differential population buildups on hairy versus glabrous plants. It is impossible to assess how much of the population differences result. In addition to the quality of food, Wellso (142) observed that larvae of *O. melanopus* fed on pubescent wheat leaves were stuffed with undigested hairs, some of which had pierced the gut wall. Trichome density on green gram pod surface, thickness of pod wall, total soluble sugar content, and phenol content of seed had a positive correlation with nymphal period and a negative correlation with body weight of nymphal instar and of adults (143). Mizukoshi and Kakizaki (144) studied the influence of trichomes on kidney bean leaves to the development of the foxglove aphid, *Aulacorthum solani*. The first instar aphids were trapped physically by trichomes and trapped aphids died from starving. Using scanning electron microscopy, trichomes appeared curved at this tips and the degree of curvature varied between cultivars. Mortality of early-instar aphids was higher on cultivars with trichomes that had heavily curved tips (144).

2.4.3 Trichomes in Relation to Oviposition

Leaf trichomes interfere with the oviposition of *Amrasca devastans* on cotton (127,145) and on okra (135,146). Effective hair length on the ventral surface of midveins showed a significant negative correlation with the number of eggs laid (147). The glabrous character of cotton has been associated with reduced oviposition in *H. zea* and *H. virescens* (148–153). Oliver and co-authors (154) stated that the ultrasmooth leaf strain D 2723 was damaged less by *Heliothis* than the other cotton strains tested in field plots. Lukefahr and his associates (155) found that a cotton strain with a glabrous character reduced a *Heliothis* population by 68%. On the other hand, Mehta (156) observed that the spotted boll worm *Earias vitella* (F.) prefers *Gossypium hirsutum* leaves over bolls for oviposition in view of the high density of hairs on the leaves. Agarwal and Katiyar (157) have recorded the lowest oviposition and damage by *E. fabia* in growing points in smooth-leaved cotton varieties compared to medium and highly pubescent varieties. Pubescent accessions support more eggs of *E. vitella* than the glabrous Deltapine smooth leaf, which indicates the nonpreference for a glabrous surface for oviposition (158). It has also been observed that in crosses in which the Deltapine smooth leaf was one of the parents, the F_1 plants were glabrous and the number of eggs deposited on them lower. Sharma and Agarwal (159) further observed that leaf hairiness correlated significantly and positively with the number of eggs laid by *E. vitella* under both field and laboratory conditions. Differences in trichome density also affect oviposition and nymphal weight of *Lygus hesperus* Knight on cotton. Females laid 28% fewer eggs on a normally hirsute line than on a pilose line and 31% fewer on a smooth leaf than on the pilose line (160). In sorghum, the presence of trichomes on the leaf surface is related to less frequency of oviposition by adults of the shootfly *A. soccata* (161,162). Pubescence in wheat greatly reduces oviposition by the cereal leaf beetle *Oulema melanopus* (163,164). Hoxie and co-authors (165) state that varieties with dense, long pubescence are highly resistant to both oviposition and larval feeding. Casagrande and Haynes (166) observed ovipositional preference for glabrous plants even in a mixed stand of glabrous and pubescent varieties. The degree of resistance of wheat cultivars can be estimated mathematically without the presence of insects (167). In lucerne, lines with long glandular hairs are mechanically resistant to the seed chalcid *Bruchophagus roddi* (Guss) because the long stiff hairs limit oviposition access to the developing seeds (168,169). The apple codling moth prefers the glabrous upper leaf surface for oviposition over the pubescent lower leaf surface (170). In cotton, of the number of factors found to affect the oviposition of *Helicoverpa armigera*, trichome length on the upper surface of the leaf, rather than density, showed a positive correlation (171).

2.4.4 Trichomes in Relation to Allelochemics

Glandular trichomes exude chemicals that are toxic to insects. The exudates of trichomes on tobacco leaves contain nicotine, anabasine, and probably nornicotine (172), which are toxic to aphids (173) and toxic on contact to the larvae of the tobacco hornworm *Manduca sexta* (L.) (174). Resistance of the wild tomato *Lycopersicum hirsutum f. glabratum* to *Helicoverpa zea* (Boddie) (175) and *Leptinotarsa decemlineata* (176) is probably due to high levels of 2-tridecanone present on the tips of trichomes. Lin and coworkers (177) studied in detail the chemicals in the glandular heads of type VI trichomes of *Lycopersicon* species. Two normal odd-chained ketones—2-undecanone and 2-tridecanone—and one unidentified sesquiterpene compose approximately 95% of the contents of the gland of *L. hirsutum f. glabratum*. In a closely related plant, *L. hirsutum,* two unidentified insecticidal sesquiterpenes account for 6% of the gland contents. In addition, small amounts of one unknown monoterpene and another unknown sesquiterpene are found in type VI glands of a commercial tomato variety (*Lycopersicum esculentum*). Bioassays, comprising the glandular exudate both by direct contact and isooctane extracts of glands on neonate larvae of the gelechiid *Keiferia lycopersicella* (Wlsm.) and the noctuid *Spodoptera exigua* (Hub.), indicated that 2-tridecanone and 2-undecanone are the major insecticidal compounds of *L. hirsutum f. glabratum* and that the two unidentified sesquiterpenes in *L. hirsutum* are also toxic to both species of insects. However, the gland contents of the commercial tomato variety serve only as a physical barrier to *K. lycopersicella* and are not detrimental to *S. exigua*. The quantities of insecticidal chemicals and density of type VI trichomes vary with plant age and location within the plants. There is an interaction between day length and light intensity and density of glandular trichomes that secrete 2-tridecanone. Mortality is higher in *M. sexta* larvae under a long-day regime than under a short-day regime. The trichome exudate from mite-resistant geraniums (*Pelargonium hortorum*) contains two anacardic acid derivatives: *O*-pentadec-7-enylsalicylic acid and *O*-heptadec-7-enylsalicylic acid. The production of these compounds has been correlated with the two complementary dominant genes reported for host resistance to spider mites (178). Clones with glandular trichomes of types A and B were the least preferred, and host selection behavior was probably influenced by volatile compounds present in the exudates of type B trichomes (179). The action of glandular hairs demonstrates the close interaction of physical and biochemical plant defenses. Although repellent to insects, glandular hairs serving as cues in host-plant finding for a few arthropod fauna associated with odorous pubescent plants is evident.

2.5 THE COMBINATION OF FACTORS

Always, in nature, resistance based on a single factor is not possible. This is particularly true if a complex of pests is involved. The picture is incomplete, but a few examples show that several factors interact in the process. Rice resistance to the stem borer *Chilo suppressalis* (Walker) results from the interaction of leaf blades with a hairy upper surface, tight leaf-sheath wrapping, small stems with ridged surface, and thicker hypodermal layers (82). Resistance in cotton to the pink bollworm, *Pectinophora gossypiella*, involves absence of bracts, glabrous leaves, cell proliferation, high gossypol content, and nectarilessness, and high-gossypol plants offer resistance to *Heliothis* spp. (150). Resistance in corn to the corn earworm, *H. zea*, has been ascribed to long husks, tight husks, blunt ear tips, flinty tip kernels, silk balling, and starchiness of kernels (58,59).

The analysis of isolated physical or chemical resistance factors provides only a partial explanation of the total syndrome manifested when an insect is exposed to a resistant plant. Resistance factors have complex complementary, synergistic, or antagonistic interactions with other plant and environmental factors. For instance, sorghums with open heads had fewer *H. zea* larvae than those with closed heads, and this probably was a consequence of larvae being more vulnerable to natural enemies (180). Similarly, certain species of thrips escaped predation within the protection of dense leaf pubescence.

There are also many possible interactions with abiotic factors. Certain plant characteristics, such as pubescence, have a profound effect on the microclimate around the plant. There are changes in reflectivity of light from pubescent surfaces that influence attractiveness to insects. The interaction among various plant species' defenses may function as an antiherbivore resource in ecological time. Monocultural cropping systems largely eliminate such defense. However, it is conceivable that some economically practical species or varietal associations enhance the interactions of defense processes. The use of trap crops is a present example of such man-manipulated strategies for plant resistance or protection.

2.6 INFLUENCE OF PLANT MORPHOLOGICAL FACTORS IN TRITROPHIC (PLANT, HERBIVORE AND CARNIVORE) INTERACTIONS

Natural enemies of herbivores benefit the plants by reducing herbivore pressure. Plants may, in turn, influence carnivores by making herbivores more vulnerable to them (181). But this positive influence does not occur in every situation; plant resistance may adversely affect natural enemies of herbivores. Price (181) tentatively classified the interactions between the three trophic levels (plant, herbivore, and carnivore) as semiochemically mediated, chemically mediated, and physically mediated interactions.

Casagrande and Haynes (166) found that selected pubescent wheat varieties did not adversely affect parasitism of the cereal leaf beetle *Oulema melanopus* (L.) by naturally occurring larval parasites. Conversely, Hulspas-Jordon and van Lenteren (182). found that the walking speed and host-searching ability of the whitefly parasite *Encarsia formosa* Gaham on surfaces of cucumber leaves were inversely proportional to leaf pubescence. Kumar and his associates (183) reported that the searching behavior of the aphid parasite *Trioxa indicus* (Hal.) is influenced by the pubescence of host plant. Parasitism of *Aphis craccivora* Koch is higher on the pigeon pea *Cajanus cajan*, which bears fine epidermal hairs, and less on *Solanum melongena*, which bears stellate trichomes. Treacy et al. (184) found that parasitism was higher on the smooth-leaved cotton, intermediate on hirsute cotton and lowest on pilose cotton. The glandular hairs on tobacco entrap the parasitoid *Lysiphlebus testaceipus* (Cress) and the tobacco hornworm parasites *Trichogramma minutun* Riley and *Apantles congregatus* (Say) (185). Rabb and Bradley (186) have noted exudates of tobacco limit the movement of *T. minutum* on plant surfaces and thus prevent parasitism, which also reduces the level of parasitism by *Telenomus sphingis* (Ashm.). Similar observations have been made by Kantanyukul and Thurston (187), Elsey and Chaplin (188), and Takada and Takenaka (189). Milliron (190) found that glandular trichomes of *Nicotiana glutinosa* entrap *Encarsia formosa*, a parasite of whiteflies. Gerling (191) has reported that the rate of parasitism by *E. formosa* is inversely related to the density of pubescence on the host plant.

Lauenstein (192) reported that pubescent leaf surfaces result in a more thorough searching behavior by the anthocorid predator *Anthocoris nemorum* (L.). Similarly, Shah (193) reported that trichomes on radish and Chinese cabbage cause larvae of the predatory two-spotted lady beetle *Adalia bipunctata* (L.) to change direction of movement frequently, resulting in more encounters with their aphid prey. In contrast, high or low trichome densities may be detrimental to predators (92,194). Larvae of the coccinellid predator *Stethorus punctillum* Weise are impaled by the hooked trichomes on the Scarlet runner bean *Phaseolus coccineus* (195,196) Similar observations have been made on potato and cotton (197,198). Sengonca and Gerlach (199) observed that the trichomes on the bean *Phaseolus vulgaris* adversely affect the predatory thrips, *Scolothrips longicornis* Priesn. Glandular trichomes on a number of crops interfere with the activity of predators (200). The searching behavior of Chrysopa *carnea* Stephens is hindered by the glandular trichomes in tobacco (201). Elsey (202) found that the searching speed of larvae of the lady beetles *Coleomegilla maculata* (Degeer) and *C. carnea* is faster on cotton than on tobacco. He concluded that the large number of trichomes on tobacco seriously reduce the speed of movement of all instars of the coccinellid *Hippodamia convergens* Guerinmeneville and are affected by the sticky exudates of leaf trichomes of tobacco cultivars. The younger the instar and the longer the larvae remain on the leaves, the greater the inhibition of their movement (203).

The activity of some predators is influenced by the general roughness of the leaf surface (204–208). Carter et al. (209) observed that the larvae of the coccinellid *Coccinealla septempunctata* L. drop more often from smoother pea leaves, compared to bean leaves, resulting in a lower rate of predation by the pea aphid *Acyrthosiphon pisum*.

The shape of cotton bracts has a profound effect on the ability of parasites to successfully attack the boll weevil, *Anthonomus grandis* (Boheman) (210,211). Cotton varieties devoid of nectaries reportedly exert an adverse effect on some parasites and predators (212,213). The type of leaf configuration (open or closed) in cabbage varieties affects the level of parasitism of *Pieris rapae* by *Phryxe vulgaris* and *Apantels glomeratus* (214). The shape of fruit sepals has a direct effect on the encyrtid parasite *Anagyrus pseudococci* (Gir.) of the citrus mealybug *Planococcus citri* (Risso) (215). The closed panicles of some sorghum cultivars reduce the effectiveness of parasites and predators against *Helicoverpa armigera* (180). Brussels sprout cultivars with glossy leaves, which appear light green in color, are more attractive to several species of aphid predators and parasites than cultivars with waxy foliage (216,217). Several other studies have also shown that parasites respond to color (218). These results show that morphological traits of plants can have a strong effect on the distribution and abundance of predators. Such predator enhancement may complement plant allelochemical defenses.

2.7 PERSPECTIVES

Host plant characteristic, including morphological, physical, and structural qualities, interfere with such insect behavior as feeding, mating, and oviposition. Thus, plant resistance mechanisms against herbivores are multidimensional (27). Genetic resistance that results from inherited characteristics is more desirable. The ultimate aim of studying different mechanisms of resistance is to develop insect-resistant varieties to minimize crop losses from insect attack. For developing insect-resistant varieties, studies on the genetics of various inherited characteristics are essential. The antixenotic mechanisms may be due to biophysical or biochemical factors or a combination of both. In this chapter, the roles of biophysical characters have been discussed in detail.

The inheritance studies for biophysical characters were done in a few cases (27). Recent developments in plant molecular biology help in identifying the inheritance of individual components of resistance. These studies increase our understanding of the mechanisms of plant resistance to insects and can facilitate the breeding process. For example, in rice, quantitative trait loci (QTLs) related to oviposition and settling behavior of brown planthopper (due to antixenosis factors) was mapped using IR64 x Azucena mapping population. They found three QTLs for each character (219). Groh et al. (220) studied the relationship of maize leaf toughness for resistance to *Diatreae* sp in two different mapping populations and located the QTLs in five different chromosomes 1, 4, 5, 7, and 8. Similarly in tomato, Nienhuis et al. (221) identified the QTL in linkage group D of tomato for type VI trichome density, which confers resistance to numerous insect pests. Maliepaard et al. (222) mapped the QTLs for type IV and type VI glandular trichome densities and greenhouse whitefly resistance. Similarly, in potato, two QTLs each for Type A and Type B trichome density were mapped (223). In soybean, QTLs for antixenosis factors conferring resistance to *H. zea* was mapped (224). So, through mapping QTLs coding for specific plant physical attributes associated with insect resistance and comparing the locations of these QTLs with those identified for the phenotypic expression of resistance to a pest species, valuable insights into the nature of resistance can be obtained. Often these insights have both basic and applied implications that can be employed to develop insect resistance crops more efficiently (225). Linking QTL mapping with current developments in genomics and proteomics will help identify expressed genes for insect resistance using different mapping populations/mutants, and map-based cloning of these genes is possible in future.

REFERENCES

1. Norris, N.M.; Dale, M.; Kogan, M. Biochemical and morphological basis of resistance. In *Breeding Plants Resistant to Insects*; Maxwell, F.G., Jennings, P.R., Eds.; John Wiley: New York, 1980; 23–62.
2. Cartier, J.J.; Varietal resistance of peas to the pea aphid biotypes under field and greenhouse conditions. J. Econ. Entomol. **1963**, *56*, 205–213.
3. Kennedy, J.S.; Booth, C.O.; Kershaw, W.J.S. Host findings by aphids in the field. III. Visual attraction. Ann. Appl. Biol. **1961**, *49*, 1–21.
4. Radcliffe, E.B.; Chapman, R.K. The relative resistance to insect attack of three cabbage varieties at different stages of plant maturity. Ann. Entomol. Soc. Am. **1965**, *58*, 897–902.
5. Radcliffe, E.B.; Chapman, R.K. Varietal resistance to insect attack in various cruciferous crops. J. Econ. Entomol. **1966**, *59*, 120–125.
6. Stephens, S.G. Sources of resistance of cotton strains to the boll weevil and their possible utilization. J. Econ. Entomol. **1957**, *50*, 415–418.
7. Dunn, J.A.; Kempton, D.P.H. Varietal differences in the susceptibility of Brussels sprouts to lepidopterous pests. Ann. Appl. Biol. **1976**, *82*, 11–19.

8. Verma, T.S.; Bhagchandani, P.M.; Narendra Singh; Lal, O.P. Screening of cabbage germplasm collections for resistance to *Brevicoryne brassicae* and *Pieris brassicae*. Indian J. Agric. Sci. **1981**, *51*, 302–305.

9. Singh, B.U.; Rana, B.S.; Rao, N.G.P. Host plant resistance to mite (*Oligonychus indicus* Hirst) and its relationship with shootfly (*Atherigona soccata* Rond.) resistance in sorghum. J. Entomol. Res. **1981**, *5*, 25–30.

10. Peregrine, W.T.H.; Catling, W.S. Studies on resistance in oats to the fruit fly. Plant Pathol. **1967**, *16*, 170–175.

11. Chiang, H.S.; Norris, D.M. "Purple stem" a new indicator of soybean stem resistance to bean flies (Diptera: Agromyzidae). J. Econ. Entomol. **1984**, *77*, 121–125.

12. Dean, P. Cotton leaf shape affects insects. Agric. Res. USA. **1982**, *30*, 16.

13. Mote, U.N.; Shahane, A.K. Biophysical and biochemical characters of sorghum varieties contributing resistance to delphacid aphid and leaf sugary exudations. Indian J. Entomol. **1994**, *56*, 113–122.

14. De Oliverira, A.P.; Castellane, P.D.; de Bortoli, S.A. Insertion angle of leaves of garlic versus thrips population. Agropecuaria catarinense **1995**, *8*, 5–6.

15. Islam, Z.; Karim, A.N.M.R. Leaf folding behavior of *Caphalocrocis medinalis* (Guenee) and *Marasmia patnalis* Bradley, and the influence of rice leaf morphology on damage inicidence. Crop Prot. **1997**, *16*, 215–220.

16. Pathak, M.D.; Andres, F.; Galacgae, N.; Raros, R. Resistance of rice varieties to striped rice borers. Tech. Bull. II. **1971**, p. 69.

17. Ukwungwa, M.N. Effects of the silica content of rice plants on the damage caused by larvae of *Chilo zacconius* (Lepidoptera: Pyralidae). ADRAO **1984**, *5*, 21–22.

18. Ukwungwa, M.N.; Odebiyi, J.A. Incidence of *Chilo zacconius* on some rice varieties in relation to plant characters. Insect Sci. Appl. **1985**, *6*, 653–656.

19. Bothe, N.N.; Pokharkar, R.N. Role of silica content in sorghum for reaction to shootfly. J. Maharashtra Agric. Univ. (India) **1985**, *10*, 338–339.

20. Swarup, V.; Chaugale, D.S. A preliminary study on resistance to stem borer, *Chilo zonellus* (Swintoe) infestation on sorghum, *Sorghum vulgare pers.* Curr. Sci. **1962**, *31*, 163–164.

21. Panda, N.; Pradham, B.; Samalo, A.P.; Prakasa Rao, P.S. Note on the relationship of some biochemical factors with the resistance in the varieties to yellow rice borer. Indian J. Agric. Sci. **1975**, *45*, 499–501.

22. Subbarao, D.V.; Perraju, A. Resistance in some rice strains to first instar of *Tryporyza incertulas* (Walker) in relation to plant nutrients and anatomical structure of the plants. IRRN. **1976**, *1*, 14–15.

23. Mishra, B.K.; Sontakke, B.K.; Mohapatra, H. Antibiosis mechanism of resistance in rice varieties to yellow stemborer, *Scirpophaga incertulas* Walker. Indian J. Plant Prot. **1990**, *18*, 81–83.

24. Singh, B. Studies on resistance in cotton to jassid, *Amarasca devastans* (Distant). Ph.D thesis, Punjab Agricultural University, Ludhiana, India, 1970.

25. Chelliah, S.; Sambandam, C.N. Evaluation of muskmelon (*Cucumis melo* L.) accessions and *Cucumis callosus* (Rottl) for resistance to the fruit fly (*Dacus cucurbitae*) Indian J. Hort. **1974**, *31*, 346–348.

26. Singh, R. Bases of resistance in okra to *Amrasca biguttula biguttula*. Indian J. Agric. Sci. **1988**, *58*, 15–19.
27. Panda, N.; Khush, G.S. Host plant resistance to Insects. CAB Int. and IRRI, Wallingford, Oxon, UK, 1995.
28. Anstey, T.H.; Moore, J.F. Inheritance of glossy foliage and cream petals in green sprouting broccoli. J. Hered. **1954**, *45*, 39–41.
29. Thompson, K.F. Resistance to the cabbage aphid (*Brevicoryne brassicae*) in *Brassica* plants. Nature. **1963**, *198*, 209.
30. Lupton, F.G.H. The use of resistance varieties in crop protection. World Rev. Pest. Control. **1967**, *6*, 47–58.
31. Jayaraj, S. Research on resistance in plants to insects and mites in Tamil Nadu. Tamil Nadu Agricultural University, Coimbatore: India, 1976; 65.
32. Atkin, D.S.J.; Hamilton, R.J. Surface of *Sorghum bicolor*. In *The Plant Cuticle*; Cutler, D.F., Alvin, K.L., Price, C.E., Eds.; Academic Press: London, 1982; 231–236.
33. Chandrasekaran, N.R.; Navakodi, K.; Shetty, B.K.; Ramaswamy, N.M. A preliminary study on the varietal resistance in castor to attack by mites. Indian Oilseeds J. **1964**, *8*, 46–48.
34. Lowe, H.J.B.; Murphy, G.J.P.; Parker, M.L. Non-glaucousness, a probable aphid resistance character of wheat. Ann. Appl. Biol. **1985**, *106*, 555–560.
35. Starks, K.J.; Jasid.; Weibel, D.E. Resistance in bloomless and sparse-bloom sorghum to greenbugs. Environ. Entomol. **1981**, *10*, 963–965.
36. Woodhead, S. Surface chemistry of *Sorghum bicolor* and its importance in feeding by *Locusta migratoria*. Physiol. Ent. **1983**, *8*, 345–352.
37. Rutherford, R.S.; Van Staden, J. Towards a rapid near-infrared technique for prediction of resistance to sugarcane borer *Eldana saccharina* Walker (Lepido Pyralidae) using stalk surface wax. Chem. Eco. **1996**, *22*, 681–694.
38. Molenaar, N.D. Genetics of thrips (*Thrips tabaci* L.) resistance and epicuticular wax characteristics of nonglossy and glossy onions (*Allium cepa* L.) Ph.D thesis, University of Wisconsin, Madison, 1984.
39. Bernays, E.A.; Chapman, R.F.; Woodhead, S. Behavior of newly hatched larvae of *Chilo partellus* (Swinhoe) (Lepidoptera Pyralidae) associated with their establishment in the host plant, sorghum. Bull. Entomol. Res. **1983**, 73, 75–83.
40. Sanford, D.E.; Espelie, K.E.; Shelton, A.M. Behaviour of neonate diamond backmoth larvae *Plutella xylostella* on leaves and on extracted leaf waxes of resistant and susceptible cabbages. J. Chem. Ecol. **1991**, *17*, 1697–1704.
41. Kingauf, F.; Nocker-Wenzel, K.; Rottger, U. Die Rolle peripherer Pflauzenwachse fur den Befall durh phytophage Insekten. Pflanzenkrankh, Z. Pflanzenschutz **1978**, *85*, 228–237.
42. Chapman, R.F.; Bernays, E.A. Insect behaviour at the leaf surface and learning aspects of host plant selection. Experientia. **1989**, *45*, 215–222.
43. Mathes, R.; Charpentier, L.J. Some techniques and observations in studying the resistance of sugarcane borer in Louisiana. Proc. Int. Soc. Sug. Cane. Technol. *II*, **1963**, 594–603.
44. Agarwal, R.A. Morphological characteristics of sugarcane and insect resistance. Entomol. Exp. Appl. **1969**, *12*, 767–776.

45. David, H. A critical evaluation of the factors associated with resistance to internode borer, *Chilo sacchariphagus indicus* (K.) in *Saccharum* sp., allied genera and hybrid sugarcane. Ph.D thesis, Calicut University, Calicut, India, 1979.

46. Sithanantham, S. Varietal incidence of the pink mealybug, *Saccharicoccus sacchari* (Ckll.) in sugarcane during different ages. Co.op. Sug. (India) **1983**, *4*, 1–4.

47. Mehta, U.K.; Jayanthi, R.; David, H. Occurrence of pink mealy bug on certain early maturing sugarcane varieties. Pestol. **1981**, *5*, 9–10.

48. Lukefahr, M.J.; Griffin, J.A. The effects of food on the longevity and fecundity of pink bollworm moths. J. Econ. Entomol. **1956**, *49*, 876–877.

49. Davis, D.D.; Ellingtan, J.J.; Brown, J.C. Mortality factors affecting cotton insects. Resistance of smooth and nectariless characters in acala cottons to *H. zea*, *P. gossypiella* and *Trichoplusia ni*. J. Environ. Qual. **1973**, *2*, 530–535.

50. Bailey, J.C.; Scales, A.L.; Meredith, W.R. Tarnished plant bug (Heteroptera: Miridae) nymph numbers decreased on caged nectariless cotton. J. Econ. Entomol. **1984**, *77*, 68–69.

51. Baloch, A.A.; Soomro, B.A.; Mallah, G.H. Evaluation of some cotton varieties with known genetic markers for their resistance/tolerance against sucking pests and bollworm complex. Turkiya Bitki Koruma Dergisi. **1982**, *6*, 3–14.

52. G.D Butler Jr, Wilson, F.D. Activity of adult whiteflies (Homoptera: Aleyrodidae) within plantings of different cotton strains and cultivars as determined by sticky-trap catches. J. Econ. Entomol. **1984**, *77*, 1137–1140.

53. Agarwal, R.A.; Singh, M.; Katiyar, K.N. Basis of resistance in cotton varieties to grey weevil. Indian J. Agric. Sci. **1976**, *44*, 608–610.

54. Lincoln, C.; Waddle, B.A. Insect resistance of frego-type cotton. Arkansas Farm Res. **1966**, *15*, 4–5.

55. Jenkins, J.N.; Parrott, N.L. Effectiveness of frego bract as a boll weevil resistance character in cotton. Crop Sci. **1971**, *11*, 739–743.

56. Thombre, M.V. Association of some morphological characters to cenfer resistance to bollworms in *Gossipium hirsutum* cotton. Andhra Agric. J. (India) **1980**, *27*, 19–20.

57. Leigh, T.F.; Hyer, A.F.; Rice, R.E. Frego bract condition of cotton in relation to insect populations. Environ. Entomol. **1972**, *1*, 390–391.

58. Luckmann, W.H.; Rhodes, A.M.; Wann, E.V. Silk balling and other factors associated with resistance of corn to corn earworm. J. Econ. Entomol. **1964**, *57*, 778–779.

59. Wiseman, B.R.; Widstorm, N.W.; McMillian, W.W. Ear characteristics and mechanisms of resistance among selected corns to corn earworm. Flo. Entomol. **1977**, *60*, 97–103.

60. Vargas, R.; Nishida, T. Evaluation of corn earworm damage and incidence of blank tip in corn grown in Hawaii and its significance in pest management. Proc. Hawaiian Ent. Soc. **1978**, *23*, 455–461.

61. Osuna Ayala, J.; De Araiyp, S.M.C.; Lara, F.M.; M-De,; Caetano, F. Analysis and selection of characters associated with resistance to the

earworm, *Heliothis zea* (Boddie, 1950) (Lepidoptera: Noctuidae) in the dent composite Jaboticabal, XI cycles of mass selection. Anais da Sociedade Entomologica do Brazl 1983.
62. Agarwal, R.A.; Verma, R.S.; Bharaj, G.S. Screening of sorghum lines for resistance against shoot bug, *Peregrinus maidis* (Ashmead) (Homoptera, Delphacidae), JNKVV Res. J. **1978**, *12*, 116.
63. Zhu, J.L. Preliminary observations on the resistance of varieties of spring wheat to *Hydrellia griseola* (Fallen). Insect Knowledge **1981**, *18*, 213–214.
64. Small, E.; Brookes, B.S. Coiling of alfalfa pods in relation to resistance against seed chalcids. Can. J. Plant Sci. **1982**, *62*, 131–135.
65. Small, E.; Brookes, B.S. Coiling of alfalfa pods in relation to resistance against seed chalcids: additional observations. Can. J. Plant Sci. **1984**, *64*, 659–665.
66. Malik, A.S.; Dhankhar, B.S.; Sharma, N.K. Variability and correlations among certain characters in relation to shoot and fruit borer infestation in brinjal. Haryana Agric. Univ. J. Res. (India) **1986**, *16*, 259–265.
67. Platt, A.W.; Farstad, C.M. The reaction of wheat varieties to wheat stem sawfly attack. Sci. Agric. **1946**, *26*, 231–247.
68. Painter, R.H. Insect Resistance in Crop plants. Macmillan: New York, **1951**, *12*, 175–185.
69. Wallace, L.E.; McNeal, F.H.; Berg, M.A. Minimum stem solidness required in wheat for resistance to the wheat stem sawfly. J. Econ. Entomol. **1973**, *66*, 1121–1123.
70. Martin, G.A.; Richard, C.A.; Hensley, S.D. Host resistance to *Diatraea saccharalis* (F)—relationship of sugarcane internode hardness to larval damage. Environ. Entomol. **1975**, *4*, 687–688.
71. Khanna, K.L.; Panje, R.R. Studies on the rind hardness of sugarcane, 1. Anatomy of the stalk and rind hardness. Indian J. Agric. Sci. **1939**, *9*, 1–4.
72. Rao, Tuljaram, J. Rind hardness as a possible factor in resistance of sugarcane varieties to the stem borer. Curr. Sci. **1941**, *8*, 365–366.
73. Rao, Tuljaram, J. Correlation between rind hardness and resistance to stem borer in sugarcane. M.Sc. thesis, Madras University, Madras, 1951.
74. Rao, D.; Seshagiri.; Puttarudriah, M.; Shastry, K.S.S. Influence of borer attack in several varieties of sugarcane grown in Visweshwarayya Canal tract, Mysore State. Proc. Int. Soc. Sug. Cane. Technol. **1956**, *9*, 895–901.
75. Rao, Siva, D.V. Studies on the resistance of sugarcane to the early shoot borer, *Chilo infuscatellus* Snell. M.Sc. thesis. Andhra University, Waltair, India, 1962.
76. Rao, Siva, D.V. Hardness of sugarcane varieties in relation to shoot borer infestation. Andhra Agric. J. (India) **1967**, *14*, 95–105.
77. Howe, W.L. Factors affecting the resistance of certain cucurbits to the squash borer. J. Econ. Entomol. **1949**, *42*, 321–326.
78. Quiras, C.T.; Stevens, M.A.; Rick, C.M.; Kokyokomi, M.K. Resistance in tomato to the pink form of the potato aphid (*Macrosiphum euphorbiae* Thomas). The role of anatomy, epidermal hairs, and foliage composition. J. Am. Soc. Hort. Sci. **1977**, *102*, 166–171.

79. Brewer, G.J.; Horber, E.; Sorensen, E.L. Potato leafhopper (Homoptera: Cicadellidae)-antixenosis and antibiosis in *Medicago* species. J. Econ. Entomol. **1986**, *79*, 421–425.

80. Tanton, M.T. The effect of leaf "toughness" on the feeding of larvae of the mustard beetle, *Phaedon cochlaariae* Fab. Entomol. Exp. Appl. **1962**, *5*, 74–78.

81. Cuthbert, F.P., Jr.; Feng, R.L.; Chambliss, U.L. Breeding for resistance to cowpea curculio in southern peas. Hort. Sci. **1972**, *9*, 69–70.

82. Patanakamjorn, S.; Pathak, M.D. Varietal resistance of rice to the Asiatic rice borer, *Chilo suppressalis* and its association with various characters. Ann. Ent. Soc. Amr. **1967**, *60*, 287–292.

83. Israel, P. Varietal resistance to rice stem borers in India. In *The Major Insect Pests of the Rice Plant. IRRI*; John Hopkins Press: Baltimore, MD, 1967, 391–403.

84. Chaudhary, R.C.; Khush, G.S.; Heinrichs, E.A. Varietal resistance to rice stem borers in Asia. Insect. Sci. Appl. **1984**, *5*, 447–463.

85. Blum, A. Anatomical phenomena in seedlings of sorghum varieties resistant to the sorghum shootfly (*Atherigona varia soccata*). Crop Sci. **1968**, *8*, 388–390.

86. Panda, N.; Mahapatra, A.; Sahoo, M. Field evaluation of some brinjal varieties for resistance to shoot and fruit borer (*Leucinodes orbonalis* Guen) Indian J. Agric. Sci. **1971**, *41*, 597–601.

87. Kadapa, S.N.; Vizia, N.C.; Patel, N.B. A note on stem tip stiffness in aphid tolerant cottons. Curr. Sci. **1998**, *57*, 265–266.

88. Uphof, J.C. Plant hairs. In *Handbuch der Pflanizenantomie 4(5)*, *Histolog*; Zimmerman, W., Ozenda, P.G., Eds.; Gebruder Borntreeger: Berlin, 1962; 1–206.

89. Hummel, K.; Staesche, K. Die Verbreitung der Haartypen in den naturlichen Verwandts-Chaftsgruppen. In *Handbuch der-Pflanzenantomie 4(5)*, *Histolog*; Zimmerman, W., Ozenda, P.G., Eds.; Gebruder Borntracger: Berlin, 1962; 207–250.

90. Johnson, H.B. Plant pubescence. An ecological perspective. Bot. Rev. **1975**, *41*, 233–258.

91. Webster, J.A. Association of plant hairs and insect resistance. An annotated bibliography. U.S. Dept. Agric. ARS Misc. Publ. 1297, **1975**, 1–18.

92. Johnson, B. The injurious effects of the hooked epidermal hairs of the French beans (*Phaseolus vulgaris* L.) on *Aphis craccivora* Koch. Bull. Entomol. Res. **1953**, *44*, 779–788.

93. Lampe, U. Examinations of the influences of hooked epidermal hairs of French beans (*Phaseolus vulgaris*) on the pea aphid *Acyrthosiphon pisum*. In *Proc 5th Int. Symp. Insect–plant Relationships*; Visser, J.H., Minks, A.K., Eds.; Wageningen: The Netherlands, 1982; 419.

94. Manglitz, G.R.; Kehr, W.R. Resistance to spotted alfalfa aphid (Homoptera:Aphididae) in alfalfa seedlings of two plant introductions. J. Econ. Entomol. **1984**, *77*, 357–359.

95. Roberts, J.J.; Foster, J.E. Effect of leaf pubescence in wheat on the bird cherry oat aphid (Homoptera: Aphididae). J. Econ. Entomol. **1983**, *76*, 1320–1322.

96. Poos, F.W.; Smith, F.F. Comparison of oviposition and nymphal development of *Empoasca fabae* (Harris) on different host plants. J. Econ. Entomol. **1931**, *24*, 361–371.

97. Lee, Y.I.; Kogan, M.; Larsen, J.R. Jr. Attachment of the potato leafhopper to soybean plant surfaces as affected by morphology of the pretarsus. Ent. Exp. Appl. **1986**, *42*, 101–107.

98. Gilbert, L.E. Butterfly–plant co-evolution: has *Passiflora adenopoda* won the selectional race with Heliconiine butterflies? Science **1971**, *172*, 585–586.

99. Rathake, B.J.; Poole, R.W. Coevolutionary race continues: butterfly larval adaptation to plant trichomes. Science **1975**, *187*, 175–176.

100. Smith, R.L.; Wilson, R.L.; Wilson, F.D. Resistance of cotton plant hairs to mortality of first instar of the pink bollworm. J. Econ. Entomol. **1975**, *68*, 679–683.

101. Wilson, R.L.; Wilson, F.D. Effects of cotton differing in pubescence and other characters on pink bollworms in Arizona. J. Econ. Entomol. **1977**, *70*, 196–198.

102. Ramalho, F.S.; Parrott, W.L.; Jenkins, J.N.; McCarty, J.C. Effects of cotton leaf trichomes on the mobility of newly hatched tobacco budworms (Lepidoptera: Noctuidae). J. Econ. Entomol. **1984**, *77*, 619–621.

103. Gibson, R.W. Glandular hairs providing resistance to aphids in certain wild potato species. Ann. Appl. Biol. **1971**, *68*, 113–119.

104. Lapointe, S.L.; Tingey, W.M. Feeding response of the green peach aphid (Homoptera: Aphididae) to potato glandular trichomes. J. Econ. Entomol. **1984**, *77*, 386–389.

105. Tingey, W.M.; Laubengayer, J.E. Defense against the green peach aphid and potato leafhopper by glandular trichomes of *Solanum berthaultii*. J. Econ. Entomol. **1981**, *74*, 721–725.

106. Lapointe, S.L.; Tingey, W.M. Glandular trichomes of *Solanum neocardenasii* confer resistance to green peach aphid (Homoptera: Aphididae). J. Econ. Entomol. **1986**, *79*, 1264–1268.

107. Kisha, J.S.A. Whitefly, *Bemisia tabaci* infestations on tomato varieties and a wild *Lycopersicon* species. Ann. Appl. Biol. **1984**, *104*, 124–125.

108. Gibson, R.W. Glandular hairs on *Solanum polyadenium* lessen damage by the Colarado beetle. Ann. Appl. Biol. **1976**, *82*, 147–150.

109. Finney, C.L. A technique for mass culture of the six-spotted mite. J. Econ. Entomol. **1953**, *46*, 712–713.

110. Huffaker, C.B. Experimental studies on predation: dispersion factors and predator–prey oscillations. Hilgardia. **1958**, *27*, 343–383.

111. Khanna, K.L.; Ramanathan, K.R. Studies on the association of plant characters and pest incidence. 1. Structure of leaf surface and mite attack in sugarcane. Proc. Natl. Inst. Sci. India. **1947**, *13*, 327–329.

112. Pallard, D.G.; Saunders, J.H. Relations of some cotton pests to jassid resistant Sakel. Emp. Cotton Gr Rev. **1956**, *33*, 197–202.

113. Palanisamy, S. Ecology and host resistance of carmine spidermite, *Tetranychus cinnabarinus* (Boisduval) in brinjal. Ph.D thesis, Tamil Nadu Agricultural University, Coimbatore, India, 1984.
114. Goonewardene, H.F.; Williams, E.B.; Kwolek, W.F.; McCabe, L.D. Resistance to European red mite, *Panonychus ulmi* (Koch) in apple. J. Am. Soc. Hort. Sci. **1976**, *101*, 532–537.
115. Massee, A.M. A new method of rearing fruit tree red spider (*Oligonychus ulmi* Koch) on apple seedlings. Rept. E Malling Res. Stn. **1974**, 122–123.
116. Kamel, S.A.; El Kassaby, F.Y. Relative resistance of cotton varieties in Egypt to spider mites, leafhoppers and aphids. J. Econ. Entomol. **1965**, *58*, 209–212.
117. Stoner, A.K.; Frank, J.A.; Gentile, G.A. The relationship of glandular hairs on tomatoes to spider mite resistance. Proc. Am. Soc. Hort. Sci. **1968**, *93*, 532–538.
118. Gentile, A.G.; Webb, R.E.; Stoner, A.K. *Lycopersicon* and *Solanum* spp. resistant to the carmine and the two-spotted spider mite. J. Econ. Entomol. **1969**, *62*, 834–836.
119. Rodriguez, J.G.; Knavel, D.E.; Aina, O.J. Studies on the resistance of tomatoes to mite. J. Econ. Entomol. **1972**, *53*, 491–495.
120. Aina, O.J.; Rodriguez, J.G.; Knavel, D.E. Characterizing resistance to *T. urticae* in tomato. J. Econ. Entomol. **1972**, *65*, 641–643.
121. Patterson, G.G.; Thurston, R.; Rodriguez, J.G. Two-spotted spider mite resistance in *Nicotiana* species. J. Econ. Entomol. **1974**, *67*, 341–343.
122. Stalk, R.S. Morphological and biochemical factors relating to spider mite resistance in the Pelargonium. Ph.D thesis, Pennsylvania State University, 1975.
123. Rodriguez, J.G.; Dabrowski, Z.T.; Stoltz, L.P.; Chaplin, C.E.; Smith, W.O. Studies in the resistance of straw-berries to mites. II. Preference and non-preference responses of *Tetranychus urticae* and *T. turkenstani* to water-soluble extracts of foliage. J. Econ. Entomol. **1971**, *64*, 383–387.
124. Dabrowski, Z.T.; Rodriguez, J.G.; Chaplin, C.E. Studies in resistance of strawberries to mites. IV. Effects of season on preference and non-preference of strawberries to *Tetranychus urticae*. J. Econ. Entomol. **1971**, *64*, 806–809.
125. Regev, S.; Cone, W.W. Chemical differences in hop varieties vs. susceptibility to the two spotted spider mite. Environ. Entomol. **1975**, *4*, 697–700.
126. Lal, K.B. Anti-jassid resistance in the cotton plant. Curr. Sci. **1937**, *6*, 88–89.
127. Hussain, M.A.; Lal, K.B. The bionomics of *Empoasca devastans* Dist. on some varieties of cotton in the Punjab. Indian J. Ent. **1940**, *2*, 123–126.
128. Afzal, M. Present position as regards breeding for jassid resistance in cotton. Indian, Central Cotton Committee Second Conference. Scientific Research Workers on Cotton in India paper No. 1, 1941.
129. Afzal, M.; Ghani, M.A. Cotton jassids in the Punjab. Sci Mongr Pakistan Assoc. Adv. Sci. **1953**, *2*, 102.
130. Sikka, S.M.; Sahni, V.M.; Butani, D.K. Studies on jassid resistance in relation to hairiness of cotton leaves. Euphytica. **1966**, *15*, 383–388.
131. Parnell, F.R. Breeding jassid resistant cottons. Emp. Cotton Gr. Rev. **1925**, *2*, 330–336.

132. May, A.W.S. Jassid resistance of the cotton plant. Q. J. Agric. Sci. **1951**, *8*, 43–68.

133. Fukuda, J. Ecological studies on the cotton leafhopper, *Chloritta biguttula* Ishida and its control. 1. On the resistance of egg plants to the cotton leafhopper. Natl TokaiKink Agr. Exp. Sta. Bull, **1952**, *1*, 159–211.

134. Parnell, F.R.; King, H.E.; Ruston, D.F. Jassid resistance and hairiness of the cotton plant. Bull. Entomol. Res. **1949**, *39*, 539–575.

135. Uthamasamy, S. Studies on host-resistance in certain Okra [*Abelmoschus esculentus* (L.) Monech] varieties to the leafhopper, *Amrasca devastans* (Dist.) (Cicadellidae–Homoptera). Ph.D thesis, Tamil Nadu Agric. Univ, Coimbatore, India, 1979.

136. Balraj Singh.; Atwal, A.S. Role of physico-chemical characteristics of the leaves of cotton varieties in imparting resistance to *Amrasca biguttula biguttula* (Ishida). Indian J. Ecol. **1976**, *3*, 172–180.

137. Uthamasamy, S. Influence of leaf hairiness on the resistance of bhendi or lady's finger, *Abelmoschus esculentus* (L.) Monech, to the leafhopper *Amrasca devastans* (Dist.) Tropical Pest. Mgmt. **1985**, *31*, 294–295.

138. Sambandam, C.N.; Chelliah, S.; Natarajan, K. A note on fox-face, *Solanum mammosum* L, a source of resistance to *Aphis gossypii* G. and *Epilachna vigintioctopunctata* F. Annamalai Agric. Res. Annu. **1969**, *1*, 110–112.

139. Khan, Z.A.; Ward, J.T.; Norris, D.M. Role of trichomes in resistance to cabbage looper, *Trichoplusia ni*. Entomol. Exp. Appl. **1986**, *42*, 109–117.

140. Schillinger, J.S.; Gallun, R.L. Leaf pubescence of wheat as a deterrent to the cereal leaf beetle, *Oulema melanopus*. Ann Entomol. Soc. Am. **1968**, *61*, 900–903.

141. Kogan, M. Intake and utilization of natural diets by the Mexican beetle *Epilachna varivestis*: a multivariate analysis. In *Insect and Mite Nutrition. Significance and Implication in Ecology and Pest Management*. Rodriguez, J.G., Ed.; North-Holland: Amsterdam, 1972; 107–126.

142. Wellso, S.D. Cereal leaf beetle: larval feeding, orientation, development and survival on four small grain cultivars in the laboratory. Ann. Entomol. Soc. Am. **1973**, *66*, 1201–1208.

143. Baruah, P.; Dutta, S.K. Association of physico-chemical character of green gram varieties with nymphal period, body weight and fecundity of *Riptortus linearis* (Fab.) (Heniptera: Alydidae). Legume Res. **1994**, *17*, 167–174.

144. Mizukoshi, T.; Kakizaki, M. Influence of trichomes on kidney bean leaves to the development of the foxglove aphid, *Aulacorthum solani* (Homoptera:Aphididae). Ann. Rep. Soc. Plant Prot. N. Jpn. **1995**, *46*, 142–146.

145. Agarwal, R.A.; Krishnananda, N. Preference to oviposition and antibiosis mechanism to jassids (*Amrasca devastans* Dist.) in cotton (*Gossypium* sp.). Symp. Biol. Hung. **1976**, *16*, 13–22.

146. Teli, V.S.; Dalaya, V.P. Varietal resistance in okra to *Amrasca biguttula biguttula* (Ishida). Indian J. Agri. Sci. **1981**, *51*, 729–731.

147. Khan, Z.R.; Agarwal, R.A. Ovipositional preference of jassid, *Amrasca biguttula* Ishida on cotton. J. Ent. Res. **1984**, *8*, 78–80.

148. Gillham, F.E.M. A study in the response of bollworm *Heliothis zea* to different genotypes of upland cotton. In *Proc 5th Annu. Cotton Improvement Conf. Dallas, Texas*, 1963; 153.

149. Lukefahr, M.J.; Martin, D.F.; Meyer, J.R. Plant resistance to five lepidoptera attacking cotton. J. Econ. Entomol. **1965**, *58*, 516–518.

150. Lukefahr, M.J.; Bottger, G.T.; Maxwell, F.G. Utilization of gossypol as a source of insect resistance. Proc Ann Cotton Dis Council. Cotton Def Phys Conf Cotton Impr Conf National Cotton Council, Memphis, Tennessee, 1966; 215–222.

151. Lukefahr, M.J.; Shaver, T.N.; Parrott, W.L. Sources and nature of resistance in *Gossypium hirsutum* to bollworm and tobacco budworm. Paper presented in Beltwide Cotton Production Research Conferences, Dallas, Texas, 1969.

152. Lukefahr, M.J.; Houghtaling, J.E.; Graham, H.M. Suppression of *Heliothis* population with glabrous cotton. J. Econ. Entomol. **1971**, *64*, 486–488.

153. Stadelbackff, R.A.; Scales, A.L. Technique for determining oviposition preference of the bollworm and tobacco budworm for varieties and experimental stocks of cotton. J. Econ. Entomol. **1973**, *66*, 418–421.

154. Oliver, B.F.; Maxwell, F.G.; Jenkins, J.N. A comparison of the damage done by the bollworm to glanded and glandless cottons. J. Econ. Entomol. **1970**, *63*, 1328–1329.

155. Lukefahr, M.J.; Houghtaling, J.E.; Gruhm, D.G. Suppression of *Heliothis* spp. with cottons containing combinations of resistant characters. J. Econ. Entomol. **1975**, *68*, 743–746.

156. Mehta, R.C. Survival and egg production of the cotton spotted bollworm, *Earias fabia* Stoll. (Lepidoptera: Noctuidae) in relation to plant infestation. Appl. Entomol. Zol. **1971**, *6*, 206–209.

157. Agarwal, R.A.; Katiyar, K.N. Ovipositional preference and damage by spotted bollworm (*Earias fabia* Stoll.) in cotton. Cotton. Dev. **1974**, *4*, 28–30.

158. Sellammal Murugesan. The spotted bollworm, *Earias vitella* (Fabricius): ecology and host resistance. Ph.D thesis, Tamil Nadu Agricultural University, Coimbatore, India, 1982.

159. Sharma, H.C.; Agarwal, R.A. Oviposition behaviour of spotted bollworm, *Earias vitella* Fab. on some cotton genotypes. Insect Sci. Appl. **1983**, *4*, 373–376.

160. Benedict, J.H.; Leigh, T.F.; Hyer, A.H. *Lygus hesperus* (Homoptera: Miridae) oviposition behaviour, growth and survival in relation to cotton trichome density. Environ. Entomol. **1983**, *12*, 331–335.

161. Maiti, R.K.; Bidinger, F.R.; Reddy, K.V.S.; Gibson, P.; Davies, J.C. Nature and occurrence of trichomes in sorghum lines with resistance to sorghum shootfly. Joint report of ICRISAT Hyderabad, India, 1980; 40.

162. Singh, S.P.; Jotwani, M.G. Mechanism of resistance in sorghum to shootfly. IV. Role of morphological characters of seedlings. Indian J. Entomol. **1980**, *42*, 806–808.

163. Gallun, R.L.; Roberts, J.J.; Finny, R.E.; Patterson, F.L. Leaf pubescence of field grown wheat: a deterrent to oviposition by the cereal leaf beetle. J. Environ. Qual. **1973**, *2*, 333–334.

164. Lampert, E.P.; Haynes, D.L.; Sawyer, A.J.; Jokinen, D.P.; Wetiso, S.G.; Gallun, R.L.; Roberts, J.J. Effects of regional releases of resistant wheats on the population dynamics of the cereal leaf beetle (Coleoptera: Chrysomelidae). Ann. Entomol. Soc. **1983**, *76*, 972–980.

165. Hoxie, R.P.; WeUso, S.G.; Webster, J.A. Cereal leaf beetle response to wheat tobacco length and density. Environ. Entomol. **1975**, *4*, 365–370.

166. Casagrande, R.A.; Haynes, D.L. The impact of pubescent wheat on the population dynamics of the cereal leaf beetle. Environ. Entomol. **1976**, 153–159.

167. Wellso, S.G. Cereal leaf beetle (Coleoptera: Chrysomelidae) and winter wheat: host plant resistance relationships. Great Lakes Entomol. **1986**, *19*, 191–197.

168. Pierce, R. Hairy sticky alfalfa fights off insects. Agric. Res. (USA) **1983**, *32*, 6–7.

169. Brewer, G.J. Resistance of pubescent Medicago species to the alfalfa seed chalcid, *Bruchophagus noddi* (Gass.). Ph.D thesis, Kansas State Univ. Manhattan, Kansas, 1985.

170. Plourde, D.F.; Goonewardene, H.F.; Kwolek, W.F. Pubescence as a factor in codling moth, oviposition and fruit entry in five apple selections. Hort. Sci. **1985**, *20*, 82–84.

171. Butter, N.S.; Surjit Singh. Ovipositional response of *Helicoverpa armigera* to different cotton genotypes. Phytoparasitica **1996**, *24*, 97–102.

172. Thurston, R.; Parr, J.C.; Smith, W.T. The phylogeny of *Nicotiana* and resistance to insects. Fourth Int Tobacco Science Congr Proc Natl Tobacco Board, Athens, Greece, 1966; 424–430.

173. Thurston, R.; Webster, J.A. Toxicity of *Nicotiana gossei* Domin to *Myzus persicae* (Sulzer). Entomol. Exp. Appl. **1962**, *5*, 223–238.

174. Thurston, R.; Smith, W.T.; Cooper, B. Alkaloid secretion by trichomes of *Nicotiana* species and resistance to aphids. Entomol. Exp. Appl. **1966**, *9*, 428–432.

175. Dimoch, M.B.; Kennedy, G.G. The role of glandular trichomes in the resistance of *Lycopersicum hirsutum* f. *glabratum* to *Heliothis zea*. Entomol. Exp. Appl. **1983**, *33*, 263–268.

176. Kennedy, G.G.; Sorenson, C.F. Role of glandular trichomes in the resistance of *Lycopersicon hirsutum* f. *glabratum* to Colorado potato beetle (Coleoptera: Chrysomelidae). J. Econ. Entomol. **1985**, *78*, 547–551.

177. Lin, S.Y.H.; Trumble, J.T.; Kumamoto, J. Activity of volatile compounds in glandular trichomes of *Lycopersicon* species against two insect herbivores. J. Chem. Ecol. **1987**, *13*, 837–850.

178. Gerhold, D.L.; Craig, R.; Mumma, R.G. Analysis of trichome exudate from mite resistant geraniums. J. Chem. Ecol. **1984**, *10*, 713–722.

179. De Moraes, J.C.; Vilela, E.F. Antixenosis to the aphid *Myzus persicae* (Sulzer), in clones of the wild potato *Solanum berthaultii* bearing glandular trichomes. Anais da Sociedade Entomologica do Brasil, **1995**, *24*, 613–618.

180. Doggett, H. A note on the incidence of American bollworm *Heliothis armigera* (Hub.) (Noctuidae) in sorghum. East Afri. Agric. For. J. **1964**, *29*, 348–349.

181. Price, P.W. Ecological aspects of host plant resistance and biological control: interactions among three trophic levels. In *Interactions of Plant Resistance and Parasitoids and Predators of Insects*; Boethel, D.J., Eikenbary, R.D., Eds.; Ellis Harwood Ltd: Chichester, 1986; 1–30.

182. Hulspas-Jordaan, P.M.; van Lenteren, J.C. The relationship between host–plant leaf structure and parasitization efficiency of the parasitic wasp *Encarsia formosa* Gahan (Hymenoptera: Aphelinidae). Med Fac Landbouww Rijksuniv Gent **1978**, *43*, 4314Q.

183. Kumar, A Tripathi, C.P.M.; Singh, R.; Pandey, R.K. Bionomics of Trioxys (Binodoxys) induces an aphid fed parasitoid of *Aphis craccivora*. II Effect of host plants on the activities of the parasitoid. Z. Angew. Ent. **1983**, *96*, 304–307.

184. Treacy, M.F.; Benedict, J.H.; Segers, J.C. Effects of smooth, hirsute and pilose cottons on the functional response of *Trichogramma pretiosum* and *Chrysopa ruflabris*. In *Proc. Beltwide Cotton Prod. Res. Conf*; Atlanta: Ga, 1984; 372–373.

185. Marcovitch, S. Experimental evidence on the value of strip farming as a method for the natural control of injurious insects with special reference to plant lice. J. Econ. Entomol. **1935**, *28*, 62–70.

186. Rabb, R.L.; Bradley, J.R. The influence of host plants on parasitism of eggs of the tobacco hornworm. J. Econ. Entomol. **1968**, *61*, 1249–1258.

187. Kantanyukul, W.; Thurston, R. Seasonal parasitism and predation of eggs in the tobacco hornworm on various host plants in Kentucky. Environ. Entomol. **1973**, *2*, 939–945.

188. Elsey, K.D.; Chaplin, J.F. Resistance of tobacco introduction 1112 to the tobacco budworm and green peach aphid. J. Econ. Entomol. **1978**, *71*, 723–725.

189. Takada, H.; Takenaka, T. Parasite complex of *Myzus persicae* (Sulzer) (Homoptera: Aphididae) on tobacco. Kontyu. **1982**, *50*, 556–568.

190. Milliron, H.E. A study of some factors affecting the efficiency of *Encarsia formosa* Gahan, an Aphelinid parasite of the greenhouse whitefly, *Trialeurodes vaporariorum*, (Westw.). Mich. State Agric. Exp. Tech. Bull. 1940; 173.

191. Gerling, D. Biological studies on *Encarsia formosa* (Hymenoptera: Aphilinidae). Ann. Entomol. Soc. Am. **1966**, *59*, 142–143.

192. Lauenstein, V.G. Zum Suchverhalten von Anthocoris nemorum L. (Het. Anthocoridae). Z. Angew. Ent. **1980**, *89*, 428–442.

193. Shah, M.A. The influence of plant surfaces on the searching behaviour of Coccinellid larvae. Entomol. Exp. Appl. **1982**, *31*, 377–380.

194. Miller, L.W. Populations of *Thrips tabaci* Lind. on bean varieties. J. Aust. Inst. Agric. Sci. **1947**, *13*, 141–142.

195. Putman, W.L. Bionomics of *Stethorus punctillum* Weise (Coleoptera: Coccinellidae) in Ontario. Can. Entomol. **1955**, *87*, 9–33.

196. Plaut, H.N. On the phenology and a control value of *Stethorus punctillum* Weise as a predator of *Tetranychus cinnabarinus* Boisd. in Israel. Entomophaga **1965**, *10*, 133–137.

197. Banks, C.I. The behaviour of individual coccinellid larvae on plants. Br. J. Anim. Behav. **1957**, *5*, 12–24.

198. Shepard, M.; Sterling, W.; Walker, J.K. Abundance of beneficial arthropods on cotton genotypes. Environ. Entomol. **1972**, *1*, 117–121.

199. Sengonca, C.; Gerlach, S. Einfuss der Blattoberflache auf die Wirksamkeit des rauberischen Thrips, *Scolothrips longicornis* (Thysan: Thripidae). Entomophaga. **1984**, *29*, 55–61.

200. Scopes, N.E.A. The potential of *Chrysopa carnea* as a biological control agent of *Myzus persicae* on glasshouse chrysanthemums. Ann. Appl. Biol. **1969**, *64*, 433–439.

201. Arzet, V.H.R. Suchverhaften der Larven vm *Chrysopa carnea* Steph. (Neuroptera: Chrysopidae). Z. Angew. Ent. **1973**, *74*, 64–79.

202. Elsey, K.D. Influence of plant host on searching speed of two predators. Entomophaga. **1974**, *19*, 3–6.

203. Belcher, D.W.; Thurston, R. Inhibition of movement of larvae of the convergent lady beetle by leaf trichomes of tobacco. Environ. Entomol. **1982**, *11*, 91–94.

204. Dixon, A.F.G. An experimental study of the searching behaviour of the predatory coccinellid beetle *Adalia decempunctata* (L.). J. Anim. Ecol. **1959**, *28*, 259–281.

205. Dixon, A.F.G. Factors limiting the effectiveness of the coccinellid beetle, *Adalia bipunctata* (L.) as a predator of the sycamore aphid, *Drepanosiphum platanoides* (Schr.). J. Anim. Ecol. **1970**, *39*, 739–751.

206. Dixon, A.F.G.; Russel, R.J. The effectiveness of *Anthocoris nemorum*, *A. confusus* (Hemiptera: Anthocoridae) as predators of the sycamore aphid, *Drepaniosiphum platanoides*. 11. Searching behavior and the incidence of predation in the field. Entomol. Exp. Appl. **1972**, *15*, 35–50.

207. Evans, G.F. The effect of prey density and host plant characteristics on oviposition and fertility in *Anthocoris confusus* (Reuter). Ecol. Entomol. **1976**, *1*, 157–161.

208. Evans, G.F. The searching behaviour of *Anthocoris confusus* (Reuter) in relation to prey density and plant surface topography. Ecol. Entomol. **1976**, *1*, 163–169.

209. Carter, M.C.; Sutherland, D.; Dixon, A.F.G. Plant structures and the searching efficiency of coccinellid larvae. Oecologia. **1984**, *63*, 394–397.

210. McGovern, W.L.; Cross, W.H. Effects of two cotton varieties on levels of boll weevil parasitism (Col: Curculionidae). Entomophaga. **1976**, *21*, 123–125.

211. Schuster, M.F.; Calderon, M. Interactions of host plant resistant genotypes and beneficial insects in cotton ecosystems. In *Interactions of Plant Resistance and Parasitoids and Predators of Insects*; Boethel, D.J., Eikenbary, R.D., Eds.; Ellis Harwood Ltd: Chichester, 1986; 84–87.

212. Henneberry, T.J.; Bariola, L.A.; Yittock, D.L. Nectariless cotton: effects on cotton leaf perforator and other cotton insects in Arizona. J. Econ. Entomol. **1977**, *70*, 797–799.

213. Adjei-Maafo, I.K.; Wilson, L.T. Factors affecting the relative abundance of arthropods on nectaried and nectariless cotton. Environ. Entomol. **1983**, *12*, 349–352.

214. Pimentel, D. An evaluation of insect resistance in broccoli, brussel sprouts cabbage, collards and kale. J. Econ. Entomol. **1961**, *54*, 156–158.

215. Berlinger, M.J.; Goliberg, A.M. The effect of the fruit sepals on the citrus mealybug population and on its parasite. Entomol. Exp. Appl. **1978**, *24*, 38–43.

216. Way, M.J.; Kurdue, G. An example of varietal variations in resistance in Brussels sprouts. Ann. Appl. Biol. **1965**, *56*, 326–328.

217. Chandler, A.E.F. Some host-plant factors affecting oviposition by *Aphidiophagous syrphidae* (Diptera). Ann. Appl. Biol. **1968**, *61*, 415–423.

218. Vinson, S.B. Host selection by insect parasitoids. Annu. Rev. Entomol. **1976**, *21*, 109–133.

219. Alam, S.N.; Cohen, M.B. Detection and analysis of QTLs for resistance to the brown planthopper, *Nilaparvata lugens* in a doubled-haploid rice population. Theor. Appl. Genet. **1998**, *97*, 1370–1379.

220. Groh, S.; Gonzalez-de-Leon, D.; Khairallah, M.M.; Jiang, C.Z.; Bergvinson, D. QTL mapping in tropical maize: III. Genomic regions for resistance to *Diatraea* spp. and associated traits in two RIL populations. Crop. Sci. **1998**, *38*, 1062–1072.

221. Nienhuis, J.; Helentjaris, T.; Solcum, M.; Ruggero, B.; Schaefer, A. Restriction fragment length polymorphism analysis of loci associated with insect resistance in tomato. Crop. Sci. **1987**, *27*, 797–803.

222. Maliepaard, C.; Bas, N.; Van Heusden, S.; Kos, J.; Pet, G. Mapping of QTLs for glandular trichome densities and *Trialeurodes vaporariorum* (green house whitefly) resistance in an F2 from *Lycopersicon esculentum* X *L. hirsutum f. glabratum*. Heredity. **1995**, *75*, 425–433.

223. Bonierbale, M.W.; Plaisted, R.L.; Pineda, O.; Tanksley, S.D. QTL analysis of trichome mediated insect resistance in potato. Theor. Appl. Genet. **1994**, *87*, 973–987.

224. Rector, B.G.; All, J.N.; Parrott, W.A.; Boerma, H.R. Identification of molecular markers linked to quantitative trait loci for soybean resistance to corn earworm. Theor. Appl. Genet. **1997**, *96*, 786–790.

225. Yencho, G.C.; Cohen, M.B.; Byrne, P.F. Applications of tagging and mapping insect resistance loci in plants. Annu. Rev. Entomol. **2000**, *45*, 393–422.

3

Nonprotein Amino Acids

3.1 INTRODUCTION

The 20 amino acids incorporated in protein are known as protein amino acids, and the amino acids that are not found in proteins are known as nonprotein amino acids. More than 900 nonprotein amino acids have been identified in plants. Many of the nonprotein amino acids resemble protein amino acids and often can be considered their structural analogues (1–5). They often derive biosynthetically from protein amino acids. Some nonprotein amino acids are simple homologues of protein amino acids. Such compounds have molecular sizes and conformations that do not differ too markedly from those of the corresponding protein amino acids and so they may act as analogue molecules, sometimes mimicking the behavior of normal molecules and sometimes acting as metabolic antagonists or inhibitors.

3.2 CHEMISTRY AND CLASSIFICATION

The nonprotein amino acids largely resemble the protein amino acids in chemical type but exhibit structural features not encountered in their protein counterparts. A variety of novel heterocycles (containing nitrogen, oxygen, sulfur, or combinations thereof) occur in these compounds. Many plants contain γ-glutamyl derivatives of certain amino acids accumulating in large concentrations. These classes of nonprotein amino acids have been reviewed earlier (1–5). However, a brief description is given here.

3.2.1   Acidic Amino Acids

Most of such compounds possess two carboxyl groups and a single amino function. A range of closely allied compounds can be regarded as substituted glutamic acids. The simplest types have hydroxy or methyl substituents on either the β- or γ-carbon atom of glutamic acid. Examples are threo-γ-hydroxy-L-glutamic acid, erythro-γ-methyl–L-glutamic acid, γ-hydroxy-γ-methylglutamic acid, γ-methylene-L-glutamic acid, γ-ethylidene- and γ-propylideneglutamic acids. Glutamine derivatives were also recorded in plants. A few amino acids characteristic of Iridaceae and Resedaceae contain a carboxyl group attached to a phenyl ring at the meta position, e.g., 3-carboxyphenylalanine, 3-carboxytyrosine, 3-carboxyphenylglycine, and 3-carboxy-4-hydroxyphenylglycine. Carboxyl groups attach to a α-cyclopropyl ring encountered in *Aesculus* species such as *cis*- and *trans*-α-(carboxycyclopropyl)glycines. The S-substituted cysteines occur in the seeds of some species of *Acacia* and *Albizzia*. These compounds include *S*-carboxyethyl and *S*-carboxyisopropyl-L-cysteines.

3.2.2   Basic Amino Acids

The basic amino acid is a diamino acid based on a linear carbon skeleton. Examples of this group of nonprotein basic amino acids are α-β-diaminopropionic acid, α,γ-diaminobutyric acid and ornithine. The β- and γ-oxalyl derivatives of diaminopropionic and diaminobutyric acids occur in *Lathyrus* species. Nonprotein amino acids containing the guanidino group include homoarginine, γ-hydroxyhomoarginine, and canavanine. The aminoimidazole derivative enduracididine occurs in *Lonchocarpus sericeus*. Basic compounds of S-substituted cysteine and S-aminoethylcysteine are also nonprotein basic amino acids.

3.2.3   Substituted Aromatic and Heterocyclic Amino Acids

Nonprotein amino acids containing a phenyl ring include 3-hydroxyphenylalanine, 3,4-dihydroxyphenylalanine (Dopa), orcylalanine, 3-hydroxymethylphenylalanine, 3-hydroxymethyl-4-hydroxyphenylalanine, and 3-aminoethylphenylalanine. Heterocyclic β-substituted alanines include compounds with a ring system having oxygen, nitrogen, or sulfur as the heteroatom. Mimosine, β-pyrazol-1-alanine, willardiine, lathyrine, and β-thiazolylalanine are a few examples.

3.2.4 Cyclopropyl and Branched-Chain Amino Acids

Hypoglycin A is a cyclopropyl derivative known as β-(methylene cyclopropyl)alanine. A lower homologue α-(methylene cyclopropyl)glycine from litchi seeds was also isolated. Branched carbon chain containing nonprotein amino acids were isolated from Sapindaceae and characterized as 2-amino-4-methylhex-5-ynoic acid and 2-amino-4-hydroxy-methylhex-5-ynoic acid.

3.2.5 Imino Acids

The lower homologue of proline, azetidine-2-carboxylic acid, was isolated from Liliaceae and Amaryllidaceae. The six-carbon heterocyclic compound pipecolic acid derivatives occur in many legume plants.

The structures of a few important nonprotein amino acids representing each of the above-mentioned classes are shown in Fig. 3.1.

3.3 OCCURRENCE

Many nonprotein amino acids occur in a large and economically important family, the Leguminosae (6). The basic amino acid is confined to species within the Papilonoideae. Cucurbitaceae represents another family producing characteristic amino acids like β-pyrazol-1-ylalanine and its γ-glutamyl derivative. The distribution of branched chain and cyclopropyl amino acids occurs within the genus *Aesculus* (7). Compounds of these structural types occurred in seed of several genera within the allied families Sapindaceae, Hippocastanaceae, and Aceraceae. The genus *Aesculus* and *Billa* occur in Hippocastanaceae.

3.4 BIOSYNTHESIS

Even though more than 900 nonprotein amino acids were isolated from plants, a detailed biosynthetic pathway has not been worked out for many of them. The pathway is known only for limited nonprotein amino acids like canavanine, canaline, homoserine, diaminobutyric, and diaminopropionic acids, as well as their oxalyl derivatives and substituted amino acids. The biosynthesis of nonprotein amino acids occur through three major routes.

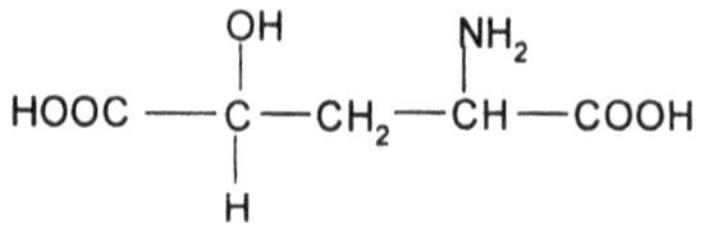

γ–Hydroxy, L-glutamic acid (Acidic)

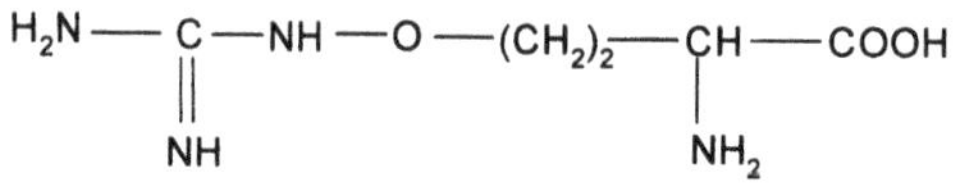

Canavanine (Basic)

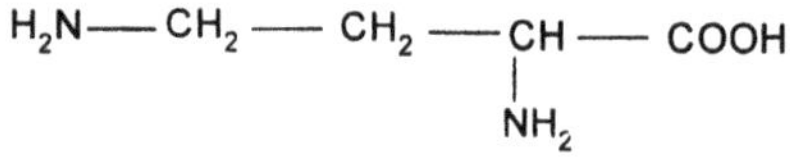

Diaminobutyric acid (Basic)

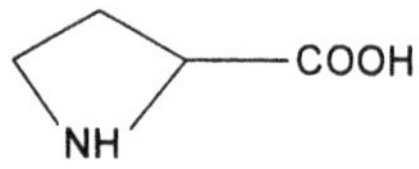

Azetidine-2-carboxylic acid (Heterocyclic)

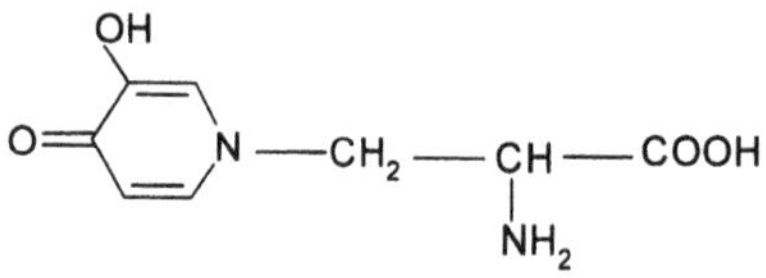

Mimosine (Substituted heterocyclic)

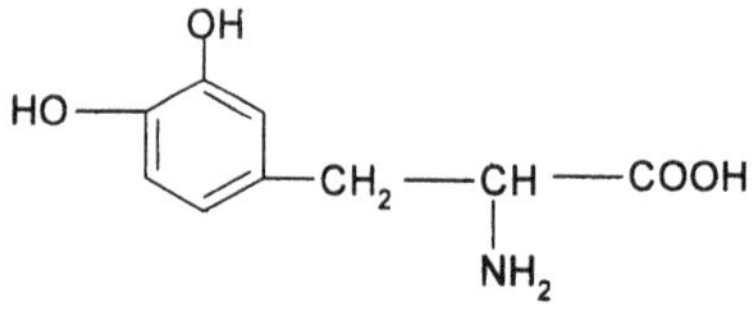

3,4- Dihydroxy phenylalanine (Substituted aromatic)

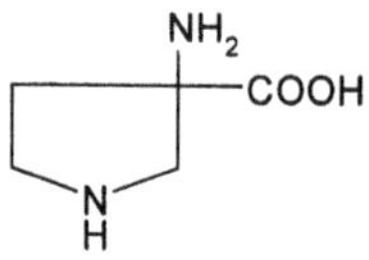

Cucurbitine (Imino)

FIGURE 3.1 Structures of nonprotein amino acids.

3.4.1 Modification of Protein Amino Acids

Cyanoalanine is synthesized from the protein amino acid L-cysteine. The enzyme β-cyanoalanine synthase (CAS) adds HCN to L-cysteine to form cyanoalanine. It has been shown that β-CAS plays an important role in cyanide metabolism in plants (7–9). The enzyme activity of CAS has been detected in a variety of plants. β-Cyanoalanine is converted to 2,4-diaminobutyric acid (DAB) or asparagine. Asparagine is converted to 2,3-diaminopropionic acid (DAP). Both DAB and DAP are oxalated to N^4-oxalyl-2,4-diaminobutyric acid and N^3-oxalyl-2,3-diaminopropionic acid, respectively (Scheme 3.1).

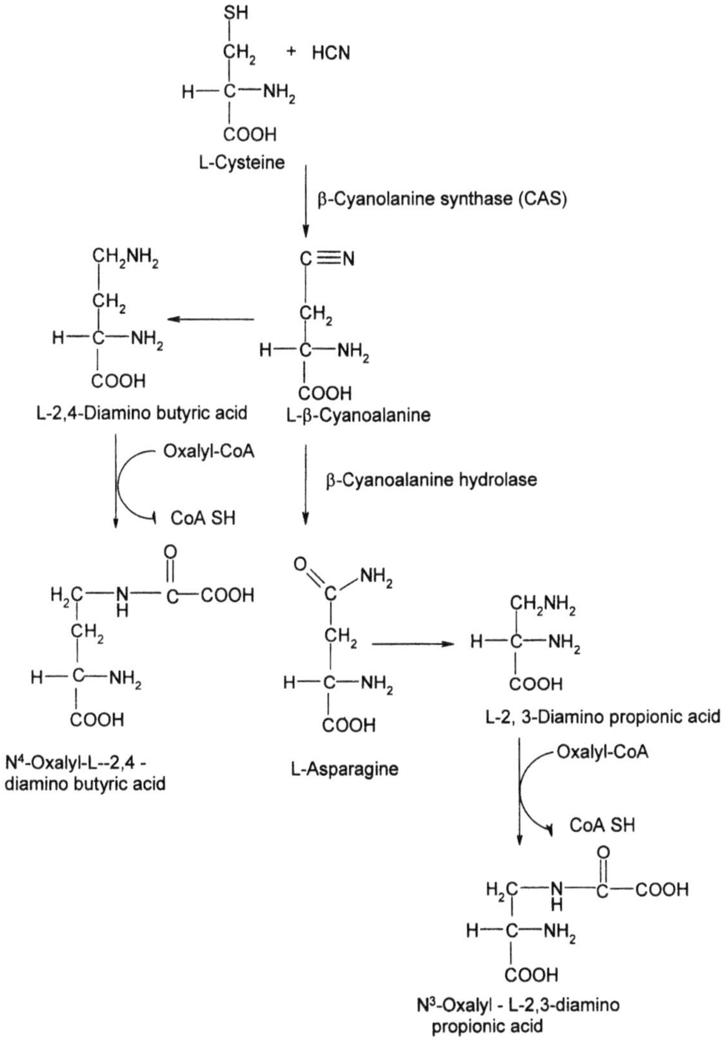

SCHEME 3.1 Biosynthesis of nonprotein amino acids in taxa of the genus Lathyrus.

3.4.2 Modification of Existing Pathway for Protein Amino Acids

m-Carboxyphenylalanine and *m*-carboxytyrosine are synthesized by modification of the shikimic acid pathway. The pathway leading to the 3-carboxyphenylalanine was documented by using doubly labeled (^{3}H and ^{14}C) shikimate to establish that these aromatic compounds were derived by an alternative of the normal shikimate aromatic pathway after the formation of chorismate, a normal intermediate in the biosynthesis of phenylalanine and tyrosine (3). A different rearrangement of the C3 moiety occurred to produce isochorismate and isoprephenate and finally *m*-carboxy-amino acids (Scheme 3.2).

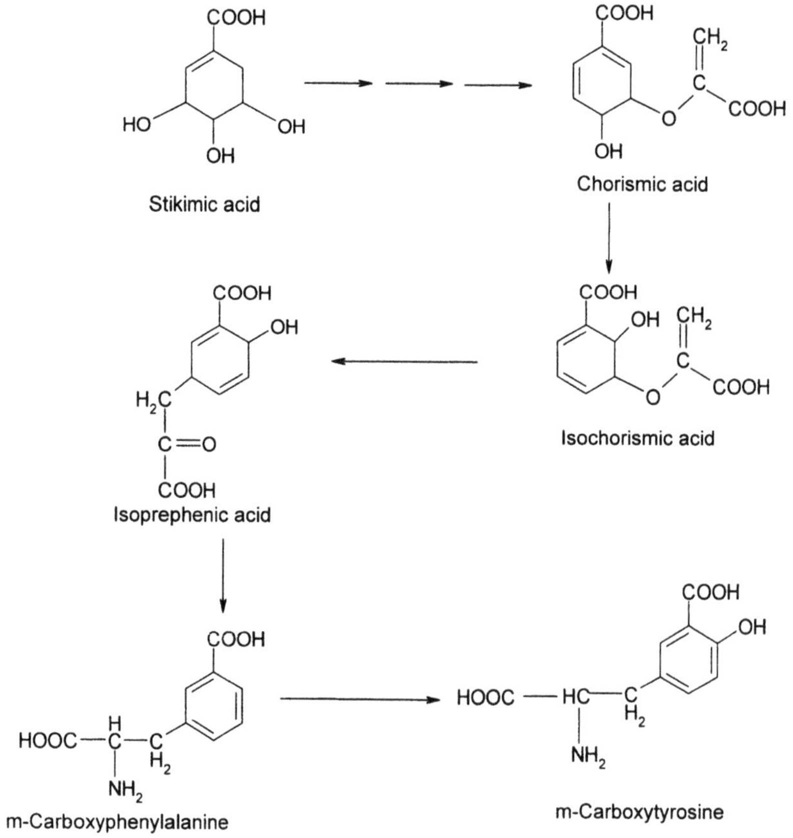

SCHEME 3.2   Biosynthesis of *m*-carboxy amino acids.

3.4.3 Novel Pathways

Canaline combines with carbamoyl phosphate to produce O-ureidohomoserine which then combines with aspartic acid in the presence of ATP to form canavaninosuccinic acid. This is cleaved into fumarate and canavanine. The canavanine is converted to canaline and urea. The canaline in the presence of NADPH is converted to homoserine by canaline reductase (Scheme 3.3). The canaline reductase (10) was isolated and purified from 10-day-old leaves of the jack bean *Canavalia ensiformis* (Leguminosae). The enzyme canaline reductase has a mass of approximately 164 kDa and is composed of 82-kDa dimers. The enzyme catalyzes a NADPH-dependent reductive cleavage of L-canaline to L-homoserine and ammonia, and is the only enzyme known to use reduced NADP to cleave an O-N bond. The cleavage of canaline to homoserine and ammonia does not appear to be reversible. The ornithine carbamoyltransferase (OCT) of canavanine-containing plants showed high canaline-dependent activities, whereas the OCT of the canavanine-deficient legume, kidney bean, shows a very low activity toward canaline. Canavanine-storing plants contain arginase which hydrolyzes L-canavanine to form the toxic metabolite L-canaline.

3.5 BIOACTIVITY

Bioactivity is not established for all nonprotein amino acids. Reported instances in which nonprotein amino acids interfering with growth and/or metabolism in plants, animals, or microorganisms are slowly increasing. Among the known nonprotein amino acid having biological activities, many interfere with the growth of vertebrates. The well-documented toxic amino acids include hypoglycin A, indospicine, mimosine, lathyrogens, and selenium-containing amino acids. Bioactivities against insects are known for a few nonprotein amino acids, which include canavanine, cyanoalanine, canaline, diaminobutyric and propionic acids, β-pyrazol-1-yl-L-alanine, mimosine, albizziine and 3,4-dopa.

3.5.1 Canavanine

Canavanine protects seeds and young foliage from insect pests. In addition to killing insects and inhibiting their reproduction, canavanine has been shown to be an nitric oxide and urea synthesis inhibitor. L-Canavanine, the guanidinooxy structural analogue of arginine, possesses significant insecticidal properties (11–13). Rosenthal (14) reviewed the biological effects and mode of action. Canavanine concentration is too weak to produce

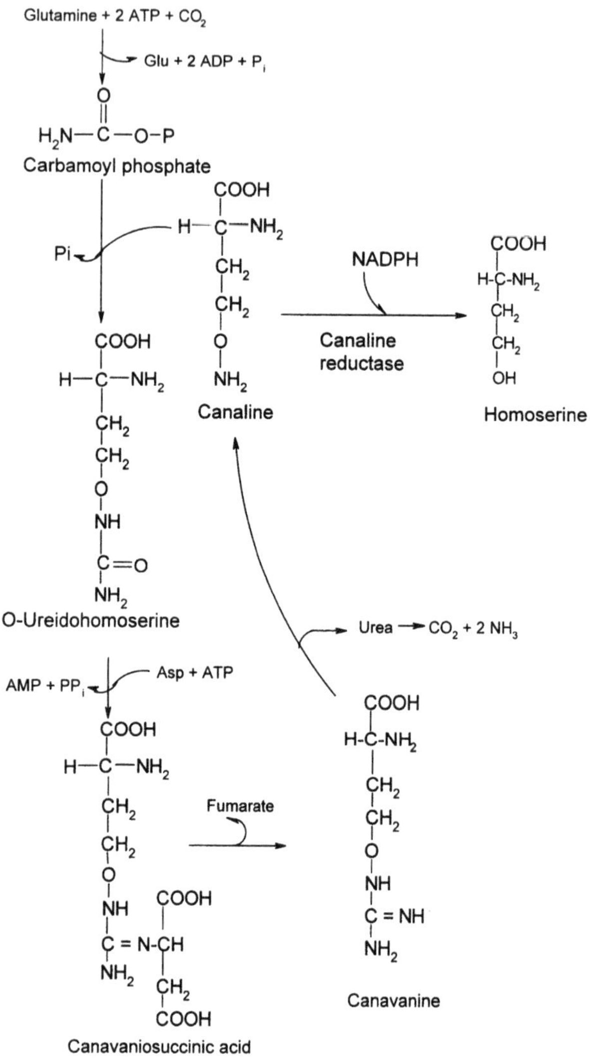

SCHEME 3.3 Biosynthesis of canaline, canavanine, and homoserine.

immediate toxic symptoms; nevertheless, it exerts deleterious effects on subsequent developmental stages (15,16). As little as 0.5 mM canavanine (88 ppm) in the larval diet of the fifth-instar tobacco hornworm, *Manduca sexta* (L.), significantly reduced the ovarial mass of the resulting adult moth (15). Palumbo and Dahlman (17) showed that canavanine reduces both the fecundity and fertility of *M. sexta*. Agar-based artificial diets containing various concentrations of L-canavanine were fed to fifth-instar tobacco hornworm larvae. The number of chorionated oocytes in the 4-day-old moths resulting from these larvae was concentration dependent between 0.75 and 2.5 mM canavanine. The mean number of chorionated oocytes from the 2.5 mM treatment was only 7.5% of that obtained from the untreated control. Fecundity and fertility of moths from 1 mM canavanine were 14% and 21% of untreated control values, respectively. Matings between treated and untreated moths show that only female reproductive capacity was significantly reduced by canavanine under these conditions. Canavanine inhibition of chorionated oocyte production may result from the alteration of oocyte composition during assembly or by retardation of the normal rate of oocyte production. This study clearly demonstrates the concentration-dependent capacity of canavanine to reduce the rate of chorionated oocyte production, but the specific biochemical explanations for this observation are not known.

Hatching failure of "apparently normal" eggs may be attributed to one or more factors. First, fertilization may not be achieved because of inviability of sperm or oocytes. Results from matings between untreated and treated moths show indirectly that adequate numbers of viable sperm are produced by canavanine–treated males. Oocytes of treated females are histologically different from those of untreated controls, but they have not been demonstrated to be incapable of fertilization. Second, even though fertilization is normal, embryogenesis may be disrupted. Eggs were not examined routinely to determine whether embryogenesis was initiated or whether it terminated at some specific point. Although the mode of action of canavanine's inhibition of fecundity and fertility is still unknown, it is important to point out that ecologically significant changes occur in the reproductive capacity of canavanine concentrations of 100–200 ppm. Such levels have been demonstrated in the seeds and foliage of legume forages such as alfalfa and red clover (18).

Although most studies on canavanine–insect interactions have used either lepidopterans or coleopterans, increased adult mortality has been demonstrated in *Drosophila melanogaster* (Meigen) exposed as larve to a canavanine-containing diet (19). However, ingested canavanine did not alter the reproductive capacity of the adult dipterous parasitoid, *Pseudosarcophaga affinis* (Fallen) (20). Dahlman et al. (21) studied the response of additional dipteran species to canavanine and evaluated the potential use of this natural product as a control measure for fly pests of cattle. These species are the housefly, *Musca domestica* L. face fly, *M. autumnalis* (De Geer), horn fly, *Haematobia irritans* (L.), and stable fly, *Stomoxyz calcitrans* (L.). Larvae of the housefly, face fly, horn fly, and stable fly were reared in media containing various concentrations of L-canavanine. Response in order of decreasing sensitivity was the stable fly, face fly, horn fly, and house fly. Only 18% housefly mortality was obtained from 8 ppm canavanine, a level that killed at least 70% of the other three species. Horn fly and face fly showed additional delayed concentration-dependent effects expressed as smaller pupal size, some degree of pupal deformation, and pupal mortality.

The current cost of canavanine prohibits its large-scale application on crops. However, its performance could be evaluated under small plot field conditions, and results of such a study were evaluated (22). L-Canavanine was sprayed on field-grown Ky 14 Burley tobacco plants at a rate of 2.25 kg/ ha. Second-instar *M. sexta* (L.) were placed on the plants, which were subsequently examined periodically for the larvae. Larvae on plants treated with canavanine experienced slower growth rates than did those on plants treated with arginine (control). The results of this field study support similar, more detailed, laboratory investigations. The significance of this study is that it illustrates the ability of a naturally occurring nonprotein amino acid to significantly alter the development rate of an insect of economic importance under field conditions. A slower growth rate exposes the insect to environmental stresses and natural enemies for an extended period. In addition, laboratory studies have shown a correlation between canavanine concentrations, growth rate, pupal and adult deformations, and egg fecundity, and fertility. Thus, the observed reduced growth rates imply that delayed effects may also occur under field conditions. The amount of canavanine applied to each plant was not measured, but each plant should have received 120 mg. If this material was distributed evenly over the leaf surface and if each plant possessed 30,000 cm^2, the plant in each cm^2 area would contain 4 μg of canavanine. A single hornworm larva consumes 1538 cm^2 of field-grown burley. Thus, each larva potentially could ingest 6.152 mg of canavanine—a dose equivalent to an artificial diet containing 1.16 mM canavanine, assuming a consumption of 30 g of diet for completion of larval development. One millimolar canavanine concentration in artificial diet resulted in delayed hornworm larval and pupal mortality, and smaller ovarial size (23). Although no data are available on most of these parameters in this field study, the observed delay in growth similar to that observed under laboratory conditions suggests that the other canavanine-related phenomena also would occur in the field.

3.5.2 Cyanoalanine

Schlesinger et al. (24) have systematically evaluated the role of various nonprotein amino acids in imparting resistance to insects in leguminous plant species in which these amino acids accumulate. *Locusta* prefers to feed on graminaceous plants more than on legumes. One of these unusual amino acids characteristic of *Vicia sativa* identified as β-cyano-L-alanine (BCNA) was toxic to *Locusta* at physiological levels when included in the diet. Toxicity is accompanied by a marked discharge of watery feces and this observation prompted them to investigate the physiological effect of BCNA on the water balance of *Locusta*. Administration of BCNA to adult locusts results in a significant decrease in hemolymph volume within a day. Dehydration becomes acute within 2–3 days and causes death within 5 days. This syndrome is attributed to impaired ability of BCNA-treated locusts to reabsorb water into the hemolymph from the rectal lumen. Chronic BCNA toxicity irreversibly disrrupts the control of water balance in locusts. Schlesinger and Applebaum (25) have shown that 2,4-diaminobutyric acid and β-cyanoalanine are only slightly toxic to the larvae of red flour beetle, *Tenebrio castaneum*. However, larvae of yellow mealworm, *Tenebrio molitor*, are much more sensitive to these neurotoxic amino acids. The treated larvae perish atttempting larval-pupal ecdysis.

3.5.3 Canaline

Canaline is a potentially poisonous nonprotein amino acid that can elicit severe larval developmental aberrations, prevent successful larval-pupal ecdysis, and cause deformities in the pupae of such canaline-sensitive insects as *M. sexta* (26). Canaline is a potent neurotoxin to the adult moth.

3.5.4 Diaminopropionic and Diaminobutyric Acid and Their Derivatives

The leaves of the legume *Lathyrus latifolius* contain α,γ-diaminobutyric acid (2,4-diaminobutanoic acid, DAB), γ-N-oxalyl-α,γ-diaminobutyric acid (2-amino-4-oxalylaminobutanoic acid, γ-ODAB), β-N-oxalyl-α,β-diaminopropanoic acid (2-amino-3-oxalylaminopropanoic acid, β-ODAP), homoserine (HS), and O-oxalylhomoserine (OHS). At the present time, major efforts are being made to develop β-ODAP-low varieties of *L. sativus* for human consumption and promote *L. sylvestris* as a fodder crop (27). There is, therefore, a need to know how the behavior of phytophagous insects is influenced by the presence of β-ODAP and other secondary compounds in *Lathyrus* species.

Bell et al. (28) described the isolation of DAB, γ-ODAB, β-ODAP, HS, and OHS from the aerial parts of *L. latifolius* and their quantitative determination in the leaves, shoots, and roots. New chromatographic data relating to the compounds are presented. The antifeedant and/or phagostimulatory effects of the individual amino acids and mixtures of amino acids on the larvae of *Spodoptera littoralis* (the African leaf worm) are described. DAB, γ-ODAB, β-ODAP, HS, and OHS were all found in the leaves and shoots of *L. latifolius*, but OHS was absent from the roots. In the insect feeding experiments all five compounds elicited significant dose-dependent responses. However, the type of activity was not uniform. Over the concentration range 0.005–0.5% (dry weight of disks), OHS and γ-ODAB showed antifeedant activity whereas HS and β-ODAP showed phagostimulant activity. DAB stimulated feeding at low concentrations but elicited significant levels of antifeedant activity at concentrations at and above 0.125%.

Mixture of the five amino acids, corresponding to the proportions found in the shoots, leaves, and roots, showed greater antifeedant activity than that obtained by summing the phagostimulatory and antifeedant activity of the individual components of the mixtures. However, the mixture was not as potent as the calculated values obtained if the activity of only those amino acids showing antifeedant activity was summed. Thus, the results cannot be explained by simple additive or synergistic activity. The responses could result from interactions among the amino acids for receptor sites on the insects' peripheral taste neurons and/or from the integration of sensory information by the central nervous system as occurs in cross-fiber patterning (29).

In order to determine the relative influence of the different amino acids on larvae feeding behavior, individual amino acids were omitted, one at a time, from the mixure that corresponded proportionately to the amino acid composition of the young leaves. Removal of OHS, β-ODAP, or DAB resulted in a significant loss of antifeedant activity. It was not surprising that the removal of OHS resulted in a loss of antifeedant amino acid. Yet when γ-ODAB, the next most potent antifeedant amino acid, was removed there was no decrease in activity. The effect on the feeding responses of the insects to the removal of DAB and β-ODAP from the mixtures was unexpected. ODAB, when tested alone, showed potent phagostimulatory activity; however, its removal from the mixture resulted in a loss of antifeedant activity. A similar decrease in activity was recorded when DAB was removed from the mixture, although this compound when tested alone showed slight phagostimulatory activity at the concentration used in the mixture. These results illustrate how difficult it is to predict how the removal of a compound from a mixture will influence the activity of that mixture. They also emphasize that the response of an organism to a complex mixture of compounds, such as a crude plant extract, cannot necessarily be predicted on the basis of its response to purified components of the mixture.

β-ODAP and γ-ODAB have been shown to have potent antifeedant activity against *Anacridium melanorohodon* at 0.5% and 5% (dry weight), respectively, and against *Locusta migratoria* at 1% and 10% (dry weight), respectively. In contrast, the study shows that β-ODAP stimulates feeding of *S. littoralis* at 0.5% (dry weight). The concentrations used in the earlier studies were not only greater than those tested but were also greater than those found in the tissues of *L. latifolius*. So it is unclear from this information whether β-ODAP has a role in deterring insects. However, the results show that γ-ODAB could act as deterrent at the concentrations found in *L. latifolius*. The leaves of *L. sylvestris*, a species closely related to *L. latifolius*, lack OHS but contain comparable concentrations of DAB, γ-ODAB, and β-ODAP.

3.5.5   β-(Pyrazol-1-yl)-L-alanine (β-PA)

Two secondary compounds of the genus *Cucumis*, pyrazole and β-(pyrazolyl)-L-alanine (β-PA), were found in its phloem sap. The effects of these compounds on attractivity, growth inhibition, and mortality were tested on aphids from cucurbit and noncucurbit hosts (30–32). β-PA had no noticeable effect on aphids tested. In contrast, pyrazol had high deleterious effects on *Acyrthosiphon pisum*, which does not feed on melon.

3.5.6   Mimosine and Other Nonprotein Amino Acids

Janzen et al. (33) found that incorporation of small doses of secondary compounds found in seeds such as nonprotein amino acids in the normal diet of the seed-eating beetle *Callosobruchus maculatus* (Bruchidae) results in toxicity. They tested L-homoarginine, D,L-pipecolic acid, albizziine, S-carboxyethylcysteine, L-tyrosine, L-djenkolic acid, L-canavanine, D,L-2,4-diaminobutyric acid, isowillardiine, γ-methylglutamic acid, tyramine, N-methyltyrosine, 5-hydroxy-L-tryptophan, *m*-carboxyphenylalanine, β-hydroxy-γ-methylglutamic acid, γ-guanidinobutyric acid, γ-aminopropionitrile fumarate, L-mimosine, β-cyano-L-alanine, azetidine-2-carboxylic acid, 5-methoxy-N,N-dimethyltryptamine by adding to the diet of *C. maculatus* larvae at 0.1%, 1.0%, and 5.0%.

From the data they obtained it was evident that the different nonprotein amino acids found in seeds were differentially toxic to cowpea weevil larvae and that there were dosage effects. Mimosine, β-cyanoalanine, and azetidine-2-carboxylic acid produced highly significant toxicity (lethal) to these larvae at the 0.1% level. Many other nonprotein amino acids produced toxicity at the 1% and 5% level. Dopa, canavanine, and homoarginine were tested for the emergence of the beetle at 0.1%, 1%, and 5% level. Dopa was lethal, and the 2–3% level was therefore the absolute tolerance limit. However, L-homoarginine showed no effect on cowpea weevils for 1–3% inclusion in the diet, a highly significant reduction at 4%, and highly significant to lethal effects at 5% depending on the trail.

At 1% concentration, 41% of the nonprotein acids were lethal and another 21% had a depressant effect on production of the adult beetles. At 5% concentration, 90% of the nonprotein amino acids were lethal and all the rest had a depressant effect on production of adult beetles. The transfer of *C. maculatus* eggs to legume seeds other than those of *V. unguiculata* confirmed that the seeds of the great majority of species examined contained toxins that killed *C. maculatus* in the egg or larval stage. The nonprotein amino acid canavanine proved lethal to the larvae of *C. maculatus* at a concentration of 5% and was markedly toxic at 1%. As this amino acid occurs in seeds of *Diocloa megacarpa* at concentrations of 8% or greater, it was of considerable interest to find that these seeds provided the larvae of a seed beetle, *Caryedes brasiliensis*, with their only source of food (5).

A group of seven naturally occurring, potentially toxic nonprotein amino acid antimetabolites—namely L-azetidine-2-carboxylic acid, L-homoarginine, L-canavanine, 5-hydroxy-L-lysine, L-albizziine, seleno- L-methionine, and 2-aminoethyl L-cysteine—were administered to terminal-instar larvae of the tobacco hornworm, *M. sexta* (Sphingidae) (34). These potentially toxic secondary metabolites were provided concurrently with bacterial cell wall fragments that induce de novo synthesis of lysozyme. Analysis of hemolymph lysozyme activity revealed that some of these compounds, such as 5-hydroxylysine, homoarginine, and albizziine, had no discernible effect on the activity of the induced lysozyme. On the other hand, providing canavanine, azetidine-2-carboxylic acid, selenomethionine, or 2-aminoethylcysteine at the time of bacterial challenge resulted in a significant loss of lysozyme activity. Concurrent administration of canavanine and 2-aminoethylcysteine markedly intensified their individual capacity to inhibit enzyme activity. The individual deleterious effect of azetidine-2-carboxylic acid and canavanine were not exacerbated when these toxicants were coadministered.

Certain legume species are conspicuously free from attack by the southern armyworm *Prodenia eridania* (35). The seeds were found to contain a number of nonprotein amino acids of which canavanine and β-hydroxy-γ-methylglutamic acid were repellent to the larvae, while 5-hydroxytryptophan was toxic at low levels but repellent at the high levels found in the seeds of *Griffonia simplicifolia*. The high level of mortality produced by a diet of *Albizzia julibrissin* was attributed to the combined action of albizziine and S-(β-carboxymethyl)cysteine. The resistance to attack of *Mucuna* seeds is probably due to high concentrations of L-3,4-dihydroxyphenylalanine. This compound inhibits tyrosinase, an enzyme required for cuticular hardening (36). Canavanine inhibits the formation of pupae from larvae of the boll weevil (37) and at a similar stage in the silkworm.

3.6   MECHANISM OF ACTION

3.6.1   General

The highest concentrations of many toxic amino acids occur in the seeds of producer species, which accounts for their possible roles in protecting the seed against attack by predatory insects. These toxic amino acids also inhibit the growth of the adjacent species after diffusion from the seed, especially during the early stages of germination. These amino acids have close structural configurations to that of protein amino acids. For example, indospicine and canavanine structurally resemble arginine, and mimosine has an aromatic character and shape similar to that of phenylalanine and tyrosine. Azetidine-2-carboxylic acid is an analogue of proline, albizziine of glutamine, and 2-amino-4-methylhex-4-enoic acid of phenylalanine (3). The metabolic antagonism may be due to analogue mimicking of the normal compound, usually in one of three principal ways.

1. The analogue molecule competes as an alternative substrate for permease systems responsible for amino acid uptake into cells.
2. One of the most important means whereby nonprotein amino acids elicit their protective effect is through their structural analogy to a protein amino acid. An important step in protein synthesis is activation and aminoacylation of protein amino acids mediated by a specific aminoacyl-tRNA synthetase. Many toxic nonprotein amino acids function as a substrate for a particular aminoacyl-tRNA synthetase and are ultimately incorporated into the protein, thus affecting its biological function.
3. The analogue acts as an inhibitor of amino acid biosynthesis either as a direct competitor of the substrate for the reactive sites of the enzyme or by mimicking the role of the end product of a biosynthetic pathway in its feedback inhibitory action on earlier key enzymes in the pathway. The mechanisms of action of a few important nonprotein amino acids are presented.

3.6.2   Canavanine

Canavanine, the guanidinooxy structural analogue of L-arginine, provided valuable insight into the biological effects and the mode of action of nonprotein amino acids (38–47). The arginyl-tRNA synthetases of numerous canavanine-free species charge canavanine and subsequently incorporated into the nascent polypeptide chain. Canavanine-containing proteins can disrupt critical reactions of RNA and DNA metabolism as well as protein synthesis. Canavanine also affects arginine uptake, formation of structural components, and other cellular processes (3). In these ways, canavanine alters essential biochemical reactions and becomes a potent antimetabolite of arginine in a wide spectrum of species. These deleterious properties of canavanine render it a highly toxic secondary plant constituent that probably functions as an allelochemic agent that deters the feeding activity of phytophagous insects and other herbivores. Canavanine, 2-amino-4-(guanidinooxy)butyric acid, and L-arginine incorporation into de novo synthesized proteins was compared by Rosenthal (38,43) in six organisms. Utilizing L-(guanidinooxy-^{14}C) canavanine and L-(guanidino-^{14}C) arginine at substrate saturation, the canavanine-to-arginine incorporation was determined in de novo synthesized proteins. *Caryedes brasiliensis* and *Sternechus tuberculatus* (canavanine-utilizing insects), *Canavalia ensiformis* (a canavanine storing plant), and, to lesser extent, *Heliothis virescens* (a canavanine-resistant insect) failed to accumulate significant canavanyl proteins (39). By contrast, *M. sexta*, a canavanine-sensitive insect, and *Glycine max*, a canavanine-free plant, readily incorporated canavanine into newly synthesized proteins. This study supports the contention that incorporation of canavanine into proteins in place of arginine contributes significantly to canavanine antimetabolic properties. The radiometric assay of L-(guanidinooxy-^{14}C)canavanine employed to study the synthesis of [^{14}C]-canavanine and [^{14}C]arginine-containing proteins from hemolymph of *M. sexta* and *H. virescens* confirmed the above (48). Moreover, physical, chemical, and immunological studies of native and canavanine-containing vitellogenin obtained from female migratory locusts (*Locusta migratoria migratorioides*) provide the first experimental evidence that canavanine can disrupt the tertiary and/or quaternary structure that yields the three-dimensional conformation unique to the protein (49). These findings enhance our understanding of the biochemical basis for canavanine antimetabolic and potent insecticidal properties.

Rosenthal and Dahlman (47) have also shown that L-canavanine is incorporated into lysozyme in response to administration of bacterial cell wall materials by canavanine-treated larvae of the tobacco hornworm *M. sexta.* Analysis of canavanine-containing lysozyme purified from these insects reveals that 21% of the arginine residues are replaced by canavanine; this residue substitution results in a loss of 49.5% of the catalytic activity. The *M. sexta* lysozyme has an arginine at positions 23, 42, and 107. The ability of incorporated canavanine to inhibit *M. sexta* lysozyme activity selectively may result from the fact that replacement of any of the three arginine residues by canavanine causes the loss of catalytic activity.

Plants producing these toxic nonprotein amino acids do not incorporate these toxicants into their own newly synthesized proteins. *Canavalia ensiformis* does not charge canavanine even though its concentration can be 200 times greater than that of arginine. Failure to charge canavanine occurs presumably because canavanine is not a substrate for the arginyl-tRNA synthetase of this legume. Similarly, insects adapted to or naturally resistant to canavanine scrupulously avoid incorporation of this natural product into newly synthesized proteins. The metabolism of L-canavanine and L-canaline were investigated in larvae of the tobacco budworm *Heliothis virescens* (Noctuidae) using L-(1,2,3,4,-^{14}C)canavanine or L-(U-^{14}C)canaline (41). This destructive insect tolerates L-canavanine and L-canaline because of its ability to reductively cleave this potentially insecticidal natural products to L-homoserine and guanidine or ammonia, respectively. *Heliothis virescens* employs a constitutive enzyme of the larval gut known as canavanine hydrolase to catalyze an irreversible hydrolysis of L-canavanine to L-homoserine and hydroxyguanidine. It represents a new type of hydrolase, one acting on oxygen–nitrogen bonds. The characteristics of this enzyme, including substrate specificity, have been described.

3.6.3 Mimosine

Mimosine has been reported to specifically prevent initiation of DNA replication in the chromosomes of mammalian nuclei (48–51). To test this hypothesis, the effects of mimosine were examined on several DNA replication systems and compared with the effects of aphidicolin, a specific inhibitor of replicative DNA polymerases. The study by Gilbert et al. (49) demonstrated that mimosine inhibits DNA synthesis in mitochondrial, nuclear, and simian virus 40 genomes to a similar extent. Mimosine also had no effect on initiation or elongation of DNA replication in *Xenopus* eggs or egg extracts containing high levels of deoxyribonucleotide triphosphates. These results showed that mimosine inhibits DNA synthesis at the level of elongation of nascent chains by altering deoxyribonucleotide metabolism.

Mosca et al. (50) provided evidence that mimosine functions by binding to an intracellular protein. The radiolabeled mimosine can be specifically cross-linked to a 50-kDa polypeptide (terminal p50) in vitro. This protein is not a channel protein and also not a cyclin. Mimosine is known to chelate iron, a cofactor for ribonucleotide reductase intracellular deoxyribonucleotide pools.

The effects of mimosine on folate metabolism were investigated in human MCF-7 mammary adenocarcinoma cells and SH-SY5Y neuroblastoma (52). It was shown that mimosine is a folate antagonist. The effect of mimosine on folate metabolism is associated with decreased cytoplasmic serine hydroxymethyltransferase expression in MCF-7 cells and reduced enzyme promoter activity over 95% (53).

3.6.4 Canaline

L-Canaline [L-2-amino(aminooxy)butyric acid] is a potent ornithine amino-transferase inhibitor. Reaction of the enzyme with [U-^{14}C]canaline produces an enzyme-bound, covalently linked, radiolabeled canaline-pyridoxal phosphate oxime. The L-[U-^{14}C]canaline-pyridoxal phosphate oxime has been isolated from canaline–treated enzyme (54). Dialysis of canaline-inactivated ornithine aminotransferase against free pyridoxal phosphate slowly reactivates the enzyme as the oxime is replaced by pyridoxal phosphate. Analysis of L-[U-^{14}C]canaline binding to ornithine aminotransferase reveals the presence of four pyridoxal phosphate molecules per enzyme. Canaline also functions as a lysine antagonist.

3.7 MOLECULAR BIOLOGY

The cDNA encoding cysteine synthase responsible for sulfur assimilation and biosynthesis of nonprotein amino acids was isolated and characterized from *Spinacea oleracea* and *Citrullus vulgaris* (55). The functional lysine residue was identified by site-directed mutagenesis. An overexpression system in *E. coli* was constructed for the bacterial production of the plant-specific nonprotein amino acids. Saito (56) made transgenic *Nicotiana tabacum* integrated with sense and antisense constructs of cysteine synthase cDNA driven by cauliflower mosaic virus 35S promoter for the purpose of genetic manipulation of biosynthetic flow of cysteine in plants.

3.8 PERSPECTIVES

Although more than 900 nonprotein amino acids have been identified in plants, only a few have been tested for bioactivity against insects. The detailed pathway of biosynthesis for these nonprotein amino acids has not been elucidated and therefore no molecular biology work related to cloning and genetic engineering of metabolic pathway enzymes involved has been attempted. There is a need to test all the known nonprotein amino acids against insect pests. Amino acids, which do not harm the mammalian system but help plants to fight against invading insects, are to be selected. Elucidation of the biosynthetic pathways of such amino acids will help to identify the rate-limiting steps and facilitate further cloning of such genes.

REFERENCES

1. Bell, E.A. Nonprotein amino acids in plants. Encycl. Plant. Physiol. New Ser. **1980**, *8*, 403–432.
2. Bell, E.A. The nonprotein amino acids occurring in plants. Prog. Phytochem. **1981**, *7*, 171–196.
3. Fowden, L. Nonprotein amino acids. In *The Biochemistry1 of Plants*; Conn, E.E., Stumpf, P.K., Eds.; Academic Press: New York, 1981; Vol. 7, 215–247.
4. Fowden, L.; Lewis, D.; Tristram, H. Toxic amino acids: their action as antimetabolites. Adv. Enzymol. **1967**, *29*, 89–163.
5. Rosenthal, G.A. Nonprotein amino acids as protective allelochemicals. In *Herbivores: Their Interaction with Secondary Plant Metabolites*; Rosenthal, G.A., Berenbaum, M.R., Eds.; Academic Press: San Diego, 1991; 1–34.
6. Bell, E.A. Nonprotein amino acids in the Leguminosae. Adv. Legume Syst. Part 2. 1981; 489–499.
7. Luckner, M. Secondary Metabolism in Microorganisms, Plants and Animals, 3rd Ed.; Springer-Verlag: Berlin, 1990.
8. Hatzfeld, Y.; Maruyama, A.; Schmidt, A.; Noji, M.; Ishi, zawa, K.; Saito, K. β-Cyanoalanine synthase is a mitochondrial cysteine synthase like protein in spinach and arabidopsis Plant. Physiol. **2000**, *123*, 1163–1172.
9. Warrilow, A.G.S.; Hawkesford, M.J. A cyanoalanine synthase and two isoforms of cysteine synthase. J. Exp. Bot. **2000**, *51*, 985–993.
10. Rosenthal, G.A. Purification and characterisation of the higher plant enzyme L-canaline reductase. Proc. Natl. Acad. Sci. USA. **1992**, *89*, 1780–1784.
11. Rehr, S.S.; Bell, E.A.; Janzen, D.H.; Feeny, P.P.; Insecticidal amino acids in legume seeds. Biochem. Syst. **1973**, *1*, 63–67.
12. Dahlman, D.L.; Rosenthal, G.A.; Nonprotein amino acid and insect interactions I. Growth effects and symptomology of L-canavanine consumption by tobacco hornworm, *Manduca sexta*. Comp. Biochem. Physiol. A **1975**, *51*, 33–36.
13. Harry, P.; Dror, Y.; Applebaum, S.W. Arginase activity in *Tribolium castaneum* and the effect of canavanine. Insect. Biochem. **1976**, *6*, 273–279.
14. Rosenthal, G.A. The biological effects and mode of action of L-canavanine, a structural analogue of L-arginine. Q. Rev. Biol. **1977**, *52*, 155–178.
15. Rosenthal, G.A.; Dahlman, D.L. Nonprotein amino acid insect interactions. II. Effects of canaline-urea cycle amino acids on growth and development of tobacco hornworm, *Manduca sexta* L.(Sphingidae). Comp. Biochem. Physiol. A **1975**, *52*, 105–108.
16. Dahlman, D.L.; Rosenthal, G.A. Further studies of the effect of L-canavanine on the tobacco hornworm, *Manduca sexta* (L.) (Sphingidae). J. Insect. Physiol. **1976**, *22*, 265–271.
17. Palumbo, R.E.; Dahlman, D.L. Reduction of *Manduca sexta* facundity and fertility by L-canavanine. J. Econ. Entomol. **1978**, *71*, 674–676.
18. Rosenthal, G.A. The interrelationship of canavanine and urease in seeds of the Lotoidae. J. Exp. Bot. **1974**, *25*, 609–613.
19. Harrison, J.; Holliday, R. Senescence and the fidelity of protein synthesis in *Drosophila*. Nature **1967**, *213*, 990–993.

20. Hegdekar, B.M. Amino acid analogues as inhibitors of insect reproduction. J. Econ. Entomol. **1970**, *63*, 1950–195.

21. Dahlman, D.L.; Herald, F.; Knapp, F.W. L-Canavanine effects on growth and development of four species of Muscidae. J. Econ. Entomol. **1979**, *72*, 678–679.

22. Dahlman, D.L. Field tests of L-canavanine for control of tobacco hornworm. J. Econ. Entomol. **1980**, *73*, 279–281.

23. Dahlman, D.L. Effect of L-canavanine on the consumption and utilisation of artificial diet by the tobacco hornworm, *Manduca sexta*. Entomol. Exp. Appl. **1977**, *22*, 123–137.

24. Schlesinger, H.M.; Applebaum, S.W.; Birk, Y. Effect of β-cyanoalanine on the water balance of *Locusta migratoria*. J. Insect. Physiol. **1976**, *22*, 1421–1425.

25. Schlesinger, H.M.; Applebaum, S.W. In *EUCARPIA/IOBL Working Group Breeding for Resistance to Insects and Mites*; Wageningen: The Netherlands, 1977; 143–147.

26. Kammer, A.E.; Dahlman, D.L.; Rosenthal, G.A. Effects of the nonprotein amino acids L-canavanine and L-canaline on the nervous system of the moth *Manduca sexta*. J. Exp. Biol. **1978**, *75*, 123–132.

27. Campbell, C.G. Threat and promise In *The Grass Pea. Proceedings of the International Network for the Improvement of Lathyrus sativus and the eradication of lathyrism*. Third World Medical Research Foundation, Spencer, P.S., Ed.; New York, 1989; 179–182.

28. Bell, E.A.; Perera, K.P.W.C.; Nunn, P.B.; Simmonds, M.S.J.; Blaney, W.M. Nonprotein amino acids of *Lathyrus latifolius* as feeding deterrents and phagostimulants in *Spodoptera littoralis*. Phytochemistry **1996**, *43*, 1003–1007.

29. Navon, A.; Bernays, E.A. Inhibition of feeding in acridids by nonprotein amino acids. Comp. Biochem. Physiol. **1978**, A*59*, 161–164.

30. Chen, Z.Q.; Rahbe, Y.; Delobel, B.; Sauvion, N.; Guillaud, J.; Febvay, G. Melon resistance to the aphid., *Aphis gossypii*: behavioural analysis and the chemical correlation with nitrogenous compounds. Entomol. Exp. Appl. **1997**, *85*, 33–44.

31. Chen, J.Q.; Delobel, B.; Rahbe, Y.; Sauvion, N. Biological and chemical characteristics of a genetic resistance of melon to the melon aphid, Entomol. Exp. Appl. **1996**, *80*, 250–253.

32. Chen, J.Q., Rahbe, Y.; Delobel, B. Effects of pyrazole compounds from melon on the aphid, *Aphis gossypii*. Phytochemistry. **1999**, *50*, 1117–1122.

33. Janzen, D.H.; Juster, H.B.; Bell, E.A. Toxicity of secondary compounds to the seed eating larvae of the bruchid beetle *Callosobruchus maculatus*. Phytochemistry **1977**, *16*, 223–227.

34. Rosenthal, G.A. The effect of nonprotein amino acid administration on the lysozyme activity of the tobacco hornworm, *Manduca sexta* (Sphingidae). Biochem. Syst. Ecol. **1998**, *26*, 257–266.

35. Janzen, D.H. Seed–eaters versus seed size, number toxicity and dispersal. Evolution. **1969**, *23*, 1–27.

36. Rehr, S.S.; Janzen, D.H.; Feeny, P.P. L-Dopa in legume seeds: chemical barrier to insect attack. Science. **1973**, *181*, 81–82.

37. Vanderzant, E.S.; Chremos, J.H. Dietary requirement of the boll weevil for arginine and the effect of arginine analogues on growth and the composition of the body amino acids. Ann. Entomol. Soc. Am. **1971**, *64*, 480–485.

38. Rosenthal, G.A. The biochemical basis for the deleterious effects of L-canavanine. Phytochemistry **1991**, *30*, 1055–1058.

39. Rosenthal, G.A.; Janzen, D.H. Avoidance of non-protein amino acid incorporation into protein by the seed predator, *Caryedes brasiliensis* (Bruchidae). J. Chem. Ecol. **1983**, *9*, 1353–1361.

40. Lenz, C.; Dahlman, D.L.; Rosenthal, G.A. The effect of L-canavanine on the hemolymph amino acid composition of the tobacco hornworm, *Manduca sexta* (L.) (Sphingidae). Arch. Insect Biochem. Physiol. **1986**, 3, 265–275.

41. Berge, M.A.; Rosenthal, G.A. Metabolism of L-canavanine and L-canaline in the tobacco budworm, *Heliothis virescens* (Noctuidae). Chem. Res. Toxicol. **1991**, *4*, 237–240.

42. Rosenthal, G.A.; Thomas, D.A. A radiometric assay for determining the incorporation of L-canavanine or L-arginine into protein. Anal. Biochem. **1985**, *147*, 428–431.

43. Rosenthal, G.A. The biological effects and mode of action of L-canavanine, a structural analogue of L-arginine. Q. Rev. Biol. **1977**, *52*, 155–178.

44. Rosenthal, G.A.; Berge, M.A.; Bleiler, J.A.; Rudd, T.P. Abberrent canavanyl protein formation and the ability to tolerate or utilize L-canavanine. Experientia **1987**, *43*, 558–561.

45. Melangeli, C.; Rosenthal, G.A.; Dahlman, D.L. The biochemical basis for L-canavanine tolerance by the tobacco budworm *Heliothis virescens*. Proc. Natl. Acad. Sci. USA. **1997**, *94*, 2255–2260.

46. Rosenthal, G.A.; Reichhart, J.M.; Hoffmann, J.A. L-Canavanine incorporation into vitellogenin and macromolecular conformation. J. Biol. Chem. **1989**, *264*, 13693–13696.

47. Rosenthal, G.A.; Dahlman, D.L. Studies of L-canavanine incorporation into insect lysozyme. J. Biol. Chem. **1991**, *266*, 15684–15687.

48. Kulp, K.S.; Vulliet, P.R. Mimosine blocks cell cycle progression by chelating iron in asynchronous human breast cancer cells. Toxicol Appl. Pharmacol. **1996**, *139*, 356–364.

49. Gilbert, D.M.; Nielson, A.; Miyazawa, H.; DePamphilis, M.L.; Burhans, W.C. Mimosine arrests DNA synthesis at replication forks by inhibiting deoxyribonucleotide metabolism. J. Biol. Chem. **1995**, *270*, 9597–9606.

50. Mosca, P.J.; Lin, H.B.; Hamlin, J.L. Mimosine, a novel inhibitor of DNA replication, binds to a 50-kDa protein in Chinese hamster cells. Nucleic Acids Res. **1995**, *23*, 261–268.

51. Kalejta, R.F.; Hamlin, J.L. The dual effect of mimosine on DNA replication. Exp. Cell. Res. **1997**, *231*, 173–183.

52. Oppenheim, E.W.; Nasrallah, I.M.; Mastri, M.G.; Stover, P.J. Mimosine is a cell-specific antagonist of folate metabolism. J. Biol. Chem. **2000**, *275*, 19268–19274.

53. Lin, H.; Falchetto, R.; Mosca, P.J.; Shabanowitz, J.; Hunt, D.F.; Hamlin, J.L. Mimosine targets serine hydroxymethyltransferase. J. Biol. Chem. **1996**, *271*, 2548–2556.

54. Rosenthal, G.A.; Dahlman D,L. Interaction of L-canaline with ornithine aminotransferase of tobacco hornworm, *Manduca sexta* (Sphingidae). J. Biol. Chem. **1990**, 265, 863–873.

55. Harms, K.; Ballmoos, P.; Brunold, C.; Holfgen, R.; Hesse, H. Expression of bacterial serine acetyl transferase in transgenic potato plants leads to increased levels of cysteine and glutamine. Plant J. **2000**, *22*, 335–343.

56. Saito, K. Molecular genetics and biotechnology in medicinal plants: studies by transgenic plants. Yakugaku Zasshi. **1994**, *114*, 1–20.

4

Lectins

4.1 INTRODUCTION

Lectins are proteins present in abundant quantity in plants and animals. The term *lectin* was first introduced in 1954 by Boyd and Shapleigh (1) and widely used presently for the designation of a sugar-binding protein of a nonimmune origin that agglutinates cells or precipitates glycoconjugates. These proteins have the ability to select human red blood cells according to their blood group. However, a very few lectins are proven to be blood group specific, and the use of lectins in blood grouping has limited scope. The agglutinating property of lectins is not restricted to red blood cells but extends to other cells, such as lymphocytes and microbial tissue cells. In lymphocytes, lectins induce mitosis and stimulate the action of antigens. Lectins interact with oligosaccharides of cell or glycoprotein surfaces. Consequently, few oligo- or monosaccharides are able to inhibit or even revert the interaction. Although lectins act by their carbohydrate binding sites, they may reinforce their interaction with cells or proteins by hydrophobic effects. Because of their binding specificity lectins may be grouped together with enzymes and antibodies. These properties have led to their widespread use for such purposes as the isolation and analysis of complex carbohydrates, cell separation, and studies of cell surface architecture (2).

Plant lectins, which are the focus of this chapter, have been proposed to have significant role in a plant's defense against insect pests, carbohydrate transport, seed germination, plant immune system, control of cell growth, and specific recognition processes (3). Plant defense is believed to be the best articulated role of vacuolar lectins. Currently, there are ample evidences to support their role in insect resistance (4,5). In addition, plant lectins mediate the interactions between rhizobia and the host plant, and they are responsible for the incompatibility of pollen or stigma in flowering plants. Recently, a lectin isolated from salt stressed rice plants highlights the possible importance of plant lectin as a stress response protein (6).

4.2   CHEMISTRY AND CLASSIFICATION

Lectins can be categorized into two main groups based on the carbohydrate group they bind, one specific for galactose and its derivatives such as fucose and N-acetyl galactosamine (GalNAc), and another specific for glucose, mannose, and their derivatives. Majority of lectins fall under the first group, which have immunological relatedness. Many galactose lectins derived from legumes are homologues as revealed by N-terminal amino acid sequence comparison.

Lectins can also be classified into two broad categories based on their evolution. One of them was evolved by the incorporation of carbohydrate binding domains into proteins of many legume species. The second one has a chitin binding domain of 43 amino acids and is found in cereals, *e.g.*, wheat germ agglutinin (WGA) and nettle lectin. Four known members of the PHA (*Phaseolus* hemagglutinin) family of polypeptides are present in common beans namely, PHA-E, PHA-L, α-amylase inhibitor, and arcelin. Lectins PHA-E and PHA-L agglutinate erythrocytes and leukocytes, respectively. Most of the scientists group lectins according to the plant family to which they belong since they share several common features and are homologous (Table 4.1).

4.2.1   Structure of Plant Lectins

The biological role of plant lectins as defense proteins is mainly based on its cell surface sugar recognition. Mechanisms for sugar recognition have evolved independently in diverse protein structural frameworks with some common features. Thus, it is vital to know the structure of lectin in order to understand its biological role. Circular dichroism studies have revealed that the β-pleated sheet is the primary structural feature of lectins (7). Complete three-dimensional structures on a high-resolution map are available for *Concanavalia ensiformis* agglutinin (ConA). The ConA subunit is dome or

TABLE 4.1 Distinguishing Characteristics of Lectins from Different Plant Families (2)

Family/source	Representative lectin/source	Carbohydrate specificity	Composition	Molecular weight (kDa)	Subunit composition
Leguminosae Subfamily: Papilionaceae Tribe: Diocleae	Concanavalin A Seeds	α-D-Mannose (D-glucose)	Rich in Asx, Ser Little Met No Cys	104	Tetramer of identical subunits
Tribe: Vicieae	Favin Seeds	α-D-Mannose (D-glucose)	Rich in Asx, Thr No Met or Cys 3% carbohydrate	53	Tetramer of 2 A chains (MW 5,571) and 2 B chains MW 20,700
Tribe: Phaseoleae	Soybean Seeds	α-D-GalNAc	Rich in Asx, Ser, Leu, Ala, Little Met No Cys 7% carbohydrate	120	Tetramer of identical or nearly identical subunits
Tribe: Loteae	*Lotus tetragonolobus* A Seeds	α-L-Fucose	Rich in Asx, Ser, Thr Little Met No Cys 8% carbohydrate	120	Tetramer of identical subunits
Tribe: Hedysareae	Peanut Seeds	β-D-Gal-(1-3) D-GalNAc	Rich in Asx, Ser, Thr	110	Tetramer of identical subunits

(continued)

TABLE 4.1 (Continued)

Family/source	Representative lectin/source	Carbohydrate specificity	Composition	Molecular weight (kDa)	Subunit composition
			Little Met No Cys No carbohydrate		
Subfamily: Caesalpinoideae	*Griffonia simplicifolia* II Seeds	β-and α-D-GlcNAc	Rich in Asx, Ser, Thr, Gly, Leu Little Met 4% carbohydrate	113	Tetramer of identical subunits
Euphorbiaceae	*Ricinus communis* agglutinin (RCA1) Seeds	β-D-Galactose	Rich in Asx, Glx, Ser, Thr, Ile, Leu Little Cys, Met, Phe 12% carbohydrate	120	Tetramer of 2 A′chains (MW 32,000) and 2 B′chains MW 36,000
Solanaceae	Potato tubers	Oligomers of β-D-GlcNAc-(1-4)-β-D-GlcNAc	Rich in hydroxyl proline, Cys, Ser, Gly No Phe 50% carbohydrate	50 100	Monomer-dimer mixture
Graminaceae	Wheat germ agglutinin Embryos	Oligomers of β-D-GlcNAc-(1-4)-β-D-GlcNAc	Rich in arabinose, Cys,Gly No carbohydrate	36	Dimer of identical subunits

gumdrop shaped with dimensions of 42 Å height, 40 Å width, and 30 Å thickness (8). The monomers are paired base to base by an exact twofold symmetry axis to form ellipsoidal dimers. Additional twofold axes to form roughly tetrahedral tetramers in turn pair the dimers. The polypeptide chain of ConA (*C. ensiformis* agglutinin) is made up of 237 amino acid residues. It recognizes oligomannosyl N-linked sugars. The active molecule is a tetramer, which has a high proportion of β-pleated sheets.

Solanaceous lectins are glycoproteins rich in hydroxyproline and arabinose. *Datura* and potato lectins have monomeric sizes of approximately 30 and 50 kDa, respectively. However, they exist as dimeric form *in vivo*. Solanaceous lectins are rich in four amino acids, namely, hydroxyproline, serine, cysteine, and glycine. Exhaustive pronase digestion followed by reductive alkylation of potato and *Datura* lectins has released 33- and 18-kDa proteins, respectively. These fractions were rich in serine and hydroxyproline but not glycine and cysteine, suggesting that the lectins have at least two distinct domains, which are cross-linked by disulfide bridges. The carbohydrate moieties contain only two sugars, namely, L-arabinose (about 90%) and D-galactose (about 10%) found exclusively in serine/hydroxyproline-rich domains.

The primary structure of a Mimosidae lectin from *Parkia platycephala* seeds has been studied recently (9). The protein contains a blocked N terminus and a single; nonglycosylated polypeptide chain composed of three tandemly arranged homologous domains. All of these domains are found to share sequence similarity with jacalin-related lectin monomers from Asteraceae, Convolvulaceae, Moraceae, Musaceae, Graminae, and Fagaceae plant families. Based on this homology, it has been predicted that each *Parkia* lectin repeat may display a β prism fold similar to that observed in the crystal structure of the lectin from *Helianthus tuberosus*. The *P. platycephala* lectin also showed sequence similarity with stress and pathogen up-regulated defense genes of a number of different plants, suggesting a common ancestry for jacalin-related lectins and inducible defense proteins.

Among the graminaceous lectins, wheat germ agglutinin (WGA) is the best characterized. Isolectins A, B, and D are encoded by their respective diploid genomes of wheat. All WGA isolectins consist of two identical 171-residue polypeptides that associate to form 36-kDa dimers (10). The WGA polypeptide shows remarkable redundancy in its structure. It is composed of four repetitive domains of 422–443 amino acids. Crystal structure analysis revealed that folding of the four independent domains is similar (Fig. 4.1B) and has disulfide bridges positioned identically. The inner core region is formed by the three-fourths of disulfides found in this lectin. Ends have a higher degree of structural flexibility and sequence variability. Two polypeptides of WGA

(A)

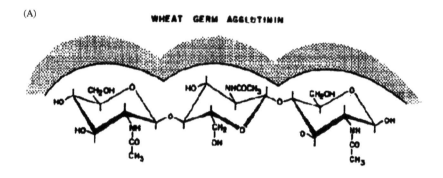

(B)

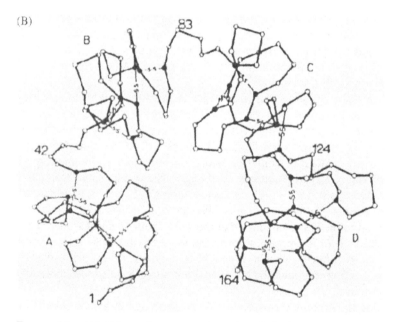

FIGURE 4.1 (A) Schematic presentation of carbohydrate binding sites of wheat germ agglutinin (15). (B) Crystal structure of wheat germ agglutinin (15).

associate in a head-to-tail fashion. Saccharide binding can occur entirely within a single domain. Amino acids involved in saccharide binding are the triad of aromatic residues at domain positions 21, 23, and 30, a serine at position 19, and a glutamate or aspartate at position 29 (Fig. 4.1A).

Snowdrop lectin or *Galanthus nivalis* L. agglutinin (GNA) was characterized by van Damme et al. (11). Edman degradation, carboxypeptidase Y digestion of mature protein, and structural analysis of the peptides obtained after chemical cleavage and modification have allowed the determination of the complete 105-amino-acid sequence of GNA. The snowdrop lectin is a pure tetrmeric protein rich in asparagine and leucine and contains three cysteine residues. The molecular weight of GNA was calculated to be 11.3 kDa. The comparison of protein with the deduced amino acid sequence from cDNA has revealed the presence of a 23-amino-acid N-terminal signal peptide and a 29-amino-acid C- terminal extension.

4.2.2 Biochemical Properties of Lectins

Biochemical properties	*Datura*	Potato
Tissue source	Seed	Tuber
Estimated mass of glycoprotein (kDa)	30	50
Isoelectric point	Not determined	~9.5
Protein (%)	60	50
Abundant amino acids	Hyp, Ser, Cys, Gly	Hyp, Ser, Cys, Gly
Carbohydrate (%)	40	50
Carbohydrate present (percentage of total carbohydrate)	84% Gal and 16% Gal	95% Ara and 5% Gal
Glycoproteins	Hyp-Ara Ser-Gal	Hyp-Ara Ser-Gal

Phytohemagglutinin (PHA) is a tetrameric lectin composed of five isoforms of polypeptides PHA-E and PHA-L in different combinations (7). Tetramer PHA-E agglutinates erythrocytes; PHA-L agglutinates leukocytes and has mitogenic activity. Lectins depend on manganese and/or calcium for sugar binding. ConA binds to α-D-mannopyranose, α-D-glucopyranose, D-fructofuranose, and their glycosides and sterically related structures. ConA also complexes with many other substances that contain the above saccharides.

A new agglutinin called CBL3 was purified recently from *Cyphomandra betacea* (Sendt.) fruits (12). CBL3 is a homodimer of 50.8 kDa with subunit 26.2 bound by disulfide linkages with a pI of 4.9. The agglutinating capacity of CBL3 is inhibited by oligomers of *N*-acetylglucosamine in the following order of potency: tetrasaccharide > trisaccharide > disaccharide. Similar to all solanaceous lectins, CBL3 is not inhibited by *N*-acetylgluco-samine. Plant lectins, when classified based on biochemical properties such as carbohydrate specificity, composition, molecular weight, and subunit composition, fall in groups that belong to single-plant families (Table 4.1).

4.3 OCCURRENCE

Lectins are ubiquitous and found in a wide range of plant species. Nearly 800 species of plants have been found to contain lectins of which 600 species belong to the family Leguminoseae. The legume species that are devoid of hemagglutinin activity were analyzed by antisera raised against one or more of the GalNAc hemagglutinins, and it was found that all of the legume lectin–like proteins cross-reacted. This suggests that lectin is present in all the legume species (3). A list of plant lectins that have insecticidal property is given below (13). *Artocarpus integrifolia* (jacalin), *Bauhinia purpurea alba* (camel's foot tree) (BPA), *Codium fragile* (CFL), *Sambucus nigra* (elderberry) bark (EBL), *Griffonia simplicifolia* lectin II (GSL), *Phytolacca americana* (PAL), *Maclura pomifera* (osage orange) (MPL), *Oryza sativa*, *Triticum vulgare* (wheat germ agglutinin, WGA), *Crocus vernus*, *Agropyrum repens* leaf, *Vicia villosa* (VVL), *Wisteria floribunda*, *Cicer arietinum*, *Cystis scoparius*, *Helix aspersa*, *Helix pomatia*, *Mycoplasma gallisepticum*, *Phaseolus vulgaris* E subunit, *Urtica dioica*, *Tulip* spp, *Eranthis hyemalis*, *Laburnum alpinum* bark (GalNAc specific lectin), *Rhizoctonia solani*, *Listera ovata*, *Aegopodium podagraria*, *Amaranthus leucocarpus*, *Bryonia dioica*, *Limax flavus*, *Epipactis helleborine*, *Cymbidium* hybrid, *Psphopcarpus scandens* (basic lectins 1 and 2), *Ceratobasidium cornigerum*, *Sambucus racemosa*, *Hordeum vulgare*, *Sedum spp*, *Agave americana*, *Dieffenbachia* spp, *Salix* spp, *Crocus vernus*, *Populus canadiensis*, *Chamaecyparis lawsoniana*, *Platanus* hybrid, *Berberis* spp, *Abies nobills*, *Abies pinsapo*, *Cedrus atlantica*, *Cedrus lebanoni*, *Juniperus sabina*, *Picea* spp needle, *Pinus* spp needle, *Sequoiadendron giganteum*, *Taxodium distichum*, *Thuya plicata*, *Agaricus bisporus* fruit bodies, *Acer pseudoplatanus* bark, *Epilobum angustifolium* leaf, *Euonymus europaeus* bark, and *Pragmites australis* leaf. The above list consists of all plant types, i.e., monocots, dicots, and lower plants like pines.

4.3.1 Cellular and Tissue Distribution

4.3.1.1 Leguminaceous Lectins

Leguminaceous lectins are abundant in seeds in which they are specifically localized in the protein bodies of the cotyledon cells (14). The amount of lectin present in other parts of the plant is negligible. In red kidney bean and soybean, low levels of lectin were detected in the roots, stems, and leaves of young seedlings (15). Lectins identified in vegetative parts are different from the seed lectins (16). Very little is known about the subcellular localization of vegetative tissue lectins. There is some evidence for the presence of lectins in root surface, especially in root hairs (14). Immunocytochemical detection with colloidal gold and antibodies to deglycosylated PHA showed that in the meristem of the primary root, PHA accumulates in vacuoles. However, in elongated root cells, PHA was found only in the cell walls, which indicates the targeting of lectin to alternate locations (17).

4.3.1.2   Graminaceous Lectins

The pattern of cereal lectin expression is species specific. Among the cereal lectins, wheat germ lectin is widely studied. In contrast to leguminous lectins, mature grains contain less lectin than developing seeds. Immunocytochemical studies using both light and electron microscopy have revealed that lectins are concentrated at the external surface of the embryo (18). Lectins are concentrated in embryonic roots, coleorhiza epiblast, root cap of the radicle, surface of the scutellum opposite from the coleoptiles, and interior surface of the grain coat. At the cellular level, wheat lectins are present at the peripheries of the protein bodies and the interface between the membrane and cell wall. *In situ* hybridization experiments and *de novo* synthesis of mRNA of lectin revealed the presence of WGA in root cap cells of wheat seedlings. The presence of lectins in protein body, vacuoles of embryonic tissues, and adult roots has been observed (19).

Other graminaceous lectins in rye, barley, and rice show variations in the tissue localization. Both the outer and inner surfaces of the coleoptile have lectin in rye. Rice lectin is present throughout the cells of coleoptile. Couch grass (*Agropyrum repens*) and common reed (*Phragmites australis*) express lectins in leaves. Barley lectin is expressed solely in the adult and embryonic roots (20). *In situ* hybridization experiments show that barley lectin mRNA is localized in the root tip of the adult plant and the analogous structure in the embryo.

4.3.1.3   Solanaceous Lectins

Solanaceous lectins form a unique group of lectins due to their immunological and hemagglutinating properties. Lectins can be detected in immature seeds, stamens, ovaries, anthers and petals of solanaceous plants. Trace levels of lectins are also detected in emerging cotyledons and roots (21). Lectins are abundant in fruits of tomato, tubers of potato, and seeds of tobacco and *Capsicum*. *Datura* seeds contain higher levels of lectins than other parts. Cellular localization by immunofluorescence data indicated that lectins are associated with plasmalemma and with intracellular organellar membranes (22). In potato, fractionation studies have shown that the lectins are found in root membranes and in cell wall.

4.3.1.4   Stinging Nettle Lectin

Stinging nettle lectin (*Urtica dioica* agglutinin UDA) is predominantly localized in rhizomes, roots, and seeds specifically in the periderm and cortex tissues of rhizome and outer periderm of roots (23).

4.3.2   Temporal Expression of Lectins

Temporal expression of lectin was studied in osage orange (*Maculura pomifera*) (15). It was reported that the lectin accumulates during early development of the seed, reaching the highest level at seed maturity and decreasing slowly during germination. However, it can be detected in low levels up to 6-month-old seedlings. In the mature tree, lectin activity is confined to the seed and is not found in the seed coat. Similarly, in red kidney bean and *Dolichus biflorus*, the decrease in seed lectin is observed during seed germination. Lectin is not found in mature plants of lentil, pea, red kidney bean, and *D. biflorus*, but it reappears upon germination. In *D. biflorus*, it is first detected on the 27th day after flowering, and in 1–2 days it attains the level present in the mature seed (24).

4.4   BIOSYNTHESIS

All lectins are synthesized as high molecular weight precursors, which are subsequently processed at the co- and/or posttranslational levels (2). Favin, a lectin from *Vicia faba*, has two subunits α and β having molecular weight 5571 and 20,700, respectively. They are synthesized as a 29-kDa precursor with both subunits and N-terminal signal sequence. The cleavage of the signal sequence occurs after translocation across the microsomal membrane. The subunits are cleaved posttranslationally. In soybean and *P. vulgaris*, the subunits are synthesized independently with an N-terminal signal peptide sequence. Glycosylation of the *P. vulgaris* lectin begins cotranslationally and the subunit assembly occurs in the endoplasmic reticulum. Subunit heterogeneity is common either because of assembly of subunits from different genes or posttranslational cleavage of a single gene product into two peptide chains.

Lectins in graminaceous plants are all 36-kDa dimers composed of two identical subunits held together by intramolecular disulfide bridges. In cultivated rice, the majority of the 18-kDa proteins undergo proteolytic cleavage, yielding two subunits of 8 and 10 kDa. WGA has three distinct portions: an N-terminal putative signal sequence, a core protein with four repetitive domains, and a C-terminal propeptide. The signal sequence has been shown to be removed cotranslationally. The C-terminal propeptide undergoes glycosylation cotranslationally and the propeptide is cleaved during transport or at vacuoles (25).

4.5 BIOACTIVITY

Earlier it was assumed that lectins have a fundamental biological role in plants because they are found in different organs and tissues of many species. Lectins serve as plant defense compounds. When eaten raw, the common bean is toxic to a variety of animals (26). Feeding trials have shown that PHA binds to intestinal mucosa of rats, resulting in lesions, disruptions, and abnormal development of microvilli. It also inhibits the absorption of nutrients across the intestinal wall and greatly increases colonization of bacteria in the small intestine.

4.5.1 Legume Lectins

The toxicity of PHA to insects was first suggested by Jansen et al. (27) and later confirmed by bioassays (28). It was reported that purified lectins inhibited the larval development of *Callosobruchus maculates*, the cowpea beetle. It was also said that the impure PHA was more effective than pure form. However, in the earlier report the PHA preparations were contaminated with α-AI, which has high toxicity to the beetle (29). α-AI shows 100% mortality to *C. maculates* (30) on transgenic plants with GNA 0.6% of total soluble proteins. Arcelin is a PHA like protein toxic to *Zabrotes subfasciatus* bean weevil (31). Feeding experiments with artificial seeds containing 10% (w/w) arcelin in the mixture reduced the percent emergence of adult *Z. subfasciatus* weevils from 76% to 18%.

A seed lectin from the African yam *Sphenostylis stenocarpa* tested on *Clavigralla tomenosicollis* showed reduced survival rate at 1–2% dry weight. At 4–8% dry weight no nymph survival was observed up to 6 days after infestation (32). In addition, significant delay in total development time was also observed. These findings, such as reduced larval growth and survival, indicate the insect resistance properties of African yam lectins. Similarly, insecticidal lectin-enriched extracts from African yam and *Lablab purpureus* were found to be toxic against three insect pests, namely, *Maruca vitrata*, *Callosobruchus maculatus*, and *Clavigralla tomentsicollis* (32). Bioassays of the above three insect pests using African yam lectin extracts showed more than 80% mortality.

Soybean lectin is found to be toxic to the developing larvae of the lepidopteron pest *Manduca sexta* (33). A glucose/mannose-specific lectin, ConA from jack bean (*Concanavalia ensiformis*), has been found to have an inhibitory effect on the insect pests from Lepidoptera and Homoptera. Artificial diet containing ConA concentrations 0.2–2.0% of total protein decreased the survival of tomato moth (*Lacanobia oleracea*) larvae up to 90%. The size of the homopteran pest, peach-potato aphid (*Myzus persicae*), was reduced up to 30% when fed with 1–9 μM concentrations of ConA in an artificial liquid diet. Feeding on transgenic potato plants expressing the ConA protein also yielded similar results. Retarded larval development, greater than a 45% reduction in the larval weight, and reduced intake of transgenic plant tissues were observed in tomato moth larvae. Similarly, ConA-expressing potato plants decreased the fecundity of *M. persicae* by up to 45% (34).

The evolution of lectin genes gives an insight to the understanding of the evolution of the insecticidal activity in these proteins (4). Lectin gene evolution is based on two principles: (a) The evolution of lectin genes within one species, *e.g.*, *Phaseolus vulgaris*, which has the following proteins (summarized below). The homologous genes presumably arose through duplication and divergence of an ancestral gene. This enabled the different plant defense properties. Some of the proteins in this family bind carbohydrates whereas others have acquired different biological properties, such as inhibition of α-amylase. (b) The evolution of lectin genes by the incorporation of carbohydrate binding domain of 43-amino-acid basic building block of WGA. This domain is also found in several other proteins that have defense properties (listed below).

Phytohemagglutinin family:
PHA-E	Toxic to mammals and birds
PHA-L	Toxic to mammals and birds
α-AI	Toxic to weevils
Arcelin	Toxic to weevils

Chitin-binding family:
WGA	Toxic to weevils, European corn borer, and Southern corn rootworm
Rice lectin	Toxic to weevils
Datura lectin	Toxic to weevils
Tomato lectin	Toxic to weevils
Nettle lectin	Toxic to weevils, inhibitory to fungi

4.5.2 *Galanthus nivalis* Agglutinin (GNA)

The effect of GNA on *Lacanobia oleracea* on larval development, growth, consumption, and survival was examined by three different bioassays (35). The first was an artificial diet with 2% w/w GNA; the second was an insect bioassay on excised leaves of transgenic potato showing 0.07% GNA; and the third was a bioassay on transgenic plants with GNA 0.6% of the total soluble proteins. Mean larval biomass was reduced by 32% and 23% in an artificial diet and excised leaves, respectively, at 21 days. Glass house trials revealed a 48% insect biomass reduction per plant after 35 days. GNA caused a 59% overall reduction in mean daily consumption of artificial diet and a significant reduction in leaf damage in glass house trials. Larval survival was decreased by approximately 40% in the glass house bioassay. The above assays confirm the insecticidal effect of GNA both *in vivo* and *in planta*. Recently, it was established that there is direct relationship between the level of GNA present in the transgenic plant and its resistance to the larva of *L. oleracea* L. (36).

GNA is also found to have resistance/tolerance to homopteran pests like rice brown planthopper (BPH) (*Nilaparvata lugens*) and rice green leafhopper (37). GNA has shown 80% corrected mortality in first-and third-instar nymphs of BPH at a concentration of 1g/L in the diet. The antifeedant activity of the GNA against BPH was measured quantitatively by the reduced honeydew excretion levels of BPH fed with an artificial diet containing 0.1% w/v of GNA.

Mannose-binding lectins—GNA, NPA (daffodil, *Narcissus pseudonarcissus*), and ASA (garlic, *Allium sativum*)—were tested for their toxic and growth inhibitory effects on nymphal development of the peach-potato aphid *Myzus persicae*. GNA was the most toxic, with 42% nymphal mortality and 50% growth inhibition at 1500 µg/mL. The other two did not show significant mortality but caused growth inhibition of 59% for NPA and 26% for ASA at 1500 µg/mL.

The impact of snowdrop lectin expressed in transgenic sugarcane on life history parameters of Mexican rice borer [*Eoreuma loftini* (Dyar)] and sugarcane borer [*Diatreaea saccharalis* (F.)] (Both Lepidoptera: Pyralidae) was evaluated (38). In the laboratory, lyophilized sugarcane leaf sheath tissue was incorporated in a meridic diet resulting in a GNA concentration of 0.47% of total protein and used for insect bioassays over two successive generations. Deleterious effects of GNA were not observed on survival weight, or developmental periods of larvae and pupae, or on adult fecundity and egg viability of *D. saccharalis*. Moreover, in the first generation, addition of transgenic sugarcane tissue to the diet enhanced larval growth in *D. saccharalis* resulting in higher larval and pupal weight compared with diet with nontransgenic sugarcane, but this effect was not observed in the second generation. In contrast, larval survival, percent adult emergence, and female fecundity of *E. loftini* were significantly reduced when fed transgenic sugarcane diet in comparison with nontransgenic sugarcane diet. In addition, a substantial reduction of female pupal weight of *E. loftini* was observed in the second generation. For both species, the only consistent effect of GNA in both generations was a reduction in adult female longevity. Life table parameters showed that GNA at the level found in the transgenic diet negatively affected development and reproduction of *E. loftini*, whereas it had a nil to positive effect on development and reproduction of *D. saccharalis*.

4.5.3 Other Lectins

Plant lectins have been implicated as antibiosis factors against insects, particularly the cowpea weevil, *Callosobruchus maculatus*. *Talisia esculenta* lectin (TEL) was tested for anti-insect activity against *C. maculatus* and *Zabrotes* subfasciatus larvae (39). TEL produced about 90% mortality to these bruchids when incorporated in an artificial diet at a level of 2% (w/w). The LD_{50} and ED_{50} for TEL was about 1% (w/w) for both insects. TEL was not disgested by midgut preparations of *C. maculatus* and *Z. subfasciatus*.

Wheat germ agglutinin produced 50% mortality in two major corn pests, European corn borer (*Ostrinia nubilalis*) and southern corn root worm (*Diabrotica undecimpunctata*), when incorporated in their diet at 0.59 and 3 mg/g, respectively (40). WGA, rice lectin, and stinging nettle lectin have deleterious effects on cowpea weevil larvae (41).

The plant defense properties of the *Amaranthus hypochondriacus* lectin (AHA) against potato-peach aphid *M. persicae* have recently been reported (42). Transgenic tobacco plants containing the AHA gene with and without intron were assayed for peach aphid. Nearly 57% and 48% reduction in insect population density was observed for intron and intronless recombinants, suggesting a favorable role for intron in the AHA expression. In some plants, even more than 85% of reduction in insect population density was observed. The above findings reveal the high toxicity level of the protein.

Lectins toxic to lepidopteran pests, European corn borer, and Southern corn borer were identified by bioassays using Stoneville diet containing 0.01–0.5% of lectin. *Triticum vulgare* lectin and WGA showed 100% mortality at 0.5% level whereas *Bauhinia purpurea alba* lectin showed 90% mortality in European corn borer. Larval growth of southern corn rootworm was reduced to 70% by lectins from *Cicer arietinum, Helix pomatia*, and by jacalin. *Griffonia simplicifolia* lectin II and *T. vulgare* lectins exhibited 60% and 50% reduction in larval growth rate, respectively (13).

A list of lectins showing insecticidal activity against pests belonging to various families is given below (5).

Insect family	Insect species	Toxic lectin
Coleoptera	*Callosobruchus maculates*	PHA, WGA, SNAII (elderberry lectin)
	Diabrotica undecimpunctata (Southern corn rootworm)	WGA, GNA
Lepidoptera	*Ostrinia nubilalis* (European corn borer)	WGA, lectins from castor and *Bauhinia purpurea*
	Manduca sexta	Soybean lectin
Homoptera	*Nilaparvata lugens* (BPH)	GNA
	Nephotettix cincticeps Ish. (rice green leafhopper)	GNA
	Empoasca fabae Harr. (potato leafhopper)	PHA, WGA
	Acyrthosiphon pisum Harr. (pea aphid)	ConA
	Myzus persicae Sulz. (potato aphid)	GNA
Diptera	*Lucilia cuprina* Weid. (blowfly)	WGA, ConA

4.6 MECHANISM OF ACTION

Lectin molecules when ingested bound to the midgut epithelial cells, which was demonstrated in an indirect immunofluorescence study using mono-specific antisera for globulin lectins (28). This mechanism of lectin toxicity is analogous to that in rats because midgut epithelial cells are disrupted. In pest species of *Phaseolus vulgaris*, such as *Acanthoscelides obtectus* (Say), the lectin molecules do not bind to epithelial cells. Therefore, no harmful effects were observed (43). Specific binding of GNA to the gut of BPH was determined by analyzing the polypeptides extracted from adult BPH guts on sodium dodecyl sulfate–polyacrylamide gel electrophoresis gels. Results indicated that GNA bound to four glycopeptides in BPH guts, of approximately 66, 43, 37, and 31 kDa (44,45).

Griffonia simplicifolia leaf lectin II (GSII) is a plant defense protein against the insect pest *Callosobruchus maculatus* F (46). Site-specific mutations were introduced in the gene to affect the *N*-acetylglucosamine (GlcAc) binding. Altered proteins were grouped into three categories based on GlcNAc binding activity (high: rGSII, rGSII-Y134F, and rGII-N196D, low: rGSII-N136D, No: rGSII-D88N, rGSII-Y134G, rGII-N1366). Insecticidal activities of the recombinant proteins were tested and a positive correlation was observed with GlcNAc binding activity. It was established that carbohydrate binding activity and biochemical stability of the proteins to digestive proteolysis are functionally linked with the insecticidal activity.

The mode of action of GNA was studied by the analysis of its pathway in (BPH) insect body. Western analysis has revealed the presence of GNA in the honeydew, gut preparation of BPH, and the crude homogenate of the whole body when BPH was fed with an artificial diet containing GNA 0.05% w/v. Within the BPH, lectin binding was concentrated on the luminal surface of the epithelial cells. This suggests that GNA binds to cell surface carbohydrate moieties in the gut. GNA is also localized in the fat bodies and hemolymph of the nymphs. Cellular disruption of the epithelial layer of midgut, deformed microvilli region, and large gaps between microvilli are evident from ultrastructural studies (45). In addition, a granular appearance in the cytoplasm with a very few mitochondria was also noticed in the GNA fed BPH midgut epithelial cells.

The mode of action in chitin-binding lectins is mediated by the binding of these lectins to chitin in the peritrophic membrane of the insect midgut. While doing so the lectins may alter the physical properties of the membrane or prevent formation of the membrane itself (47).

An aphidophagous predator (*Adalia bipunctata*) was found to show reduced longevity, fecundity, and egg viability after feeding on aphids that were fed with transgenic potato plants expressing snowdrop lectin (48). This suggests that the lectin in the aphid's body is stable and able to retain its deleterious effect even after passing through the digestive system.

4.7 MOLECULAR BIOLOGY

Closely related molecular forms, which are otherwise called isolectins, are found in a single species. This corresponds to multiple genes, which might have evolved by gene duplication. In the polyploid wheat, each genome codes for a single molecular species of lectin subunit. Each of these isolectins was correlated to a genome in alloploid wheat. In soybean, two lectin genes, *L1* and *L2*, coding for *N*-acetylgalactosamine and galactose specific seed lectin, respectively, have been identified. In ricin, a fusion of tandemly duplicated genes may have occurred during the evolution of the β chain of ricin as evidenced by homologies in amino acid sequence between NH_2 and

COOH terminal halves of this lectin as well as presence of structurally similar domains (49). Multigene family means altered patterns of expression in different tissues of the same plant.

PHA-L and PHA-E are encoded by two tandemly linked genes *dlec-1* and *dlec-2* (50). These two genes have 90% homology at the nucleotide level and 82% at the amino acid level. Along with them, the genes for α-amylase inhibitor (α -AI) and arcelin are also linked. α-AI is a 27-kDa protein and the gene encoding this protein shows 82% homology to *dlec-1* and *dlec-2* genes. However, the carbohydrate-binding domain of PHA is missing in this gene (31). Arcelin is another tandemly arranged gene with the above cluster having 78% homology to PHA. It shows 58–61% identity at the amino acid level to PHA-E and PHA-L.

The mRNA level of lectins in *P. vulgaris*, *G. max*, and *P. sativum* increases during midmaturation of seeds, which also corresponds to higher production of lectins suggesting a transcriptional level of regulation in lectins. Abscisic acid was found to act as a regulatory agent in the synthesis of lectins. In wheat, it has been observed that the hormone shift in the developing embryo to maturation phase in which stored reserves accumulate and then into a quiescent state prevents the precarious germination of seeds. Rice and wheat embryos when removed from endosperm and cultured in the absence of abscisic acid they stop lectin synthesis and begin to germinate. Presence of abscisic acid in the growth medium blocks germination and stimulates the synthesis of lectins.

Genes coding for pea lectin (p-Lec) have been expressed at high levels in transgenic tobacco plants with CaMV-35S promoter by *Agrobacterium* transformation. The resultant P-Lec transgenic plants had enhanced levels of resistance/tolerance to *Heliothis virescens*. Computer image analysis showed a significant reduction in the larval biomass and leaf damage in transgenic (P-Lec) plants (51). Transgenic tobacco plants with both cowpea trypsin inhibitor gene (CpTI) and P-Lec gene were obtained by cross-breeding plants derived from two independently transformed lines. These plants showed additive effect of 89% insect biomass reduction compared with 50% when expressing the genes individually. This is the first example of a transgenic plant showing enhanced insect resistance and additive protective effect of different plant–derived insect resistance genes.

Transgenic rice plants were developed using a phloem-specific promoter (52) obtained from rice. A gene construct containing phloem specific promoter obtained from rice sucrose synthase gene (RSs1) when fused to the GNA gene expressed the protein in phloem tissues. Transgenic rice plants expressing this insecticidal protein against sap-sucking insects were tested using antibodies raised against GNA. Phloem-specific expression was confirmed (53). Constitutive expression of GNA in transgenic rice plants was obtained using maize ubiquitin 1 (Ubi 1) gene promoter. The transgenic GNA plants reduced the nymphal survival and fecundity, and reduced growth and development of BPH. A multimechanistic action with antibiosis and antixenosis mechanism of GNA was found to hamper the growth of BPH (44).

Similarly in wheat, GNA expression using a phloem-specific promoter has been shown to affect the aphid population feeding on them (54). At GNA levels greater than 0.4% of the total soluble protein, it was found to decrease the fecundity of grain aphids (*Sitobion avenae*).

Two lectins, designated as HTAI and HTAII and their corresponding cDNAs were isolated from *Helianthus tuberosus* (55). The sequence of the cDNA consisted of 432 nucleotides coding for a polypeptide of 143 amino acids. Active HTAI synthesized using *Escherichia coli* had a hemagglutination property. The deduced amino acid sequence has homology to other lectins and jasmonate-induced proteins; accordingly, its expression levels increased in a dose-dependent manner. Recently, the lectin gene from *Amaranthus hypochondriacus* (AHA) has been characterized (42). The AHA gene consists of 2453 base pairs including a 1538 intron and two exons of 212 bp and 703 bp, respectively. The presence of intron has a favorable effect on the expression of AHA protein in transformed plants.

4.8 PERSPECTIVES

Virtually all known plants contain lectins, usually as one of a number of proteins in seeds but often also distributed among the other parts of plants, including leaves, roots, and storage organs. However, the insecticidal activity of lectins has been recognized only in a handful of plant species. Among them, GNA, AHA, ConA, and P-lec have been genetically engineered, and the transgenic plants were also found to be toxic as expected (34,42,51,54). Unlike *Bacillus thuringiensis* (Bt) toxins, resistance to a wide spectrum of pests can be achieved by exploiting plant-derived insect control genes such as lectins. However, the scale of effects produced by these genes has not been convincing enough to lead to serious attempts at commercialization. The limitations could be that these proteins produce chronic rather than acute effects, and many serious pests are simply not susceptible to lectins. In addition, few lectins such as the lectin found in red kidney beans (PHA) are known to damage the cell structure of the small intestine or affect the pancreas and production of digestive enzymes when eaten by human or animals (56,57). To date, only the snowdrop lectin (GNA) is found to be harmless to animals other than insects (56). For this reason, GNA is being highly exploited to develop transgenic plants in several species.

In future, to increase the protective efficacy, spectrum of activity and durability of resistance, it is envisaged that "packages" of different genes could be introduced into crops. The components of such packages should each act on different targets within the insect, thus mimicking the multiple mechanism of resistance that occurs in nature. If identification of insecticidal lectins with no toxic effect to human is difficult, manipulation or engineering the protein for specificity can be done. Attempts have already been made to modify the amino groups of PHA lectin to alter its biological properties. One such experiment has led to the inactivation of agglutinating activity but not its toxicity (58). Similar approaches to improving the insecticidal activity and specificity of the toxins will be a boon to the biotechnologist. Lectins also have other beneficial effects for plants. Recent research findings suggest that lectins are defense proteins for abiotic stresses like salt and frost injury (6,59). Molecular analysis combined with suitable physiological techniques may provide information about the role of lectin as a stress-induced defense protein.

REFERENCES

1. Boyd, W.C.; Shapleigh, E. Antigenic relations of blood group antigens as suggested by tests with lectins. J. Immunol. **1954**, *73*, 226–231.
2. Etzler, M.E. Plant lectins: molecular and biological aspects. Annu. Rev. Plant Physiol. **1985**, *36*, 209–234.
3. Shannon, L.M. Structural relationships and properties of legume lectins. In *Lectins Biology, Biochemistry, Clinical Biochemistry*; Bog-Hansen, T.C., Spengler, G.A., Eds.; Walter de Gruyter: New York, 1983; Vol. 3, 573–581.
4. Chrispeels, M.J.; Raikhel, N.V. Lectins, lectin genes, and their role in plant defense. Plant Cell **1991**, *3*, 1–9.
5. Gatehouse, A.M.R.; Gatehouse J.A. Identifying proteins with insecticidal activity: use of encoding genes to produce insect-resistant transgenic crops. Pest. Sci. **1998**, *52*, 165–175.
6. Zhang, W.; Peumans, W.J.; Barre, A.; Astoul, C.H.; Rovira, P.; Rouge, P.; Proost, P.; Truffa-Bachi, P.; Jalali, A.A.; Van Damme, E.J.M. Isolation and characterization of a jacalin-related mannose-binding lectin from salt-stressed rice (*Oryza sativa*) plants. Planta **2000**, *210*, 970–978.
7. Goldstein, I.J.; Hayes, C.E. The lectins: carbohydrate-binding proteins of plants and animals. Adv. Carbohydr. Chem. Biochem. **1978**, *35*, 127–340.
8. Edelman, G.M.; Cunningham, B.A.; Reeke, G.N.; Becker, J.W.; Waxdal, M.J.; Wang, J.L. The covalent and three-dimensional structure of concanavalin A. Proc. Natl. Acad. Sci. USA. **1972**, *69*, 2580–2584.
9. Mann, K.; Farias, C.M.; Delsol, F.G.; Santos, C.F.; Grangeiro, T.B.; Nagano, C.S.; Cavado, B.S.; Calvete, J.J. The amino acid sequence of the glucose/mannose-specific lectin isolated from *Parkia platycephala* seeds reveals three tandemly arranged jacalin-related domains. Eur. J. Biochem. **2001**, *268*, 4414–4422.

10. Raikhel, N.V.; Lee, H.I.; Brockaert, W.F. Structure and function of chitin-binding proteins. Annu. Rev. Plant Physiol. Plant Mol. Biol. **1993**, *44*, 591–615.

11. Van Damme, E.J.M.; Kaku, H.; Perini, F., Goldstein, I.J.; Peeters, B.; Yagi, F.; Decock, B.; Peumans, W.J. Biosynthesis, primary structure and molecular cloning of snow drop (*Galanthus nivalis* L.) lectin. Eur. J. Biochem. **1991**, *202*, 23–30.

12. Sampietro, A.R.; Isla, M.I.; Quiroga, E.N.; Vattuone, M.A. An N-acetylglu-cosamine oligomer binding agglutinin (lectin) from ripe *Cyphomandra betacea* Sendt. fruits. Plant Sci. **2001**, *160*, 659–667.

13. www.nal.usda.gov/bic/Biotech_Patents/1995patents/05407454.html.

14. Etzler, M.E.; MacMillan, S.; Scates, S.; Gibson, D.M.; James, D.W.; Cole, D.; Thayer, S. Subcellular localization of two *Dolichos biflorus* lectins. Plant Physiol. **1984**, *76*, 871–878.

15. Lis, H.; Sharon, N. Lectins in higher plants. In *The Biochemistry of Plants: A Comprehensive Treatise*; Marcus, A., Ed.; Academic Press: New York, 1981; Vol. 6, 371–447.

16. Gade, W.; Schmidt, E.L.; Wold, F. Evidence for the existence of an intracellular root lectin in soybeans. Planta **1983**, *158*, 108–110.

17. Kjemtrup, S.; Borkhsenious, O.; Raikhel, N.V.; Chrispeels, M.J. Targeting and mistargeting of vacuolar phytohemagglutinin E released from roots of bean seedlings. Plant Physiol. **1995**, *109*, 603–610.

18. Mishkind, M.L.; Raikhel, N.V.; Palevitz, B.A.; Keegstra, K. Immuno-cytochemical localization of wheat germ agglutinin in wheat. J. Cell. Biol. **1982**, *92*, 753–764.

19. Raikhel, N.V.; Mishkind, M.L.; Palevitz, B.A. Characterization of a wheat germ agglutinin like lectin from adult wheat plants. Planta **1984**, *162*, 55–61.

20. Raikhel, N.V.; Bednarek, S.Y.; Wikins, T.A. Cell-type specific expression of wheat germ agglutinin gene in embryos and young seedlings of Triticum aestivum. Planta **1988**, *176*, 406–414.

21. Kilpatrick, D.C.; Jeffree, C.E.; Lockhart, C.M.; Yeoman, M.M. Immunological evidence for structural similarities among lectins from species of the solanaceae. FEBS Lett. **1980**, *113*, 129–133.

22. Jeffree, C.E.; Yeoman, M.M.; Kilpatrick, D.C. Immunoflourescence studies on plant cell: localization of enzymes, lectins and other proteins in plants. Int. Rev. Cytol. **1982**, *80*, 231–265.

23. Lerner, D.R.; Raikhel, N.V. The gene for stinging nettle lectin (*Urtica dioica* agglutinin) encodes both a lectin and a chitinase. J. Biol. Chem. **1992**, *267*, 11085–11091.

24. Pueppke, S.G.; Bauer, W.D.; Keegstra, K.; Ferguson, A. The role of lectins in plant–microorganism interactions. II. Distribution of soybean lectin in tissues of *Glycine max* (L) Merr. Plant Physiol. **1978**, *61*, 779–784.

25. Bednarek, S.Y.; Raikhel, N.V. The barley lectin carboxyl-terminal propeptide is a vacuolar protein sorting determinant in plants. Plant Cell **1991**, *3*, 1195–1206.

26. Pusztai, A.; Clarke, E.M.W.; King, T.P. The nutritional toxicity of *Phaseolus vulgaris*. J. Nutr. Soc. **1979**, *38*, 115–120.

27. Janzen, H.; Juster, H.B.; Liener, I.E. Insecticidal action of the phytohemagglutinin in black bean on a bruchid beetle. Science **1976**, *192*, 795–796.

28. Gatehouse, A.M.R.; Dewey, F.M.; Dove, J.; Fenton, K.A.; Pusztai, A. Effect of seed lectin from *Phaseolus vulgaris* on the development of larvae of *Callosobruchus maculatus*: mechanism of toxicity. J. Sci. Food Agric. **1984**, *35*, 373–380.

29. Murdock, L.L.; Huesing, J.E.; Nielsen, S.S.; Pratt, R.C.; Shade, R.E. Biological effects of plant lectins on the cowpea weevil. Phytochemistry **1990**, *29*, 85–89.

30. Ishimoto, M.; Kitamura, K. Identification of the growth inhibitor on azuki bean weevil in kidney bean (*Phaseolus vulgaris* L.). Jpn. J. Breeding **1998**, *38*, 367–370.

31. Osborne, T.C.; Burow, M.; Bliss, F.A. Purification and characterization of arcelin seed protein from common bean. Plant Physiol. **1988**, *86*, 399–405.

32. Omitogun, O.G.; Jackai, L.E.N.; Thottappilly, G. Isolation of insecticidal lectin-enriched extracts from African yam bean, (*Sphenostylis stenocarpa*) and other legume species. Entomol. Exp. Appl. **1999**, *90*, 301–311.

33. Shukle, R.H.; Murdock, L.L. Lipoxygenase, trypsin inhibitor and lectin from soybeans: effects on larval growth of *Manduca sexta* (Lepidoptera: Sphingidae). Environ. Entomol. **1983**, *12*, 787–791.

34. Sauvion, N.; Rahbe, Y.; Peumans, W.J; van Damme, E.J.M.; Gatehouse, J.A.; Gatehouse, A.M.R. Effects of GNA and mannose binding lectins on development and fecundity of the peach-potato aphid, *Myzus persicae*. Entomol. Exp. Appl. **1996**, *79*, 285–293.

35. Fitches, E.; Gatehouse, A.M.R.; Gatehouse, J.A. Effects of snowdrop lectin (GNA) delivered via artificial diet and transgenic plants on the development of tomato moth (*Lacanobia oleracea*) larvae in laboratory and glasshouse trials. J. Insect Physiol. **1997**, *43*, 727–739.

36. Down, R.E.; Ford, L.; Bedford, S.J.; Gatehouse, L.N.; Newell, C.; Gatehouse, J.A.; Gatehouse, A.M.R. Influence of plant development and environment on transgene expression in potato and consequences for insect resistance. Transgenic Res. **2001**, *10*, 223–236.

37. Powell, K.S.; Gatehouse, A.M.R.; Hilder, V.A.; Gatehouse, J.A. Antimetabolic effects of plant lectins and plant and fungal enzymes on the nymphal stages of two important rice pests, *Nilaparvata lugens* and *Nephotettix cincticeps*. Entomol. Exp. Appl. **1993**, *75*, 61–65.

38. Setamou, M.; Bernal, J.S.; Legaspi, J.C.; Mirkov, T.E.; Legaspi, B.C. Evaluation of lectin-expressing transgenic sugarcane against stalkborers (Lepidoptera: Pyralidae): effects on life history parameters. J. Econ. Entomol. **2002**, *95*, 469–477.

39. Macedo, M.L.; Freire M.G.M.; Novello, J.C.; Marangoni, S. *Talisia esculenta* lectin and larval development of *Callosobruchus maculatus* and *Zabrotes subfasciatus* (Coleoptera: Bruchidae). Biochem. Biophys. Acta **2002**, *157*, 83–88.

40. Czapla, T.H.; Lang B.A. Effect of plant lectins on the larval development of European corn borer (Lepidoptera: Pyralidae) and Southern corn root worm (Coleoptera: Chrysomelidae). J. Econ. Entomol. **1990**, *83*, 2480–2485.

41. Huesing, J.E.; Murdock, L.L.; Shade, R.E. Effect of wheat germ isolectins on development of cowpea weevil. Phytochemistry **1991**, *30*, 785–788.

42. Zhou, Y.G.; Tian, Y.G.; Mang, K.Q. Cloning of AHA gene from *Amaranthus hypochondriacus* and its aphid inhibitory effect in transgenic tobacco plants. Sheng Wu Gong Cheng Xue Bao **2001**, *17*, 34–39.

43. Gatehouse, A.M.R.; Shackley, S.J.; Fenton, K.A.; Bryden, J.; Pusztai, A. Mechanism of seed lectin tolerance by a major insect storage pest of *Phaseolus vulgaris, Acanthoscelides obtectus.* J. Sci. Food Agric. **1989**, *47*, 269–280.

44. Bharathi, M.; Gatehouse, A.M.R.; Gatehouse, J.A.; Rao, K.V.; Hodges, T.K.; Sudhakar, D.; Christou, P. Enhancing the resistance of rice to brown planthopper. Poster presented at the general meeting of the International Programme on Rice Biotechnology of the Rockefellor Foundation, Malacca, Malaysia, Sept 15–19, Rockefellor Foundation: New York, 1997; 347.

45. Powell, K.S.; Spence, J.; Bharathi, M.; Gatehouse, J.A.; Gatehouse, A.M.R. Immunohistochemical and developmental studies to elucidate the mechanism of action of the snowdrop lectin on the rice brown planthopper *Nilaparvata lugens* (Stal). J. Insect Physiol. **1998**, *44*, 529–539.

46. Zhu-salzman, K.; Shade, R.E.; Koiwa, H.; Salzaman, R.A.; Narasimhan, M.; Bressan, R.A.; Hasegarva, P.M.; Murdoc, L.L. Carbohydrate binding and resistance to proteolysis: control insecticidal activity of *Griffonia simplicifolia* lectin II. Proc. Natl. Acad. Sci. USA **1998**, *95*, 15123–15128.

47. Richards, A.G.; Richards, P.A. The pleiotrophic membranes of insects. Annu. Rev. Entomol. **1991**, *2*, 219–240.

48. Birch, A.N.E.; Geoghegan, I.E.; Majerus, M.E.N.; McNicol, J.W.; Hackett, C.A.; Gatehouse, A.M.R.; Gatehouse, J.A. Tritrophic interactions involving pest aphids, predatory 2-spot lady birds and transgenic potatos expressing snowdrop lectin for aphid resistance. Mol. Breed. **1999**, *5*, 75–83.

49. Vasil, I.K.; Hubbel, D.H. The effect of lectins on cell division in tissue culture of soybean and tobacco. In *Cell Wall Biochemistry Related to Specificity in Host–Plant Pathogen Interactions*; Solheim, B., Raa, J. Eds.; Universitets forlaget: Tromso-Oslo-Bergen, 1977; 361–367.

50. Hoffman, L.M.; Donaldson, D.D. Characterization of two *Phaseolus vulgaris* phytohemagglutinin genes closely linked on chromosome. EMBO J. **1985**, *4*, 883–889.

51. Boulter, D.; Edwards, G.A.; Gatehouse, A.M.R; Gatehouse, J.A.; Hilder, V.A. Additive protective effects of incorporating two different higher plant derived insect resistance genes in transgenic tobacco plants. Crop Prot. **1990**, *9*, 351–354.

52. Shi, Y.; Wang, M.B.; Powell, K.S.; van Damme, E.J.M.; Hilder, V.A.; Gatehouse, A.M.R.; Boulter, D.; Gatehouse, J.A. Use of the rice sucrose synthase-1 promoter to direct phloem-specific expression of B-glucuronidase and snowdrop lectin genes in transgenic tobacco plants. J. Exp. Bot. **1994**, *45*, 623–631.

53. Sudhakar, D.; Fu, X.; Stoger, E.; Williams, S.; Spence, J.; Brown, D.P.; Bharathi, M.; Gatehouse, J.A.; Christou, P. Expression and immunolocalisation of the snowdrop lectin, GNA in transgenic rice plants. Transgenic Res. **1998**, *5*, 371–378.

54. Stoger, E.; Williams, S.; Christou, P.; Down, R.E.; Gatehouse, J.A. Expression of the insecticidal lectin from snowdrop (*Galanthus nivalis* agglutinin GNA) in transgenic wheat plants: effect on predation by the grain aphid *Sitobion avenae*. Mol. Breeding **1999**, *5*, 65–73.

55. Nakagawa, R.; Yasokawa, D.; Okumura, Y.; Nagashima, K. Cloning and sequence analysis of cDNA coding for a lectin from *Helianthus tuberosus* callus and its jasmonate-induced expression. Biosci. Biotechnol. Biochem. **2000**, *6*, 1247–1254.

56. Pusztai, A.; Ewen, S.W.B.; Grant, G.; Brown, D.S.; Stewart, J.C.; Peumans, W.J.; van Damme, E.J.M.; Bardocz, S. Antinutritive effect of wheat germ agglutinin and other N-acetylglucosamine specific lectins. Br. J. Nutr. **1993**, *70*, 313–321.

57. Pusztai, A.; Ewen, S.W.B.; Grant, G.; Peumans, W.J.; van Damme, E.J.M.; Rubio, L.; Bardocz, S. The relationship between survival and binding of plant lectins during small intestine passage and their effectiveness as growth factors. Digestion **1990**, *46*, 310–316.

58. Figueroa, M.O.R.; Lajolo, F.M. Effect of chemical modifications of *Phaseolus vulgaris* lectins on their biological properties. J. Agric. Food Chem. **1997**, *45*, 639–643.

59. Hincha, D.K.; Pfuller, U.; Schmitt, J.M. The concentration of cryoprotective lectins in mistletoe (*Viscum album* L.) leaves is correlated with leaf frost hardiness. Planta **1997**, *203*, 140–144.

5

Enzyme Inhibitors

5.1 INTRODUCTION

Plants have evolved general insect defense mechanisms that are sufficient for plant survival but not always effective enough to keep the damage to a level that would be acceptable for crop plants. One such defense mechanism is the use of plant proteinaceous inhibitors. The three important enzymes required for the digestion of food materials by insects are proteinases, amylases, and lipases. Polypeptides and proteins that are inhibitors of proteolytic enzymes are present in cells of almost all life forms and may be ubiquitous in nature. Natural proteinase or protease inhibitors are often found as major components in plants and are concentrated in seeds and tubers, particularly in those of the Graminae, Leguminosae (Fabaceae), and Solanaceae families. Proteinases that are inhibited by plant inhibitors are primarily found in animal, bacterial, and fungal species and only occasionally in plants. Most plant proteinase inhibitors are specific for serine endopeptidases. A few inhibitors from plants that inhibit metallocarboxypeptidases, acid proteinases, and sulfhydryl proteinases are known, but inhibitors of plant aminopeptidases have not yet been identified. In 1946, Kunitz crystallized an inhibitor of trypsin from soybeans (1) that still bears his name, the Kunitz soybean trypsin inhibitor (STI). The role of plant proteinase inhibitors in plant metabolism, their possible contribution to the natural protective system of plants, and their evolutionary relationship have been reviewed (2–12).

Protein inhibitors of mammalian α-amylases have been purified and characterized from many plant seeds. These inhibitors are abundant in cereal grains and are also present in legumes and other seeds. These inhibitors also inhibit insect α-amylases. The natural concentrations of α-amylase inhibitors (α-AIs) in seeds are related to resistance to seed pests such as bruchid beetles. The α-amylases are the main carbohydrate-digesting enzymes in insect gut; resistance is achieved presumably by interfering with assimilation of carbohydrate from the diet. Unlike proteinase inhibitors, induced synthesis of amylase inhibitors by insect attack has not been observed; therefore, the precise physiological role of these proteins remains open to speculation. However, those purified from wheat and common bean have been shown to be insecticidal to nonpest species when tested in artificial diet. There is no detailed report available on proteinaceous inhibitors of lipases except one research publication (13).

5.2 CHEMISTRY AND CLASSIFICATION

Based on the hydrolytic function, enzyme inhibitors are broadly classified as follows:

1. Protease/proteinase inhibitors
2. Amylase inhibitors
3. Lipase inhibitors

5.2.1 Proteinase Inhibitors

The terms *proteinase* and *protease* are often confused in the literature. In this chapter the definition by Barret (14) is followed. Protease refers to both endo- and exopeptidases, where as proteinase describes endopeptidase only. Since most of the plant inhibitors target mainly endopeptidases, the term *proteinase inhibitor* is used here.

In general, most plant proteinase inhibitors are specific for the serine endopeptidases trypsin, chymotrypsin, or enzymes with similar specificities. Few plant inhibitors are specific for sulfhydryl endopeptidases although several inhibitors of sulfhydryl enzymes have been isolated from animals. A new class of inhibitors in potato tubers that specifically inhibit animal metalloxopeptidases, carboxypeptidase A and B, was discovered. Inhibitors of acid proteinases are also rare, and only one such type has been isolated from plants (15). Inhibitors of aminopeptidases have not yet been reported from plants or animals, but have been found in microorganisms.

5.2.1.1 Serine Proteinase Inhibitors

Serine proteinase inhibitors found in plants are typically polypeptides and proteins composed entirely of L-amino acids linked through peptide bonds. They generally contain high percentages of half-cystine residues, present as disulfide cross-links, and it is not uncommon for inhibitors to have one half-cystine for every 8–10 amino acid residues. Plant inhibitors are typically low in or devoid of methionine, histidine, and tryptophan but are often rich in aspartic acid, glutamic acid, serine, and lysine residues. No plant proteinase inhibitor has yet been identified as a glycoprotein. This contrasts with the situation in animals, in which several glycoprotein proteinase inhibitors occur.

Sizes of plant serine proteinase inhibitors vary from 4 to 60 kDa. It is not unusual to find that gene duplication–elongation has taken place, resulting in inhibitors having two nearly identical halves, each with an active site and a "double-headed" inhibitor capable of simultaneously inhibiting two molecules of enzyme per molecule of inhibitor. This double-headedness is not shared by all plant proteinase inhibitors but is commonly found in plant inhibitors with molecular weight (MW) near 10 kDa.

Plant proteinase inhibitors generally are quite stable molecules and are often resistant to heat, to pH extremes, and to proteolysis by proteinases, even by those they do not inhibit. This stability has been attributed in part to the high degree of cross-linking through disulfide bridges. It is clear, however, that other non-covalent interactions also contribute significantly to the stability of the inhibitors. For example, inhibitor I from potatoes is a powerful chymotrypsin inhibitor stable in solution at 80°C for several minutes. Yet it contains but one disulfide bond per monomer unit (MW ∼ 8.3 kDa) that can be reduced and carboxymethylated without loss of inhibitory activity. More than 100 serine proteinase inhibitors of plant origin have been reported. These include the widely studied Bowman-Birk and Kunitz inhibitors that are major components in seeds of legumes such as soybean.

5.2.1.2 Bowman-Birk Inhibitor

Bowman-Birk proteinase inhibitor was first identified in the genus *Glycine* (16) and characterised by Birk in 1985 (17). Since then many members of this inhibitor family have been isolated and characterized mostly from members of the family Leguminosae such as cowpea (18). The Bowman-Birk inhibitors (BBIs) are relatively small proteins with a molecular weight of about 8 kDa. The inhibitor contains 71 amino acids. It is especially rich in cystine residues in that it has seven disulfide bonds, but it is devoid of glycine and tryptophan. It is a double-headed inhibitor with independent binding sites for chymotrypsin and trypsin. It exhibits marked stability toward heat, acid, and alkali, a property most likely attributable to the stabilizing effect of the disulfide bonds on the structure of the protein. The trypsin reactive site (Lys 16–Ser 17) as well as the chymotrypsin reactive site (Leu 44–Ser 45) lie within nonapeptide loops formed by a single disulfide bond. The sequence of amino acids surrounding these two reactive sites are in fact remarkably similar not only to each other but to the corresponding active sites of other legume inhibitors (19–21).

By taking advantage of a single methionine residue at position 27, which could be cleaved with CNBr, Odani and Ikenaka (22) succeeded in separating the inhibitor into two active fragments: one (38 residues) contained the trypsin reactive site and inhibited only trypsin, the other (29 residues) contained the chymotrypsin reactive site and was effective only toward chymotrypsin. Odani and Ikenaka (22) also reported the synthesis of the nonapeptide loop encompassing the trypsin reactive site (Cys 14–Cys 22). Although this peptide did not form a stable complex with trypsin, it did display significant inhibitory activity toward the esterase and peptidase activities of this enzyme.

A considerable number of BBIs were sequenced (23–26). All the sequenced inhibitors consist of two tandem homology regions on the same polypeptide chain, each with a reactive site. However, significant differences exist among the BBIs, where interhomology region disulfide bonds link homology regions. When the interhomology region disulfides are split, the resulting two-homology regions are separately active but are weaker inhibitors than the parent molecule, presumably due to partial loss of rigidity.

In most Bowman-Birk-type inhibitors the P_1 residue in the first (NH_2 terminal) homology region is Lys and trypsin is inhibited, whereas the P_1 residue in the second homology region is Leu and chymotrypsin is inhibited (20–26). However, these residues vary a great deal. Particularly interesting is the soybean inhibitor D-II, in which both P_1 residues are Arg and which is double headed for trypsin. This inhibitor can be viewed as an ancestor of other inhibitors. Soybean inhibitor C-II and garden bean inhibitor II, both of which have Ala and Arg as their P_1 residues, inhibit both elastase and trypsin (24).

Many homologous BBIs are isolated from each species of beans or even from a single bean. Some of these forms appear to differ only in the length of their NH_2 terminal sequence, but many represent significant differences in sequence. They have a tendency to form homo- or heterodimers and trimers (27–29).

5.2.1.3   Kunitz Inhibitor

The first plant inhibitor to be well characterized was soybean trypsin inhibitor (Kunitz). Its isolation and crystallization and that of its complex with trypsin by Kunitz is one of the classic achievements of inhibitor chemistry (30). Kunitz soybean trypsin inhibitor is often confused with BBI, but the two are strikingly different. The polypeptide chain in single-headed Kunitz soybean inhibitor has 181 residues and only 2 disulfide bridges while Bowman-Birk is typically double headed, has 7 disulfide bridges, and only about 71 residues (31). The reactive site is located at residues Arg63 and Ile64. On the basis of this sequence and theoretical considerations it seems that this molecule have very little α-helical structure and is largely in the form of a random coil. This was verified by X-ray crystallographic data (32). Kunitz inhibitor is resistant to denaturing agents such as urea and guanidinium chloride. Although the disulfide bond encompasses the small loop of the molecule, and Cys136 and Cys145 may be selectively cleaved without loss in activity, rupture of both disulfide bonds causes inactivation.

X-ray crystallographic studies of soybean Kunitz inhibitor and subtilisin inhibitor (33) permit comparison of the three-dimensional structure of these inhibitors. The overall folding is dramatically different; thus, they belong to separate families, strongly supporting the notion that the inhibitor families arose by convergent evolution (34). The amino acid sequences surrounding the reactive sites of the various families are summarized in Table 5.1.

The reactive site residue, P_1, generally corresponds to the specificity of the cognate enzyme. Thus, inhibitors with P_1 Lys and Arg tend to inhibit trypsin and trypsin-like enzymes (6); those with P_1 Tyr, Phe, Trp (artificial), Leu, and Met inhibit chymotrypsin and chymotrypsin-like enzymes; and

TABLE 5.1   Amino Acid Residues in Sequences Surrounding the Reactive Site of Inhibitors

Inhibitor family	Reactive site amino acid							
	P_4	P_3	P_2	P_1	P'_1	P'_2	P'_3	P'_4
Kunitz inhibitor	P	S	Y	R	I	R	F	I
Bowman-Birk inhibitor	A	C	T	K	S	N	P	P
Pancreatic trypsin inhibitor (Kazal)	G	C	P	R	I	Y	N	P
Pancreatic trypsin inhibitor (Kunitz)	G	P	C	K	A	R	I	I
Potato inhibitor I	P	V	T	L	D	Y	R	C
Potato inhibitor II	A	S	Y	K	S	V	C	E

those with P_1 Ala and Ser inhibit elastase-like enzymes. Those that inhibit chymotrypsin often, but not always, inhibit subtilisin.

In many instances, strong inhibitors of trypsin with P_1 Arg and Lys inhibit chymotrypsin on the same reactive site. In other cases, related inhibitors do not inhibit chymotrypsin, and failure to inhibit is not yet fully predictable. The most striking feature of P_1 specificity is that exchange of Lys for Arg at this position, either by actual mutation or by semisynthetic replacement, leaves the inhibitor specificity and strength approximately the same. The exchange of Lys or Arg for a chymotrypsin-specific residue generally changes the inhibitor from a good trypsin inhibitor to a good chymotrypsin inhibitor. An odd exception is the strange behavior of Phe63 soybean trypsin inhibitor.

5.2.1.4   Cysteine Proteinase Inhibitors

Proteinaceous cysteine proteinase inhibitors of animal origin have evolved from a cognate ancestral gene and from the cystatin superfamily, which is further classified into three different families on the basis of their molecular structures. Family 1 cystatins lack disulfide bonds, family 2 cystatins contain two disulfide bonds, and family 3 cystatins conserve the sequence Gln-Val-Val-Ala-Gly or its homologues that might be involved in the inhibition of cysteine proteinases (35).

Plant cysteine proteinase inhibitors are usually small, about 12–18 kDa, and do not have disulfide bonds (36–40). The rice cysteine proteinase inhibitor oryzacystatin has been isolated, characterized, and has shown significant homology to type II of the mammalian cystatins (37). It contains the sequence Gln-Val-Val-Ala-Gly, a highly conserved sequence throughout the cystatin superfamily. Oryzacystatin (OC) in rice is composed of 102 residues (Met 1–Ala 102). Two forms of oryzacystatin, I and II, are known. Oryzacystatin shows significant sequence homology with animal cystatins. It bears no disulfide bond and could be classified as family 1. However, in terms of the numbers and location of identical amino acid residues, it seems that OC more closely resembles family 2 than family 1 cystatins. For example, the sequences Phe-Ala-Val (residues 29–31) and Phe-Trp-Met (residues 83–85) of oryzacystatin are common to the family 1 cystatins. In any event, it is likely that OC and animal cystatins must have evolved from a cognate ancestral gene. Incidentally, it is noteworthy that the conserved sequence Gln-Val-Val-Ala-Gly exists in oryzacystatin (38). The mRNA of OC contains 598 base pairs. The amino acid sequence of OC deduced from the cDNA sequence was significantly homologous to those of mammalian cystatins, especially family 2 cystatins.

A nearly full-length cDNA clone encoding 102 amino acid residues was obtained.

5.2.1.5 α-Amylase Inhibitor

A series of structural studies has been made on a white kidney bean (*P. vulgaris*) α-amylase inhibitor (α-A1), which is composed of two kinds of glycopolypeptide subunits, α and β (41). These studies disclosed for the first time that a legume α-amylase inhibitor has a tetrameric structure, $\alpha^2\beta^2$, which is essential for the inhibitory activity. In this study, with a view to gaining an insight into the common structural features of two leguminous α-amylase inhibitors, Nakaguchi et al. (42) undertook a study to determine the structure of an α-amylase inhibitor (α-AI-2) from a wild common bean (*P. vulgaris*), that has an inhibitory specificity quite different from that of α-AI-1. The amino acid sequences established for the two kinds of subunits, α and β, of α-AI were highly homologous to those of α-AI-1. Polypeptide molecular weight of α-AI-2 determined by size exclusion chromatography (SEC)/light scattering technique, considered together with the sequence molecular weights of the subunits α and β, demonstrated that α-AI-2 is also a tetrameric complex, $\alpha^2\beta^2$, analogous to α-AI-1, suggesting that the tetrameric structure is common to leguminous α-amylase inhibitors. Further more, a comparison of the amino acid sequences of these inhibitors and other *P. vulgaris* defense proteins suggested that the posttranslational processing of the precursors to form an active tetramer needs an Arg residue close to the processing site (42).

The polypeptide molecular weights of α and β subunits of α-AI-2 were estimated to be about 8 and 15 kDa, respectively. No tryptophan was detected in the α subunit and no cysteine in the β subunit. Valine and lysine contents were much higher in the β subunit. The α subunit contained about 27% by weight of carbohydrate, whereas the β subunit had only 4%. The sugar compositions of these subunits suggested the possibility that the α subunit has an abundance of high-mannose oligosaccharides and the β subunit contains one short xylomannose oligosaccharide chain. The amino acid sequence of α-AI-2 was deduced from the nucleotide sequence of the cloned gene (43). Therefore, to establish the complete sequences of the α and β subunits, these subunits were subjected to terminal sequence analyses. The α and β subunits were finally found to comprise of 72 and 135 amino acids, respectively, and a comparison of α-AI-2 with α-AI-1 revealed a high degree of identity (78%) in their amino acid sequences (42). Based on light scattering technique and size exclusion chromatographic elution patterns on Sephadex 200, the molecular weight of α-AI-2 was calculated as 48.6 kDa. This value is nearly twice the sequence molecular weight of the heterodimer α and β, i.e., 23,122 (corrected for isotopic composition), demonstrating that α-AI-2 is a tetrameric complex composed of $\alpha^2\beta^2$ analogous to the structure recently elucidated for α-AI-1.

The ragi (*Eleucine coracana* Gaertneri; Indian finger millet) bifunctional α-amylase/trypsin inhibitor (RATI) is the prototype of a cereal inhibitor superfamily (44). It has two independent sites for the cognate enzymes. It inhibits α-amylases from various sources. The crystal structure of RATI and its nuclear magnetic resonance (NMR) structure have also been reported (45). It is a monomer of 122 amino acids with a pI of 10.3. The globular fold of RATI comprises of four α helices with a simple up-and-down topology and two short antiparallel β strands. It is a very stable molecule, resistant to urea, guanidine hydrochloride, and thermal denaturation (46).

The amaranth α-amylase inhibitor (AAI) is a 32-residue peptide isolated from the seeds of the Mexican crop plant *Amaranthus hypochondriacus* (47). The peptide strongly inhibits α-amylase activity of insect larvae and does not inhibit human and mammalian α-amylases. AAI's sequence contains three cystine bridges. The sequence is not closely related to any known protein; however, its disulfide topology and residue conservation pattern are similar to those found in members of the squash family of proteinase inhibitors. This group of proteins was named "Knottins" because the disulfides form a topological knot (5). AAI is the shortest of the peptide α-amylase inhibitors described so far, which makes it an attractive candidate for conferring pest resistance to transgenic plants. Another interesting feature of the AAI molecular architecture is that it can harbor different biological activities within a small and compact molecular framework, so that it may be used to design novel enzyme inhibitors of therapeutic relevance. The amino acid sequence and the cysteine bridges are shown below.

5.3 OCCURRENCE

5.3.1 Proteinase Inhibitors

Serine proteinase inhibitors are distributed throughout plants. They are found primarily in storage tissues such as seeds and tubers where they often represent several percent of the total proteins. Serine proteinase inhibitors also occur in other vegetative tissues but are generally present in lower concentrations than in storage organs.

Inhibitors of the BBI type have been found in the seeds of many members of Leguminoseae (9), i.e., soybean, garden bean, azuki bean, mung bean, cowpea, lima bean, peanut, navy bean, kidney bean, black-eyed pea, runner bean, lentil, *Macrotyloma axillare*, *Lonchocarpus capassa*, *Pterocarpus angolensis*, and *Vicia angustifolia*.

Kunitz-type inhibitors have been identified in a wide range of species (9). They were found in Papilionideae: winged bean, *Erythrina latissima*, *E. crystagalli*, *Glycine maxima*; Cesalpinoideae: *Peltophorum africanum*; *Mimosidae*: *Adenanthera pavonina*, *Albizzia julibrissin*, *Acacia elata*, *A. sieberana*; Gramineae: *Hordeum vulgare*, *Triticum aestivum*, and *Secale cereale*.

Most cysteine proteinase inhibitors have been found in animals, but several have been isolated from plant species as well, including pineapple (48), potato (49), corn (50), rice (51), cowpea (52), mung bean (53), tomato (54), wheat, barley, rye, and millet (55). Their presence in plants may be ubiquitous.

A novel inhibitor of the acid proteinase was isolated from potatoes (56,57). Metallocarboxypeptidase inhibitors have been identified in potato tubers and wounded tomato plants (58).

5.3.2 Amylase Inhibitors

Amylase inhibitors are abundant in cereal grains and are also present in legumes and other seeds. Insect-specific α-amylase inhibitors have been found in wheat (59), barley (60), maize (61), rice (62), ragi (63), rye (64), sorghum (65), triticale (66), and legumes (9) such as common bean, red and white kidney bean, cowpea, chickpea, wild legumes, azuki beans, lentils, mung bean and lima bean. It was also found in tubers (67) such as colocasia, dioscoria, Job's tears, local tubers of Sri Lanka, yams, and coleus.

5.3.3 Trypsin/α-Amylase Bifunctional Inhibitors

Heterologous α-amylase/trypsin bifunctional inhibitor has been found in wheat, barley, rye (68,69), and ragi (44).

5.4 BIOSYNTHESIS

The mRNA of proteinase inhibitors accumulates during the mid-maturation stage of the developing seeds and reaches a steady state later in development. After synthesis, the inhibitor undergoes a specific and extensive proteolytic processing (6,9). The fact that cycloheximide, but not chloramphenicol, inhibits accumulation of the inhibitor suggests that cytoplasmic ribosomes may be involved in the synthesis, but not chloroplast or mitochondrial ribosomes (6).

α-Amylase inhibitor is synthesized on the endoplasmic reticulum (ER) as a preprotein of 25–28 kDa. After removal of the signal peptide followed by glycosylation at two or more sites, the glycoprotein of 32–36 kDa is formed which is transported to the storage vacuoles. The Golgi apparatus mediates this transport where some of the N-linked glycans are modified (70). Arrival of α-AI in the protein storage vacuoles is followed by removal of a short carboxy terminus and proteolytic cleavage at the carboxyl side of Asn77, resulting in the formation of two subunits with estimated molecular weights of 10 and 14.6 kDa, respectively. Further assembly of the subunits results in the formation of the matured active α-amylase inhibitor.

5.5 BIOACTIVITY

Members of the serine and cysteine proteinase inhibitor families have been more relevant to the area of plant defense than metallo- and aspartyl proteinase inhibitors, since only a few of these latter two families of inhibitors have been found in plants. Therefore, recent studies concerning the possible roles for serine and cysteine proteinase inhibitors as well as amylase inhibitors in natural plant defense systems are emphasized below.

Since the inhibitory activities of most known proteinase inhibitors are primarily directed against digestive proteinases found in animals and microorganisms, and only a few are known that inhibit plant proteinases, one major role might be plant protection by arresting the digestive proteolytic enzymes of invading pests.

5.5.1 Proteinase Inhibitors

Plant proteinase inhibitors have been thought to be involved in a protective role against plant pests. Applebaum (71) proposed that proteinase inhibitors in legumes probably evolved as a defense mechanism against insects. He suggested that protein digestion of insects should be considered an important factor in host selection.

5.5.1.1 Serine Proteinase Inhibitors

Serine proteinases have been identified in extracts from the digestive tracts of insects from many families, particularly those of the Lepidoptera (72), and many of these enzymes are inhibited by proteinase inhibitors. In the order Lepidoptera, which includes a number of crop pests, the pH optima of the guts are in the alkaline range of 9–11 where serine proteinases and metalloexopeptidases are most active. In addition, serine proteinase inhibitors have antinutritional effects against several lepidopteran insect species.

Studies on the effects of proteinase inhibitors on insect diets first began in the 1950s when Lipke et al. (73) found that a protein fraction from soybean inhibited growth, as well as proteolytic activity *in vitro*, of the mealworm, *Tribolium confusum*. Growth of larvae of a related organism, *Tribolium castaneum*, was also inhibited by purified fractions from soybean (74). When soybean Bowman-Birk trypsin inhibitor (SBTI) and some weaker inhibitors of bovine trypsin isolated from corn were fed to larvae of European corn borer at 2–5% of diets, SBTI inhibited growth of the larvae and delayed pupation, whereas the corn inhibitors had no effect on growth or pupation (75).

Broadway and Duffey (76) compared the effects of purified SBTI and potato inhibitor II (an inhibitor of both trypsin and chymotrypsin) on the growth and digestive physiology of larvae *Helicoverpa zea* and *Spodoptera exigua*. At levels of about 10% of the proteins in the diet, growth of larvae was inhibited. Trypsin inhibitors at 10% of the diet were toxic to larvae of the *Callosobruchus maculatus* (77) and *Manduca sexta* (78). The negative response to the presence of the inhibitors could be reversed by the addition of methionine to the diets. The authors suggest that inhibition of protein digestion alone does not only cause the adverse effects of the inhibitors but, in addition, causes hyperproduction of digestive enzymes and enhances the loss of sulfur amino acids by the insects.

Blood-sucking insects are also affected by the addition of proteinase inhibitors from plant origins to blood supplied to the insects as food. For examples, horn flies (*Haematobia irritans* L.) fed blood containing 1.02% w/v leupeptin or 0.1% w/v SBTI experienced severe reduction in fecundity (79), and stable flies (*Stomoxys calcitrans*) fed SBTI encapsulated in erythrocytes experienced total elimination of egg production and 50% mortality (80). Transgenic tobacco plants expressing SBTI show a higher level of protection against *H. virescens* than that conferred by cowpea trypsin inhibitor (81,82).

Broadway et al. (83) demonstrated that leaves from wounded tomato plants induced by wounding to accumulate more than 200 µg/g leaf tissue of potato inhibitors I and II produced severely reduced growth of *S. exigua* larvae. Similar experiments were performed by Edwards et al. (84) in which wounded tomato leaves, fed to larvae of Spodoptera, caused a change in feeding patterns and an inhibition of growth. Although an increase in proteinase inhibitors in tomato leaves could be correlated with levels of resistance to larvae of Spodoptera, the direct contributions of the proteinase inhibitors to the observed effects could not be firmly established since other induced defensive chemicals could have contributed to the results of these experiments.

Hilder et al. (85) demonstrated the role of proteinase inhibitors in plant leaves to defend against insects. Transformed tobacco plants expressing a foreign cowpea trypsin inhibitor gene and producing about 1% of the leaf proteins as the inhibitor were more resistant to feeding by larvae of *Heliothis virescens* than untransformed control plants or transformed plants that did not express the gene. The pure cowpea inhibitor is an antinutrient agent against a wide range of insects including larvae of *Heliothis, Spodoptera, Diabrotica,* and *Tribolium,* all agronomically important insect pests (77).

Gatehouse et al. (86) found that a single variety of cowpea obviously selected as the only resistant variety toward the insect *Callosobruchus maculatus* out of 5000 varieties tested contains significantly higher proteinase inhibitors than any other. The antimetabolic nature of the cowpea trypsin inhibitor was confirmed by feeding trials. When the albumin fraction containing the inhibitors was fed to larvae at a 10% level, the larvae died. When the trypsin inhibitors were selectively removed from the albumin fraction, it was no longer toxic to the insects. This is the clearest evidence to date that the increased resistance of one seed variety over another toward insect attack can be attributed to elevated trypsin inhibitor content.

A novel class of serine proteinase inhibitors that inhibit enterokinase was reported to be present in peanuts, tomato leaves, and potato tubers (87). The presence in the food of insects of an inhibitor capable of preventing activation of secreted digestive proteinases could be a significant factor in arresting digestion.

Table 5.2 shows the range of the insect pests that are susceptible to cowpea trypsin inhibitor. These include many field and storage pests. Cowpea trypsin inhibitor has no deleterious effects on mammals and is a small polypeptide of 80 amino acids.

Studies on several inhibitors of the midgut proteinases of *Tenebrio molitor*, a common pest that also consumes stored grains, have been undertaken (88). Lima bean inhibitor, soybean inhibitor, and purified BBI inhibited *Tenebrio* larval trypsin. The proteinases of *T. molitor* were studied in detail by Zwilling (89). Besides the presence of trypsin and chymotrypsin, previously reported by Applebaum et al. (88) two new proteinases were reported, called α and β proteases. α-Protease had molecular weight (24 kDa) similar to that of trypsin and chymotrypsin but did not hydrolyze typical ester substrates and was inhibited by any natural proteinase inhibitors. The β-protease had a molecular weight of about 60 kDa and was inhibited by five naturally occurring plant and animal proteinase inhibitors of trypsin and chymotrypsin. This work suggested that insect

TABLE 5.2   Insect Pests Affected by the Cowpea Trypsin Inhibitors (CpTIs)

Order	Pests	Crops attacked
Lepidoptera	*Heliothis virescens*	Tobacco, cotton
	Helicoverpa zea	Cotton, maize, tobacco, beans
	Helicoverpa armigera	Cotton, maize, sorghum, beans
	Spodoptera littoralis	Rice, cotton, tobacco, maize
	Chilo partellus	Rice, maize, sorghum, sugarcane
	Autographa gamma	Potato, cabbage, lettuce, beans, sugarbeet
	Manduca sexta	Potato, tomato, tobacco
Orthoptera	*Locusta migratoria*	Many crops, especially cultivated grasses
Coleoptera	*Diabrotica undecimpunctata*	Maize, clover
	Costelytra zealandica	Cotton
	Callosobruchus maculatus	Cowpea, soybean
	Tribolium confusum	Most flours

protein digestion might be more complex than that of higher animals due to the variability in the kinds of digestive endopeptidase that are present.

The possible involvement of proteinase inhibitors as protective agents toward insects received strong support when Green and Ryan (90) found that wounding of the leaves of potato or tomato plants by adult Colorado potato beetles or their larvae induced a rapid accumulation of a proteinase inhibitor throughout the aerial tissues of the plants. The effect of insect damage could be stimulated by mechanically wounding the leaves. More than 0.2 mg inhibitor protein accumulated per milliliter of leaf juice within 48 h of wounding. A factor identified as systemin was released from the damaged leaves and spread rapidly throughout the plants. In response to the factor, the level of proteinase inhibitor in both damaged and adjacent leaves rose strikingly within a few hours. Wounds inflicted on leaf tissues near midveins were more effective in inducing inhibitor in distal leaves than wounds near the outer edge of the leaves.

The half-time of transmission of the inducing factor from wounded leaves was found to be about 3.5 h. The overall response was light and temperature dependent. Inhibitor accumulation increased linearly as a function of light intensity through 300 foot candles (fc) and approached a full response at intensities in excess of 500 fc (91). The temperature dependence was unusual. Between 22°C and 32°C the amount of inhibitor I that accumulated in leaves per unit time due to wounding increased ninefold. The optimal temperature for the response was 36°C. In response to wounding, the inhibitor has been demonstrated to accumulate in the central vacuole of the cell where it remained for long time. The accumulation of a high concentration of proteinase inhibitor in leaves as a result of insect attack demonstrated that the insects can rapidly influence the protein-synthesizing machinery in the entire plant by attacking a single leaf. The overall system is reacting in the manner of a primitive immune response (92).

The *Medicago scutellata* trypsin inhibitor (MsTI) from seeds of snail medic (*M. scutellata*) has been purified by ion-exchange chromatography, gel filtration chromatography, and reversed-phase high-performance liquid chromatography (HPLC). The protein inhibits the catalytic activity of bovine β-trypsin, with an apparent K_d of 1.8×10^{-9}, but exhibits no activity toward bovine α-chymotrypsin. Moreover, MsTI inhibits the trypsin-like proteinase activity present in larvae of the crop pests *Adoxophyes orana*, *Hyphantria cunea*, *Lobesia botrana*, and *Ostrinia nubilalis* (93).

5.5.1.2 Cysteine Proteinase Inhibitor

Cysteine proteinases are not secreted as intestinal digestive enzymes in higher animals but are found in insect midguts of several families of Hemiptera (94) and Coleoptera (95) where they appear to have important roles in the digestion of food proteins. These particular insects characteristically have mildly acidic pH in their midguts near the pH optima of cysteine proteinases (pH ~ 5). Several members of the order Coleoptera are seed- and leaf–eating insects that are important pests of agricultural crops (96). None of these particular insects appears to use, or even require, serine proteinases to digest food proteins, but they employ cysteine proteinases as major digestive enzymes. Thus, larvae of the cowpea weevil, for example, can consume cowpea tissues containing several percent of their proteins as proteinase inhibitors without adverse effects on digestion. Similarly, larvae of Colorado potato beetle (another Coleoptera species) can readily consume potato and tomato tissues. These insects can consume tissues that have been induced to accumulate serine proteinase inhibitors to high concentrations in response to the chewing damage (90). Isolation of the midgut cysteine proteinases have been reported from of the larvae of the cowpea weevil *C. maculatus* (96), the bruchid beetle *Zabrotes subfaceatus* (97), the flour beetle *Tribolium castaneum*, the Mexican beetle *Epilachna varivestis* (94), as well as the bean weevil *Acanthocelides obtectus* Say (98). Both synthetic and naturally occurring cysteine proteinase inhibitors inhibited cysteine proteinases isolated from insect larvae. In a study of the proteinases from midguts of several members of the Coleoptera order, 10 of 11 beetle species representing 11 different insect families had gut proteinases that were inhibited by a potent sulfhydryl reagent, indicating that the proteinases were of the cysteine mechanistic class. Plant cysteine proteinase inhibitors (CPIs) (phytocystatins) have been implicated as defensive molecules against coleopteran and hemipteran insect pests.

Larvae of the coleopterans *Callosobruches maculatus* (cowpea weevil), *Zabrotes subfasciatus* (Mexican bean beetle), *Leptinotarsa decemlineata* (Colorado beetle). *Sitophilus oryzae* (rice weevil), *Diabrotica undecimpunctata howardii* (Southern corn rootworm), *Tenebrio molitor* (yellow mealworm), and *Tribolium castaneum* (red flour beetle) have digestive cysteine proteinases that are inhibited by cystatin inhibitors from rice, cowpea, or soybean (99–102).

Recent work suggests that cysteine proteinase inhibitors may also be used to provide protection against other invertebrate pests. Adult females of the potato cyst nematode (*Globodera pallida*) contain high levels of cysteine proteinase activity (103), leading Urwin et al. (104) to explore the use of cystatins to confer resistance. It was shown that wild-type and mutated forms of oryzacystatin I expressed in *E. coli* inhibited the cysteine proteinase from the free-living nematode *Caenorhabditis elegans* (again expressed in *E. coli*) and the development of the larvae of this species. The cystatins were then engineered into hairy roots of tobacco, where they inhibited the growth and development of *G. pallida* (103,104). The inhibitory activity against both the gcp-1 protease and the growth of *G. pallida* larvae was greater with a mutant form in which a single aspartate residue was deleted than with the wild-type enzyme, indicating that protein engineering should prove to be a powerful tool to increase the activity and specificity of such inhibitors.

Feeding bioassay results established that the soybean cysteine proteinase inhibitor N (soyacystatin N, scN) substantially inhibits growth and development of Western cornrootworm (WCR), by attenuating digestive proteolysis (105). Recombinant scN was inhibitory than the potent and broad specificity cysteine proteinase inhibitor E-64. WCR digestive proteolytic activity was separated by mildly denaturing SDS-PAGE into two fractions and in-gel assays confirmed that the proteinase activities of each were largely scN sensitive.

Cystatin CsC, a cysteine proteinase inhibitor from chestnut (*Castanea sativa*) seeds, has been purified and characterized. Recombinant CsC inhibited papain (Ki 29 nM), chymopapain (Ki 366 nM), and cathepsin B (Ki 473 nM). By contrast with most cystatins, it was also effective toward trypsin (Ki 3489 nM). CsC is active against digestive proteinases from the insect *Tribolium castaneum* and the mite *Dermatophagoides farinae*, two important agricultural pests (106). Its effects on the cysteine proteinase activity of two closely related mite species revealed the high specificity of the chestnut cystatin.

Three distinct digestive protease systems were induced in larvae of the herbivorous pest, Colorado potato beetle (CPB; *Leptinotarsa decemlineata*

Say), and used as a model to assess the ability of the proregion of papaya proteinase IV (PPIV; glycyl endopeptidase) to act as an inhibitor of insect digestive cysteine proteinase. As shown by gelatin/PAGE and complementary inhibition assays, a recombinant form of the proregion produced in *E. coli* inhibited a fraction of the insect proteases also inhibited by the well-characterized inhibitor of cysteine proteinases, oryzacystatin I (OCI). In contrast with OCI, the inhibitory potency of the proregion was effected by an increase of the temperature, suggesting a certain alteration of its structural integrity by the insect nontarget proteases (107).

Two soybean cystatins, soyacystatin N (scN) and soyacystatin L (scL), have 70% sequence identity, but scN is a much more potent inhibitor of papain, vicilin peptidohydrolase, and insect gut proteinases. When these cystatins were displayed on phage particles, papain binding affinity and cysteine proteinase inhibitory (CPI) activity of scN were substantially greater than those of scL, in direct correlation with their relative CPI activity as soluble recombinant proteins. Futhermore, scN substantially delayed cowpea weevil [*Callosobruchus maculatus* (F.)] growth and development in insect feeding bioassays, whereas scL was essentially inactive as an insecticide (108). Papain biopanning selection of phage-displayed soyacystatins resulted in a 200- to 1000-fold greater enrichment for scN relative to scL. These results establish that binding affinity of cystatins can be used in phage display biopanning procedures to select variants with greater insecticidal activity.

The use of oryzacystatins I and II, two cysteine proteinase inhibitors naturally produced in rice grains, represents an attractive approach to the control of Coleoptera insect pests. The study was done to analyze the inhibitory effect of recombinant oryzacystatins produced in *E. coli* as fusion proteins against digestive proteinases of the major pest, Colorado potato beetle (*Leptinotarsa decemlineata* Say). Both inhibitors had a significant effect on total proteolytic activity, but maximal inhibitions ranged from 20% to 80% for pHs varying from 5.0 to 7.0, respectively (109). This pH-dependent efficiency of plant cystatins was due to the selective inactivation of potato beetle cathepsin H, as demonstrated by the use of inhibitors with different specificities against cathepsins B and H. These results demonstrate the importance of having an adequate knowledge of insect proteinases specifically recognized by the inhibitors to be used in pest control strategies.

High levels of proteinase inhibitors are induced in potato leaves by wounding. These inhibitors, when ingested by Colorado potato beetle (*L. decemlineata* Say) larvae, induce expression of specific proteolytic activities in the gut. Induced proteinase activities cannot be inhibited by potato inhibitors and thus enable the insects to overcome this defense

mechanism of potato plants. The induced aminopeptidase and endoproteolytic activities both have the characteristics of cysteine proteases. Twenty-
one protein inhibitors of different structural types have been examined for
their ability to inhibit these activities *in vitro* (110). Members of the cystatin
superfamily were found to be poor inhibitors of the induced endoproteolytic
activities, except for the third domain of human kininogen, which was a
fairly strong inhibitor (75% inhibition). The strongest inhibition (85%) of
induced endoproteolytic activity was obtained using structurally different
thyroglobulin type 1 domain–like inhibitors, i.e., equistatin and maps histo-
compatibility complex (MHC) class II-associated p41 invariant fragment.
Experiments performed using three synthetic substrates for endoproteases
gave similar results and indicate the existence of at least different endoproteolytic enzymes resistant to potato inhibitors. Only the stefin family
of inhibitors in cystatin superfamily can inhibit the induced aminopeptidase
activity. Colorado potato beetle larvae fed on equistatin-coated potato
leaves were strongly retarded in their growth and almost 50% died after 4
days. These results demonstrated the potential of equistatin to protect crops
from insect attack.

5.5.2 Amylase Inhibitors

Gatehouse et al. (111) found that α-amylase inhibitors purified from wheat
were potent inhibitors of amylases found in the midguts of the larvae from
two species of insects, *C. maculatus* and *T. confusum*, which are storage pests
of legumes. However, when incorporated into diets of the two larvae, the
inhibitors could inhibit the development and increase the mortality of larvae
of only *T. confusum*. This suggested either that larvae of *C. maculatus* were
not inhibited by the inhibitor or that the gut contents could detoxify the
amylase inhibitors. If the latter is occurring, then the presence of proteinase
inhibitors of gut proteinases might significantly enhance the activities of the
amylase inhibitors that were ineffective alone. Wheat α-amylase inhibitors are
also potent inhibitors of the α-amylase present in several stored-grain
insects, including *Tenebrio*, *Tribolium*, *Sitophilus*, and *Oryzaephilus* (112–115).
    Four distinct α-amylases were isolated from the midguts of *C.
macualtus*. None of the enzyme fractions was inhibited by an α-amylase
inhibitor from bean, *P. vulgaris*, that inhibits α-amylase from the insect guts
of *Augusta kuhnilella*, *Tribolium castaneum*, *Tribolium confusum*, and *Tenebrio
molitor* (116). However, three of the four forms of α-amylases from *C.
maculatus* were inhibited by an α-amylase inhibitor isolated from wheat
kernels. Wheat amylase inhibitors also inhibited amylases from the rice
weevil, *S. oryzae*. However, there is considerable variation in the specificities
of the different forms (117). Thus even within the Coleoptera, enzymes from

T. molitor (yellow mealworm) and *Sitophilus oryzae* (rice weevil) are more sensitive to the monomeric wheat inhibitor, while *L. decemlineata* (Colorado beetle) and *Oryzaephilus surinamensis* (grain beetle) enzymes are more sensitive to dimeric forms. In contrast, lepidopteran (*Ephestia keuhniella*, Mediterranean flour moth) α-amylases are more strongly inhibited by the tetrameric wheat inhibitor (118). The monomeric trypsin inhibitor of barley, also called CMe, shows little or no activity against α-amylases, in contrast to the homologues from maize and finger millet that are bifunctional (119). Carbonero et al. (117) reported that transgenic tobacco plants showing constitutive expression of the wheat monomeric α-amylase inhibitor or the barley trypsin inhibitor were lethal to lepidopteran larvae (*Agrotis ipsilon*, black cut worm and *Spodoptera littoralis*, common leaf worm) in leaf disc assays.

Rajendran and Thayumanavan (120) have shown that the α-amylase inhibitor from *Panicum sumatrens* Roth inhibited the larvae midgut α-amylase from tomato fruit borer (*Helicoverpa armigera*), brinjal shoot and fruit borer (*Leucinodes orbonalis*), and cucurbit fruit fly (*Dacus cucurbitae*).

5.6 MODE OF ACTION

5.6.1 Proteinase Inhibitors

On the surface of each inhibitor molecule lies at least one (more in multiheaded inhibitors) peptide bond called the reactive site, which specifically interacts with the active site of the cognate enzyme. The value of K_{cat}/K_m for the hydrolysis of this peptide bond by the cognate enzyme at neutral pH is very high, 10^4–$10^6 M^{-1} s^{-1}$, compared to a typical value for normal substrates of about $10^3 M^{-1} s^{-1}$ (121). However, for inhibitors, the values of K_{cat} and K_m are many orders of magnitude lower than the values for normal substrates. At typically used concentrations and neutral pH, therefore, their hydrolysis is extremely slow, and the system behaves as if it were a simple equilibrium between the enzyme and free inhibitor on the one hand and the complex on the other. The equilibrium constant for the association is extremely high (in the range 10^7–$10^{13} M^{-1}$). An additional property of the inhibitory reactive sites is that their hydrolysis does not proceed to virtual completion. Instead, at neutral pH, the equilibrium constant between modified inhibitor (reactive site peptide bond hydrolyzed) and virgin inhibitor (reactive site peptide bond intact) is near unity. Since the same stable complex is formed between the enzyme and either modified or virgin inhibitor, both are thermodynamically equally strong inhibitors of the cognate enzyme. However, in cases so far examined, the rate of complex formation from modified inhibitor and the enzyme is much lower than from virgin inhibitor and the enzyme. In a few cases, this difference is so great that it led to the assumption that the modified inhibitor was inactive, while longer incubation was seen to be an error.

The detailed molecular events at the active site of the enzyme and at the reactive site of the inhibitor are not as clear as some facile statements imply. The work of the group led by Huber (122), summarized in several reviews, is presented below. The scissile peptide bond (P_1–P_1') of the inhibitor is intact in the complex. However, the carbonyl carbon of the reactive site peptide bond (P_1 residue) is not fully trigonal but is distorted about half way toward tetrahedral. Since the carbon is 2.6 Å (much longer than the 1.4 Å covalent C-O bond, but shorter than a van der Waals contact) from the oxygen in the catalytic Ser195 of the enzyme, it was believed that it was this oxygen that caused the tetrahedral distortion of the carbonyl carbon. This view was abandoned when it was realized that anhydrotrypsin (where the catalytic Ser195 residue is dehydrated to dehydroalanyl195) forms stable inhibitor complexes of comparable strength with essentially the same geometry, and with tetrahedral distortion precisely the same as in a complex with trypsin (123). Therefore, the partial tetrahedral distortion is currently believed to result from the attraction between the carbonyl oxygen of the inhibitor and the oxyanion hole of the enzyme, the NH's of Gly193 and Ser195. The carbonyl carbon is now poised for nucleophilic attack by oxygen of Ser195, but in the stable complex the attack has not yet taken place.

The X-ray crystallographic studies do not explain as yet why inhibitors are inhibitors, not just excellent substrates. This is a complex problem because strong inhibitors of bovine trypsin are simply very good substrates for a homologous trypsin 1 from the starfish *Dermasterias imbricata* (124). Human cationic trypsin is intermediate in behavior between these two enzymes. Thus, the high k_{cat}/K_m value of the reactive site of an inhibitor applies to many homologous enzymes. Some of these partition this value into very low k_{cat} and K_m, and thus are strongly inhibited; others show more conventional k_{cat} and K_m values, and the inhibitor serves them as a substrate.

Katz et al. (125) describe a new serine proteinase inhibition motif in which binding is mediated by a cluster of very short hydrogen bonds (< 2.3 Å) at the active site. This protease–inhibitor binding paradigm is observed at high resolution in a large set of crystal structures of trypsin, thrombin, and urokinase-type plasminogen activator (uPA) bound with a series of small-molecule inhibitors [2-(2-phenol)indoles and 2-(2-phenol)-benzimidazoles]. In each complex there are eight enzyme–inhibitor or enzyme–water–inhibitor hydrogen bonds at the active site, three of which are very short. These short hydrogen bonds connect a triangle of oxygen atoms comprising OSer195, a water molecule cobound in the oxyanion hole (H_2Ooxy), and the phenolate oxygen atom of the inhibitor (O6'). Two of the other hydrogen bonds between the inhibitor and active site of the trypsin and uPA complexes become short in the thrombin counterparts, extending the three-centered short hydrogen-bonding array into a tetrahedral array of atoms (three oxygen and one nitrogen) involved in short hydrogen bonds. In the uPA complexes, the extensive hydrogen bonding interactions at the active site prevent the inhibitor S1 amidine from forming direct hydrogen bonds with Asp189 because the S1 site is deeper in uPA than in trypsin or thrombin. Ionization equilibria at the active site associated with inhibitor binding are probed through determination and comparison of structures over a wide range of pH (3.5–11.4) of thrombin complexes and of trypsin complexes in three different crystal forms. The high-pH trypsin-inhibitor structures suggest that His57 is protonated at pH values as high as 9.5. The pH-dependent inhibition of trypsin, thrombin, uPA, and factor Xa by 2-(2-phenol)benzimidazole analogues in which the pK_a of the phenol group is modulated is shown to be consistent with a binding process involving ionization of both the inhibitor and the enzyme. These data further suggest that the pK_a of His57 of each protease in the unbound state in solution is about the same, approximately 6.8.

Three new analogues of trypsin inhibitor CMTI-III were synthesized by the solid-phase method: [Lys5]CMTI-III, [Orn5]CMTI-III, and [Dab5]CMTI-III (126). Only one analogue with lysine residue in position P1 showed inhibitory activity of the same order of magnitude as did wild CMTI-III. Two remaining analogues were completely inactive. A conclusion was drawn that the distance between the basic group of the amino acid residue's side chain in position P1 of the trypsin inhibitor CMTI-III and Asp189 in the substrate pocket of trypsin plays an essential role for the trypsin–inhibitor interaction.

McBride et al. (127) described a template-assisted combinatorial peptide library based on the antitryptic reactive site loop of a BBI. Sequences that displayed inhibitory activity redirected to chymotrypsin were found to have a consensus-binding motif, with their most striking feature being that exclusively threonine was found at the P2 position. The study investigated the reason for this surprising specificity by maintaining the binding motif but systematically varying the P2 residue. From analysis of 26 variants, it was found that the requirements for inhibitory activity at P2 are finely tuned and, in agreement with the library work, threonine at P2 provides optimal inhibition. In addition, peptides with threonine at P2 are significantly less susceptible to hydrolysis. Examination of all available BBI sequences shows that threonine is very highly conserved at P2, which implies that the functional requirement extends to the full-length BBI protein. Their results are consistent with a dual requirement for hydrophobic recognition within the S2 pocket and maintenance of an inhibitory conformation via hydrogen bonding within the reactive site loop. As the isolated peptide loop reproduces the active region of full-length BBI, these results explain why threonine is well conserved at P2 in this class of inhibitor. Furthermore, they illustrate that proteinase inhibitor specificity can have characteristics that are not easily predicted from information on the substrate preferences of a proteinase.

Two peptides, isolated from the hemolymph of *Schistocerca gregaria* SGCI and SGTI, inhibited chymotrypsin and trypsin, respectively (128). Their primary structure were found to be identical with SGP-2 and SGP-1, two of a series of peptides isolated from ovaries of the same species. All these peptides are composed of 35–36 amino acid residues and contain three homologous disulfide bridges. The residues imparting specificity to SGCI and SGTI were identified as Leu30 and Arg29, respectively. The pepetides were synthesized by solid-phase peptide synthesis, and the synthetic ones displayed the same inhibition as the natural forms: SGCI is a strong inhibitor of chymotrypsin ($K_i = 6.2 \times 10^{-12}$ M), and SGTI is a rather weak inhibitor of trypsin ($K_i = 2.1 \times 10^{-7}$ M). The replacement of P_1 and P_1' residues of SGCI with trypsin–specific residues increased affinity for trypsin 3600- and 1100-fold, respectively; thus, SGCI was converted to a strong trypsin inhibitor ($K_i = 5.0 \times 10^{-12}$ M) that retained some inhibitory affinity for chymotrypsin ($K_i = 3.5 \times 10^{-8}$ M). The documented role of both P_1 and P_1' highlights the importance of $S_1'-P_1'$ interactions in enzyme–inhibitor complexes.

Complementary DNA coding for the channel-forming α subunit of a large-conductance Ca^{2+}-activated K^+ channel was cloned from bovine aortic smooth muscle cells (129). Their results lead to the conclusion that the α subunit of maxi KCa channels contains a conserved proteinase inhibitor binding site. It is hypothesized that this site corresponds to a C-terminal domain of the channel protein that struturally resembles serine proteinases.

5.6.2 Cysteine Proteinase Inhibitors

The biochemical interactions between digestive proteinases of the Coleoptera pest black vine weevil (*Otiorynchus sulcatus*) and two plant cysteine proteinase inhibitors, oryzacystatin I (OCI) and oryzacystatin II (OCII), were assessed using gelatin-PAGE, OCI affinity chromatography, and recombinant forms of the two plant inhibitors (130). The insect proteinases were resolved in gelatin-containing polyacrylamide gels as five major bands, only three of them being totally or partially inactivated by OCI and OCII. The maximal inhibitory effect of both OCs at pH 5.0 was estimated at 40%, and the inhibition was stable with time despite the presence of OC-sensitive proteinases, indicating the stability of the OCI and OCII effects. After removing OC-sensitive proteinases from the insect crude extract by OCI affinity chromatography, the effects of the insect cystatin–insensitive proteinases on the structural integrity of the free OCs were analyzed. While OCI remained stable, OCII was subjected to limited proteolysis leading to its gradual transformation into an approximately 10.5 kDa unstable intermediate, OCIIi. As shown by the degradation pattern of a glutathione S-transferase (GST)/OCII fusion protein, the appearance of OCIIi resulted from the C-terminal truncation of OCII. Either free or linked to GST, OCIIi was as active against papain and human cathepsin H as OCII near its active (inhibitory) site. It is also suggested that a general conformational destabilization of this inhibitor following its initial cleavage subsequently leads to its complete hydrolysis. This apparent susceptibility of OCII to proteolytic cleavage by the insect proteinases could have major implications when planning the use of this plant cystatin for insect pest control.

5.6.3 Amylase/Trypsin Bifunctional Inhibitor

Amylase inhibitors show remarkable structural variety leading to different modes of inhibition and different specificity profiles against diverse α-amylases. Of particular interest are some bifunctional inhibitors with additional favorable properties (131). Studies on the substrate–inhibitor interactions establish that under the assay conditions, the ragi amylase/trypsin bifunctional inhibitor (RATI) binds to the soluble substrate (44). Since cytochrome c, which is also a basic protein with a pI close to that of RATI, does not bind to the soluble starch, the interactions of RATI with substrate are not based predominantly on the positive charges of RATI and are not completely nonspecific in nature. It is known that nonpolar and H-bond interactions are involved in the binding of proteins to oligosaccharides and polysaccharides. While RATI is highly charged, 40% of its sequences composed of nonpolar amino acids. It is possible that a combination of nonpolar and uncharged polar residues makes up a sugar binding region. Equally important is the loss of inhibitory activity upon binding, which is relevant to kinetics of amylase inhibition. The inhibitor molecules are bound to the substrate, leading to less inhibition than expected. Since less inhibition is observed even with preincubation of enzyme and inhibitor, the added substrate shifts the binding equilibrium toward dissociation of enzyme and inhibitor. The concentration of the substrate is 3–4 orders of magnitude more than the inhibitor concentration in the assay conditions. Even if the interactions are weak, the high effective concentration of the substrate will render the depletion of inhibitor highly significant. Hence, complete inhibition of α-amylase with proteinaceous inhibitors is generally not observed. It was expected that larger polysaccharide would be more efficient in binding with the inhibitor. The kinetic scheme of α-amylase inhibition using larger substrate is as follows:

$$
\begin{array}{ccccc}
E + S & \rightleftharpoons & ES & \rightleftharpoons & E + P \\
+ & & + & & \\
I & & I & & \\
\uparrow\downarrow & & \downarrow & & \\
EI & & SI & &
\end{array}
$$

Since no inhibition was observed when SI was added to α-amylase, the binding site of RATI was not accessible to α-amylase. It implies that the complex formed between substrate (starch) and the RATI is a stable complex. It has already been reported the large substrates occupy the catalytic site and the surface site (a second sugar binding site) of α-amylase. Structural data on the complexes of α-amylases and their inhibitors have clearly shown that the inhibitor interactions do not cover the surface site. In view of these observations, it seems unlikely that a ternary complex, (ESI) is formed. This kind of impairment of the function of a proteinaceous inhibitor has not been reported so far in enzyme kinetics. Thus, in the absence of any evidence of the complex inhibition mode, the observed loss of inhibition should be explained by the substrate–inhibitor interactions. Since it could be estimated as to how much inhibitor was lost by way of binding with the substrate, the amount of substrate bound to inhibitor was deducted from the added inhibitor amount. A Dixon plot of the data established a competitive mode of inhibition with starch as a substrate, when the corrected RATI concentrations were used. The calculated K_i value of 0.1 nM is comparable to that in other reported studies.

It has been shown by Alam et al. (44) that the N-terminal fragments of RATI interact with the catalytic region of α-amylase. The intact RATI has a K_I value of 15 ± 2 nM whereas the peptides have 10–40 times more than this value. The nonamer peptide is more effective than the heptamer or elevenmer among the homologous synthetic peptides. The sequence -Cys6-Ile7-Pro8-Ala9- forms a type II β-turn in the intact RATI. This stable conformation might act as an anchor to place the first six amino acids appropriately in the binding region of α-amylase. In the case of heptapeptide (P7), the β-turn structure is not available. On the other hand, in the case of P11, the turn conformation may be available, but since residues 10 and 11 do not contribute significantly to interactions with α-amylase they are not required for the inhibition, thus making P11 a poorer inhibitor of α-amylase and a well-defined stable C-terminal conformation. P9 shows more affinity to α-amylase. The critical role of the side chain of terminal residue is brought out in the Ser1-Ala1 substitution. The side chain of Ser1 is involved in the H bonding with His299 and Asp300 in subsites 3 and 4 of the substrate-binding region of the enzyme. The terminal nitrogen of Ser1 is involved in H bonding with Asp197 and Glu233.

5.7 MOLECULAR BIOLOGY

As proteinases are fundamental in insect digestion, expression of proteinase inhibitors in crop plants should, in theory, give protection from insect attack. To date, most attempts to genetically engineer insect resistance have concentrated on the insertion of proteinase inhibitors. Around 15 different proteinase inhibitors have been introduced experimentally into crop plants, with some achieving high degrees of protection.

5.7.1 Proteinase Inhibitors

The most widely used inhibitor to date has been a serine proteinase inhibitor, the cowpea trypsin inhibitor (CpTI). This has been inserted into crops such as tobacco (132,133), oilseed rape, potato, sweet potato, rice (134), cabbage (135), and strawberry (136), resulting in increased resistance in most cases. In glass house or growth–room studies, for example, transgenic cabbage demonstrated improved resistance to the lepidopteran caterpillars *Pieris rapae L.* and *Helicoverpa armigera* (135). In tobacco, a 50% reduction in biomass of *Spodoptera litura* was demonstrated from CpTI-expressing plants (133). This reduction in biomass has also been demonstrated in potato on the tomato moth, *Lacanobia oleracea* (137).

Other serine proteinase inhibitors have also been evaluated. The potato inhibitors have also demonstrated significant protection when expressed in a range of crops. Potato proteinase inhibitor 2 (PIN2) conferred significant protection in several japonica rice varieties against the pink stem borer (*Sesamia inferens*) (138). In transgenic poplar lines, PIN2 demonstrated protection from the willow leaf beetle (*Plagiodera vesicolora*) (139). Potato proteinase inhibitors I and II have also been shown to increase mortality of a range of cereal aphids including the Russian wheat aphid (*Diuraphis noxia*) and the greenbug (*Schizaphis graminum*), whereas other inhibitors, such as soybean trypsin inhibitor, demonstrated a species-specific effect, while a lima bean proteinase inhibitor had no effect (140).

An insect proteinase inhibitor, antielastase from *M. sexta* when placed under the control of the cauliflower mosaic virus (CaMV) 35S promoter and inserted into alfalfa reduced the onset of thrip predation (141). Insect anti-trypsin, antichymotrypsin, and antielastase expressed in *Nicotiana tabacum* L. reduced insect reproduction by as much as 98% compared to controls (141).

A cDNA clone, encoding a trypsin inhibitor was placed under the control of a CaMV 35S promoter and introduced into tobacco (142). The transgenic plants exhibited resistance to infestation by insects (*Heliothis virescens*). The protection afforded by cowpea trypsin inhibitor has subsequently been demonstrated for other lepidopteran pests including *Helicoverpa zea, Spodoptera littoralis*, and *M. sexta*. Similarly, the tomato inhibitor II gene, when expressed in the same plant at higher levels, has also been shown to confer insect resistance. The tomato and potato inhibitor II gene encodes a trypsin inhibitor and expression of this gene in tobacco resulted in increased levels of protection against the larvae of *M. sexta*. Widespread use of these genes to produce transgenic plants of economically important species is yet to take place.

The mustard trypsin inhibitor MTI-2 is a potential tool in the study of interactions between insect pests and plants (143). It can be applied to study the adaptations of digestive proteinases in insect pests. The effects of MTI-2 expressed at different levels in transgenic tobacco, *Arabidopsis* and oilseed rape lines have been evaluated against three different lepidopteran insect pests (144). The levels of insect gut proteolytic activities of the larvae still alive at the end of a 7-day feeding bioassay were usually higher than in the controls, but no new proteinases were expressed in any case. The combined results demonstrated the relevance of a case-by-case analysis in a PI-based strategy for plant protection.

Nicotiana tabacum plants were transformed with the cDNA of barley trypsin inhibitor BTI-CMe under the control of the 35S CaMV promoter (145). Although the transgene was expressed and the protein was active in the homozygous lines selected, growth of *Spodoptera exigua* (Lepidoptera: Noctuidae) larvae reared on transgenic plants was not affected. The protease activity in larval midgut extracts after 2 days feeding on transformed tobacco leaves from the highest expressing plant showed a reduction of 25% in the trypsin–like activity compared to that from insects fed on nontransformed controls. The susceptibility of digestive serine proteases to inhibition by BTI-CMe was confirmed by activity staining gels. This decrease was compensated with a significant induction of leucine aminopeptidase-like and carboxypeptidase A–like activities, whereas chymotrypsin, elastase, and carboxypeptidase B–like proteases were not affected.

MCTI-II (*Momordica charantia* trypsin inhibitor II) isolated from bitter gourd (*Momordica charantia* Linn.) seeds is one of the serine proteinase inhibitors of the squash family. Sato et al. (146) cloned cDNA that encodes MCTI-II and constructed an expression system for MCTI-II by using a baculovirus vector. The inhibitory activities of MCTI-II expressed toward trypsin were examined in terms of the K_i value.

Transgenic tobacco plants expressing both *Bacillus thuringiensis* (Bt) insecticidal protein and cowpea trypsin inhibitor (CpTI) genes were used to evaluate the insecticidal activity on the cotton bollworm (*Helicoverpa armigera*) in comparison with transgenic tobacco with Bt insecticidal protein gene alone (147). First- to third-instar larvae fed continuously on transgenic tobacco with the two genes could not survive until pupation. Second-instar larvae were fed transgenic tobacco plants for 3 days and then transferred to an artificial diet. The efficacy of transgenic two genes tobacco on mortality, larvae weight, percentage of pupation, and time to pupation were significantly higher than that of transgenic tobacco with Bt insecticidal protein gene alone. The results with transgenic tobacco as model plant were valuable for other crops both for the enhancement of insecticidal efficacy and for the delay of insect adaption to transgenic Bt crops.

The gene encoding for a pumpkin (*Curcubita maxima*) trypsin inhibitor CMTI-V was synthesized chemically (148). The synthetic gene was prepared from four overlapping oligonucleotides by overlapping extension. The synthetic gene was amplified by PCR, cloned into a T7 expression vector, and expressed in *E. coli* as a fusion protein. The clone, namely 70-1, encoded a fusion protein containing 7 amino acid residues of the N terminus of the bacterial protein 10 and the entire 68 residues of CMTI-V. The wild-type fusion protein constituted approximately 15% of the total bacterial protein mass and was purified to homogeneity in a single step by antibody affinity chromatography. The wild-type fusion protein possesses inhibitory activity toward trypsin and factor XIIa, but to a lesser extent when compared with the natural CMTI-V. A mutant, T43A, in which threonine at position 43 (P2 position) was replaced by alanine was constructed. This mutant showed considerably lower specific inhibitory activity toward both trypsin and factor XIIa.

A two-domain portion of the proteinase inhibitor precursor from *Nicotiana alata* (NaProPI) has been expressed and its structure determined by NMR spectroscopy (149). The expressed protein comprises residues 25–135 of NaProPI and encompasses the first two contiguous structural domains, namely, the chymotrypsin inhibitor CI and the trypsin inhibitor TI, joined by a five-residue linker, and is referred to as CI-TI. The tertiary structure of each domain in CI-TI is identical to that found in the isolated inhibitors. The lack of strong interdomain association is likely to be important for the function of individual inhibitors by ensuring that there is no masking of reactive sites upon release from the precursor.

To express transgenes in plant cells, appropriate promoter sequences have to be introduced alongside the gene to ensure efficient transcription of mRNA (150,151). The same promoter, CaMV 35S (or derivatives of it), has been used in the majority of insect-resistant transgenic plants (152). This promoter originates from the cauliflower mosaic virus and, although not completely constitutive, produces continuous gene expression in most tissues of the plant. However, levels of gene expression have been reported to vary between different species of plant and different parts of the plant. In transgenic cotton, for example, CaMV 35S resulted in gene expression in all tissues except mature petals and pollen, and maize CaMV lines expressed high levels of toxin in pith and root, moderate levels in the kernal, and no detectable toxin in the pollen or anthers. However, continuous gene expression in all plant tissues is likely to increase the risk of the pests developing resistance and may also result in yield penalties as the plant directs more resources than necessary to its defense. Considerable research effort is now directed to concentrating expression in those parts of the plant attacked by insects. An example is the deployment of a phloem-specific promoter for genes providing resistance to phloem-sucking insect pests such as aphids. There is also potential for the use of wound-induced promoters, which lead to gene expression only when the plant is actually attacked. Other specific promoters that have been used with insect resistance genes include seed-specific and pollen-specific promoters, and promoters especially suited for use in monocotyledonous plants.

Selectable marker genes were introduced alongside the insect resistance gene to allow separation of plant cells that have incorporated the new genes from untransformed cells. The majority of insect-resistant transgenic plants produced today about which sufficient information has been published contain antibiotic resistance genes as the selectable marker genes. The most commonly used marker is the bacterial neomycin-phosphotransferase II gene [nptII or APH (3′) II]. A second antibiotic resistance marker gene, used for rice and soybean transformation, was the hygromycin-phosphotransferase gene (hpt, hph, or APH IV) (153,154). Recent problems with the approval of transgenic maize in the European Community were based on risks connected with the antibiotic resistance marker gene rather than the insect resistance gene itself. Therefore, the current trend is either to not use antibiotic resistance marker genes or to deliver the marker gene in a different locus so that it can later be bred away. Herbicide tolerance genes are an alternative to antibiotic resistance markers, and genes coding for 5-enolpyruvylshikimate-3-phosphate synthase (EPSPS) and phosphinothricin-N-acetyltransferase (PAT) provide tolerance to glyfosate and glufosinate-ammonium herbicides, respectively, and have been used as selectable markers in maize. However, there remains a shortage of efficient selectable marker genes, and it is difficult to carry out repeated transformations using the same marker genes to introduce more than one desirable gene into the same plant line (155).

The open reading frame and terminator region of a wound-inducible tomato inhibitor I gene, regulated by the CaMV 35S promoter, was stably integrated into the genomes of nightshade (*Solanum nigrum*), tobacco (*Nicotiana tabacum*), and alfalfa (*Medicago sativa*), using an Agrobacterium-mediated transformation system (156). The expression of the foreign inhibitor I gene in leaves of each species was studied at the mRNA and protein levels. The levels of inhibitor I protein present in leaves of each species correlated with the levels of mRNA. The average levels of both mRNA and inhibitor I protein were highest in leaves of transgenic nightshade plants (over 125 µg of inhibitor I per gram of tissue), less in tobacco plants (about 75 µg/g tissue), and lowest in leaves of transgenic alfalfa plants (below 20 µg/g tissue). Inhibitor I protein was observed in all tissues throughout transgenic plant species, but inhibitor concentration per gram of tissue was 2–3 times higher in young developing leaf tissues and floral organs. The differences in the expression of the CaMV-tomato inhibitor I gene among the different plant genera suggest that either the rate of transcription of the foreign gene or the rate of degradation of the nascent inhibitor I mRNA varies among genera. Using electron microscopy techniques, the newly synthesized pre-pro-inhibitor I protein was shown to be correctly processed and stored as a mature inhibitor I protein within the central vacuoles of leaves of transgenic nightshade and alfalfa. The results of these experiments suggest that maximal expression of foreign proteinase inhibitor genes, and perhaps other foreign defense genes, may require gene constructs that are fashioned with promoters and terminators that allow maximal expression in the selected plant species.

Most effort to introduce proteinase inhibitor genes have been concentrated on serine proteinase inhibitors from the plant families Fabaceae, Solanaceae, and Poaceae, which were targeted mainly against lepidopteran species but also against some coleopteran and orthopteran pests (157). The most active inhibitor identified to date is CpTI, which has been transferred into at least 10 other plant species. Experiments with transgenic plants and artificial diets have shown that CpTI affects a wide range of lepidopteran and coleopteran species; CpTI-transformed tobacco, field tested in California, caused significant larval mortality of cotton bollworm (*Helicoverpa zea*) larvae, but the protection provided by CpTI was less pronounced and consistent than that of tobacco containing a truncated Bt toxin gene. The serine-proteinase inhibitors KTi3, C-II, and PI-IV (from soybean) resulted in up to 100% mortality of first-instar cotton leaf worms (*Spodoptera littoralis*) when expressed in tobacco. However, resistance levels achieved by the same constructs in potato were much lower, resulting mainly in retarded growth of *S. littoralis* rather than direct mortality. To date, no crop expressing proteinase inhibitor transgenes has been commercialized.

5.7.2 Cysteine Proteinase Inhibitors

There are fewer examples of transgenic plants expressing cysteine proteinase inhibitors. There are reports of gene insertion into poplar trees (158), potato (159), and rice (160). In rice, the effect of the inhibitor was examined for nematode resistance with a 55% reduction in egg production observed. A number of cysteine proteinase inhibitors have been evaluated in artificial diets and in most cases their usefulness for insect control was demonstrated. Oryzacystatins I and II caused growth retardation of *Callosobruchus chinensis* (Coleoptera) and *Riptortus clavatus* (Hemiptera) (161). Cystatins identified in cereal have been cloned and their antiviral and antipest effects evaluated. Characterization of proteinases in target pest species will allow more effective deployment of transgenes expressing the most effective inhibitors.

Potato cysteine proteinase inhibitors (PCPIs) represent a distinct group of proteins because they show no homology to any other known cysteine proteinase inhibitor superfamilies but belong to the Kunitz-type soybean trypsin inhibitor family. cDNA clones for five PCPIs have been isolated and sequenced (162). Amino acid substitutions occurring in the limited regions forming loops on the surface of these proteins suggest further classification of PCPIs into three subgroups. Accumulation of PCPIs was observed in vacuoles of stems after treatment with jasmonic acid (JA) using immunocytochemical localization, implying that these inhibitors are part of a potato defense mechanism against insects and pathogens. Genomic DNA analysis show that PCPIs form a multigene family and suggest that their genes do not possess any introns.

Studies of the effects of insect-resistant transgenic plants on beneficial insects have, to date, concentrated mainly on small-scale "worst case scenario" laboratory experiments or on field trials. Schuler et al. (163) presented a laboratory method using large population cages that represent an intermediate experimental scale, allowing the study of ecological and behavioral interactions between transgenic plants, pests, and their natural enemies under more controlled conditions than is possible in the field. Previous studies have also concentrated on natural enemies of lepidopteran and coleopteran target pests. However, natural enemies of other pests, which are not controlled by the transgenic plants, are also potentially exposed to the transgene product when feeding on hosts. Reduction in the use of insecticides on transgenic crops could lead to increasing problems with such nontarget pests, normally controlled by sprays, especially if there are any negative effects of the transgenic plant on their natural enemies. This study tested two lines of insect-resistant transgenic oilseed rape (*Brassica napus*) for side effects on the hymenopteran parasitoid *Diaeretiella rapae* and its aphid host, *Myzus persicae*. One transgenic line expressed the delta-endotoxin Cry1Ac from *Bacillus thuringiensis* and a second expressed the proteinase inhibitor oryzacystatin I from rice. These transgenic plant lines were developed to provide resistance to lepidopteran and coleopteran pests, respectively. No detrimental effects of the transgenic oilseed rape lines on the ability of the parasitoid to control aphid populations were observed. Adult parasitoid emergence and sex ratio were also not consistently altered on the transgenic oilseed rape lines compared with the wild-type lines.

Corn cystatin (CC), a phytocystatin, shows a wide inhibitory spectrum against various cysteine proteinases. Irie et al. (164) produced transgenic rice plants by introducing CC cDNA under CaMV 35S promoter as a first step in obtaining a rice plant with insecticidal activity. This attempt was based on the observation that many insect pests, especially Coleoptera, have cysteine proteinases, probably digestive enzymes, and also that oryzacystatin, an intrinsic rice cystatin, shows a narrow inhibition spectrum and is present in ordinary rice seeds in insufficient amounts to inhibit the cysteine proteinases of rice insect pests. The transgenic rice plants generated contained high levels of CC mRNA and CC protein in both seeds and leaves, with the CC protein content of the seed reaching about 2% of the total heat-soluble protein. The CC activity from seeds was also recovered, and it was found that the CC fraction efficiently inhibited both papain and cathepsin H, whereas the corresponding fraction from nontransformed rice seeds showed much lower or undetectable inhibitory activities against these cysteine proteinases. Furthermore, CC prepared from transgenic rice plants showed potent inhibitory activity against proteinases that occur in the gut of the insect pest, *Sitophilus zeamais*.

A chimeric gene containing a cDNA clone of rice cystatin (oryzacystatin I; OC-I), the cauliflower mosaic virus 35S promoter, and the nopaline synthase 3′ region was introduced into tobacco plants through *Agrobacterium tumefaciens* (165). The presence of chimeric gene in transgenic plants was detected by a PCR-amplified assay, and transcriptional activity was shown by RNA blot analysis. Heated extract from transgenic tobacco plants, as well as from progeny obtained by selfing a primary transformant, contained protein bands that corresponded in molecular mass to OC-I and reacted with antibodies raised against rOC, a recombinant OC-I protein produced by *E. coli*. Similar bands were absent in extracts from untransformed control plants. OC-I levels reached 0.5% and 0.6% of the total soluble proteins in leaves and roots, respectively, of some progeny.

On a fresh weight basis, the OC-I content was higher in leaves (50 µg/g) than in roots (30 µg/g). OC-I was partially purified from protein extracts of rice seeds and from transgenic tobacco leaves by affinity to anti-rOC antibodies. OC-I from both sources was active against papain.

    A cDNA clone that encodes oryzacystatin, a cysteine protease inhibitor from rice, was isolated and expressed in *E. coli* BL-21 (DE3) using an expression plasmid under the control of a T7 RNA polymerase promoter (166). The construct pT7OC 9b encoded a fusion protein containing 11 amino acid residues of the NH_2 terminus of the bacterial protein phi 10 and 79 residues of oryzacystatin lacking 23 NH_2-terminal residues of the wild-type protein. Recombinant oryzacystatin (ROC) constituted approximately 10% of the total bacterial protein mass and was purified in a single step by anion-exchange chromatography. The inhibitory activity of ROC toward papain ($K_i = 3 \times 10^{-8}$ M) was comparable with that of the naturally occurring protein isolated from rice. Caseinolytic activity in midgut homogenates from seven species of stored product insects was inhibited from 18% to 85% by ROC, whereas the same activity was inhibited from 14% to 69% by the serine proteinase inhibitor phenylmethylsulfonyl fluoride. Midguts of stored product insects apparently contain both cysteine proteinases and serine proteinases, but the relative amount varies with the species. When fed to the red flour beetle, *Tribolium castaneum*, 10% ROC in the diet suppressed growth approximately 35% relative to that of the control group of insects.

    A cDNA clone for a cysteine proteinase inhibitor of rice (oryzacystatin) was isolated from a lambda gt10 cDNA library of rice immature seeds by screening with synthesized oligonucleotide probes based on partial amino acid sequences of oryzacystatin (167). A nearly full-length cDNA clone was obtained which encoded 102 amino acid residues. The amino acid sequence of oryzacystatin deduced from the cDNA sequence was significantly homologous to those of mammalian cystatins, especially family 2 cystatins. Oryzacystatin contained the sequence Gln-Val-Val-Ala-Gly conserved among most members of the cystatin superfamily. The gene for oryzacystatin was transcribed into a single mRNA species of about 700 nucleotides. The content of mRNA reached its highest level 2 weeks after flowering and then gradually decreased to undetectable levels at 10 weeks. This feature of transient expression is coordinate with that of glutelin (a major storage protein), although the expression of oryzacystatin precedes that of glutelin by about a week.

    Misaka et al. (168) isolated both genomic DNA and cDNA clones from soybean that encode a cystatin consisting of 245 amino acid residues (soyacystatin). The genomic DNA encoding soyacystatin is also unique in that it consists of four exons with three introns in its coding regions. The mRNA for soyacystatin is distinctly expressed in soybean seeds 2 weeks after flowering. Soyacystatin purified from mature soybean seeds had a molecular mass of about 26 kDa on SDS-PAGE, suggesting that it contains the extension sequences. Papain inhibition experiments demonstrate that this endogenous soyacystatin has almost the same inhibitory activity as that of its deletion mutant (102 amino acid residues) recombinantly produced by truncation of the amino and carboxy terminal extensions, indicating that the occurrence of the extensions does not affect the cystatin activity. Immunohistochemical experiments reveal that soyacystatin is expressed nearly uniformly in the cotyledons.

Abe et al. (169) constructed an expression plasmid containing a full-length oryzacystatin cDNA at the multicloning site of pUC 18 and produced a lacZ'-oryzacystatin fusion protein in *E. coli*. The partially purified expressed protein efficiently inhibits papain activity assayed using N-benzoyl-D,L-arginine-2-naphthylamide as a substrate. Expression plasmids lacking the 5' and 3' regions of cDNAs that encode NH_2- and COOH terminally truncated oryzacystatins was also constructed. An N-truncated oryzacystatin lacking Gly5 and retaining Gln53-Val54-Val55-Ala56-Gly57 inhibited papain as efficiently as the full-length oryzacystatin, although both Gly5 and Gln53-Gly57 (oryzacystatin numbering) are conserved among members of most cystatin superfamilies. However, another N-truncated oryzacystatin lacking the NH_2 terminal 38 residues was almost completely inactive. On the other hand, a COOH terminally truncated oryzacystatin lacking the COOH terminal 11 residues shows much less inhibitory activity, although it retains the two well-conserved features Gly5 and Gln53-Gly57. These results indicate that the NH_2 terminal 21 residues containing Gly5 and the COOH terminal 11 residues are not essential, suggesting that a portion of the polypeptide segment containing Gln53-Gly57 is necessary for oryzacystatin to elicit its papain inhibitory activity efficiently.

The mRNA for oryzacystatin I is expressed maximally at 2 weeks after flowering and is not detected in mature seeds, whereas the mRNA for oryzacystatin II is constantly expressed throughout the maturation stages and is clearly detected in mature seeds (170). Western blot analysis using antibody to oryzacystatin II showed that, as is the case with oryzacystatin I, oryzacystatin II occurs in mature rice seeds. Thus, two oryzacystatin species are believed to be involved in the regulation of proteolysis caused by different proteinases.

A full–length cDNA clone for a cysteine proteinase inhibitor (cystatin) was isolated from a lambda gt10 cDNA library of immature corn kernels by screening with a mixture of cDNA inserts for oryzacystatins I and II (171). The cDNA clone spans 960 bp, encoding a 135-amino-acid protein containing a signal peptide fragment. The protein, named corn cystatin I, is considered to be a member of the cystatin superfamily because it contains the commonly conserved Gln-Val-Val-Ala-Gly region that exists in most known cystatins as a probable binding site and is significantly similar to other cystatins in its overall amino acid sequences. Corn cystatin I expressed in *E. coli* showed a strong papain inhibitory activity. Northern blot analysis showed that the amount of mRNA for corn cystatin I reaches a maximum 2 weeks after flowering and then decreases gradually.

5.7.3   Amylase Inhibitors

Spencer and Hodge (172) constructed an oligonucleotide probe corresponding to the N-terminal sequences of an α-AI-like protein from cocoa beans and isolated the corresponding cDNA from a library made from poly(A) + RNA from immature cocoa beans. Suzuki et al. (173) isolated and determined the sequence of an 852-nucleotide cDNA, designated as α-AI2, and found it to contain a 720-bp open reading frame. *Liu* et al. (174) found that ABA did not affect BASI (bifunctional α-amylase/substilisin inhibitor from barley) mRNA translation. Nuclear run-on assays demonstrated that ABA had no effect on transcriptional activity. They suggested that the increase in steady-state levels of BASI mRNA due to exogenous ABA might be due to its influence on the stability of BASI mRNA through synthesis of a short-lived protein that protects the message.

Grosset et al. (175) isolated the gene encoding barley CMd protein (subunit of the tetrameric α-AI from barley) from a genomic library using a cDNA probe that encoded the wheat CM3 protein. Analysis of the promoter sequences revealed motifs found in genes specifically expressed in endosperm and aleurone cells, as well as TATA and other putative functional boxes. Jones et al. (176) cloned the barley amy-1 gene (encoding the bifunctional α-amylase/subtilisin inhibitor, BASI) into a pMAL vector that expressed the fusion protein. The purified fusion protein was cleaved with a specific protease to release the native BASI protein. Sparvoli et al. (177) isolated a cDNA clone corresponding to an α-AI-like (AIL) protein from lima bean. The clone showed 93.7% nucleotide identity with another clone corresponding to an arcelin-like protein.

Shade et al. (178) published the first report on the successful application of an α-AI gene for the production of insect-resistant transgenic plants. They have transformed peas (*Pisum sativum*) with the α-AI gene from common bean (*Phaseolus vulgaris*) through *Agrobacterium*-mediated transformation. The levels of α-AI protein expressed in the transgenic pea seeds were comparable to those normally found in common bean seeds, and the peas became resistant to cowpea weevil and azuki bean weevil.

Double-gene constructs containing different combinations of the genes encoding bean chitinase (BCH), snowdrop lectin (GNA), and wheat α-AI (WAI) were used by Gatehouse et al. (179) for transforming potato cv. Desire. Constitutive expression of the proteins in transgenic plants carrying the double constructs WAI/GNA and BCH/GNA had significant influence on fecundity and mortality of the peach-potato aphid, *Myzus persicae*. Masoud et al. (180) reported the successful transformation and subsequent expression of a corn bifunctional inhibitor of serine proteinases and insect α-AIs in tobacco plants. The level of constitutive expression of the inhibitor in young leaves of the transgenic tobacco plants was to the tune of 0.05% of total leaf proteins. However, the use of a double 35S promoter did not enhance the protein accumulation indicating that posttranscriptional events had a major role in the low accumulation of the protein in tobacco.

Titarenko and Chrispeels (181) reported the characterization and cDNA cloning of two α-amylase isozymes from larvae of the Western corn rootworm (*Diabrotica virgifera virgifera* LeConte). Larvae raised on artificial media have very low levels of amylase activity, and much higher levels are found in larvae raised on maize seedlings. At pH 5.7, the optimal pH for enzyme activity, α–amylase is substantially but not completely inhibited by amylase inhibitors from the common bean (*Phaseolus vulgaris*) and from wheat (*Triticum aestivum*). Using the reversed-transcriptase polymerase chain reaction (RT-PCR), they cloned two cDNAs with 83% amino acid identity that encode α-amylase-like polypeptides. Expression of one of the two cDNAs in insect cells with a baculovirus vector showed that this cDNA encodes an active amylase with a mobility that corresponds to that of one of the two isozymes present in larval extracts. The same two inhibitors substantially inhibited the expressed enzyme. They also showed that expression in *Arabidopsis* of the cDNA that encodes the amylase inhibitor AI-1 of the common bean results in the accumulation of active inhibitor in the roots, and the results are discussed with reference to the possibility of using amylase inhibitors as a strategy to genetically engineer maize plants that are resistant to Western corn rootworm larvae.

5.8 PERSPECTIVES

Currently, there are two major groups of plant-derived genes used to confer insect resistance on crops, namely, inhibitors of digestive enzymes and lectins. These have been transferred into crop plants without major alteration and expression has been at a similar level to codon-optimized

Bt toxins. However, this approach has not resulted in the same high levels of insect control.

Most effort was concentrated on serine-proteinase inhibitors from the plant families Fabaceae, Solanaceae, and Poaceae, which are targeted against lepidopteran species and with limited report on coleopteran and orthopteran pests. The most efffective inhibitor identified to date is the cowpea trypsin inhibitor. CpTl-transformed tobacco (when field tested in California, United States) caused significant larval mortality of cotton bollworm larvae, but the protection provided by CpTl was less pronounced and consistent than that of tobacco containing the Bt toxin gene (182). In addition to serine proteinase inhibitors, cysteine-proteinase inhibitor OC-1 from rice has been introduced in several other crops (183). To date, no crop expressing proteinase or amylase inhibitor genes has been commercialized.

More research is also needed to find specific promoters such as inducible promoters to replace, at least partly, the CaMV 35S promoter. Efforts are also needed to resolve the possible environmental factors or endogenous processes influencing the stability of expression and at overcoming the limitations of conventional marker genes.

Resistance has not been widely described, but Jongsma et al. (184) demonstrated that feeding larvae of *Spodoptera exigua* on tobacco leaves expressing the trypsin/chymotrypsin inhibitor II of potato resulted in an increased level of inhibitor-insensitive enzyme, from about 12% of the total tryptic activity in insects fed control plants to 72% in those fed transgenic plants. Such resistance is less likely to develop if multiple protective proteins are employed.

Resistance-breaking insects capable of overcoming selected *Bacillus thuringiensis* endotoxin genes are emerging (185), and resistance to proteinase inhibitors has already been reported (186). Cultural practices, such as non-transgenic refugia for nonresistant pest genotypes (187) or use of seed mixtures, are being evaluated as strategies to delay the emergence of resistance-breaking population of insects. An interesting approach to increasing the durability of transgenes is gene pyramiding, in which two, three, or more proteinase inhibitors and/or lectins (with different modes of action) could be expressed together in the same plant. Not only would this decrease the chance of selecting for resistance-breaking insects but it could also increase efficacy and broaden the spectrum of insect species controlled by any one transgenic line. Transgenic crops expressing anti-insect genes of enzyme inhibitor will almost certainly be most effective and useful when incorporated into integrated pest management schemes, when they can be used to reduce the input of chemical control agents and reduce our overdependance on the few commercial sources of plant resistance genes currently available.

REFERENCES

1. Kunitz, M. Crystalline soybean trypsin inhibitor. J. Gen. Physiol. **1946**, *29*, 149–154.
2. Ryan, C.A. Proteinase inhibitors. In *The Biochemistry of Plants, Proteins and Nucleic Acids*; Marcus, A., Ed.; Academic Press: New York, 1981; Vol. 6, 351–370.
3. Shewry, P.R.; Lucas, L.A. Plant proteins that confer resistance to pests and pathogens. Adv. Bot. Res. **1997**, *26*, 135–192.
4. Gatehouse, A.M.R.; Gatehouse, J.A. Identifying proteins with insecticidal activity—use of encoding genes to produce insect-resistant transgenic crops. Pest. Sci. **1998**, *52*, 165–175.
5. Liener, I.E.; Kakade, M.L. Proteinase inhibitors. In *Toxic Constituents of Plant Foodstuffs*; Liener, I.E., Ed.; Academic Press: New York, 1980; 7–71.
6. Ryan, C.A. Proteolytic enzymes and their inhibitors. Annu. Rev. Plant Physiol. **1973**, *24*, 197–224.
7. Laskowski, M.; Kato, I. Protein inhibitors of proteinases. Annu. Rev. Biochem. **1980**, *49*, 593–626.
8. Ryan, C.A. Protease inhibitors in plants: genes for improving defenses against insects and pathogens. Annu. Rev. Phytopathol. **1990**, *28*, 425–449.
9. Garcia-Olmedo, F.; Salcedo, G.; Sanchez-Monge, R.; Gomez, L.; Royo, J.; Carbonero, P. Plant proteinaceous inhibitors of proteinases and α-amylases. Ox Surv. Plant Mol. Cell Biol. **1987**, *4*, 275–334.
10. Watt, K.; Graham, J.; Gorden, S.C.; Woodhead, M.; McNicol, R.J. Current and future transgenic control strategies to vine weevil and other insect resistance in strawberry. J. Hort. Sci. Biotechnol. **1999**, *74*, 409–421.
11. Xavier-Filho, J.; Campus, F.A.P. Proteinase inhibitors. In *Toxicants of Plant Origin, Proteins and Amino Acids*; Cheeke, P.R., Ed.; CRC Press: Boca Raton, 1989; Vol. 3, 1–27.
12. Richardson, M. Seed storage proteins: the enzyme inhibitors. Meth. Plant Biochem. **1991**, *5*, 259–305.
13. Funatsu, M.; Shimoda, T.; Kim, B.M. Lipase inhibitors from green pepper, *Caspsicum annum* Lin. I. separation and some properties of crude lipase inhibitor. J. Fac. Agric. Kyushu Univ. **1977**, *21*, 1–8.
14. Barret, J.A. The classes of proteolytic enzymes. In *Plant Proteolytic Enzymes*; Dalling, M.J., Ed.; CRC Press: Boca Raton, 1986; 1–6.
15. Keilova, H.; Tomasek, V. Isolation and properties of cathepsin D inhibitor from potatoes. Collection Zechoslov Chem. Commun. **1976**, *41*, 489–497.
16. Bowman, D.E. Fractions derived from soybeans and navy beans, which retard tryptic digestion of casein. Proc. Soc. Environ. Physiol. Med. **1944**, *57*,139–140.
17. Birk, Y. The Bowman-Birk inhibitor. Trypsin and chymotrypsin inhibitor from soybeans. Int. J. Peptide Prot. Res. **1985**, *25*, 113–131.
18. Hilder, V.A.; Barker, R.F.; Samour, R.A.; Gatehouse, A.M.R.; Gatehouse, J.A.; Boulter, D. Protein and cDNA sequence of Bowman-Birk proteinase inhibitors from the cowpea (*Vigna unguiculata* Walp). Plant Mol. Biol. **1989**, *13*, 701–710.

19. Belzunces, L.P.; Lenfant, C.; Pasquale, S.D.; Marc Edourc, C. In vivo and in vitro effects of wheat germ agglutunin and Bowman-Birk soybean trypsin inhibitor, two potential transgene products, on midgut esterase and protease activities from *Aphis mellifera*. Comp. Biochem. Physiol. **1995**, *109*, 63–69.

20. Odani, S.; Ikenaka, T. Studies on soybean trypsin inhibitors IV. Complete amino acid sequence and the antiproteinase sites of Bowman-Birk soybean proteinase inhibitor. J. Biochem. **1972**, *71*, 839–848.

21. Odani, S.; Ikenaka, T. Studies on soybean trypsin inhibitors VIII. Disulfide bridges in soybean Bowman-Birk inhibitor. J. Biochem. **1973**, *74*, 697–715.

22. Odani, S.; Ikenaka, T. Scission of soybean Bowman-Birk proteinase inhibitor into two small fragments having either trypsin or chymotrypsin inhibitory activity. J. Biochem. **1973**, *74*, 857–860.

23. Odani, S.; Ikenaka, T. The amino acid sequences of two soybean double-headed proteinase inhibitors and evolutionary consideration of the legume proteinase inhibitors. J. Biochem. **1976**, *80*, 641–643.

24. Odani, S.; Ikenaka, T. Studies on soybean trypsin inhibitors XI. Complete amino acid sequence of a soybean trypsin-chymotrypsin-elastase inhibitor, C-II. J. Biochem. **1977**, *82*, 1523–1531.

25. Odani, S.p.; Koide, T.; Ono, T. The complete amino acid sequence of barley trypsin inhibitor. J. Biol. Chem. **1983**, *258*, 7998–8003.

26. Odani, S.; Koide, T.; Ono, T. Wheat germ trypsin inhibitors. Isolation and structural characterisation of single-headed and double-headed inhibitors of the Bowman-Birk type. J. Biochem. **1986**, *100*,978–983.

27. Wilson, K.A.; Laskowski, M. The partial amino acid sequence of trypsin inhibitor II from garden bean, *Phaseolus vulgaris*, with location of trypsin and elastase-reactive sites. J. Biol. Chem. **1975**, *250*, 4261–4267.

28. Yamamoto, M.; Hara, S.; Ikenaka, T. Amino acid sequences of two trypsin inhibitors from winged bean seeds. J. Biochem. **1983**, *94*, 849–863.

29. Yoshikawa, M.; Kiyohara, T.; Iwasaki, T.; Ishii, Y.; Kimura, N. Amino acid sequences of proeinase inhibitors II and II′ from azuki beans. Agric Biol. Chem. **1979**, *43*, 787–796.

30. Kunitz, M. Crystalline soybean trypsin inhibitor 2. General properties. J. Gen. Physiol. **1947**, *30*, 291–307.

31. Koide, T.; Tsunazava, S.; Ikenaka, T. Studies on soybean trypsin inhibitor. 2. Amino acid sequence around the reactive site of soybean trypsin inhibitor. Eur. J. Biochem. **1973**, *32*, 408–416.

32. Blow, D.M.; Janin, J.; Sweet, R.M. Mode of action of soybean trypsin inhibitor (Kunitz) as a model for specific protein–protein interactions. Nature **1974**, *249*, 54–57.

33. Sweet, R.M.; Wright, H.T; Janin, J.; Chotina, C.H; Blow, D.M. Crystal structure of the complex of porcine trypsin with soybean trypsin inhibitor (Kunitz) at 2.6 Å resolutions. Biochemistry **1974**, *13*, 4212–4228.

34. Mitsui, Y.; Satow, Y.; Wanatabe, Y.; Hirono, S.; Iitaka, Y. Crystal structures of *Streptomyces subtilisin* inhibitor and its complex with subtilisin. Nature **1979**, *277*, 447–452.

35. Barrett, A.J. The cystatins: a new class of peptidase inhibitors. Trends Biochem. Sci. **1987**, *12*, 193–196.
36. Waldrom, C.; Wegrich, L.M.; Owens, P.A.; Walsh, T.A. Characterization of genomic sequence coding for potato multicystatin, an eight-domain cysteine proteinase inhibitor. Plant Mol. Biol. **1993**, *23*, 801–812.
37. Abe, A.; Emori, Y.; Kondo, H.; Suzuki, K.; Arai, S. Molecular cloning of cysteine proteinase inhibitor of rice (oryzacystatin). J. Biol. Chem. **1987**, *262*, 16793–16797.
38. Arai, S.; Watanabe, H.; Kondo, H.; Emori, Y.; Abe, K. Papain-inhibitory activity of oryzacystatin, a rice seed cysteine protease inhibitor depends on the central Gln-Val-Val-Ala-Gly region conserved among cystatin superfamily members. J. Biochem. **1991**, *109*, 294–298.
39. Abe, K.; Kondo, H.; Arai, S. Purification and characterization of a rice cysteine protease inhibitor. Agric. Biol. Chem. **1987**, *51*, 2763-2768.
40. Kondo, H.; Abe, K.; Nishimura, I.; Watanabe, H.; Emori, Y.; Arai, S. Two distinct cystatin species in rice seeds with different specificities against cysteine proteinases. J. Biol. Chem. **1990**, *265*, 15832–15837.
41. Kasahara, K.; Hayashi, K.; Arakawa, T.; Philo, J.S.; Wen, J.; Hara, S.; Yamaguchi, H. Complete sequence, subunit structure and complexes with pancreatic α-amylase of an α-amylase inhibitor from *Phaseolus vulgaris* white kidney beans. J. Biochem. **1996**, *120*, 177–183.
42. Nakaguchi, T.; Arakawa, T.; Philo, J.S.; Wen, J.; Ishimoto, M.; Yamaguchi, H. Structural characterization of an α-amylase inhibitor from a wild common bean (*Phaseolus vulgaris*): insight into the common structural features of leguminous α-amylase inhibitors. J. Biochem. **1997**, *121*, 350–354.
43. Suzuki, K.; Ishimoto, M.; Kitamura, K. cDNA sequence and deduced primary structure of an α-amylase inhibitor from a bruhid-resistant wild common bean. Biochem. Biophys. Acta. **1994**, *1206*, 289–291.
44. Alam, N.; Gourinath, S.; Dey, S.; Srinivasan, A.; Singh, T.P. Substrate–inhibitor interactions in the kinetics of α-amylase inhibition by ragi α-amylase/trypsin inhibitor (RATI) and its various N-terminal fragments. Biochemistry **2001**, *40*, 4229–4233.
45. Strobl, S.; Muhlhahn, P.; Bernstein, R.; Wiltscheck, R.; Maskos, K.; Wunderlich, M.; Huber, R.; Glockshuber, R.; Holak, T.A. Determination of the three-dimensional structure of the bifunctional α-amylase/trypsin inhibitor from ragi seeds by NMR spectroscopy. Biochemistry **1995**, *34*, 8281–8293.
46. Alagiri, S.; Singh, T.P. Stability and kinetics of a bifunctional amylase/trypsin inhibitor. Biochem. Biophys. Acta **1993**, *1203*, 77–84.
47. Lopez, A.C.; Labra, A.B.; Patthy, A.; Sanchez, R.; Pongor, S. A novel alpha amylase inhibitor from amaranth (*Amaranthus hypocondriacus*) seeds. J. Biol. Chem. **1994**, *269*, 23675–23680.
48. Reddy, M.N.; Keim, P.S.; Heinrikson, R.L.; Kezdy, F.J. Primary structural analysis of sulfhydryl protease inhibitors from pineapple stem. J. Biol. Chem. **1975**, *250*, 1741–1750.

49. Rodis, P.; Hoff, J.E. Naturally occurring protein crystals in the potato. Plant Physiol. **1984**, *74*, 907–911.
50. Abe, M.; Whitaker, J.R. Purification and characterization of a cysteine proteinase inhibitor from the endosperm of corn. Agric. Biol. Chem. **1988**, *52*, 1583–1584.
51. Abe, K.p; Kondo, H.; Arai, S. Purification and characterization of a rice cysteine proteinase inhibitor. Agric. Biol. Chem. **1987**, *51*, 2763–2768.
52. Rele, M.V.; Vartak, H.G.; Jagannathan, V. Proteinase inhibitors occurring from *Vigna unguiculata* sub sp *cylindrica* I. Occurrence of thiol proteinase inhibitors in plants and purification from *Vigna unguiculata* sub sp *cylindrica*. Arch. Biochem. Biophys. **1980**, *204*, 117–128.
53. Baumgartner, B.; Chrispeels, M.J. Partial characterization of a protease inhibitor, which inhibits the major endopeptidase present in cotyledons of mung beans. Plant Physiol. **1976**, *58*, 1–6.
54. Akers, C.P.; Hoff, J.E. Simultaneous formation of chymopapain inhibitor activity and cubical crystals in tomato leaves. Can. J. Bot. **1980**, *58*, 1000–1003.
55. Fossum, K. Proteolytic enzymes and biological inhibitors. III. Naturally occurring inhibitors in some animal and plant materials and their effect upon enzymes of various origin. Acta Pathol. Microbiol. Scand. Sect. B. Microbiol. **1970**, *78*, 741–754.
56. Rancour, J.M.; Ryan, C.A. Isolation of a carboxypeptidase B inhibitor from potatoes. Arch. Biochem. Biophys. **1968**, *125*, 380–382.
57. Hass, G.M.; Ryan, C.A. Carboxypeptidase inhibitor from potatoes. Meth. Enzymol. **1981**, *80*, 778–791.
58. Graham, J.S.; Ryan, C.A. Accumulation of metallocarboxypeptidase inhibitor in leaves of wounded potato plants. Biochem. Biophys. Res. Commun. **1981**, *101*, 1164–1170.
59. Chen, M.S.; Feng, G.; Zen, K.C.; Richardson, M.; Rodriguez, S.V.; Reeck, G.R.; Kramer, K.J. Alpha amylases from three species of stored grain coleoptera and their inhibition by wheat and corn proteinaceous inhibitors. Insect Biochem. Mol. Biol. **1991**, *22*, 261–268.
60. Weselake, R.J.; Alexander, W.; Macgregor, W.; Hill, R.D. Effect of endogenous barley α-amylase inhibitor on hydrolysis of starch under various conditions. J. Cereal Sci. **1985**, *3*, 249–259.
61. Blanco-Labra, Iturbe-Chinase, F.A. Purification and characterization of an α-amylase inhibitor from maize (*Zea mays*). J. Food Biochem. **1981**, *5*, 1–17.
62. Feng, G.H.; Chen, M.S.; Kramer, K.J.; Reeck, G.R. Alpha amylase inhibitors from rice: fractionation and selectivity toward insect, mammalian and bacterial α-amylase. Cereal Chem. **1991**, *68*, 516–521.
63. Shivaraj, B.; Pattabiraman, T.N. Natural plant enzyme inhibitors. VIII. Purification and properties of two α-amylase inhibitors from ragi (*Eleusine coracana*) grains. Ind. J. Biochem. Biophys. **1980**, *17*, 181–185.
64. Taefal, A.; Bohn, H.; Flamme, W. Protein inhibitors of α-amylase in mature and germinating grain of rye (*Secale cereale*). J. Cereal Sci. **1997**, *25*, 367–273.

65. Mulimani, V.H.; Supriya, D. Alpha amylase inhibitor in sorghum (*Sorghum bicolor*). Plant Foods Hum. Nutrn. **1993**, *44*, 261–266.

66. Zawistowski, U.; Langstaff, J.; Friesen, A.D. Purification and characterization of two double-headed triticale isoinhibitors of endogenous α-amylase and subtilisin. J. Food Biochem. **1989**, *13*, 235–239.

67. Prathibha, S.; Nambisan, B.; Leelamma, S. Enzyme inhibitors in tuber crops and their thermal stability. Plant Foods Hum. Nutrn. **1995**, *48*, 247–257.

68. Cantagalli, P.; Giorgio, G.D.; Morisi, G.; Pocchiari, F.; Silano, V. Purification and properties of three albumins from *T. aestivum* seeds. J. Sci. Food Agric. **1971**, *22*, 256–259.

69. Mikola, J.; Kirsi, M. Differences between endospermal and embryonal trypsin inhibitors in barley, wheat and rye. Acta Chem. Scand. **1972**, *26*, 787–795.

70. Pueyo, J.J.; Hunt, D.C.; Chrispeels, M.J. Activation of bean α-amylase inhibitor requires proteolytic processing of the pro-protein. Plant Physiol. **1993**, *101*, 1341–1348.

71. Applebaum, S.W. Physiological aspects of host specificity in the bruchidae. I. General considerations of developmental compatability. J. Insect Physiol. **1964**, *10*, 783–788.

72. Applebaum, S.W. Biochemistry of digestion. In *Comprehensive Insect Physiology, Biochemistry and Pharmacology*; Kerkut, G.A., Gilbert, L.I., Eds.; Pergamon press: New York, 1985; Vol. 4, 279–311.

73. Lipke, H.; Fraenkel, G.S. Liener, I. Effect of soybean inhibitor on growth of *Tribolium confusum*. Food Chem. **1954**, *2*, 410–414.

74. Birk, Y.; Applebaum, Y. Effect of soybean trypsin inhibitor on the development and midgut proteolytic activity of *Tribolium castaneum* larvae. Entomolgia **1960**, *22*, 318–326.

75. Steffens, R.; Fox, F.R.; Kassel, B. Effect of trypsin inhibitor on growth and metamorphosis of corn borer larvae *Ostrinia nubilalis*. J. Agric. Food Chem. **1978**, *26*, 170–175.

76. Broadway, R.M.; Duffey, S.S. Plant proteinase inhibitors: mechanism of action and effect on the growth and digestive physiology of larvae *Heliothis zea* and *Spodoptera exigua*. J. Insect Physiol. **1986**, *32*, 827–833.

77. Gatehouse, A.M.R.; Boulter, D. Assessment of antimetabolic effects of trypsin inhibitor from cowpea (*Vigna unguiculata*) and other legumes on development of the bruchid beetle *Callosobruchus maculatus*. Entomol. Exp. Appl. **1983**, *39*, 279–286.

78. Shukle, R.H.; Murdock, L.L. Lipoxygenase, trypsin inhibitor and lectin from soybeans: effects on larval growth of *Manduca sexta* (Lepidoptera: Sphingidae). Environ. Entomol. **1983**, *12*, 787–791.

79. Spates, G.E.; Harris, R.L. Reduction of fecundity, egg hatch and survival in adult horn flies fed protease inhibitors. Southwest Entomol. **1984**, *4*, 399–403.

80. Deloach, J.R.; Spates, G.E. Effect of soybean trypsin inhibitor-loaded erythrocytes on fecundity and midgut protease and haemolysis activity of stable flies. J. Econ. Entomol. **1980**, *73*, 590–594.

81. Gatehouse, A.M.R.; Shi, Y.; Powell, K.S.; Brough, C.; Hilder, V.A.; Hamilton, W.D.O.; Newell, C.A.; Merryweather, A.; Boulter, D.; Gatehouse, J.A. Approaches to insect resistance using transgenic plants. Philos. Trans. Roy. Soc. London B. **1993**, *342*, 279–296.

82. Johnson, K.A.; Brough, M.J.; Hilder, V.A.; Gatehouse, A.M.R.; Gatehouse, J.A. Protease activity in the larval midgut of *Heliothis virescens*: evidence for trypsin and chymotryspin like enzymes. Insect Biochem. Mol. Biol. **1995**, *25*, 375–383.

83. Broadway, R.M.; Duffey, S.S.; Pearce, G.; Ryan, C.A. Plant proteinase inhibitors: a defense against herbivorous insects? Entomol. Exp. Appl. **1986**, *41*, 33–38.

84. Edwards, P.J.; Wratten, S.D.; Cox, H. Wound induced changes in the acceptability of tomoto to larvae of *Spodoptera littoralis*: a laboratory bioassay. Ecol. Entomol. **1985**, *10*, 155–158.

85. Hilder, V.A.; Gatehouse, A.M.R.; Sheerman, S.E.; Barker, R.F.; Boulter, D. A novel mechanism of insect resistance engineered into tobacco. Nature **1987**, *330*, 160–163.

86. Gatehouse, A.M.R.; Gatehouse, J.A.; Dobie, P.; Kilminster, A.M.; Boulter, D. Biochemical basis of insect resistance in *Vigna unguiculata*. J. Sci. Food Agric. **1979**, *30*, 948–958.

87. Lau, A.; Ako, H.; Washburne, M.W. Survey of plants for enterokinase inhibitors. Biochem. Biophys. Res. Commun. **1980**, *92*, 1243–1249.

88. Applebaum, S.W.; Birk, Y.; Harpaz, I.; Bondi, A. Comparative studies on pro-teolytic enzymes of *Tenebrio molitor* L. Comp. Biochem. Physiol. **1964**, *2*, 85–103.

89. Zwilling, R. On the evolution of endopeptidase. IV. Alpha- and beta-protease from *Tenebrio molitor*. Z. Physiol. Chem. **1968**, *349*, 326–332.

90. Green, T.R.; Ryan, C.A. Wound inducted proteinase inhibitors in plant leaves. a possible defense mechanism against insects. Science **1972**, *175*, 776–777.

91. Green, T.R.; Ryan, C.A. Wound-induced proteinase inhibior in tomato leaves-some effects of light and temperature on the wound response. Plant Physiol. **1972**, *51*, 19–21.

92. Ryan, C.A.; Moura, D.S. Wound-inducible proteinase inhibitors in pepper. Differential regulation upon wounding, systemin and methyl jasmonate. Plant Physiol. **2001**, *136*, 289–298.

93. Ceciliani, F.; Tava, A.; Iori, R.; Mortarino, I.; Odoardi, M.; Ronchi, S. A trypsin inhibitor from snail medic seeds active against pest proteases. Phytochemistry **1997**, *44*, 393–398.

94. Houseman, J.G.; Downe, A.E.R. Cathepsin D like activity in the posterior midgut of hemipteran insects. Comp. Biochem. Physiol. B. **1983**, *75*, 509–512.

95. Murdock, L.L.; Brookhart, G.; Dunn, D.E.; Foard, D.E.; Kelley, S. Cysteine digestive proteinases in Coleoptera. Comp. Biochem. Physiol. B. **1987**, *87*, 783–787.

96. Kitch, L.W.; Murdock, L.L. Partial characterization of a major proteinase from larvae of *Callosobruchus maculatus* (F.). Arch. Insect Biochem. Physiol. **1986**, *3*, 561–576.

97. Lemos, F.J.A.; Xavier-Filho, J.; Campos, F.A.P. Proteinases of the midgut of *Zabrotes subfasciatus* larvae. Arq. Biol. Technol. **1987**, *30*, 46.

98. Wieman, K.F.; Nielsen, S.S. Isolation and partial characterization of a major gut proteinase from larval *Acanthocelides obtectus* Say (Coleoptera: Bruchidae). Comp. Biochem. Physiol. B. **1988**, *89*, 419–426.

99. Hines, M.E.; Osuala, C.L.; Nielsen, S.S. Screening for cysteine proteinase inhibitor activity in legume seeds. J. Sci. Food Agric. **1992**, *59*, 555–557.

100. Liang, C.; Brookhart, G.; Feng, G.H.; Reeck, G.R.; Kramer, K.J. Inhibition of digestive proteinases of stored grain Coleoptera by oryzacystatin, a cysteine proteinase inhibitor from rice seed. FEBS Lett. **1991**, *278*, 139–142.

101. Lemos, F.J.A.; Campos, F.A.P.; Silva, C.P.; Xavier-Filho, J. Proteinase and amylases of larval midgut of *Zabrotes subfasciatus* reared on cowpea seeds. Entomol. Exp. Appl. **1990**, *56*, 219–227.

102. Edmonds, H.S.; Gatehouse, L.N.; Hilder, V.A.; Gateshouse, J.A. The inhibitor effect of cysteine protease inhibitor oryzacystatin on digestive proteases and on survival and development of the southern corn rootworm (*Diabrotica undecimpunctata* Howard). Entomol. Exp. Appl. **1996**, *78*, 83–94.

103. Koritsas, V.M.; Atkinson, H.J. Proteinases of females of the phytoparasite *Globodera pallida* (potato cyst nematode). Parasitology **1994**, *109*, 357–365.

104. Urwin, P.E.; Atkinson, H.J.; Waller, D.A.; McPherson, M.J. Engineered oryzacystatin-I expressed in transgenic hairy roots confers resistance to *Globodera pallida*. Plant J. **1995**, *8*, 121–131.

105. Koiwa, H,; Shade, R.E.; Zhu-Salzman, K.; D'Urzo, M.P.; Murdock, L.L.; Bressan, R.A.; Hasegawa, P.M. A plant defensive cystatin (soyacystatin) targets cathepsin L like digestive cysteine proteinases in the larval midgut of western corn rootworm (*Diabrotica virgifera virgifera*). FEBS Lett. **2000**, *471*, 67–70.

106. Pernas, M.; Sanchez-Monge, R.; Gomez, L.; Salcedo, G. A chestnut seed cystatin differentially effective against cysteine proteinases from closely related pests. Plant Mol. Biol. **1998**, *38*, 1235–1242.

107. Visal, S.; Taylor, M.A.; Michaud, D. The proregion of papaya proteinase IV inhibits Colorado potato beetle digestive cysteine proteinases. FEBS Lett. **1998**, *434*, 401–405.

108. Koiwa, H.; Shade, R.E.; Zhu-Salzman, K.; Subramanian, L.; Murdock, L.L.; Nielson, S.S.; Bressan, R.A.; Hasegawa, P.M. Phage display selection can differentiate insecticide activity of soybean cystatins. Plant J. **1998**, *14*, 371–379.

109. Michaud, D.; Nguyen-Quoc, B.; Yelle, S. Selective inhibition of Colorado potato beetle cathepsin H by oryzacystatins I and II. FEBS Lett. **1993**, *331*, 173–176.

110. Gruden, K.; Strukelj, B.; Popovic, T.; Lenarcic, B.; Bevec, T.; Brzin, J.; Kregar, I.; Herzog-Velikonja, J.; Stiekema, W.J.; Bosch, D.; Jongsma, M.A. The cysteine protease activity of Colorado potato beetle (*Leptinotarsa decemlineata* Say) guts, which is insensitive to potato protease inhibitors, is inhibited by thyroglobulin type I domain inhibitors. Insect Biochem. Mol. Biol. **1998**, *28*, 549–560.

111. Gatehouse, A.M.R.; Fenton, K.A.; Jepson, I.; Pavey, D.J. The effects of α-amylase inhibitors on insect storage pests: inhibition of α-amylase in vitro and effects on development in vivo. J. Sci. Food Agric. **1986**, *37*, 727–734.

112. Silano, V.; Furia, M.; Gianfreda, L.; Macri, A.; Palescandolo, R. Inhibition of amylases from different origins by albumins from the wheat kernel. Biochim. Biophys. Acta **1975**, *391*, 170–178.

113. Powers, J.R.; Culbertson, J.D. Interaction of a purified bean (*Phaseolus vulgaris*) glycoprotein with an insect amylase. Cereal Chem. **1982**, *60*, 427–429.

114. Baker, J.E. Purification of an α-amylase inhibitor from wheat (*Triticum aestivum* and its interaction with α-amylase from the rice weevil, *Sitophilus oryzae* (Coleoptera: Curculionidae). Insect Biochem. **1987**, *18*, 107–116.

115. Yetter, M.A.; Saunders, R.M.; Boles, H.P. α-Amylase inhibitors from wheat kernels as factors in resistance to post harvest insects. Cereal Chem. **1979**, *56*, 243–244.

116. Ishimoto, M.; Kitamura, K. Idenitfication of the growth inhibitor on azuki bean weevil in kidney bean (*Phaseolus vulgaris*) Jpn. J. Breed. **1989**, *38*, 367–370.

117. Carbonero, P.; Salcedo, G.; Sanchez-Monge, R.; Garcia-Maroto, F.; Royo, J.; Gomez, L.; Mena, M.; Medina, J.; Diaz, I. A multigene family from cereals which encodes inhibitors of trypsin and heterologous α-amylases In *Innovations in Proteases and Their Inhibitors*; Aviles, F.X., Ed.; Walter de Gruyter: Berlin, 1993; 333–348.

118. Gutierrez, C.; Garcia-Casado, G.; Sanchez-Monge, R.; Gomez, L.; Castanera, P.; Salcedo, G. Three inhibitor types from wheat endosperm are differentially active against α-amylase of Lepidoptera pests. Entomol. Exp. Appl. **1993**, *66*, 47–52

119. Moralejo, M.A.; Garcia-Casado, G.; Sanchez-Monge, R.; Lopez-Otin, C.; Romagosa, I.; Molina-Cano, J.L.; Salcedo, G. Genetic variants of the trypsin inhibitor from barley endosperm show different inhibitory activities. Plant Sci. **1993**, *89*, 23–29.

120. Rajendran, P.; Thayumanavan, B. Purification of α-amylase inhibitor from seeds of little millet (*Panicum sumatrens* Roth) J. Plant Biochem. Biotechnol. **2000**, *9*, 89–94.

121. Estell, D.A.; Wilson, K.A.; Laskowski M. Jr. Thermodynamics and kinetics of the hydrolysis of the reactive-site peptide bond in pancreatic trypsin inhibitor (Kunitz) by *Demasterias imbricata* trypsin I. Biochemistry **1980**, *19*, 131–137.

122. Huber, R.; Kukla, D.; Bode, W.; Schwager, P.; Bartels, K.; Deisenhofer, J.; Steigemann, W. Structure of the complex formed by bovine trypsin and bovine pancreatic trypsin inhibitor II. Crystallographic refinement at 1.9 Å resolution. J. Mol. Biol. **1974**, *89*, 73–101.

123. Hunkapiller, M.W.; Forgac, M.D.; Yu, E.H.; Richards, J.H. 13C NMR studies of the binding of soybean trypsin inhibitor to trypsin. Biochem. Biophys. Res. Commun. **1979**, *87*, 25–31.

124. de la Sierra, I.L.; Quillien, L.; Flecker, P.; Gueguen, J.; Brunie, S. Dimeric crystal structure of a Bowman-Birk protease inhibitor from pea seeds. J. Mol. Biol. **1999**, *285*, 1195–1207.

125. Katz, B.A.; Elrod, K.; Luong, C.; Rice, M.J.; Mackman, R.L.; Sprengeler, P.A.; Spencer, J.; Hataye, J.; Janc, J.; Link, J.; Litvak, J.; Rai, R.; Rice, K.; Sideris, S.; Verner, E.; Young, W. A novel serine protease inhibition motif involving a multi-centered short hydrogen bonding network at the active site. J. Mol. Biol. **2001**, *307*, 1451–1486

126. Jakiewicz, A.; Lesner, A.; Roycki, J.; Rodziewicz, S.; Rolka, K.; Ragnarsson, U.; Kupryszewski, G. Distance between the basic group of the amino acid residues side chain in position P1 of trypsin inhibitor CMTI-III and Asp 189 in the substrate pocket of trypsin has an essential influence on the inhibitor activity. Biochem. Biophys. Res. Commun. **1997**, *240*, 869–871.

127. McBride, J.D.; Brauer, A.B.E.; Nievo, M.; Leatherbarrow, R.J. The roles of threonine in the P2 position of Bowman-Birk proteinase inhibitors: studies in P2 variation in cyclic peptides encompassing the reactive site loop. J. Mol. Biol. **1998**, *282*, 447–457.

128. Malik, Z.; Amir, S.; Pal, G.; Buzas, Z.; Varallyay, E.; Antal, J.; Szilagyi, Z.; Vekey, K.; Asboth, B.; Patthy, A.; Graf, L. Proteinase inhibitors from desert locust *Schistorcerca gregaria*: engineering of both P(1) and P(1)' residues converts a potent chymotrypsin inhibitor to a potent trypsin inhibitor. Biochem. Biophys. Acta. **1999**, *1434*, 143–150.

129. Moss, G.W.; Marshall, J.; Morabito, M.; Howe, J.R.; Moczydlowski, E. An evolutionarily conserved binding site for serine proteinase inhibitors in large conductance calcium-activated potassium channels. Biochemistry **1996**, *35*, 16024–16035.

130. Michaud, D.; Cantin, L.; Vrain, T.C. Carboxy- terminal truncation of oryzacystatin II by oryzacystatin-insensitive insect digestive proteinases. Arch. Biochem. Biophys. **1995**, *322*, 469–474.

131. Franco, O.L.; Rigden, D.J.; Melo, F.R.; Grossi-De-Sa, M.F. Plant α-amylase inhibitors and their interactions with insect α-amylases. Eur. J. Biochem. **2002**, *269*, 397–412.

132. Hilder, V.A.; Gatehouse, A.M.R.; Sheerman, S.E.; Barker, R.; Boulter, D. A novel mechanism of insect resistance engineered in tobacco. Nature **1987**, *330*, 160–163.

133. Sane, V.A.; Nath, P.; Aminuddin, P.; Sane, P.V. Development of insect-resistant transgenic plants used in plant genes: expression of cowpea trypsin inhibitor in transgenic tobacco plants. Curr. Sci. **1997**, *72*, 741–747

134. Xu, D.P.; Xue, Q.Z.; McElroy, D.; Mawal, Y.; Hilder, V.A.; Wu, R. Constitutive expression of a cowpea trypsin inhibitor gene CpTi in transgenic rice plants confers resistance to two major rice pests. Mol. Breed. **1996**, *2*, 167–173.

135. Hao, Y.; Ao, G.M. Transgenic cabbage plants harbouring cowpea trypsin inhibitor (CpTi) gene showed improved resistance to two major insect pests, *Pieris rapae* L. and *Heliothis armigera*. FASEB **1977**, *11*, 68.

136. Graham, J.; Gordon, S.; McNicol, R.J. The effect of the CpTi gene in strawberry against attack by vine weevil (*Otiorhynchus sulcatus* F. Coleoptera: Curculionidae). Ann. Appl. Biol. **1997**, *131*, 133–139.

137. Gatehouse, L.N.; Shannon, A.L.; Burgless, E.P.J.; Christeller, J.T. Characterisation of major midgut proteinase cDNAs from *Helicoverpa armigera* larvae and changes in gene expression in response to four proteinase inhibitors in the diet. Insect Biochem. Mol. Biol. **1997**, *2*, 929–944.

138. Duan, X.L.; Li, X.G.; Xue, Q.Z.; Abcelsaad, M.; Xu, D.P.; Wu, R. Transgenic plants harboring an introduced potato proteinase inhibitor II gene are insect resistant. Nat. Biotechnol. **1996**, *14*, 494–498.

139. Klopfenstein, N.B.; Allen, K.K.; Avila, F.J.; Heuchelin, S.A.; Martinez, J.; Carman, R.C.; Hall, R.B.; Hart, E.R.; McNabb, H.S. Proteinase inhibitor II gene in transgenic poplar. Chemical and biological assays. Biomass and Bioenergy **1997**, *12*, 299–311.

140. Tran, P.; Cheesebrough, T.M.; Keicheffer, R.W. Plant proteinase inhibitors are potential anticereal aphid compounds. J. Econ. Entomol. **1997**, *90*, 1672–1677.

141. Thomas, J.C.; Adams, D.G.; Keppenne, V.D.; Wasmann, C.C.; Brown, J.K.; Kanost, M.R.; Bohnert, H.J. *Manduca sexta*-encoded proteinase inhibitors expressed *Nicotiana tabacum* provide protection against insect. Plant Physiol. Biochem. **1995**, *33*, 611–614

142. Johnson, R.; Narvaez, J.; An, G.; Ryan, C.A. Expression of proteinase inhibitors I and II in transgenic tobacco plants: effects on natural defense against *Manduca sexta* larvae. Proc. Natl. Acad. Sci. USA **1989**, *86*, 9871–9875.

143. Volpicella, M.; Ceci, L.R.; Gallerani, R.; Jongsma, M.A.; Beekwilder, J. Functional expression on bacteriophage of the mustard trypsin inhibitor MTI-2. Biochem. Biophys. Res. Commun. **2001**, *280*, 813–817.

144. Leo, F.D.; Bottino, M.B.; Ceci, L.R.; Gallerani, R.; Jouanin, L. Effects of a mustard trypsin inhibitor expressed in different plants on three lepidopteran pests. Insect Biochem. Mol. Biol. **2001**, *31*, 593–602.

145. Lara, P.; Ortego, F.; Hidalgo, E.G.; Castanera, P.; Carbonero, P.; Diaz, I. Adaptation of *Spodoptera exigua* (Lepidoptera: Noctuidae) to barley trypsin inhibitor BTI-CMe expressed in transgenic tobacco. Transgenic Res. **2000**, *9*, 169–178.

146. Sato, S.; Kamei, K.; Taniguchi, M.; Sato, H.; Takano, R.; Mori, H.; Ichida, M.; Hara, S. Cloning and expression of the *Mimordica charantia* trypsin inhibitor II gene in silkworm by using a baculovirus vector. Biosci. Biotechnol. Biochem. **2000**, *64*, 393–398.

147. Fan, X.; Shi, X.; Zhao, J.; Zhao, R.; Fan, Y. Insecticidal activity of transgenic tobacco plants expressing both Bt and CpTI genes on cotton bollworm (*Helicoverpa armigera*). Chin. J. Biotechnol. **1999**, *15*, 1–5.

148. Wen, L.; Kim, S.S.; Tinn, T.T.; Huang, J.K.; Krishnamoorthi R.; Gong, Y.X.; Lwin, Y.N.; Kyin, S. Chemical synthesis, molecular cloning, over expression and site-directed mutagenesis of the gene coding for pumpkin (*Curcubita maxima*) trypsin inhibitor CMTI-V. Protein Expr. Purif. **1993**, *4*, 215–222.

149. Schirra, H.J. Scanlon, M.J. Lee, M.C.S.; Anderson, M.A.; Craik, D.J. The solution structure of CI-TI a two-domain proteinase inhibitor derived form a circular precursor protein from *Nicotiana alata*. J. Mol. Biol. **2001**, *306*, 69–79.
150. Webb, K.J.; Morris, P. In *Plant Genetic Manipulation for Crop Protection*; Gatehouse, A.M.R., Hilder, V.A., Boulter, D., Eds.; CAB Int., 1992; 7–43 pp.
151. Finch, R.P. In *Molecular Biology in Crop Protection*. Marshall, G., Walters, D., Eds.; Chapman and Hall: London, 1994; 1–37.
152. Roush, R. In *Advances in Insect Control: The Role of Transgenic Plants*; Carozzi, N., Koziel, M., Eds.; Taylor and Francis: London, 1997; 271–294.
153. Wunn, J.; Klotic, A.; Burkhardt, P.K.; Biswas, G.C.; Launis, K.; Iglisias, V.A.; Potrykus, I. Transgenic indica rice breeding line IR 58 expressing a synthetic Cry IA (b) gene from *Bacillus thuringiensis* provides effective insect pest control. Biotechnology **1996**, *14*, 171–176.
154. Stewart Jr, C.N.; Adang, M.J.; All, J.N.; Boerma, H.R.; Cardineau, G.; Tucker, D.; Parrott, W.A. Genetic transformation recovery, and characterization of fertile soybean transgenic for a synthetic *Bacillus thuringiensis* Cry IAc gene. Plant Physiol. **1996**, *112*, 121–129.
155. Ebinuma, H.; Sugita, K.; Matsunaga, E.; Yamakado, M. Selection of marker-free transgenic plants using the isopentenyl transferase gene. Proc. Natl. Acad. Sci. USA **1997**, *94*, 2117–2121.
156. Ryan, C.A.; Orozco-Cardenas, M.L.; Narvaez-Vasquez, J. Differential expression of a chimeric CaMV-tomato proteinase inhibitor I gene in leaves of transformed night shade, tobacco and alfalfa plants. Plant Mol. Biol. **1992**, *20*, 1149–1157.
157. Schuler, T.H.; Poppy, G.M.; Kerry, B.R.; Denholm, I. Insect resistant transgenic plants. Trends Biotechnol. **1998**, *16*, 168–175.
158. Leple, J.C.; Bottino, M.B.; Augustin, S.; Pilare, G.; Dumonois, L.T.V.; Delplanque, A.; Cornu, D.; Jouanin, L. Toxicity to *Chrysomela tremulae* (Coleoptera: Chrysomelidae) of transgenic poplars expressing a cysteine proteinase inhibitor. Mol. Breed. **1995**, *1*, 319–328.
159. Benchkroun, A.; Michaud, D.; Nguyen-Quoc, B.; Overney, S.; Desjardins, Y.; Yelle, S. Synthesis of active oryzacystatin I in transgenic potato plants. Plant Cell Rep. **1995**, *14*, 585–588.
160. Vain, P.; Worland, B.; Clarke, M.; Richard, G.; Beavis, M.; Liu, H.; Kohli, A.; Leech, M.; Snape, J.; Christou, P.; Atkinson, H. Expresson of an engineered cysteine proteinase inhibitor (oryzacystatin-1 Delta D86) for nematode resistance in transgenic rice plants. Theor. Appl. Genet. **1998**, *96*, 267–271.
161. Kuroda, M.; Ishimoto, M.; Suzuki, K.; Abe, K.; Kitamura, K.; Arai, S. Oryzacystatins exhibit growth inhibitor and lethal effects on different species of bean insect pests, *Callosobruchus chinensis* (Coleoptera) and *Riptortus clavatus* (Hemiptera). Biosci. Biotechnol. Biochem. **1996**, *6*, 209–212.
162. Gruden, K.; Strukelj, B.; Ravnikar, M.; Prijatelj, M.P.; Mavric, I.; Brzin, J.; Pungercar, J.; Kregar, I. Potato cysteine proteinase inhibitor gene family: molecular cloning, characterization and immunocytochemical localization studies. Plant Mol. Biol. **1997**, *34*, 317–323.

163. Schuler, T.H.; Denholm, I.; Jouanin, L.; Clark, S.J.; Clark, A.J.; Poppy, G.M. Population-scale laboratory studies of the effect of transgenic plants on nontarget insects. Mol. Ecol. **2001**, *10*, 1845–1853.

164. Irie, K.; Hosoyama, H.; Takeuchi, T.; Iwabuchi, K.; Watanabe, H.; Abe, M.; Abe, K.; Arai, S. Transgenic rice established to express corn cystatin exhibits strong inhibitory activity against insect gut proteinases. Plant Mol. Biol. **1996**, *30*, 149–157.

165. Masoud, S.A.; Johnson, L.B.; White, F.F.; Reeck, G.R. Expression of a cysteine proteinase inhibitor (oryzacystatin-I) in transgenic tobacco plants. Plant Mol. Biol. **1993**, *21*, 655–663.

166. Chen, M.S.; Johnson, B.; Wen, L.; Muthukrishnan, S.; Kramer, K.J.; Morgan, T.D.; Reeck, G.R. Rice cystatin: bacterial expression, purification, cysteine proteinase inhibitory activity and insect growth suppressing activity of a truncated form of the protein. Protein Expr. Purif. **1992**, *3*, 41–49.

167. Abe, K.; Emori, Y.; Kondo, H.; Suzuki, K.; Arai, A. Molecular cloning of a cysteine proteinase inhibitor of rice (oryzacystatin). Homology with animal cystatins and transient expression in the ripening process of rice seeds. J. Biol. Chem. **1987**, *262*, 16793–16797.

168. Misaka, T.; Kuroda, M.; Iwabuchi, K.; Abe, K.; Arai, S. Soyacystatin, a novel cysteine proteinase inhibitors in soybean, is distinct in protein structure and gene organisation from other cystatins of animal and plant origin. Eur. J. Biochem. **1996**, *240*, 609–614.

169. Abe, K.; Emori, Y.; Kondo, H.; Arai, S.; Suzuki, K. The NH2-terminal 21 amino acid residues are not essential for the papain-inhibitory activity of oryzacystatin, a member of the cystatin superfamily. Expression of oryzacystatin cDNA and its truncated fragments in *Escherichia coli*. J. Biol. Chem. **1988**, *263*, 7655–7659.

170. Kondo, H.; Abe, K.; Nishimura, I.; Watanabe, H.; Emori, Y.; Arai, S. Two distinct cystatin species in rice seeds with different specificities against cysteine proteinases. Molecular cloning, expression and biochemical studies on oryzacystatin II. J. Biol. Chem. **1990**, *265*, 15832–15837.

171. Abe, M.; Abe, K.; Kuroda, M.; Arai, S. Corn kernel cysteine proteinase inhibitor as a novel cystatin superfamily member of plant origin. Molecular cloning and expression studies. Eur. J. Biochem. **1992**, *209*, 933–937.

172. Spencer, M.E.; Hodge, R. Cloning and sequencing of the complementary DNA encoding the major albumin of *Theobroma cacao*: identification of the protein as a member of the Kunitz protease inhibitor family. Planta **1991**, *183*, 528–535.

173. Suzuki, K.; Ishimoto, M.; Kitamura, K. cDNA sequence and deduced primary structure of an α-amylase inhibitor from a bruchid resistance wild common bean. Biochem. Biophys. Acta **1994**, *1206*, 289–191.

174. Liu, J.H.; Hill, R.D.; Liu, J.H. Post transcriptional regulation of bifunctional α-amylase/subtilisin inhibitor expression in barley embryos by abscisic acid. Plant Mol. Biol. **1995**, *29*, 1087–1091.

175. Grosset, J.; Alary, R.; Gautier, M.F.; Menossi, M.; Martinez-Izquierdo, J.A.; Jourdrier, P. Characterisation of barley gene coding for an amylase inhibitor

subunit (CMd protein) and analysis of its promoter in transgenic tobacco plants and in maize kernels by microprojectile bombardment. Plant Mol. Biol. **1997**, *34*, 331–338.

176. Jones, M.E.; Vickers, J.E.; De-Jersey, J.; Henry, R.J.; Simons, M.H.; Marschke, R.J. Bacterial expression of the bifunctional α-amylase/subtilisin inhibitor from barley. J. Inst. Brew. **1997**, *103*, 31–33.

177. Sparvoli, F.; Gallo, A.; Marinelli, D.; Santucci, A.; Bollini, R. Novel lectin related proteins are major components in lima bean (*Phaseolus lunatus*) seeds. Biochim. Biophys. Acta **1998**, *1382*, 311–323.

178. Shade, R.E.; Schroeder, H.E.; Peuyo, J.J.; Tabe, L.M.; Murdock L.L.; Higgins, T.J.V.; Chrispeels, M.J. Transgenic pea seeds expressing the α-amylase inhibitor of the common bean are resistant to bruchid beetles. Biotechnol. **1994**, *13*, 793–796.

179. Gatehouse, A.M.R.; Down, R.E.; Powell, K.S.; Sauvion, N.; Rahbe, Y.; Newell, C.A.; Merryweather, A.; Hamilton, W.D.O.; Gatehouse, J.A. Transgenic potato plants with enhanced resistance to the peach potato aphid *Myzus persicae*. Entomol. Exp. Appl. **1996**, *79*, 295–307.

180. Masoud, S.A.; Ding, X.; Johnson, L.B.; White, F.F.; Reeck, G.R. Expression of a corn bifunctional inhibitor of serine proteinases and insect α-amylase in transgenic tobacco plants. Plant Sci. **1996**, *115*, 59–69.

181. Titaranko, E.; Chrispeels, M.J. cDNA cloning, biochemical characterization and inhibition by plant inhibitors of the amylases of the Western corn rootworm *Diabrotica virgifera virgifera*. Insect Biochem. Mol. Biol. **2000**, *30*, 979–990.

182. Hoffmann, M.P.; Zalom, F.G.; Wilson, L.T.; Smilanick, J.M.; Malyi, L.D.; Kiser, J.; Hilder, V.A.; Barnes, W.M. Field evaluation of transgenic tobacco containing genes encoding *Bacillus thuringiensis* delta-endotoxin or cowpea trypsin inhibitor efficacy against *Helicoverpa zea* (Lepidoptera: Noctuidae). J. Econ. Entomol. **1992**, *85*, 2516–2522.

183. Hilder, V.A.; Gatehouse, A.M.R.; Sheerman, S.E.; Barker, R.F.; Boulter, D. A novel mechanism of insect resistance engineered into tobacco. Nature **1987**, *330*, 160–163.

184. Jongsma, M.A.; Bakker, P.L.; Peters, J.; Bosch, D.; Stiekma, W.J. Adaptation to Spodoptera exigua larvae to plant proteinase inhibitors by induction of gut proteinase activity insensitive to inhibition. Proc. Natl. Acad. Sci. USA **1995**, *92*, 8041–8045.

185. Perez, C.J.; Shelton, A.M. Resistance of *Plutella xylostella* (Lepidoptera: Plutellidae) to *Bacillus thuringiensis* Berliner in Central America. J. Econ. Entomol. **1997**, *90*, 87–93.

186. Girard, C.; Le Metayer, M.; Bonade-Bottino, M.; Pham-Delegue, M.H.; Jouanin, L. High level of resistance to proteinase inhibitors may be conferred by proteolytic cleavage in beetle larvae. Insect Biochem. Mol. Biol. **1998**, *28*, 229–237.

187. Hyde, J.; Martin, M.; Preckel, P.V.; Dobbins, C.L. Edwards, C.R. An economic analysis from non-Bt corn refuges. Crop Prot. **2001**, *20*, 167–171.

6

Cyanogenic Glycosides

6.1 INTRODUCTION

Cyanogenic glycosides (CGs) are nitrogen-containing secondary metabolites whose biosynthetic origin is similar to that of glucosinolates. They are intermediately polar, water-soluble compounds that often accumulate in the vacuoles of plant cells (1–13). These compounds are usually O-β-glycosides of cyanohydrins (α-hydroxynitriles). They are optically active because of chirality of the hydroxylated C-atom 2, which occurs in either the S or the R configuration. Cyanogenic glycosides are formed in the cytoplasm but stored in the central vacuole. Storage of cyanogenic glycosides is tissue specific and takes place in epidermal vacuoles in sorghum. The enzymes that degrade cyanogenic glycosides, namely, β-glucosidase and hydroxynitrile lyase, are present in the adjacent mesophyll cells, which is safely away from the cyanogenic glycosides. More than 75 different cyanogenic glycosides occur among plants. A closely related nitrile glycosides and cyanolipids also occur in plants, but their distribution is restricted. Cyanolipids are found only in Sapindaceae and Hippocastanaceae (14). Several reviews on cyanogenic compounds have been published (1–13). All plants produce cyanide; however, in most cases cyanide is present in extremely small quantities. A level of 10 mg HCN per kg is the minimum for a plant to be considered cyanogenic. Cyanogenic glycosides are stored in plant cell vacuoles and enzyme in the cytosol. When the plant tissues are damaged due to herbivory, trampling, intense heat, or frost, cyanide release occurs. Since nonruminants have more acidic stomach, they are not sensitive to cyanogenic glycosides. Tiger beetles, millipedes, and centipedes are able to use cyanogenic glycosides as a defense against predators. These insects sequester the cyanide in their cells. When an insect is attacked by a predator, cyanide is released as a defense.

It is estimated that between 3000 and 12,000 plant species produce cyanogenic glycosides. Many important crop plants are cyanogenic, including sorghum, almond, lima bean, and white clover. Many bacteria and fungi are cyanogenic. The cyanogenic compounds of these organisms are usually labile. Cyanogenic glycosides are present in food and forage plants such as cassava and sorghum.

6.2   OCCURRENCE

Cyanogenic glycosides have been reported from at least 2650 plants belonging to more than 550 genera and 130 families. Cyanogenic plant species include ferns, gymnosperms, monocotyledonous and dicotyledonous plants (15–39). Despite the widespread occurrence of cyanogenesis in plants, the compounds responsible for producing HCN have been isolated and identified only from limited species. The important families in which cyanogenic glycosides commonly occur are Araceae, Asteraceae, Euphorbiaceae, Fabaceae, Flacourtiaceae, Maleesherbia, Proteaceae, Rosaceae, Sapindaceae, Turneraceae, Compositae, Gramineae, and Leguminoseae (1–13) The UN Food and Agriculture Organisation has estimated that 23 crops produced in the largest tonnage worldwide are listed as cyanogenic. Cyanogenesis is not known to occur in the genera *Lycopersicon esculentum* Miller (tomato), *Helianthus annus* L. (sunflower), and *Ananas comosus* Miller (pineapple). Major food plants in which cyanogenic glycosides have been identified are listed in Table 6.1. Plants reported to be cyanogenic are shown in Table 6.2. The compounds responsible for cyanogenesis have been isolated from about 475 species. In some instances different glycosides occurred in different plant parts. For example, prunasin is found in the vegetative portions and amygdalin is found in the seeds of *Prunus* species. A number of cases have been reported in which several cyanogenic glycosides co-occur in the same plant tissues. Both mono- and diglycosides co-occur in flax (*Linum usitatissimum*) and in rubber (*Hevea brasiliensis*) (23,30). Complex mixtures of di-, tri-, and tetraglycosides occur in members of the family Asteraceae.

The linamarin and lotaustralin have a relatively broad distribution in the plant kingdom, having been demonstrated in the following plant

TABLE 6.1   Food Plants Containing Cyanogenic Glycosides

Crop	Scientific name	Cyanogenic glycosides
Wheat	*Triticum aestivum*	Dhurrin
	T. monococcum	Linamarin, lotaustralin, epilotaustralin
Sorghum	*Sorghum bicolor*	Dhurrin
Cassava	*Manicot esculenta*	Linamarin, lotaustralin
Lima bean	*Phaseolus lunatus*	Linamarin, lotaustralin
French bean	*P. vulgaris*	Linamarin, lotaustralin
Ragi	*Elusine coracana*	Triglochinin
Barley	*Hordeum vulgare*	Epiheterodendrin
Oat	*Avena sativa*	Linamarin
Rye	*Secale cereale*	Dhurrin
Apple	*Malus pimila*	Amygdalin, prunasin
Taro	*Colocasia esculenta*	Triglochinin
Almond	*Prunus dulcis*	Amygdalin
Peach	*P. persica*	Amygdalin
Sweet cherry	*P. avium*	Amygdalin, prunasin
Sour cherry	*P. cerasus*	Amygdalin, prunasin
Papaya	*Carica papaya*	Tetraphyllin, prunasin
Passion fruit	*Passiflora edulis*	Prunasin
Sapote	*Pouteria sapota*	Lucumin
Bamboo shoots	*Bombusa vulgaris*	Taxiphyllin
Giant taro	*Alocasia macrorhiza*	Isotriglochinin, triglochinin

families: Compositae, Euphorbiacae, Linaceae, Papaveraceae, and Fabaceae (Leguminosae). A similar wide distribution has been observed for prunasin in six families (Polypodiasaceae, Myrtaceae, Rosaceae, Saxifragaceae, Scrophulariaceae, and Myoporaceae). Sambunigrin, vicianin, amygdalin, all of which are closely related to prunasin, have been demonstrated in three families (Caprifoliaceae, Mimosaceae, Oleaceae), two families (Polypodiaceae, Fabaceae), and one family (Rosaceae), respectively. The more common distribution pattern is that a particular cyanogenic compound will occur in one or two families (40). Conversely, it is generally true that, with few exceptions, only one or two characteristic glycosides will occur in a given plant family (Poaceae: dhurrin; Compositae: linamarin; Polypodiaceae: prunasin and vicianin; Rosaceae: amygdalin and prunasin). The cyanogenic compounds of plants belong undoubtedly to secondary plant metabolites that have or can have a chemotaxonomic character. The majority of these families belong to the Angiospermatophyta, but there are some exceptions (Polypodiaceae/Pteridophyta, Taxaceae/Gymnospermatophyta). Both the class Dicotyledonopsida and Monocotyledonopsida have plant families with

TABLE 6.2 Classification of Cyanogenic Glycosides Based on Biosynthetic Origin

Biosynthetic precursor	Cyanogenic glycoside	Family
a. Phenylalanine	1. Amygdalin	Rosaceae
	2. Prunasin	Compositae, Leguminosae, Caprifoliaceae, Myrtaceae, Rosaceae
	3. Samunigrin	Caprifoliaceae, Leguminosae
	4. Lucumin	Sapotaceae
	5. Vicianin	Leguminosae, Polypodiaceae, Fabaceae
	6. Epilucumin	Asteraceae
	7. Perilla glycoside	Lamiaceae
	8. Anthemis glycoside A	Asteraceae, Fabaceae, Rosaceae
	9. Anthemis glycoside B	Asteraceae, Fabaceae, Rosaceae
	10. Holacalin B	Leguminosae, Liliaceae
	11. Zierin	Caprifoliaceae, Rutaceae
b. Tyrosine	1. Dhurrin	Gramineae, Proteaceae
	2. Nandinin	Berberidaceae
	3. Triglochinin	Araceae, Liliaceae, Magnoliaceae, Platanaceae
	4. Taxiphyllin	Graminae, Taxaceae, Euphorbiaceae, Cupressaceae
	5. 4-Gluosyloxymandelo-nitrile	Leguminosae, Berberidaceae
	6. Proteacin	Proteaceae, Ranunculaceae
c. Leucine	1. Cardiospermin	Sapindaceae
	2. Heterodendrin	Leguminosae, Sapindaceae
	3. Epiheterodendrin	Sapindaceae, Poaceae
	4. 3-Hydroxyheterodendrin	Sapindaceae
	5. Proacacipetalin	Leguminosae
	6. Epiproacacipetalin	Leguminosae
d. Isoleucine, valine	1. Linamarin	Compositae, Euphorbiaceae, Leguminosae, Liliaceae, Papaveraceae
	2. Lotaustralin	Compositae, Euphorbiaceae, Liliaceae, Papaveraceae
	3. Epilotaustralin	Poaceae
	4. Linustatin	Passifloraceae
	5. Neolinustatin	Passifloraceae
	6. Sarmentosin epoxide	Crassulaceae

(*continued*)

TABLE 6.2   (Continued)

Biosynthetic precursor	Cyanogenic glycoside	Family
e. 2-(2-Cyclopentenyl) glycine	1. Tetraphyllin A	Passifloraceae
	2. Tetraphyllin B	Passifloraceae
	3. Tetraphyllin B sulfate	Passifloraceae, Turneraceae
	4. Tetraktophyllin	Flacourtiaceae
	5. Eqivolkenin	Flacourtiaceae
	6. Volkenin	Flacourtiaceae
	7. Deidaclin	Passifloraceae
	8. Gynocardin	Flacourtiaceae
	9. Passisuberosin	Flacourtiaceae
	10. 6'-O-rhamnopyranosyl tetraktophyllin	Flacourtiaceae
	11. Passicapsin	Passifloraceae
	12. Passibiflorin	Passifloraceae
	13. Passitrifasciatin	Passifloraceae
f. Leucine and fatty acids	1. 1-Cyano-2-methyl-prop-2-en-1-ol ester	Sapindaceae, Boraginaceae
	2. 1-Cyano-2-hydroxymethyl prop-2-en-1-ol diester	Sapindaceae, Boraginaceae
g. Nicotinic acid	1. Acalypin	Euphorbiaceae
h. Nitroacids, nitroalcohols	1. 3-Nitropropionic acid	Fabaceae
	2. Cibarian	Violaceae
	3. Coronarian	Malpighiaceae
i. Nitrile glycoside	1. Simmondsin	Crassulaceae
	2. Bauhinin	Simmondsiaceae
	3. Lithosperoside	Simmondsiaceae, Aquifoliaceae

cyanogenic compounds, but most families belong to the dicots. The families Saxifragaceae, Rosaceae, Minsaceae, Fabaceae, Myrtaceae, Linaceae, and Euphorbiaceae are in the subclass Rosidae; other cyanogenous families are in subclasses Ranunculudae (family Papaveraceae), Lamiidae (families Caprifoliaceae, Rosaceae, Oleaceae), and Asteridae (family Compositae). The occurrence or omission of the cyanogenic compounds has probable chemotaxonomic importance too, but these relations have not been sufficiently documented or investigated.

CGs have been reported from many members of the three subfamilies of Fabaceae, but some of these reports should be reconfirmed (41).

Cyanogenic members of the Papilionideae have been reported from 18 tribes (the ability to produce HCN upon hydrolysis), but the compounds responsible have been isolated from only a few of these tribes. Linamarin and lotaustralin have been found in many species of the tribes Loteae and Trifoliae, and probably occur in the Coronilleae. Vacianin occurs in seeds of several *Vicia* species. At least one member of the Crotalarieae appears to contain prunasin. Cyanogenic compounds from the Galegeae have been reported, but their chemical nature is unclear. Species of the Indigofereae appear to contain prunasin or sambunigrin, whereas those of the Phaseoleae contain linamarin and lotaustralin.

The new, more developed analytical methods have been now used to isolate and identify new cyanogenic compounds, or new derivates of known compounds. The 6-trans-2-butenoyl ester of prunasin was isolated as a new CG from *Centaurea aspear* var. *subinermis* (42), the purshianin from the *Purshia tridentata* (Rosaceae), and the multifidin from the latex of *Jatropa multifida* (Euphorbiaceae) (43). Another new CG (esterified with iridoid glycoside) was isolated from *Canthium schimperianum* (44). In other cases, CGs (or HCN) were identified and determined from plant species or varieties: in seeds of *Hevea brasiliensis*, in leaves and callus cultures of *Schlecterina mitostemmatoides*/Passifloraceae (45), in different morphological parts of *Moringa oleifera* (Moringaceae) (46) and in some legumes, grasses, and other plants (15). The common food plants containing cyanogenic glycosides are shown in Table 6.1.

6.3 CHEMISTRY AND CLASSIFICATION

Cyanogenic glycosides have the general structure (Fig. 6.1) with α-hydroxynitriles as their aglycone. The sugar moiety is usually D-glucose linked by an O-β-glucosyl bond. The important exceptions are amygdalin, vicianin, lucumin, linustatin, and neolinustatin, which have disaccharides as their sugar component. R1 and R2 in the general structure shown may be aliphatic or aromatic substituents (or hydrogen). The carbinol carbon of the aglycone is usually chiral since R1 and R2 are not identical. This introduces the possibility of epimeric compounds. Few important and commonly occurring cyanogenic glycosides are shown in Fig. 6.1.

6.3.1 Classification

Cyanogenic glycosides are classified according to the chemical nature of their aglycones, such as aliphatic, aromatic, or alicyclic, but the most preferred classification is based on the biosynthetic origin of the glycone.

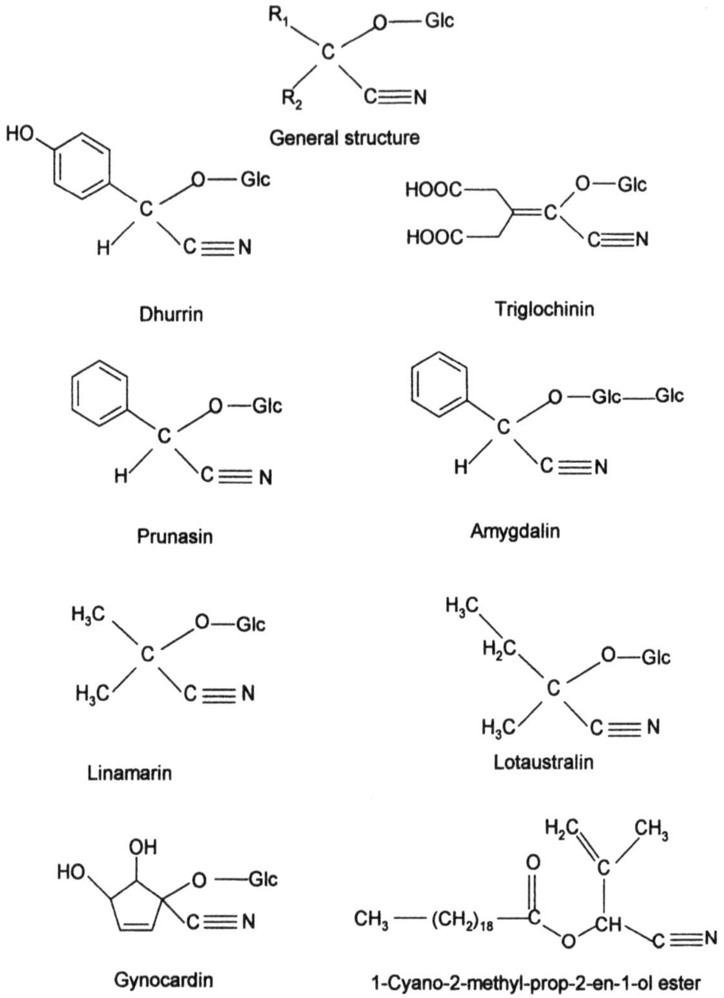

FIGURE 6.1   Structure of some important cyanogenic glycosides.

Most cyanogenic glycosides are derived from five protein amino acids, others from the nonprotein amino acid 2-(2-cyclopentenyl)glycine and from nicotinic acid (Table 6.2). The important classes of cyanogenic glycosides include those derived from

  1. Phenylalanine
  2. Tyrosine
  3. Leucine

4. Valine and isoleucine
5. Cyclopentenylglycine
6. Leucine and fatty acids (cyanolipids)
7. Nicotinic acid
8. Nitroacids and nitroalcohol
9. Tyrosine nitrile glycosides

6.3.1.1   Cyanogens Derived from Phenylalanine

Several cyanogenic glycosides are derived from L-phenylalanine (Table 6.2). These cyanogens are found in the Rosidae and Asteridae. The important compound of this group is amygdalin, which is widespread in seeds of members of the Rosaceae, such as apples, peaches, cherries, and apricots. These glycosides include monoglycosides and the diglycosides. Complex glycosides containing several sugar moieties and *p*-hydroxycinnamate residues are found in the Asteraceae, Fabaceae, and Rosaceae (25). A quite unusual iridoid-monoterpene containing cyanogen is found in the family Rubiaceae.

6.3.1.2   Cyanogens Derived from Tyrosine

Cyanogens derived from tyrosine such as dhurrin, taxiphyllin, and triglochinin are widespread in nature. Tyrosine-derived glycosides most commonly occur in monocotyledonous angiosperms and in the Magnolidae. Dhurrin, the well-known cyanogenic glycoside, make up to 30% of the dry weight of the leaves and coleoptiles of etiolated sorghum seedlings.

6.3.1.3   Cyanogens Derived from Leucine

Several cyanogenic glycosides arising from leucine have restricted distribution. Glycosides such as proacacipetalin, epiproacacipetalin, and pro-acaciberin are known only from the genus *Acacia*. Heterodendrin and epiheterodendrin occur only in the Sapindaceae and in the Poaceae (grasses) (25).

6.3.1.4   Cyanogens Derived from Valine and Isoleucine

The most widespread cyanogenic glycosides in this group are linamarin and lotaustralin. Including the other less common glycosides epilotaustralin, linustatin, neolinustatin, and sarmentosin epoxide, which are mostly encountered in the species of the Asteraceae, Euphorbiaciae, Fabaceae, and Linaceae (25,30). The gentibiose derivative linustatin and neolinustatin are found in flax seed and *Rossiflora* species. These compounds are transport forms in rubber tree and cassava.

6.3.1.5   Cyanogens Derived from Cyclopentenylglycine

The nonprotein amino acid 2-(2-cyclopentenyl)glycine yields several cyanogenic glycosides that contain a cyclopentenoid ring structure (21,23,25,33–
35). These compounds include decidaclin, tetraphyllin A, tetraphyllin B,
volkenin, gynocardin, and other cyanogens. These compounds occur in the
families Flacourtiaceae, Turneraceae, Passifloraceae, Malesherbiaceae, and
Achariaceae. *Passiflora* species contain cyclopentenoid glycosides as well as
those derived from valine and isoleucine.

6.3.1.6   Cyanolipids

Cyanolipids are derivatives of a leucine-derived cyanohydrin esterified
with a long-chain fatty acid found in the seed oils of several species of the
family Sapindaceae, with limited distribution in Hippocastanaceae and
Boraginaceae (4,11). The fruit of *Allophylus cobbe* contains the cyanogen
cardiospermin and mixture of cyanolipids. Cardiospermin is also found in
the extract of whole-insect *Leptocoris isolata.*

6.3.1.7   Cyanogens Derived from Nicotinic Acid

The cyanogenic lipid acalyphin from *Acalypha indica* from Euphorbiaceae is
derived from nicotinic acid (25).

6.3.1.8   Nitroglycosides

Several plant families contain the glycoside of nitroacids and nitroalcohol
(5). Several esters of 3-nitropropionic acid occur in the genera *Astragalus,
Coronilla, Indigofera* (Fabaceae), *Heteropteris, Hiptage* (Malpighiaceae),
Viola (Violaceae), and *Corynocarpus* (Corynocarpaceae). The 3-nitropro-
pionic acid is derived from aspartic acid. The carbon–nitrogen bond of
aspartic acid is preserved during the formation. Few poisonous compounds
to several insect and other animals are derived from 3-nitropropanol.

6.3.1.9   Nitrile Glucosides

Nitrile glucosides are noncyanogenic compounds having structural similarities to intermediates in cyanogenic glycoside biosynthesis. These compounds are found in members of the Simmondsiaceae, Aquifoliaceae,
Menispemaceae, Fabaceae, Boraginaceae, Ranunculaceae, and Crassula-
ceae families (5).

6.4   BIOSYNTHESIS

The pathway for the different cyanogenic glycosides is thought to follow a
common biosynthetic scheme (47–71). Biosynthetic studies with radioactively

labeled precursors and trapping experiments in which unlabeled putative intermediates were included in the microsomal reaction mixtures identified N-hydroxyamino acids, N,N-dihydroxyamino acid, aldoximes, nitriles, and cyanohydrins as key intermediates (Scheme 6.1). The biosynthesis of the L-tyrosine-derived cyanogenic glycoside dhurrin has been elucidated in sorghum [*Sorghum bicolor* (L.) Moench] (54–56,59,61–63). Studies carried out using microsomes prepared from etiolated cassava seedlings demonstrate the involvement of the same classes of intermediates as in sorghum (52,60,66,70). All the intermediate compounds except N,N-dihydroxyamino acid have been chemically synthesized and are metabolized by the microsomal system. The extreme lability of the N,N-dihydroxyamino acid makes it impossible to investigate directly since it can be neither chemically synthesized nor isolated. Microsomal activity is dependent on the presence of oxygen and NADPH. The conversion of tyrosine to the p-hydroxymandelonitrile proceeds with the consumption of three molecules of oxygen, indicating the involvement of three hydroxylation reactions. Two molecules of oxygen are consumed in the conversion of tyrosine to the aldoxime, whereas a single oxygen molecule is consumed in the conversion of the aldoxime to the cyanohydrin (Scheme 6.1). Biosynthetic experiments using stable isotopes have also helped elucidate the nature of the intermediates involved. If the hydrogen atom at the α-carbon atom of the amino acid is labeled with deuterium, it is conserved in the aldoxime. Biosynthetic experiments with $^{18}O_2$ also revealed that the aldoxime was labeled with ^{18}O in the hydroxylamine function.

The whole pathway from the parent amino acid to the cyanogenic glycoside is catalyzed by three enzymes. Two are multifunctional cytochrome P450s and the third a UDPG-glucosyltransferase. The first committed steps in the biosynthesis of cyanogenic glycosides are the conversion of amino acids to the corresponding oximes. A multifunctional cytochrome P450 enzyme, designated as P450tyr, has been isolated from etiolated sorghum seedlings. It catalyzes the conversion of tyrosine to (Z) p-hydroxyphenylacetaldoxime. Cytochrome P450 from cassava (*Manihot esculenta* Crantz) catalyzing the first steps in the biosynthesis of the linamarin and lotaustralin has been isolated.

A cytochrome P450, designated P450ox, that catalyzes the conversion of (Z)-P-hydroxyphenylacetaldoxime to p-hydroxymandelonitrile in the biosynthesis of the cyanogenic glycoside dhurrin has been isolated from microsomes prepared from etiolated seedlings of sorghum [*Sorghum bicolor* (L.) Moench] (68). P450ox is multifunctional, catalyzing the dehydration of (Z)-oxime to p-hydroxyphenylacetonitrile and C-hydroxylation of p-hydroxyphenylacetonitrile to nitrile. P450ox is extremely labile compared with the P450s previously isolated from sorghum (68).

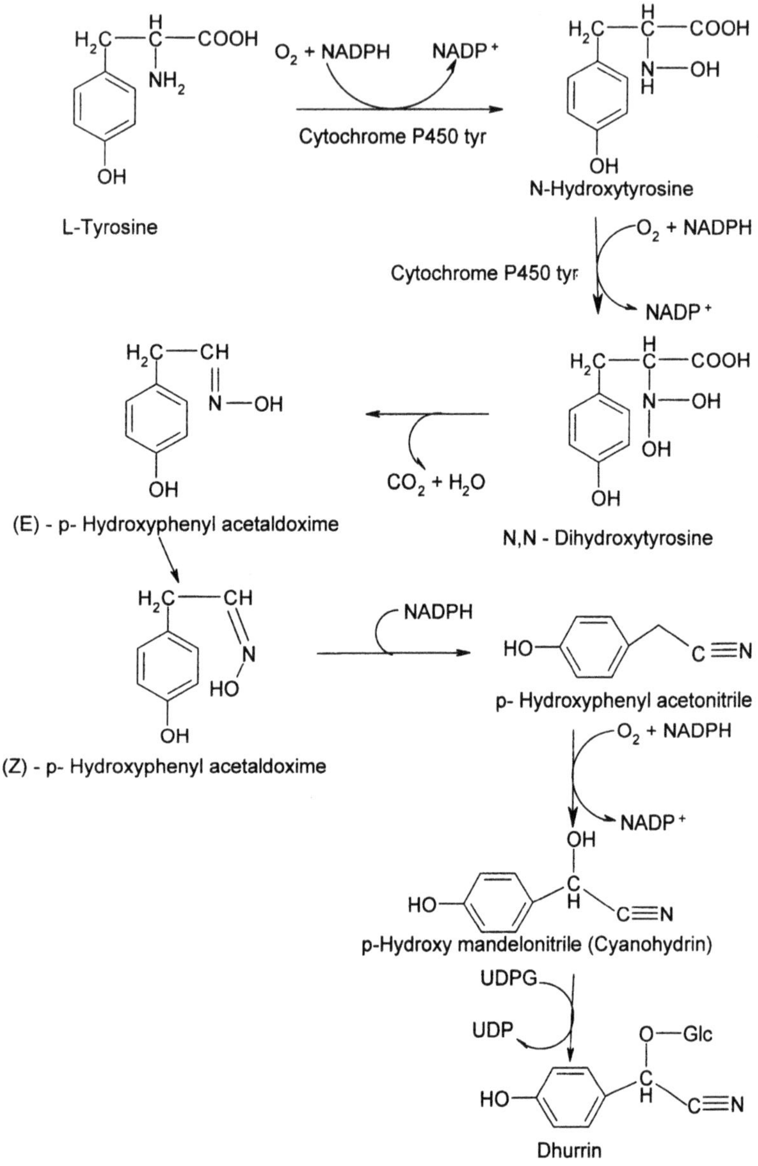

SCHEME 6.1 The biosynthetic pathway of dhurrin in sorghum.

The final step in the biosynthesis of dhurrin in sorghum is transformation of labile cyanohydrin into a stable storage form by glucosylation of (S)-p-hydroxymandelonitrile at the cyanohydrin function. The UDP-glucose:p-hydroxymandelonitrile-O-glucosyltransferase was isolated from etiolated seedlings of sorghum (63,69). Glucosyltransferases have been isolated and purified from a few cyanogenic species (69). They have an absolute specificity for UDP glucose and are less specific toward cyanohydrins.

6.5 BIOACTIVITY

One of the most probable explanations of the biological role of cyanogenic glycosides in some plants is the participation in defense mechanisms against different phytopathogens, despite the fact that both plants and animals possess the ability to detoxify cyanide. So the presence of CGs in plants is not necessarily inimical to herbivory. Larvae of *Spodoptera eridania* prefer to graze on CG-containing plants and grow better when cyanide is present in their diet (72). Studies have shown that CGs can act as feeding deterrents or phagostimulants, depending on the insect species (73). Malagon and Garrido (74) concluded that bitter almond plants are resistant to larvae of buprestid *Capnodis tenebrionis*, owing to the high concentration of CGs. Fourteen sorghum varieties were investigated for their susceptibility to fly species *Atherigona soccata* and *Chilo partellus* (75). The CG dhurrin was in greater quantities in susceptible cutivars CSH-1, Swarna, and IS10795, and it was suggested that dhurrin acts as an oviposition activator for the pests. Clear examples of the protective effect of cyanogenesis do exist, even against apparently specialist pests. Thus, the CG of cassava tubers improved its resistance to the cassava root borer (76).

Both cyanide and the aglycones resulting from the hydrolysis of cyanogenic glycosides are toxic to nonadapted herbivores. Cyanide release from dhurrin of field-grown sorghum is a potential plant defense compound against acridid grasshoppers in West Africa and India (5).

Amygdalin has been reported to serve as a phagostimulant for *Malacosoma americana* larvae. The normal host of the insect, black cherry (*Prunus serotina*), is strongly cyanogenic. The cyanogenic compound present in it is prunasin. Larval feeding was extensive when the level of glycosides dropped well below the maximum observed in immature leaves. The moth *Yponomeuta evonymellus* does not respond to prunasin, but other members of this genus are both sensitive to the presence of prunasin and deterred from feeding on the plant (77). The armyworm *Spodoptera exempta*, feeds preferentially on *Cynodon plectostachya* (Poaceae). The level of cyanide in the leaves of the grass, especially after defoliation by armyworm larvae, were high enough to kill cattle, although the larvae themselves appeared unaffected.

Although it is generally believed that the major source of the toxicity of cyanogenic compounds is HCN, attention should also be given to the aldehydes and ketones that are simultaneously released (78). Acetone and butanone released by hydrolysis of linamarin and lotaustralin are actual deterrents to feeding on *L. corniculatus* and *T. ripens* by molluscs. There is negative correlation of cyanogenic glycosides (79). Cyanolipids are toxic in the diet of *Callosobruchus maculatus* at both the 1% and 5% (w/w) level and were toxic when incorporated into the larval diet of the European corn borer, *Ostrinia nubilalis*. Contact with oils containing cyanolipids temporarily paralyzed grain beetles, *Oryzaephilus surinamensis* (80). A number of esters of 3-nitropropionic acid occur in members of the genera *Astragalus, Coronilla*, and others. 3-Nitropropionic acid is toxic to the bruchid *C. maculatus* at 0.1% (w/w) when incorporated into artificial diets. 3-Nitropropanol compounds are poisonous to livestock and to several insects species (81).

The relationship between cyanogenesis in bracken fern (*Pteridium aquilinum*) and the insect fauna feeding on the plant was investigated over a 3-year period (82,83). The most common insects between May and July, when cyanide levels were high, were the sawflies *Strongylogaster impressata* Provancher, *S. multicincta* Norton, *Aneugmenus flavipes* (Norton), the aphid *Macrosiphum euphorbiae* (Thomas), and a microlepidopteran species of *Monochroa*. Collections of insects from cyanogenic and acyanogenic fronds showed significantly fewer sawflies on the cyanogenic fronds. The aphid and the microlepidopteran were randomly distributed with respect to cyanogenicity. Feeding tests for two of the sawfly species showed that larvae grew more slowly and had a higher mortality when raised on cyanogenic fronds than on acyanogenic ones. Field-collected cyanogenic bracken fronds were found to have sustained less damage from chewing herbivores compared with acyanogenic fronds. The cyanoganic glycoside present in bracken fern is prunasin. It was shown that prunasin was at its highest concentration when the fronds are young, declining to low levels as the season progresses and the fronds mature.

The concentration of prunasin in bracken also varies between clones and habitats. Bracken is polymorphic for the ability to produce cyanide when its tissues are damaged. Lawton (84) examined the diversity and abundance of arthropods on bracken and found that both diversity and abundance increased over the season.

Numerous studies have evaluated the impact of specific polyphagous herbivorous species on cyanogenic as compared with acyanogenic phenotypes of plants. Grasshoppers, aphids, various slugs and snails, and deer have been found to prefer acyanogenic as compared with cyanogenic phenotypes of a given plant species (85).

Sorghum leaves contain the cyanoglycoside dhurrin located in the cell vacuole. The degree of liberation of HCN in leaves is dependent on the age and variety of plant, as well as on certain environmental factors. Leaves of young sorghum are often rejected at the first bite by the graminivorous locust *Locusta migratoria*, but other sorghum is eaten in large quantity (86). The change in palatability is related to the rate at which HCN is released from the leaf at the time of biting. Concentration as low as 0.01 mM was sufficient to inhibit feeding. The amount of HCN initially released from young sorghum (8 days old) was calculated to be equivalent to concentration to a solution between 0.5 mM and 1.0 mM hydrocyanic acid and thus would be effective as a deterrent.

Invertebrate herbivores avoid feedings or feed at reduced levels on cyanogenic white clovers (87). Whitman (88) also found that damage by unspecified herbivores to white clover was lowest on cyanogenic types in unmanaged habitats. Mowat and Shakeel (89) associated reduced density of larvae and eggs of root-feeding weevils, *Sitona spp.* and lower leaf damage by adult weevils with high levels of cyanogenesis in white clover. Insect response to cyanogenic white clover may vary seasonally. Raffaelli and Mordue (90) observed preferred feeding on acyanogenic types during late summer and fall and on cyanogenic types during spring. Alfalfa weevils *Hypera postica* (Gyllenhal) were the dominated herbivorous insect species present in sweep-net collections from damaged white clover. Damage by alfalfa weevil larvae appeared consistently lower on 'Louisiana S-1', a cultivar known to be cyanogenic, compared with other cultivars. Ellsbury et al. (87) also observed that the cyanogenic cultivar Louisiana S-1 was least damaged among eight cultivars rated visually for weevil feeding in naturally infested plots. Variation in leaf damage also was observed among 104 white clover plant introduction and cultivars rated for number of leaves showing insect damage in field plantings. Larvae of alfalfa weevils and adults of clover head weevils, *Hypera meles* (F.), exhibited preference for leaflets from acyanogenic genotypes offered in free-choice feeding tests. They suggested that levels of cyanogenic glycosides could be selected in white clover to improve resistance to insects. White clover contains linamarin or lotaustralin. Reduced weevil damage in Louisiana S-1 plots suggests either that weevils oviposited less in those plots or that some antibiotic effect, possibly from cyanogenesis, affected larvel populations after eggs were laid.

Evidence is presented that female *Euptoieta hegesia* do not show preference for host plants on the basis of their cyanogenesis concentration but do prefer *Turnera ulmifolia* over equally cyanogenic, closely related, secondary host plant species (*Passiflora* sp). Similarly, cyanogenesis in *T. ulmifolia* had little effect on the food preference, growth or development of the larvae (90). The potential host range of *E. hegesia* is limited, even within the genus *Turnera*, but this does not appear to be due to host plant cyanogenesis. Pupae suffer very high mortality levels in the wild that are not associated with host plant cyanogenesis although these studies indicated that larvae are capable of sequestering cyanogenic glycosides from their host plants and possibly of synthesizing these or similar compounds. Evidence is presented that the presence of sequestered cyanogenic glycosides in the larvae protects them from terrestrial-based predators, such as *Anolis* lizards.

Sudan grass, such as common sorghum (*Sorghum bicolor*), contains the glycoside dhurrin, which can be hydrolyzed to yield cyanide. Because activity of fractionated Sudan grass extracts against *Meloidogyne hapla* is associated with the presence of cyanide in the fractions, it seems likely that dhurrin is involved in the mode of action of Sudan grass on *M. halpa* (91).

Cassava roots (*Manihot esculenta*, Euphorbiaceae) contain various quantities of cyanogenic glucosides, particularly linamarin. When cells are damaged, enzymes hydrolyze these glucosides, releasing cyanide via cyanohydrin intermediates. Manipueira, a liquid formed during processing of cassava roots, has been utilized for nematode control for decades in Brazil.

6.6   MECHANISM OF ACTION

Plants containing cyanogenic glycosides usually contain β-glycosidases capable of removing glucose or sugar from cyanogenic glycoside to produce a cyanohydrin (aglycone). A second enzyme, hydroxynitrile lyase, catalyzes the dissociation of the cyanohydrin to a carbonyl compound (an aldehyde or a ketone) and hydrogen cyanide (92,93) (Scheme 6.2). Naturally the substrate and enzymes are compartmentalized within the plant and the cyanide release occurs only when the plant is damaged.

The HCN produced is a highly toxic substance that inhibits cytochrome oxidase and other respiratory enzymes. Studies have shown that cyanogenic glycosides can act as either feeding deterrents or phagostimulants, depending on the insect species. It has also been shown that dhurrin, a cyanogenic glycoside, act as an oviposition activator for the pests.

Some insects produce cyanogenic glycosides de novo while a few other insects sequester cyanogenic compounds from plants (94) Some insects feed on cyanogenic plants but contain cyanogens different from those of the host plant. Many insects possess significant levels of 3-cyanoalanine synthase, an enzyme capable of converting HCN and cysteine to 3-cyanoalanine. Asparagine is further synthesized from 3-cyanoalanine catalyzed by

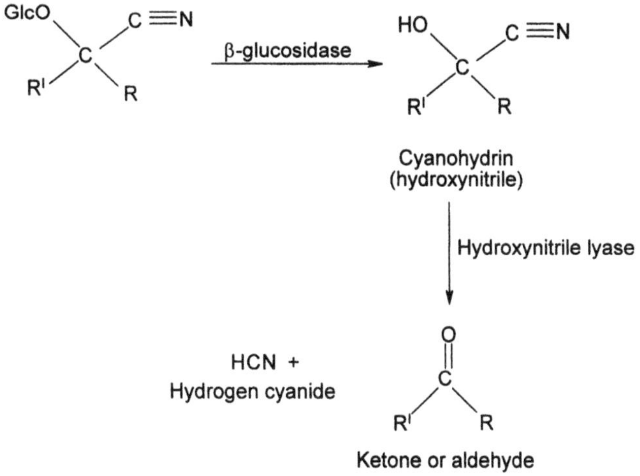

SCHEME 6.2   Hydrolysis of cyanogenic glycosides to hydrogen cyanide.

3-cyanoalanine hydrolase. 3-Cyanoalanine is found in several families of Lepidoptera. The genus *Heliconius*, which feeds on *Passiflora* species, has evolved alternate β-glycosidases that inhibit hydrolysis of the cyanogenic glycoside. Spencer proposed that β-glycosidases of insect origin bind with a plant glycosidase–CGs complex either during or after complex formation, in a competitive manner (78). Hence, many adapted insects use mechanisms such as sequesteration of CGs, detoxification of the HCN produced from CGs using 3-cyanoalanine synthase, or production of β-glycosidases that compete with the plant β-glycosidases.

6.7   MOLECULAR BIOLOGY

The biosynthesis from amino acids of the aglycone part of cyanogenic glucosides has been shown to involve P450 enzymes in *S. bicolor* (55), *M. esculenta* (60), and *Triglochin maritima* (67) (Scheme 6.3). In *S. bicolor*, two multifunctional P450 enzymes have been shown to catalyze all of the membrane-associated steps leading to the aglycone of dhurrin. Tyrosine is converted to (Z)-*p*-hydroxyphenylacetaldoxime by CYP79A1 (65,68), and (Z)-*p*-hydroxyphenylacealdoxime is subsequently converted to *p*-hydroxy-mandelonitrile by CYP71E1 (71). In *M. esculenta*, a CYP79 homologue, CYP79D1, has recently been identified and characterized (70). Also, the biosynthetic pathway of at least some glucosinolates involves P450 enzymes

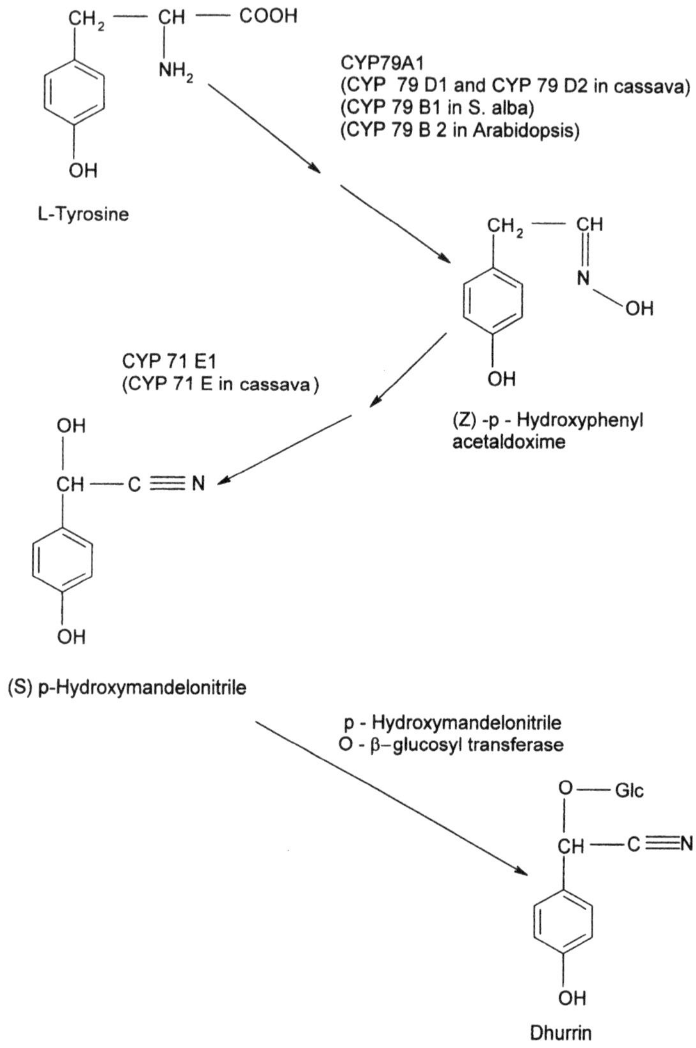

SCHEME 6.3 Molecular biology of the dhurrin biosynthetic pathway.

in the conversion of the parent amino acid to the corresponding aldoxime (66). CYP79 homologues have been identified in the two glucosinolate-producing plants, *S. alba* (CYP79B1) and *Arabidopsis* (CYP79B2), but so far no successful expression of the isolated clones has been achieved (95).

The CYP79B1 clone was isolated by screening an *S. alba* cDNA library with an expressed sequence tag (CYP79B2) probe from *Arabidopsis* (95). Both plants are members of the Brassicaceae family but contain glucosinolates derived from different substrates (52,66). Despite this difference, the two full-length clones showed 89% identity at the amino acid level. Accordingly, the first attempts to isolate a CYP79 homologue from *T. maritima* was carried out using heterologous probes from *S. bicolor* to take advantage of the fact that both plants are monocotyledonous and use Tyr as substrate. However, no CYP79 homologue could be isolated using this strategy. Instead, they used a polymerase chain reaction (PCR) approach, taking advantage of the presence of some highly conserved CYP79-specific regions covering part of the l-helix β-6-1 and β-1-4. These regions are involved in substrate recognition (SRS) as well as heme binding, and accordingly were thought to be less variable than other SRSs and therefore suitable for the design of degenerate primers. Conserved regions within groups and families of P450s have previously been used successfully to design degenerate primers for the amplification of specific P450 clones (95,96).

The major PCR product obtained using cDNA and genomic DNA revealed 62–70% identity to other members of the CYP79 family. This confirmed that the primers were highly specific toward the CYP79 family, as illustrated with primer 1F (DNPSNA) covering a region that in the vast majority of other P450s contains a highly conserved Thr residue instead of Asn. Only one of the degenerate primers covering FN (V/L) PHVA did not anneal perfectly with the sequences subsequently obtained from *T. maritima*, having a single nucleotide difference at the 5' end of the primer. Here the Ala codon in the primer was replaced by a Ser codon in the *T. maritima* sequences. The PCR fragment was used as a probe to screen the cDNA library made from *T. mariima* flowers and fruits. Microsomal preparations isolated from these tissues had previously been shown to have high catalytic activity compared with other tissues (51,67), thus raising the abundance of the putative mRNA. Two partial CYP79 homologues were isolated and used to isolate a full-length and a nearly full-length clone, designated CYP79E1 and CYP79E2, respectively, as the first members of a new subfamily. These clones show 94% identity. CYP79E1 showed 45–49% identity to other members of the CYP79 family. In sequence alignment to other members of the CYP79 family, CYP79E1 showed the highest identity and similarity to CYP79A1. In contrast, CYP79A1 showed higher identity/similarity toward the three other sequences—CYP79B1, CYP79B2, and CYP79D1—that were all isolated from dicotyledonous plants. This is surprising considering that CYP79B1 and CYP79B2 are putatively involved in the biosynthesis of glucosinolates, whereas CYP79D1 catalyzes the conversion of aliphatic amino acids. The very high degree of identity between the three clones involved in the biosynthesis of cyanogenic glucosides, explaining the lack of success using a heterologous probe from *S. bicolor* for screening the *T. maritima* library.

Cytochrome P450s are thought to possess the same tertiary structure (97,98); nevertheless, they exhibit only one highly conserved sequence, FXXGXRXCXG (X being any amino acid), harboring the heme-binding Cys residue. When restricted groups of families of P450s are aligned, additional conserved amino acid residues are found, such as the proposed highly conserved A-group heme-binding consensus sequence PFGXGRRXCXG (98) and the CYP79-specific sequence SFSTG (K/R) RGC (A/I) A (99). The CYP79 family diverges from the A-group heme-binding consensus sequence and, within the CYP79 family, CYP79E diverges in several positions from otherwise conserved CYP79 residues. A notable difference is that CYP79E1, CYP79E2, and CYP79A2 (GenEMBL No. AB010692 comp (11,000–13,200 region) contain the generally highly conserved Gly residues positioned two amino acids downstream from the heme-binding Cys residue (amino acid 479 in CYP79E1), whereas all the other CYP79s contain Ala at this position. This Gly residue is positioned closely to the heme plain and allows a sharp turn from the Cys pocket to the α helix (97).

In the highly conserved PERF region within microsomal P450s (448–451 in CYP79E1), the Phe residue has been replaced by His, as is also observed in other CYP79s (87). The K helix contains a partly conserved region, KETLR (392–396 in CYP79E1), in many P450s with the Glu and the Arg residues being particularly conserved (97). In the CYP79 family, Lys has been replaced by Arg, Thr has been replaced by Ala, and Leu has been replaced with Phe, with the exception that CYP79A2 still contains Lys and CYP79Es still contains Leu. Only minor differences are observed between the two different clones isolated from *T. maritima*. The region of positive charges preceding the Pro-rich region contains a repeat of 3x KS in CYP79E1, whereas CYP79E2 contains the sequence KPKS in the same region. Finally, the amino acids EGR (296–298 in CYP79E1) are deleted in CYP79E2, even though the Gly is conserved in all other CYP79s. The differences both within the CYP79 family and in relation to residues that are more or less conserved within the A group or microsomal P450s in general illustrate how difficult it is, based on sequence alignments, to predict amino acids important for different functions.

To verify the catalytic activity of the two CYP79 homologues from *T. maritima*, they were functionally expressed in *E. coli*. Codon usage in pro- and eukaryotic genes is different, and most highly expressed *E. coli* genes have a high AT content in the 5′ end of the gene. Therefore, modification of the 5′ end of most eukaryotic P450 genes is necessary to achieve high expression. Chimeric CYP79E2, containing the first 24 amino acids of the bacterial *lacZ* gene, showed a dramatically increased expression level and turnover of Tyr per milligram of protein compared with CYP79E1 expressed with a few silent mutations or in a truncated form containing a modified P4502E1 N terminus devoid of the membrane-spanning anchor. A partial CYP79E1 clone in frame with *lacZ* at the same nucleotide position as CYP79E2 showed the same high expression level at *lacZ*-modified CYP79E2, indicating the importance of the N-terminal sequence and choice of expression vector for efficient expression. Surprisingly, the CYP79E1 construct, containing a modified P45017 (N terminus with a stretch of 13 hydrophobic amino acids for membrane insertion) resulted in no detectable expression and no activity. The same type of construct [TYR(1–25)$_{bov}$] resulted in the highest level of expression of CYP79A1. All constructs showed higher expression level and activity when expressed in the *E. coli* strain JM109 compared with XL-1Blue.

Triglochin maritima contains the two cyanogenic glucosides triglochinin and taxiphyllin, which are both derived from Tyr. Triglochinin biosynthesis requires a ring cleavage step that supposed takes place after an additional hydroxylation of the phenolic side chain. To further study this pathway, the catalytic activities of the two isolated CYP79Es from *T. maritima* were tested using other putative substrates. Neither Phe nor dihydroxyphenylalanine (dopa) was metabolized by the P450s, which strongly suggests that the ring cleavage reaction leading to triglochinin formation proceeds after *p*-hydroxyphenylacetonitrile formation. The ring cleavage reaction could be dependent on hydroxylation carried out by a 2-oxoglutarate-dependent dioxygenase.

(Z)-*p*-Hydroxyphenylacetaldoxime is converted to *p*-hydroxymande-lonitrile by CYP71E1 in *S. bicolor* (99). Oximes have not been found to accumulate in cyanogenic plants. The membrane bound enzymes involved in the biosynthesis of cyanogenic glucosides are therefore believed to be closely associated. Indeed, reconstitution experiments including native or 2E1 constructs of CYP79E1 or the CYP71E1 from *S. bicolor*, revealed formation of considerable amounts of *p*-hydroxybenzaldehyde and the presence of low amounts of *p*-hydroxyphenylacetaldoxime. CYP71E1 is a labile enzyme that in its isolated state releases considerable amounts of *p*-hydroxyphenylaceto-nitrile from the active site before the subsequent C-hydroxylation reaction proceeds. Accordingly, *p*-hydroxyphenylacetonitrile is also observed to accumulate in the reaction mixtures, including the reaction mixture composed solely of *S. bicolor* CYP79A1 and CYP71E1.

In biosynthetic experiments using *S. bicolor* microsomes, no *p*-hydroxyphenylacetonitrile is observed while *p*-hydroxyacetonitrile accumu-lates in *T. maritima* (100). In the reconstitution experiments, the relative levels of *p*-hydroxyacetonitrile and *p*-hydroxybenzaldehyde were the same in all experiments, indicating that the activity of CYP71E1 is not differentially affected by the presence of CYP79A1 from *S. bicolor* or by native and truncated forms of CYP79E1 and CYP79E2 from *T. maritima*. This indicates that other membrane-associated regions of CYP79E1 are sufficient to retain correct orientation and association with CYP71E1. The efficient interaction between enzymes from *T. maritima* and *S. bicolor* indicates a high conservation of the enzymes involved in the biosynthesis of cyanogenic glucosides and suggests the presence of a CYP71E1 homology in *T. maritima*. *In vivo*, it is conceivable that, in contrast to CYP71E1 from *S. bicolor*, the CYP71E1 homologue present in *T. maritima* releases sufficient amounts of *p*-hydroxyphenylacetonitrile to permit the formation of triglochinin, the major cyanogenic glucoside in *T. maritima*.

Two cDNA clones encoding cytochrome P450 enzymes belonging to the CYP79 family have been isolated from *T. maritima* (100). The two proteins show 94% sequence identity and have been designated CYP79E1 and CYP79E2. Heterologous expression of the native and truncated forms of the two clones in *E. coli* demonstrated that both encode multifunctional *N*-hydroxylases catalyzing the conversion of tyrosine to *p*-hydroxyphenyl-acetaldoxime in the biosynthesis of the two cyanogenic glucosides taxi-phyllin and triglochinin in *T. maritima*. This renders CYP79E functionally identical to CYP79A1 from *S. bicolor* and unambiguously demostrates that cyanogenic glucoside biosynthesis in *T. maritima* and *S. bicolor* is catalyzed by analogous enzyme systems with *p*-hydroxyphenylacetaldoxime as a free intermediate. In *S. bicolor* CYP71E1 catalysed the subsequent conversion of *p*-hydroxyphenylacetaldoxime to *p*-hydroxymandelonitrile when CYP79E1 from *T. maritima* was reconstituted with CYP71E1 and NADPH-P450 oxidoreductase from *S. bicolor*, efficient conversion of tyrosine to *p*-hydroxymandelonitrile was observed.

Novel cyanogenic plants have been generated by the simultaneous expression of the two multifunctional sorghum [*Sorghum bicolor* (L.) Moench] P450 enzymes CYP79A1 and CYP71E1 in tobacco (*Nicotiana tabacum* cv Xanthi) and *Arabidopsis* under the regulation of the constitutive 35S promoter (101). CYP79A1 and CYP71E1 catalyze the conversion of the parent amino acid tyrosine to *p*-hydroxymandelonitrile, the aglycone of the cyanogenic glycoside dhurrin. CYP79A1 catalyzes the conversion of tyrosine to *p*-hydroxyphenylacetaldoxime and CYP71E1, the subsequent conversion to *p*-hydroxymandelonitrile. *p*-Hydroxymandelonitrile is labile and dissociates into *p*-hydroxybenzaldehyde and hydrogen cyanide, the same products released from dhurrin upon cell disruption as a result of pest or herbivore attack. In transgenic plants expressing CYP79A1 as well as CYP71E1, the activity of CYP79A1 is higher than that of CYP71E1, resulting in the accumulation of several p-hydroxyphenylacetaldoxime-derived products in addition to those derived from p-hydroxymandelonitrile. Transgenic tobacco and *Arabidopsis* plants expressing only CYP79A1 accumulate the same p-hydroxyphenylacetaldoxime-derived products as transgenic plants expressing both sorghum cytochrome P450 enzymes. In addition, the transgenic CYP79A1 *Arabidopsis* plants accumulate large amounts of p-hydroxybenzylglucosinolate. In transgenic *Arabidopsis* expressing CYP71E1, this enzyme and the enzymes of the preexisting glucosinolate pathway compete for the p-hydroxyphenylacetaldoxime as substrate, resulting in the formation of small amounts of p-hydroxybenzyl-glucosinolate. Cyanogenic glucosides are phytoanticipins, and the study demonstrates the feasibility of expressing cyanogenic compounds in new plant species by gene transfer technology to improve pest and disease resistance. The entire pathway for synthesis of the tyrosine-derived cyano-genic glucoside dhurrin has been transferred from *S. bicolor* to *Arabidopsis thaliana*. Tattersall et al. (102) documented that genetically engineered plants are able to synthesize and store large amounts of new natural products. The presence of dhurrin in the transgenic *A. thaliana* plants confers resistance to the flea beetle *Phyllotreta nemorum*, which is a natural pest of other members of the crucifer group, demonstrating the potential utility of cyanogenic glucoside in plant defense.

Sorghum has two isozymes of the cyanogenic β-glucosidase dhurri-nase-1 (Dhr-1) and dhurrinase-2 (Dhr-2). A nearly full-length cDNA encoding dhurrinase was isolated from 4-day-old etiolated seedlings and sequenced. The cDNA has a 1695-nucleotide-long open reading frame, which codes for a 565-amino-acid-long precursor and a 514-amino-acid-long mature protein, respectively (103). Deduced amino acid sequence of the sorghum Dhr showed 70% identity with two maize (*Zea mays*) β-glucosidase isozymes. Southern blot data suggested that β-glucosidase is enclosed by a small multigene family in sorghum. Nouthern blot data indicated that the mRNA corresponding to the cloned etiolated seedlings but at low levels in the root only in the zone of elongation and the tip region. Light grown seedling parts have lower levels of Dhr mRNA than those of etiolated seedlings. Immunoblot analysis performed using maize anti-β-glucosidase sera detected two distinct dhurrinases (57 and 62 kDa) in sorghum. The distribution of Dhr activity in different plant parts supports the mRNA and immunoreactive protein data, suggesting that the cloned cDNA corresponds to the Dhr-1 (57 kDa) isozyme and that the dhr-1 gene shows organ-specific expression.

A cDNA encoding CYP79B1 has been isolated from *Sinapis alba*. CYP79B1 from *S. alba* shows 54% sequence identity and 73% similarity to sorghum CYP79A1, and 95% sequence identity to the *Arabidopsis* T42902, assigned CYP79B2 (99). The high identity and similarity to sorghum CYP79A1, which catalyzes the conversion of tyrosine to p-hydroxyphenyl-acetaldoxime in the biosynthesis of the cyanogenic glucoside dhurrin, suggests that CYP79B1 similarly catalyzes the conversion of amino acid(s) to aldoxime(s) in the biosynthesis of glucosinolates. Within the highly conserved 'PERF' and the heme binding region of A-type cytochromes, the sequences are FXP(E/D)RH and SFSTG(K/R)RGC(A/T)A, respectively. Sequence analysis of PCR products generated with CYP79B subfamily-specific primers identified CYP79B homologues in *Tropaeolum majus*, *Carica papaya*, *Arabidopsis*, *Brassica napus*, and *S. alba*. The five glucosinolate-producing plants showed a CYP79B amino acid consensus sequence KPERHLNECSEVTLTENDLRFISFSTGKRGC. The unique substitutions in the *'PERF'* and the heme binding domain and the high sequence isolation of CYP79B1, CYP79B2, and CYP79A1, together with the presence of Caricaceae and Brassicaceae in the Capparales order, show that the initial part of the biosynthetic pathway of glucosinolates and cyanogenic glycosides is catalyzed by evolutionarily conserved cytochromes P450. This confirms that the appearance of glucosinolates in Capparales is based on a cyanogen "predisposition." Identification of CYP79 homologues in glucosinolate-producing plants provides an important tool for tissue-specific regulation of the level of glucosinolates to improve nutritional value and pest resistance.

Three higher plant β-glucosidase genes have been cloned as cDNA. Analysis of the deduced amino acid sequence of the cyanogenic β-glucosidase from white clover and from cassava and a noncyanogenic β-glucosidase from white clover reveals considerable homology between the three genes and also with several cloned prokaryotic β-glucosidase enzymes (104). Both of the plant cyanogenic β-glucosidases are posttranslationally modified by signal peptide cleavage and glycosylation. The white clover cyanogenic enzyme has an extracellular location but the cassava enzyme is intracellular, being synthesized and subsequently located specifically in the latex vessels of this plant.

Hydrocyanic acid (HCN) is released from all tissues of cassava (*Manihot esculenta* Crantz) following mechanical damage (105,106). HCN is produced during the breakdown of two structurally related cyanogenic glucosides by the sequential action of a β-glucosidase (linamarase) and an α-hydroxynitrile lyase. The final stage in the biosynthesis of the cyanogluco-sides is catalyzed by UDP-glucosyltransferase. The cloning and expression of these three soluble enzymes have been approached by a variety of methods. A cDNA library prepared from cassava cotyledons has been screened for both the β-glucosidase and glucosyltransferase genes. A linamarase cDNA clone from white clover was used to select the equivalent cyanogenic enzyme from cassava. The β-glucosidase clone has been characterized and sequenced, and gene expression has been localized by nonisotopic *in situ* hybridization of mRNA. Several glucosyltransferase clones have been isolated and sequenced using a flavonoid glucosyltransferase clone from *Antirrhinum majus* as a heterologous probe. Comparisons have been made between expression of these genes and enzyme potential of cyanogenic glucosyltransferase clones which appear to represent single-copy genes. Peptide sequences for HNL have been obtained preparatory to cloning. It has been confirmed that this enzyme is not glycosylated in cassava.

A linamarase cDNA clone (pCAS5) was isolated from a cotyledon cDNA library using a white clover β-glucosidase heterologous probe (104). The nucleotide and derived amino acid sequence is reported and five putative *N*-asparagine glycosylation sites are identified. Concanavalin A affinity chromatography and endoglycosides H digestion demonstrate that linamarase from cassava is glycosylated having high-mannose-type *N*-asparagine linked oligosaccharides. Consistent with this structure and the extracellular location of the active enzyme is the identification of an *N*-terminal signal peptide on the deduced amino acid sequence of pCAS5.

6.8 PERSPECTIVES

The biosynthetic pathway for cyanogenic glucosides involves mainly three enzymes of which two are multifunctional cytochrome P450s and a third a UDPG-glucosyltransferase. Because these three enzymes each contain only a single subunit, three structural genes are enough to enclose all the enzymes needed to synthesize cyanogenic glycosides. Through gene technology, it is now feasible to insert the pathway for cyanogenic glycoside synthesis in acyanogenic plants. This will allow detailed studies on the protective mechanisms of cyanogenic glycosides toward generalist feeders (105). On the other hand, it is also possible to reduce the level of cyanogenic glycosides in cassava by antisense or cosuppression techniques by blocking the synthesis of cyanogenic glycoside biosynthesis either in whole plant or in tissue specifically. These studies will also pave the way to investigate the effect of cyanogenic glycosides on herbivores, insect pests, and pathogens. If the transgenic plants devoid of cyanogenic glycosides turn out to be more susceptible to attack, a new defense mechanism based on less harmful substances should be introduced in the plants (106).

The biosynthesis of cyanogenic glycosides and glucosinolates involves common intermediates. Thus, attempts to produce unnatural, novel products through metabolic engineering will pave the way for new strategies for pest control. The similarity between the cytochrome P450s catalyzing the cyanogenic glycoside biosynthesis in plants and certain insects may be addressed. The other important topics to be delineated related to cyanogenic glycosides are their transport mechanisms and their site of storage and protection from degradative enzymes.

The cyanogenic glycosides in cassava tubers constitute a potential health hazard for the millions of poor people who are dependent on these roots as their staple food. Isolation of homologous cDNAs from other agriculturally important cyanogenic crops will allow tissue-specific regulation of the level of cyanogenic glycosides to optimize food safety and pest resistance.

Currently no tertiary structure of UDPG-glucosyltransferase, an enzyme involved in the conversion of aglycone to cyanogenic glycosides, is available. The availability of a crystal structure would serve to identify residues determining substrate specificity and provide clues on how to use site-directed mutagenesis to alter the substrate binding site toward aglycones and thus to alter the profiles of cyanogenic glycosides produced.

For long-distance transport to be effective and to avoid toxic releases of HCN, the cyanogenic glycosides should be transported via compartments not containing β-glucosidases or be converted to transport forms not recognized as substrates by the β-glucosidases generally present. Diglucosides may constitute such transport forms. The expression of this gene is likely to be required for transport of cyanogenic glycosides and is therefore an obvious target for metabolic engineering of cyanogenic plants.

REFERENCES

1. Moller, B.L.; Poulton, J.E. Cyanogenic glycosides. In *Methods in Plant Biochemistry, Enzymes of Secondary Metabolism*; Lea, P.J., Ed.; Academic Press: London, 1993; Vol. 9, 183–207.
2. Moller, B.L.; Seigler, D.S. Biosynthesis of cyanogenic glycosides, cyanolipids, and related compounds. In *Plant Amino Acids Biochemistry and Biotechnology*; Singh, B.K., Ed.; Marcel Dekker: New York, 1999; 563–609.
3. Sibbesen, O.; Koch, B.M.; Rouze, P.; Moller, B.L.; Halikier, B.A.; Biosynthesis of cyanogenic glucosides. Elucidation of the pathway and characterisation of the cytochromes P-450 involved. In *Amino Acids and Their Derivatives in Higher Plants*; Wallsgrove, R.M., Ed.; Cambridge University Press: London, 1995; 227–241.
4. Seigler, D.S. Cyanide and cyanogenic glycosides. In *Herbivores: Their Interactions with Secondary Plant Metabolites*; Rosenthal, G.A., Berenbaum, M.R., Eds.; Academic Press: New York, 1991; Vol. 1, 35–77.
5. Bernays, E.A. Nitrogen in defense against insects. In *Nitrogen as an Ecological Factor*; Lee, J.A., McNeill, S., Rorison, I.H., Eds.; Blackwell Scientific: London, 1983; 321–344.

6. Conn, E.E. Cyanogenic glycosides. Annu. Rev. Plant Physiol. **1980**, *31*, 433–451.

7. Davis, R.H. Cyanogens. In *Toxic Substances in Crop Plants*; D'Mello, J.P.F., Duffus, C.M., Eds.; Royal Society of Chemistry: Cambridge, 1991; 202–225.

8. Jones, D.A. Cyanogenic glycosides and their function. In *Phytochemical Ecology*; Harborne, J.B., Ed.; Academic Press: London, 1972; 103–124.

9. Poulton, J.E. Cyanogenic compounds in plants and their toxic effects. In *Handbook of Natural Toxins*, Keeler, R.F., Tu, W.T., Eds.; Marcel Dekker: New York, 1983; Vol. 1, 117–157.

10. Poulton, J.E. Cyanogenesis in plants. Plant Physiol. **1990**, *94*, 401–405.

11. Conn, E.E. Cyanogenic glycosides. In *International Review of Biochemistry*: *Biochemistry of Nutrition IA*, Neuberger, A., Jukes, T.H., Eds.; University Park Press: Baltimore, 1979; Vol. 27, 21–43.

12. Conn, E.E. Cyanogenic glycosides. In *The Biochemistry of Plants*: *Secondary Plant Products*, Stumpf, P.K., Conn, E.E., Eds.; Academic Press: New York, 1981, Vol. 7, 479–500.

13. Nahrstedt, A. Cyanogenic compounds as protecting agents for organisms. Plant Syst. Evol. **1985**, *150*, 35–47.

14. Ahmad, I.; Ansari, A.A.; Osman, S.M.; Cyanolipids of Boraginaceae seed oils. Chem. Ind. **1978**, 16, 626–627.

15. Aikman, K.; Bergman, D.; Ebinger, J.E.; Seigler, D. Cyanogenic variation in some midwestern US plant species. Biochem. Syst. Ecol. **1996**, *63*, 637–645.

16. Aritomi, M.; Kumori, M.T.; Kawasaki, T.T. Cyanogenic glycosides in leaves of *Perilla frutescens* var. *acuta*. Phytochemistry **1985**, *24*, 2438–2439.

17. Akazawa, T.; Miljanich, P.; Conn, E.E. Studies on cyanogenic glucoside of *Sorghum vulgare*. Plant Physiol. **1960**, *35*, 535–538.

18. Erb, N.; Zinsmeister, D.; Lehmann, G.; Nahrstedt, A. A new cyanogenic glucoside from *Hordeum vulgare*. Phytochemistry **1979**, *18*, 1515–1517.

19. Forslund, K.; Johnson, L. Cyanogenic glucosides and their metabolic enzymes in barley in relation to nitrogen levels. Physiol. Plant **1997**, *101*, 367–372.

20. Jaroszewski, J.W.; Andersen, J.V.; Billeskov, I. Plants as a course of chiral cyclopentenes: taraktophyllin and epivolkenin, new cyclopentenoid cyanohydrin glucosides from Flacourtiaceae. Tetrahedron **1987**, *43*, 2349–2354.

21. Jaroszewski, J.W.; Bruun, D.; Clausen, V.; Cornett, C. Novel cyclopentenoid cyanohydrin rhamnoglucosides from Flacourtiaceae. Planta Med. **1988**, *54*, 333–337.

22. Jenses, S.R.; Nielsen, B.J. Gynocardin in *Ceratiosicyos laevis* (Achariaceae), Phytochemistry **1986**, *25*, 2349–2350.

23. Lieberei, R.; Selmar, D.; Biehl, B. Metabolization of cyanogenic glucosides in *Hevea brasiliensis*. Plant Syst. Evol. **1985**, *150*, 49–63.

24. Lykkesfeldt, J.; Moller, B.L. Cyanogenic glucosides in cassava, *Manihot esculenta* Crantz. Acta Chem. Scand. **1994**, *48*, 178–180.

25. Nahrstedt, A. Recent developments in chemistry, distribution, and biology of the cyanogenic glycosides. In *Biologically Active Natural Products*; Hostettmann, K.; Lea, P.J., Eds.; Clarendon Press: Oxford, 1987; 213–234.

26. Nahrstedt, A.; Sattar, E.A.; El-Zalabani, S.M.H. Amygdalin acyl derivatives, cyanogenic glycosides from the seeds of *Merremia dissecta*. Phytochemistry **1990**, *29*, 1179–1181.
27. Pourmohseni, H.; Ibenthal, W.D.; Machinek, R.; Remberg, G.; Wray, V. Cyanogenic glucosides in the epidermis of *Hordeum vulgare*. Phytochemistry **1993**, *33*, 295–298.
28. Rockenbach, J.; Nahrstedt, A.; Wray, V. Cyanogenic glycosides from *Psydrax* and *Oxyanthus* species. Phytochemistry **1992**, *31*, 567–570.
29. Seigler, D.S.; Brinker, A.M. Characterization of cyanogenic glycosides, cyanolipids, nitroglycosides, organic nitro compounds and nitrile glucosides from plants. In *Modern Methods in Plant Biochemistry*; Dey, P.M., Harborne, J.B., Eds.; Academic Press: London, 1993, Vol. 8, 51–131.
30. Selmar, D.; Lieberei, R.; Junqueira, R.; Biehl, B. Changes in cyanogenic glucoside content in seeds and seedlings of *Hevea* species. Phytochemistry **1991**, *30*, 2135–2140.
31. Selmar, D.; Irandoost, Z.; Wray, V. Dhurrin-6'-glucoside, a cyanogenic diglucoside from *Sorghum bicolor*. Phytochemistry **1996**, *43*, 569–572.
32. Shimomura, H.; Sashida, Y.; Adachi, T. Cyanogenic and phenylpropanoid glucosides from *Prunus grayana*. Phytochemistry **1987**, *26*, 2365–2366.
33. Spencer, K.C.; Seigler, D.S. Cyanogenic glycosides of *Malesherbia*. Biochem. Syst. Ecol. **1985**, *13*, 23–24.
34. Spencer, K.C.; Seigler, D.S. Cyanogenic glycosides and the systematics of the Flacourtiaceae. Biochem. Syst. Ecol. **1985**, *13*, 421–431.
35. Spencer, K.C.; Seigler, D.S. Passibiflorin, epipassibiflorin, and passitrifasciatin: cyclopentenoid cyanogenic glycosides from *Passiflora*. Phytochemistry **1985**, *24*, 981–986.
36. Spencer, K.C.; Seigler, D.S.; Fraley, S.W. Cyanogenic glycosides of the Turneraceae. Biochem. Syst. Ecol. **1985**, *13*, 433–435.
37. Spencer, K.C.; Seigler, D.S.; Nahrstedt, A. Linamarin, lotaustralin, linustatin and neolinustatin from *Passiflora* species. Phytochemistry **1986**, *25*, 645–647.
38. Swenson, W.; Dunn, J.E.; Conn, E.E. Cyanogenesis in *Acacia sutherlandii*. Phytochemistry **1987**, *26*, 1835–1836.
39. Bokanga, M. Distribution of cyanogenic potential in cassava germplasm. Acta. Hortic. **1994**, *375*, 117–123.
40. Vetter, J. Plant cyanogenic glycosides. Toxicon **2000**, *38*, 11–36.
41. Seigler, D.S.; Maslin, B.R.; Conn, E.E. Cyanogenesis in the Leguminosae. Monographs in Systematic Botany for the Missouri Botanical Garden **1989**, *29*, 645–672.
42. Cardona, L.; Fernandez, I.; Pedro, J.R.; Vidal, R. Polyoxygenated terpenes and cyanogenic glucosides from *Centaurea alpera* var *subinermis*. Phytochemistry **1992**, *31*, 3507–3509.
43. Berg van den, A.J.J.; Horsten, S.F.A.J.; Kettenes, J.J.; Kroes, B.H.; Labadie, R.P.; Multifidin, a cyanoglucoside in the latex of *Jatropha multifida*. Phytochemistry **1995**, *40*, 597–598.

44. Schwartz, B.; Wray, V.; Proksch, P. A cyanogenic glycoside from *Canthium schimperianum*. Phytochemistry **1996**, *42*, 633–636.
45. Jager, A.K.; McAlister, B.C.; Staden, J. Cyanogenic glycosides in leaves and callus cultures of *Schlechterina mitostemmatoides*. S. Afr. J. Bot. **1995**, *61*, 274–275.
46. Makkar, H.P.S.; Becker, K. Nutrients and antiquality factors in different morphological parts of the *Moringa oleifera* tree. J. Agric. Sci. **1997**, *128*, 311–322.
47. McFarlane, I.J.; Lees, E.M.; Conn, E.E. The in vitro biosynthesis of dhurrin, the cyanogenic glycoside of *Sorghum bicolor*. J. Biol. Chem. **1975**, *250*, 4708–4713.
48. Conn, E.E. Biosynthesis of cyanogenic glycosides. Biochem. Soc. Symp. **1973**, *8*, 277–302.
49. Conn, E.E. Cyanogenic glycosides; a possible model for the biosynthesis of natural products. In *The New Frontiers in Plant Biochemistry*; Akazawa, T., Ashai, T., Imaseki, H., Eds.; Japan Scientific Societies Press: Tokyo, 1983, 11–22.
50. Conn, E.E. Biosynthetic relationship among cyanogenic glycosides, glucosinolates, and nitro compounds. In *Biologically Active Natural Products: Potential Use in Agriculture. Am. Chem. Soc. Symp. Ser*, Culter, H.G., Ed.; Americian Chemical Society: Washington, DC, 1988; Vol. 380, 143–154.
51. Cutler, A.J.; Hosel, W.; Sternberg, M.; Conn, E.E. The in vitro biosynthesis of taxiphyllin and the channeling of intermediates in *Triglochin maritima*. J. Biol. Chem. **1981**, *256*, 4253–4258.
52. Du, L.; Bokanga, M.; Moller, B.L.; Halkier, B.A. The biosynthesis of cyanogenic glucosides in roots of cassava. Phytochemistry **1995**, *39*, 323–326.
53. Halkier, B.A.; Moller, B.L. Biosynthesis of the cyanogenic glucoside dhurrin in seedlings of *Sorghum bicolor* (L.) Moench and partial purification of the enzyme involved. Plant Physiol. **1989**, *90*, 1552–1559.
54. Halkier, B.A.; Moller, B.L. The biosynthesis of cyanogenic glucosides in higher plants. Identification of three hydroxylation steps in the biosynthesis of dhurrin in *Sorghum bicolor* (L.) Moench and the involvement of 1-aci-nitro-2-(p-hydroxyphenyl)ethane as an intermediate. J. Biol. Chem. **1990**, *265*, 21114–21121.
55. Halkier, B.A.; Moller, B.L. Involvement of cytochrome P-450 in the biosynthesis of dhurrin in *Sorghum bicolor* (L.) Moench. Plant Physiol. **1991**, *96*, 10–17.
56. Halkier, B.A.; Olsen, C.E.; Moller, B.L. The biosynthesis of cyanogenic glucosides in higher plants The (E)- and (Z)-isomers of p-hydroxyphenylacetaldehyde oxime as intermediates in the biosynthesis of dhurrin in *Sorghum bicolor*. J. Biol. Chem. **1989**, *264*, 19487–19494.
57. Halkier, B.A.; Scheller, H.V.; Moller, B.L. Cyanogenic glucosides; the biosynthetic pathway and the enzyme system involved. In *Cyanide Compounds in Biology. Ciba Symposium*; Evered, D., Harnett, S., Eds.; John Wiley and Sons: Chichester, 1988; Vol. 140, 49–66.
58. Hosel, W.; Nahrstedt, A. In vitro biosynthesis of the cyanogenic glucoside taxiphyllin in *Triglochin maritima*. Arch. Biochem. Biophys. **1980**, *203*, 753–757.

59. Kahn, R.A.; Bak, S.; Svendsen, I.; Halkier, B.A.; Moller, B.L. Isolation and reconstitution of cytochrome P450ox and in vitro reconstitution of the entire biosynthetic pathway of the cyanogenic glucoside dhurrin from sorghum. Plant Physiol. **1997**, *115*, 1661–1670.

60. Koch, B.; Nielsen, V.S.; Halkier, B.A.; Olsen, C.E.; Moller, B.L. The biosynthesis of cyanogenic glucosides in seedlings of cassava (*Manihot esculenta* Crantz). Arch. Biochem. Biophys. **1992**, *292*, 141–150.

61. Moller, B.L.; Conn, E.E. The biosynthesis of cyanogenic glucosides in higher plants. N-Hydroxytyrosine as an intermediate in the biosynthesis of dhurrin by *Sorghum bicolor* (Linn) Moench. J. Biol. Chem. **1979**, *254*, 8575–8583.

62. Moller, B.L.; Conn, E.E. The biosynthesis of cyanogenic glucosides in higher plants. Channeling of intermediates in dhurrin biosynthesis by a microsomal system from *Sorghum bicolor* (L.) Moench. J. Biol. Chem. **1980**, *255*, 3049–3056.

63. Reay, P.F.; Conn, E.E. The purification and properties of a uridine diphosphate glucose: aldehyde cyanohydrin-glucosytransferase from sorghum seedlings. J. Biol. Chem. **1974**, *249*, 5826–5830.

64. Schuler, M.A. Plant cytochrome P450 monooxygenase. Crit. Rev. Plant Sci. **1996**, *15*, 235–284.

65. Sibbesen, O.; Koch, B.; Halkier, B.A.; Moller, B.L. Cytochrome P-450TYR is a multifunctional heme-thiolate enzyme catalyzing the conversion of l-tyrosine to p-hydroxyphenylacetaldehyde oxime in the biosynthesis of the cyanogenic glucoside dhurrin in *Sorghum bicolor* (L.) Moench. J. Biol. Chem. **1995**, *270*, 3506–3511.

66. Du, L.; Bokanga, M.; Moller, B.L.; Halkier, B.A. The biosynthesis of cyanogenic glucosides in roots of cassava. Phytochemistry **1995**, *39*, 323–326.

67. Nielsen, J.S.; Moller, B.L. Biosynthesis of cyanogenic glucosides in *Triglochin maritima* and the involvement of cytochrome P 450 enzymes. Arch. Biochem. Biophys. **1999**, *368*, 121–130.

68. McMahon, J.M.; White, W.L.B.; Sayre, R.T. Cyanogenesis in Cassava (*Manihot esculenta* Crantz). J. Exp. Bot. **1995**, *46*, 731–741.

69. Jones, P.R.; Moller, B.L.; Hoj, P.B. The UDP-glucose:p-hydroxymandelonitrile-O-glucosyl transferase that catalyse the last step in synthesis of the cyanogenic glucoside dhurrin in *Sorghum bicolor*. J. Biol. Chem. **1999**, *274*, 35483–35491.

70. Andersen, M.D.; Busk, P.K.; Svendsen, I., Moller, B.L. Cytochromes P-450 from cassava (*Manihot esculenta* Crantz) catalysing the first steps in the biosynthesis of the cyanogenic glucosides linamarin and lotaustralin. J. Biol. Chem. **2000**, *275*, 1966–1975.

71. Kahn, R.A.; Fahrendorf, T.; Halkier, B.A.; Moller, B.L. Substrate specificity of the cytochrome P450 enzymes CYP79A1 and CYP71E1 involved in the biosynthesis of the cyanogenic glucoside dhurrin in *Sorghum bicolor* (L.) Moench. Arch. Biochem. Biophys. **1999**, *363*, 9–18.

72. Brattsen, L.B.; Samuelian, J.H.; Long, K.Y.; Kincaid, S.A.; Evans, K. Cyanide as a feeding deterrent for the southern armyworm, *Spodoptera eridania*. Ecol. Entomol. **1991**, *8*, 125–132.

73. Poulton, J.E. Toxic compounds in plant foodstuffs: cyanogens. In *Food Proteins*. Kinsella, J.E.; Soucie, W.G., Eds.; The American Oil Chemists' Society; Champaign, Illinois, 1989; 381–401.

74. Malagon, J.; Garrido, A. Relation between cyanogenic glycosides content and the resistance to *Capnodis tenebrionis* (L.) in stone fruits. Boletin de Sanidad Vegetal. **1990**, *16*, 499–503.

75. Alborn, H.; Stenhagen, G.; Leuschner, K.; Rizvi, S.J.H. Biochemical selections of sorghum crop varieties resistant to sorghum shoot fly (*Atherina soccata*) and stem borer (*Chilo partellus*): role of allelochemicals. In *Allelopathy: Basic and Applied Aspects*. Rizvi, V., Ed.; Chapman and Hall: London, 1992; 101–117.

76. Belloti, A.C.; Arias, B. The possible role of HCN on the biology and feeding behaviours of the cassava burrowing bug (*Cyrtomenus bergi*, Froeschner). In *Proceedings of the First International Scientific Meeting of the Cassava Biotechnology Network*. Roca, W.M., Thro, A.M., Eds.; (Centro International de Agricultura Tropical). CIAT: Cali, Columbia, 1994; 406–409.

77. Stadler, E. Contact chemoreception. In *Chemical Ecology of Insects*; Bell, W.J., Carde, R.T., Eds.; Chapman and Hall: New York, 1984; 3–35.

78. Spencer, K. In *Chemical Mediation of Coevolution*; Spencer, K., Ed.; Academic Press: Orlandos, 1988; 167–240.

79. Jones, D.A. Cyanogenesis in animal–plant interactions. In *Cyanide Compounds in Biology: Ciba Symposium*; Evered, D., Harnett, S., Eds.; John Wiley and Sons: Chichester, 1988; Vol. 140, 151–170.

80. Mikolajczak, K.L.; Madrigal, R.V.; Smith, C.R.; Reed, D.K, Insecticidal effects of cyanolipids on three species of stored product insects, European corn borer (Lepidoptera: Pyralidae) larvae and striped cucumber beetle (Coleoptera: Chysomelidae). J. Econ. Entomol. **1984**, *77*, 1144–1148.

81. Janzen, D.H.; Juster, H.B.; Bell, E.A. Toxicity of secondary compounds to the seed eating larvae of the bruchid beetle *Callosobruchus maculatus*. Phytochemistry **1977**, *16*, 223–227.

82. Schreiner, I.; Nafus, D.; Pimentel, D. Effects of cyanogenesis in bracken fern (*Pteridium aquilinum*) on associated insects. Ecol. Entomol. **1984**, *9*, 69–79.

83. Cooper-Driver, G.A.; Finch, S.; Swain, T.; Bernays, E. Seasonal variation in secondary plant compounds in relation to palatability of *Pteridium aquilinum*. Biochem. System Ecol. **1977**, *5*, 177–183.

84. Lawton, J.H. Host-plant influences on insect diversity; the effects of space and time. In *Insect Faunas. RES Symposium 9*; Mound, L.A.; Waleff, N., Eds.; Blackwell Scientific: Oxford, 1978; 105–125.

85. Dritschilo, W.; Krummel, J.; Nafus, D.; Pimentel, D. Herbivorous insects colonizing cyanogenic and acyanogenic *Trifolium repens*. Heredity **1979**, *42*, 49–56.

86. Woodhead, S.; Bernays, E. Changes in release rates of cyanide in relation to palatability of sorghum to insects. Nature **1977**, *270*, 235–236.

87. Ellsbury, M.M.; Pederson, G.A.; Fairbrother, T.E. Resistance to foliar-feeding hypergine weevils (Coleoptera: Curculionidae) in cyanogenic white clover. J. Econ. Entomol. **1992**, *85*, 2467–2472.

88. Whitman, R.J. Herbivore feeding and cyanogenesis in *Trifolium repens*. L. Heredity **1973**, *30*, 241–245.
89. Mowat, D.J.; Shakeel, M.A. The effect of different cultivars of clover on numbers of and leaf damage by some invertebrate species. Grass Forage Sci. **1989**, *44*, 11–18.
90. Raffaelli, D.; Mordue, A.J. The relative importance of molluscs and insects as selective grazers of acyanogenic white clover (*Trifolium repens*). J. Molluscan Stud. **1990**, *56*, 37–45.
91. Chitwood, D.J. Phytochemical based strategies for nematode control. Annu. Rev. Phytopathol. **2002**, *40*, 221–249.
92. Hughes, J.; Pacro, A.; Brown, K.; Haysom, H.; McCartney, H.; Ketty, K.; Fletcher, M.; Hughes, M.A. Molecular studies of cyanogenesis in cassava. In *Centro International de Agricultural Tropical Cali*; Roca, W.M.; Thro, A.M., Eds.; Columbia (CIAT) 1993; 384–389.
93. Li, HQ.; Sautter, C.; Potrukus, I.; Puonti-Kaerlas, J. Genetic transformation of cassava (*Manihot esculenta* Crantz). Nature Biotechnol. **1996**, *14*, 736–740.
94. Nahrstedt, A. Cyanogenesis and role of cyanogenic glycosides in insects. In *Cyanide Compounds in Biology. Ciba Symposium*; Evered, D.; Harnett, S., Eds.; John Wiley and Sons: Chichester, 1988; Vol. 140, 131–150.
95. Bak, S.; Kahn, R.A.; Nielson, H.L.; Moller, B.L.; Halkier, B.A, Cloning of three A-type cytochrome P450 CYP71E1, CYP98, and CYP99 from *Sorghum bicolor* (L.) Moench by a PCR approach and identification by expression in *Escherichia coli* of CYP71E1 as a multifunctional cytochrome P 450 in the biosynthesis of the cyanogenic glucoside dhurrin. Plant Mol. Biol. **1998**, *36*, 393–405.
96. Frank, M.R.; Deyneka, J.M.; Schuler, M.A. Cloning of wound induced cytochrome P450 monooxygenases expressed in pea. Plant Physiol. **1996**, *110*, 1035–1046.
97. Hasemann, C.A.; Kurumbail, R.G.; Boddupalli, S.S.; Peterson, J.A.; Deisenhofer, J. Structure and function of cytochromes P450; a comparative analysis of three crystal structures. Structure **1995**, *3*, 41–62.
98. Peterson, J.A.; Graham, S.E. A close family of resemblance: the importance of structure in understanding cytochromes P450. Structure **1998**, *6*, 1079–1085.
99. Bak, S.; Nielsen, H.L.; Halkier, B.A. The presence of CYP79 homologous in glucosinolate-producing plants shows evolutionary conservation of the enzymes in the conversion of amino acid to aldoxime in the biosynthesis of cyanogenic glucosides and glucosinolates. Plant Mol. Biol. **1998**, *38*, 724–734.
100. Nielsen, J.S.; Moller, B.L. Cloning and expression of cytochrome P450 enzymes catalysing the conversion of tyrosine to p-hydroxyphenyl acetaldoxime in the biosynthesis of cyanogenic glucosides in *Triglochin maritima*. Plant Physiol. **2000**, *122*, 1311–1321.
101. Bak, S.; Olsen, C.E.; Halkier, B.A.; Moller, B.L, Transgenic tobacco and arabidopsis plants expressing the two multifunctional sorghum cytochrome P450 enzymes, CYP79A1 and CYP71E1, are cyanogenic and accumulate metabolites derived from intermediates in dhurrin biosynthesis. Plant Physiol. **2000**, *123*, 1437–1448.

102. Tattersall, D.B.; Bak, S.; Jones, P.R.; Olsen, C.E.; Nielsen, J.K.; Hansen, M.L.; Hoj, P.B.; Moller, B.L. Resistance to an herbivore through engineered cyanogenic glucoside synthesis. *Science* **2001**, *293*, 1826–1828.

103. Cicek, M.; Esen, A, Structure and expression of a dhurrinase (β-glucosidase) from sorghum. *Plant Physiol.* **1998**, *116*, 1476–1478.

104. Hughes, M.A.; Brown, K.; Pacoro, A.; Murray, B.S.; Oxtoby, E.; Hughes, J. A molecular and biochemical analysis of the structure of the cyanogenic β-glucosidase (linamarase) from cassava (*Manihot esculenta* Cranz). Arch. Biochem. Biophys. **1992**, *295*, 273–274.

105. Cases, A.M.; Kononowicz, A.J.; Zehr, U.B.; Tomes, D.T.; Axtell, J.D.; Butler, L.G.; Bressan, R.A.; Hasegava, P.M. Transgenic sorghum plants via microprojectile bombardment. Proc. Natl. Acad. Sci. USA. **1993**, *90*, 11212–11216.

106. Estruch, J.J.; Carozzi, N.B.; Desai, N.; Duck, N.B.; Warren, G.W.; Koziel, M.G. Transgenic plants: an emerging approach to pest control. Nature Biotechnol. **1997**, *15*, 137–141.

7

Glucosinolates

7.1 INTRODUCTION

Glucosinolates are amino acid–derived, secondary plant products containing a sulfate and thioglucose moiety. They are a uniform class of naturally occurring hydrophilic, nonvolatile, water-soluble, anionic compounds. These group of compounds were initially called thioglucosides or mustard oil glucosides. The glucosinolates were first isolated more than a century ago. The first compound isolated was derived from white mustard seeds (*Sinapis alba* L.) and named sinalbin. Many of the individual substances of this class are still known by such trivial names as sinalbin and sinigrin (isolated from black mustard seeds, *Brassica nigra* Koch.) In a subsequent naming system adopted by Kjaer (1), the prefix *gluco* was attached to a part of the Latin species name from which the compound was first isolated (e.g., glucobrassicin, glucoiberin).

A detailed discussion of the early work on the structure of mustard oil glucosides is given in a monograph written by Challenger (2). A more well-adopted system that relates to the chemical structure was suggested by Ettlinger and Kjaer (3). In this system, all members of the class are called glucosinolates, to which is added a prefix that chemically describes a variable part of the molecule, e.g., methyl glucosinolate (methyl-GS) or benzyl glucosinolate (benzyl-GS).

When hydrolyzed, glucosinolates liberate D-glucose, sulfate, and an unstable aglucone, which undergo rearrangement to yield isothiocyanate (commonly called mustard oil) as the main product or thiocyanate or a nitrile (organic cyanide) as secondary products. Because of the presence of the glucose moiety and sulfate group (ionic), glucosinolates are hydrophilic and nonvolatile. On the other hand, isothiocyanates are generally volatile and are chemically very active. More than 100 glucosinolates have been isolated from plants and all of them have a similar general structure. A number of reviews have been written about glucosinolates and their products (4–15). Many of the reviews address specific agricultural, food, biochemical, or pest control aspect.

7.2 Occurrence

Glucosinolates are known to occur in dicotyledonous plants. They occur mainly in the order Capparales. They have been detected in every species studied of the Cruciferae, Capparaceae, Tovariaceae, Resedaceae, and Moringaceae. Other families in which they occur in one or more taxa include Limnanthaceae, Caricaceae, Tropaeolaceae, Gryostemonaceae, Salvadoraceae, and Euphorbiaceae (4–6,11,16). The glucosinolates in the Cruciferae are of special interest because a number of important vegetable and agricultural crops belong to this family.

Glucosinolate types in plant species are highly variable. The main glucosinolate in radish seed (*Raphanus sativus*) is 4-methylsulfinyl-3-butenyl glucosinolate, whereas the mustard seed (*Brassica juncea*) is dominated by allyl glucosinolate. Rapeseed (*Brassica napus*) contains four major glucosinolates, namely, 2-hydroxy-3-butenyl, 3-butenyl, 4-pentenyl, and 2-hydroxy-4-pentenyl, whereas cabbage seed (*Brassica oleracea*) contains mainly allyl and 2-hydroxy-3-butenyl glucosinolates (13).

Benzyl glucosinolate has a broad family distribution and methyl glucosinolate rarely occurs outside the Capparaceae (family specific). Three ketoalkyl glucosinolates found in South American species of the genus *Capparis* have not been detected in the species of the same genus from other part of the world.

There is also difference in the distribution based on biosynthetic origin. Glucosinolates derived from chain-extended amino acids are found primarily in the Cruciferae and Resedaceae. Indole glucosinolates are characteristic of the first four listed families of the order Capparales. The distribution of glucosinolates in families, genera, and species has often been used in chemotaxonomic considerations.

Glucosinolate concentrations vary within a single species and fluctuations occur with plant age (17). Glucosinolate types and quantities also vary within plant tissues of an individual plant. The major glucosinolates in radish seed is not detected in radish leaves or roots, whereas allyl glucosinolate is found in all *B. juncea* tissues. Various plant parts of a given crucifer (roots, stems, leaves, inflorescences, fruits, and seeds) quantitatively display the same pattern, but changes occur in each glucosinolate from the roots to green organs and seeds (18). Numerous analyses of concentrations in vegetables and other crops have been reported (19). Generally, the levels in fresh plant parts are about 0.1% or less based on fresh weight, whereas levels in seeds may be up to 10% of the dry weight. Only limited knowledge is available on the glucosinolate concentration in individual plant cells, but in seeds the endosperm is the site of accumulation. The glucosinolates in horseradish root cells have been localized in the vacuoles. Environmental and cultural factors, such as spacing, moisture regime, and nutrient availability, affect the concentration of glucosinolates (20,21).

7.3 CHEMISTRY AND CLASSIFICATION

7.3.1 Structure

The general structure and structures of few important glucosinolate are shown in Fig. 7.1. Fenwick et al. (22) have proposed a structure for glucosinolate based on the degradative and physicochemical investigations. Glucosinolates are considered to possess the Z. configuration shown around C=N double bond and are called (Z)-*cis*-N-hydroximinosulfate esters, possessing a side chain R and sulfur-linked D-glucopyranose moiety. Natural glucosinolates contain exclusively a β-D-glucopyranosyl linkage. This structure is assumed to be valid for all glucosinolates. These glucosinolates invariably occur in nature in the anionic form because of the low pK value of the sulfonic acid group. The various glucosinolates are derived by naming the side chain R as a prefix.

7.3.2 Classification

The classification of glucosinolates are based on the structure, biosynthesis, and distribution pattern. Based solely on structure, three main classes of glucosinolates are known:

1. Aliphatic
2. Aromatic
3. Indole

Classification could also be done by selecting the groups of glucosinolates containing side chains conferring special reaction possibilities

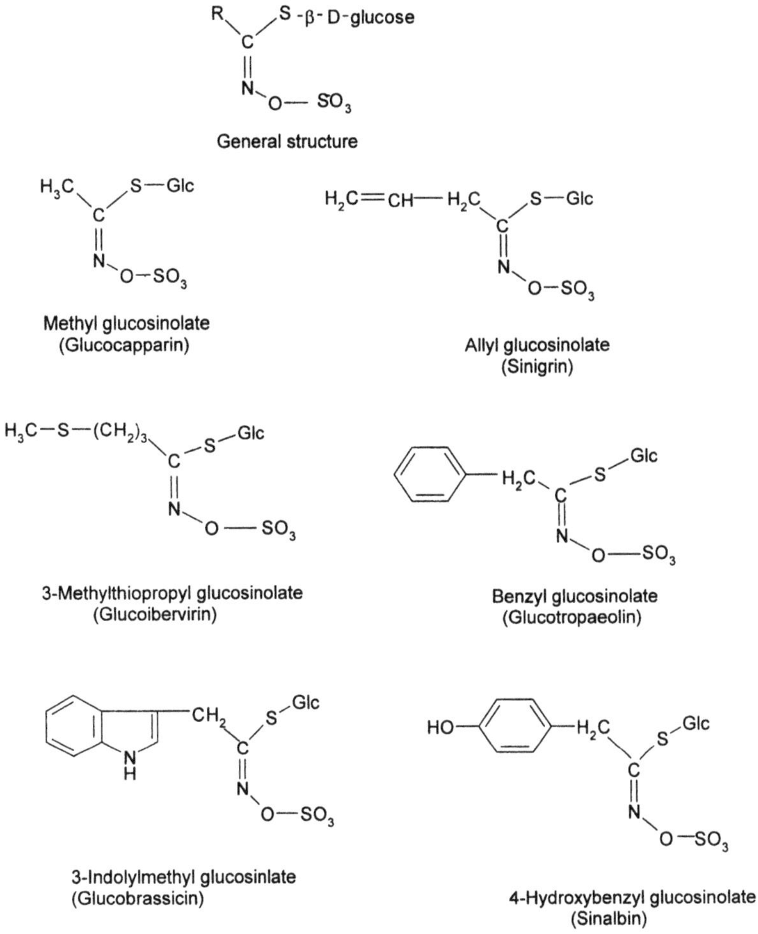

FIGURE 7.1 Structures of some glucosinolates.

on the glucosinolates or their decomposition products. Classification of glucosinolates based on structure is given in Table 7.1 (4,6).

7.3.3 Degradation Products of Glucosinolates

Intact glucosinolates are nontoxic but upon enzyme hydrolysis they yield an array of physiologically active cleavage products (6). The cleavage is catalyzed by a thioglucoside glucohydrolase named myrosinase (EC 3.2.3.1). The myrosinase is also called thioglucosidase. Myrosinase (23,24) catalyzes the cleavage of glucosinolate molecule to an unstable thiohydroxamate-O-sulfonate

TABLE **7.1** Classification of Glucosinolate

Structure type	Chemical name
I. Aliphatic	
1. Saturated without functional group	Methyl GS
2. With double bond	Allyl GS, but-3-enyl GS
3. With alcohol group	1-Methyl-2-hydroxyethyl GS
	2-Hydroxypent-4-enyl GS
4. With keto group	4-Oxoheptyl GS
5. With methylthio group	3-Methylthiopropyl GS
	4-Methylthiobutyl GS
6. With methylsulfinyl group	3-Methylsulfinylpropyl GS
7. With methylsulfonyl group	3-Methylsulfonylpropyl GS
	4-Methylsulfonylbutyl GS
8. With esterified carboxyl group	3-Methyloxycarbonylpropyl GS
II. Aromatic	
1. Without functional group	Benzyl GS
	2-Phenylethyl GS
2. With phenol groups, free, methylated, or glycosylated	p-Hydroxybenzyl GS
3. With alcohol group	2-Hydroxy-2-phenylethyl GS
III. Indole	
1. Without functional group	Indol-3-ylmethyl GS
2. With methoxyl group	1-Methoxyindol-3-ylmethyl GS
	4-Methoxyindol-3-ylmethyl GS
3. With hydroxyl group	4-Hydroxyindol-3-ylmethyl GS
4. With sulfonyl group	N-Sulfoindol-3-ylmethyl GS

GS, glucosinolate.

and other aglycones (Scheme 7.1). The bond between sulfur and glucose is cleaved under neutral conditions, the aglycone moiety gives rise to sulfate and, by a Loessen-type molecular rearrangement, yields isothiocyanate. Formation of a nitrile, also known as organic cyanide, does not require rearrangement and involves sulfur loss from the molecule. Nitrile formation is favored over isothiocyanate at low pH. The presence of Fe^{2+} or thiol compounds and epithio specifier protein increases the likelihood of nitrile formation. Epithionitrile formation requires the same conditions as those for nitriles plus the presence of terminal unsaturation of the R group. Another product, organic thiocyanate, although not a common occurrence is sometime produced. Thiocyanate production is controlled by the presence of specific R group. Indole and 4-hydroxybenzyl glucosinolate yield thiocyanate. The formation of thiocyanate from indole glucosinolate occurs over

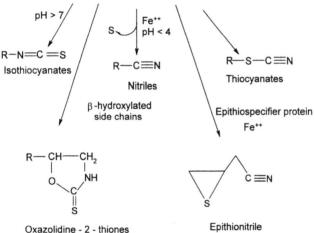

SCHEME 7.1 Hydrolytic products of glucosinolates.

a wide pH range, whereas 4-hydroxybenzyl glucosinolates yield, thiocyanate only at a more basic pH (25). Glucosinolates with β-hydroxylated side chains spontaneously cyclize to yield the oxazolidine-2-thiones.

Myrosinase and glucosinolates are separated from each other in intact plant tissues. Myrosinases are localized in the myrosin cells scattered throughout most plant tissues. Within these cells the enzyme is stored inside myrosin grains (26). Glucosinolates are probably contained in vacuoles of various types of cells in the tissue. Disruption of cellular tissues allows mixing of glucosinolates and myrosinase resulting in the rapid release of glucosinolate degradation products.

7.3.4   Characteristics of Hydrolytic Products

Isothiocyanates are often volatile with pungent flavors or odors (27). The presence of allyl isothiocyanates in mustards and horseradish is responsible for much of the flavor and hence are sometimes called mustard oils. Compounds with strongly polar and large side chains have restricted volatility. Isothiocyanates carrying a hydroxyl group in the β position spontaneously cyclize to yield substituted oxazolidone-2-thiones. The goitrogenic oxazolidone thus formed are major product from rapeseed (*Brassica napus*). Isothiocyanates derived from the indole glucosinolates are unstable in neutral or alkaline solutions and produce thiocyanate ions. Toxicity of other organic products, nitriles, epithionitrile, and thiocyanates is likely related to the cyano group. Toxicities of organic nitriles vary but are generally lower than those of cyanide salts and hydrogen cyanide. Toxicity seems particularly pronounced toward the liver and kidney. Unsaturated nitriles appear less toxic than saturated nitriles. Thiocyanate and epithionitrile toxicities appear similar to nitrile toxicities. Alkyl thiocyanates are generally more active than aromatic thiocyanate and toxicities decrease with increasing molecular weight. Thiocyanates are generally less toxic than isothiocyanates and organic cyano compounds.

7.4   BIOSYNTHESIS

The substrates for the biosynthesis of glucosinolates are L-amino acids. The synthesis of glucosinolates from amino acid was first demonstrated by incorporation of [β^{14}C]tryptophan into indol-3-yl methyl glucosinolate in *Brassica oleracea* (28) and by the incorporation of ^{14}C-labeled phenylalanine into benzylglucosinolate in *Tropaeolum majus* (29). The activity from [α^{14}C] phenylalanine was located in the thiohydroximate carbon, whereas that from [β^{14}C]phenylalanine was found in the benzyl carbon (29). No activity was present in the glucosinolate when [carboxyl-^{14}C]phenylalanine was fed. With labeling in specific positions and subsequent degradation was carried out to show that the isotope was present in the expected positions. During the biosynthesis of glucosinolates from amino acid, the nitrogen is preserved. This was demonstrated by using double-labeled L-[^{14}C,^{15}N]phenylalanine, which was converted to benzylglucosinolate in *T. majus* (30). The result also proved that the amino nitrogen and the carbon skeleton of phenylalanine were incorporated intact into benzylglucosinolate and the carboxyl carbon was not incorporated into the glucosinolate. The readers may refer the lattest reviews on biosynthesis of glucosinolates for more details (31,32).

Previous *in vivo* studies using seedlings or excised tissues have shown that amino acids, *N*-hydroxyamino acids, nitro compounds, oximes, thiohydroximates, and desulfoglucosinolates are precursors or intermediate compounds in the biosynthetic pathway of glucosinolates (10) (Scheme 7.2). The biosynthesis of glucosinolates comprises three important stages.

Stage 1. Synthesis of chain elongated amino acids
Stage 2. Development of the glucone moiety
Stage 3. Side chain modification

7.4.1   Chain Elongation of Amino Acids

The protein amino acids undergo transamination reaction to produce the corresponding α-ketoacids. This is followed by condensation with acetyl-CoA and then a second transamination to recover the amino group (33–37). The formation of chain elongated amino acids was noticed after *in vivo* administration of amino acids and acetate to plants. The methionine chain elongation pathway in the biosynthesis of glucosinolates in *Eruca sativa* (Brassicaceae) was demonstrated by Graser et al. (35) The acetyl-CoA condenses with 2-oxoacid derived from methionine and then undergoes oxidative decarboxylation to yield a new oxoacid with one additional methyl group. Glucosinolate biosynthesis through α-ketoacid elongation in *Arabidopsis thaliana* was demonstrated by deQuiros et al. (37) These chain lengthening processes, in some cases, are also followed by transformations in the side chains. The chain lengthening process involving the incorporation of acetyl moiety is shown in Scheme 7.3. It has been shown that labeled carbon atoms from acetate and phenylalanine are incorporated in the expected positions in 2-phenylethyl glucosinolate. The chain lengthening process not only produces glucosinolates with one additional carbon atom but is also responsible for the production of the entire homologous series of glucosinolates. Studies conducted in *Brassica campestris* with labeled methionine and acetate showed the incorporation of label in expected positions of 3-butenyl glucosinolate. The most common glucosinolates derived are 2-propenyl, but-3-enyl, and pent-4-enyl glucosinolates. Chain elongation of aromatic amino acids was shown by Dornemann et al. (36).

The protein amino acids which are directly involved in the synthesis of glucosinolates are alanine, valine, isoleucine, leucine phenylalanine, tyrosine, and tryptophan.

Alanine   – – – – – – – – –→   Methyl GS

Valine   – – – – – – – – –→   Isopropyl GS

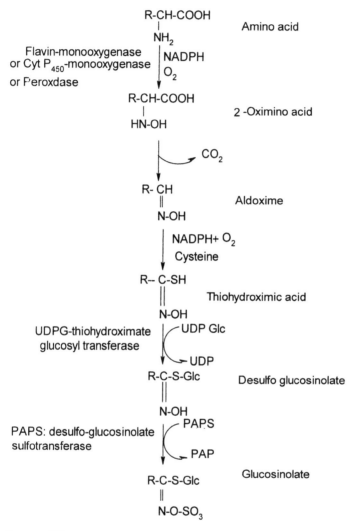

SCHEME 7.2 Biosynthesis of glucosinolate.

Isoleucine	– – – – – – – – →	*sec*-Butyl-GS
Leucine	– – – – – – – – →	Isobutyl GS
Phenylalanine	– – – – – – – – →	Benzyl GS
Tyrosine	– – – – – – – – →	Hydroxybenzyl GS
Tryptophan	– – – – – – – – →	Inol-3-ylmethyl GS

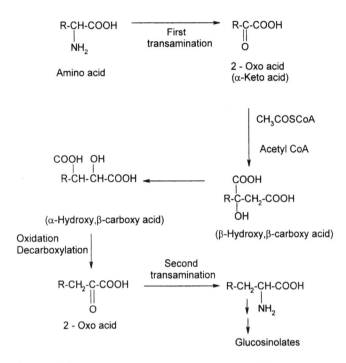

SCHEME 7.3 Side-chain elongation of amino acids.

7.4.2 Development of Glucone Moiety

The first committed step in the formation of glucone moiety of glucosinolates is the amino acid-to-oxime conversion (38). It has been proposed that homologous enzyme systems catalyze the conversion of amino acids to oximes in the glucosinolate and cyanogenic glucoside pathways (10).

From our current knowledge of the enzymes that catalyze the conversion of amino acids to oximes in the glucosinolate pathway, three different types of enzymes appear to be involved (Scheme 7.4).

1. Cytochrome P_{450s} (39)
2. Flavin-containing monooxygenases (40)
3. Peroxidases (41)

Based on homology with the biosynthetic pathway of cyanogenic glycosides, the cytochrome P_{450s}-dependent monooxygenase in glucosinolate biosynthesis is believed to catalyze two consecutive N-hydroxylations

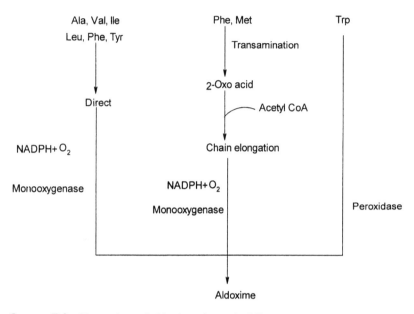

SCHEME 7.4   Formation of aldoxime through different routes.

followed by a dehydration and decarboxylation reaction. Using microsomes isolated from young green leaves of rapeseed, the involvement of flavin-containing monooxygenases in the conversion of chain elongated amino acids to their corresponding oximes was demonstrated (42,43). The reaction is dependent on oxygen and NADPH. No detailed mechanism has been proposed for the reaction catalyzed by flavin-containing monooxygenases. The conversion of tryptophan to indole acetaldoxime in seedlings of Chinese cabbage (*Brassica compestris*) is catalyzed by a plasma membrane–bound peroxidase. In the peroxidase-catalyzed conversion of tryptophan to indole acetaldoxime, it has been proposed that the amino nitrogen is oxidized to H_2O_2 followed by dehydrogenation and decarboxylation reactions.

The involvement of three different enzymes in the conversion of amino acids to oximes suggests convergent evolution. *Sinapis alba* contains a cytochrome P_{450}-dependent monooxygenase for its tyrosine-derived gluco-sinolate, a peroxidase for its indolyl glucosinolate, and the flavin-containing monooxygenase for the minor glucosinolates derived from chain elongated amino acids.

The conversion of oxime to thiohydroximate is a poorly understood step in the biosynthesis of glucosinolates. The oxidation of the oxime to an aci-nitro compound that subsequently functions as the acceptor for an appropriate thiol donor was proposed (44). The sulfur donor for the conversion of aci-nitro intermediate to thiohydroximate is not clearly known. It has been proposed that cysteine is possibly conjugated to aci-nitro compound to produce S-alkylthiohydroximate with the help of a glutathione-S-transferase. This intermediate is cleaved by a C-S lyase to yield the thiohydroximate (42). The next step is conversion of thiohydroximate to desulfoglucosinolate by an S-glucosylation reaction in the presence of UDP-glucose and a soluble UDPG-thiohydroximate glucosyltransferase (45). The desulfoglucosinolate is subsequently sulfated by a soluble 3'-phosphoadenosine 5'-phosphosulfate: desulfoglucosinolate sulfotransferase (46). This enzyme has been isolated and partially purified from rapeseed, which showed high substrate specificity for thiohydroximate but no specificity for the structure of the side chain (46).

7.4.3 Side Chain Modifications

Secondary modifications of the side chain may occur following glucone development. In brassicacious plants, a family of glucosinolates derived from methionine occurs. This family contains side chain that can be collectively expressed as $MeS(CH_2)_n$ with n ranging from 3 to 11. Additional complexity is introduced via later-stage oxidations that produce the corresponding sulfoxides or sulfones. The important modified side chains derived from methionine are (a) methylthioalkyl (CH_3-S-$(CH_2)_n$), methylsulfinylalkyl (CH_3-SO-$(CH_2)_n$), methylsulfonyl alkyl (CH_3-SO_2-$(CH_2)_n$), alkenyl (CH_2=CH-$(CH_2)_n$), hydroxyalkenyl (CH_2=CH-CH(OH)-$(CH_2)_n$), and hydroxyalkyl (HO-$(CH_2)_n$).

Based on genetic studies, a model for side chain modifications of aliphatic glucosinolates has been proposed (47). The initial products after glucone formation are likely to be methylthioalkyl glucosinolates. Further oxidation result in the formation of methylsulfinylalkyl and methylsulfonylalkyl side chains. The production of alkenyl homologues occurs by removal of the methylthio group followed by insertion of a double bond. The desaturation and addition of a hydroxyl group to methylsulfinylpropyl glucosinolate occurs.

7.5 BIOACTIVITY

Glucosinolates are nonvolatile and can therefore be recognized by insects only on contact. Isothiocyanates, the most important hydrolytic product, are generally volatile and can induce olfactory responses. The biological activity of glucosinolates and isothiocyanates seems to be determined by the nature of the side chain, by the concentration of the compound, and by the type of pest. Glucosinolates have generally been considered to interfere with the insects. Host plant selection at two levels: preingestive and postingestive. In the former phase, the insects are repelled by the pungent odor of mustard oil and in the latter phase the insects are confronted with the toxic component of the mustard oil glycoside. These compounds act as plant defense chemicals against generalized insects including aphids, grass-hoppers, and many noncrucifer-feeding lepidopterans. Glucosinolates or their hydrolysis products have been implicated in a variety of interactions between plants and their potential herbivores. Normally, glucosinolates interfere with the oviposition and egg hatchability, and interact with larval chemoreceptors. Glucosinolates have profound effects on larval and adult beetles. They also function as fumigant, parasite attractant, and nematicidal compounds.

7.5.1 Oviposition Response

Glucosinolates stimulate oviposition in insects associated with species of cruciferae. Also, isothiocyanates have been shown to induce oviposition in a number of cases. Adult females of *Pieris brassicae* possess tarsal contact chemoreceptors that are sensitive to glucosinolates and are probably involved in oviposition (48). Adult females of *Delia brassicae* flies also possess tarsal chemoreceptors that respond to sinigrin with a thresholds of 10^{-5} to 10^{-4} M, but not to the corresponding isothiocyanate (49).

$P. brassicae$ organisms lay eggs on non host surfaces painted with allyl glucosinolate or if their tarsi are brushed with the compound. Egg-laying *Pieris napi* adults distinguish among groups of plant species that show consistent difference in glucosinolates profiles, suggesting that the glucosi-nolates may be used as oviposition cues. Allyl glucosinolate generally stimulated oviposition activity in *P. rapae* and *P. napi* but did not promote discrimination among plants (50). A positive egg laying response to glucosinolates in leaves was also shown by *P. brassicae*. Whole-plant extracts were more stimulatory than any fraction. Laboratory experiments suggest that glucosinolates are involved but not sufficient in themselves to explain completely the oviposition behavior of the pierids (51).

Huang and Renwick (52) showed that allyl and indol-3-ylmethyl glucosinolate are the stimulants in oviposition. No differences were noted in the stimulatory effect in *P. rapae* between phenethyl and 2-hydroxy-2-phenylethyl glucosinolates when compared to indol-3-ylmethyl glucosi-nolate, and the insect was less sensitive to aliphatic glucosinolates, while *P. napi oleracea* was strongly stimulated by the aliphatic group (53). The 3-methylsulfinylpropyl and 3-methylsulfonylpropyl glucosinolates were identified as the most active stimulants to *P. napi oleracea* (54), while 3-methylsulfinylpropyl was recognized to be as strong as 2-propenyl glucosinolate in stimulating oviposition by this species. Among glucosinolates, the indoles have been demonstrated to be the most powerful oviposition kairomones for *P. brassicae* and *P. rapae* (55).

The cabbage root fly, *Delia radicum* (L.)/*Delia brassicae* (Wiedemann), attacks a wide range of brassicacious plants, suggesting that several volatile chemicals are probably involved in attracting the flies and in stimulating them to lay (56). Oviposition stimulation was found in species yielding different types of isothiocyanate, while in plants lacking volatile glucosinolate breakdown products no stimulation of oviposition was found. It was observed that the cabbage root fly can locate sources of allyl isothiocyanate solely by olfactory cues, providing the basis for very effective yellow water traps using mixtures of volatile isothiocyanates (57,58). Roessingh et al. (59) found that the *D. sensilla* on segments 3 and 4 of the tarsus of *D. radicum* females contain a receptor cell sensitive to glucosinolates, especially to indol-3-ylmethyl, phenethyl, and pent-4-enyl glucosinolates. The turnip root fly *Delia floralis* (Fall) was also shown to differentiate between glucosinolates, the type D tarsal sensilla being identified as the most responsive to pent-4-enyl, but-3-enyl, and indol-3-ylmethyl glucosinolate (60). However, differentiation by both *Delia* species has been attributed not only to glucosinolate composition of the leaf surface on which the flies walk, but also to an unidentified nonglucosinolate compounds (61).

Reed et al. (62) reported indol-3-ylmethyl glucosinolate to be as stimulatory as aromatic and aliphatic glucosinolates in oviposition of *Plutella xylostella*. Several other studies have shown these compounds to be at least partially responsible for host-plant selection by *Plutella maculipennis* (Curtis) and the cabbage aphid *Brevicoryne brassicae* (L.) (63).

Traynier (64) reported oviposition preferences of *Hylemia brassicae* (Bouche) to be correlated with the presence of glucosinolates in hosts. He suggested that chemostimuli provided by plants might be more potent than the physical characteristics of hosts in eliciting oviposition and found allyl glucosinolate to be less effective than juice squeezed from the root of swede (*B. napus* Napobrassica group), a favored host of *H. brassicae*, pointing to the involvement of other compounds. Nair and McEwen (65) studied the ovipositional behavior of this insect and confirmed that allyl, *p*-hydroxybenzyl, and benzyl glucosinolates induce oviposition equally, while 3-methylsulfinylpropyl and 3-methylsulfonylpropyl were significantly less effective. Allyl isothiocyanate was shown to stimulate flies into greater activity and attracted them to its source. Because oviposition may be dependent on the presence of glucosinolates, development of plant varieties containing a very low concentration of the most active glucosinolates might confer some degree of resistance, but the "glucosinolate pattern" of a given species is more important than the total glucosinolate content in determining the acceptability of a plant.

7.5.2 Egg Hatching

Eggs of herbivorous insects are susceptible to isothiocyanates produced by wounded tissue near oviposition sites. Benzyl isothiocyanate is toxic to eggs and first instars of three fruit fly pests of *Carica papaya* at levels comparable to those produced by ripening fruit (66). *Brassica juncea* and *Brassica nigra* are less suitable hosts of the midge *Dasineura brassicae* due to the presence of allyl and phenylethyl glucosinolates (67). Allyl glucosinolate was the most toxic, lethal at 10 ppm. This concentration is within the range of the corresponding parent allyl glucosinolate in *Brassica juncea* cultivars. The effect of allyl glucosinolate and 2-phenylethyl glucosinolates, including their hydrolysis products isothiocyanates and an epithionitrile, was studied on eggs of the midge *D. brassicae*. Oviposition sites for these midges may accumulate hydrolysis products because the midge used holes made by a seed weevil (68).

Control of the black vine weevil, *Otiorhynehus sulcatus* (T.), with allelochemicals produced from glucosinolates may be possible; however, plant-derived isothiocyanates are not readily available for bioassay. Their objective was to predict the toxicity of plant-derived isothiocyanates using a model developed with commercially available compounds. Contact toxicities of 12 organic isothiocyanates were determined by dipping black vine weevil eggs into isothiocyanate solutions (69). Quantitative relationships between the molecular structure of the isothiocyanates and their toxicities were estimated by regressing lethal concentrations against the compound's respective physiochemical parameters. Isothiocyanate polarity (log octanol/water partition coefficient) had the most significant effect on observed toxicities, whereas electronic and steric characteristics were unimportant. Using this linear structure–activity relationship, they predict that the highest contact toxicities to black vine weevil eggs will result from glucosinolates producing isothiocyanates with higher numbers of carbon atoms or those bearing sulfinyl, thio, or aromatic moieties.

7.5.3 Larval Responses

Electrophysiological study on lepidopterous larvae shows that contact chemoreceptors in the mouthparts respond to glucosinolates but are only weekly responsive to isothiocyanates. In *Pieris brassicae* and *P. rapae*, lateral and medial maxillary sensilla of each species show different response profiles toward a range of glucosinolates tested (48).

Glucosinolates are the active contact stimuli in the feeding specificity of cruciferous insects. Lepidopteran larvae of the crucifer-feeding moth *Plutella maculipennis* were fed cellulose-agar gels containing a variety of glucosinolates isolated from Cruciferae and Tropaeoaceae. All of the glucosinolates stimulated feeding activity in these cabbage-reared larvae. Minimal concentrations that elicit feeding responses were reported as 8 ppm (70), a concentration comparable to thresholds observed in electrophysiological studies of another crucifer specialist, *Pieris brassicae*. Exogenous methyl and allyl glucosinolates in semisynthetic diets of the *P. brassicae* similarly elicits larval feeding in the absence of crucifer leaves. Larval lepidopteran behavior differs toward glucosinolates with different R groups. In *Plutella maculipennis*, allyl glucosinolate was the most highly stimulatory at the lowest concentration (< 400 ppm) whereas other glucosinolates, notably 2-hydroxy-3-butenyl, elicited greater response at higher concentrations (70). These results suggest that glucosinolates could provide a basis for caterpillar preference among crucifer parts containing several glucosinolates in different proportions. Short-chain and unsaturated allyl R groups gave the strongest response in electrophysiological studies of larvae of another lepidopteran crucifer specialist, *Mamestra brassicae*. Glucosinolates and their breakdown products are implicated as the active contact stimuli in larval feeding. Larvae of diamondback moth, *Plutella xylostella*, were induced to feed on agar fortified with various glucosinolates (54). However, allylisothiocyanate did not stimulate feeding. Glucosinolates stimulated larval feeding by *Pieris brassicae* and *P. rapae*. Larval *P. brassicae* were highly sensitive to concentration and composition of glucosinolate. The concentration of allylglucosinolate tested at 0.01 mM against *P. rapae* was well below average concentration in cabbage (12,70,71). Glucosinolate also affects larval development.

7.5.4   Phagostimulants

Glucosinolates function as gustatory stimulants for the immature adapted insects. The stimulatory effect varies among the glucosinolates or hydrolytic products evaluated (72). Methyl glucosinolate was the strongest stimulant to *Pieris brassicae* among the compounds tested. Benzylglucosinolate was most stimulatory to *Pieris cruciferae*, and 3-methylsulfonylpropyl glucosinolate was more stimulatory than benzyl or allyl glucosinolate.

Glucosinolates and their degradation products are considered to function as phagostimulants and to have a major role in host plant location and colonization by many phytophagous insects specifically adapted to brassicacious plants (73). The nature of side chain R in the glucosinolates and isothiocyanates, their concentration in the host plant, and the type of pest determine their biological activity. A plant lacking glucosinolates was not recognized as a host plant by brassicacious specialist insects and insects like *Phyllotreta* spp. were able to distinguish between cultivars on the basis of their glucosinolate content. Pawlowski et al. (74) have demonstrated that grasshoppers (*Melanoplus sanguinipes*) discriminated cultivars based on isothiocyanate levels. The cultivars containing lower isothiocynate levels were greatly preferred for their feeding. Several workers have shown stimulation of the flea beetle *Phyllotreta cruciferae* (Goeze) and the striped flea beetle *P. striolata* (F.) in the presence of glucosinolates and their hydrolysis products. Allyl glucosinolate had a strong stimulating effect on *Phyllotreta armoraciae* (Koch). Benzyl, *p*-hydroxybenzyl, 2-phenethyl, and 2-butyl glucosinolates were also recognized as stimulants (75).

Larsen et al. (76) identified 6-methylsulfinylhexyl (glucohesperalin), 3,4-dihydroxybenzyl (glucomatronalin), and 3-*O*-apiosylglucomatronalin, as the most powerful feeding stimulants for *Ceutorhynchus inaffectatus* Gyllenhal, when compared to allyl, 3-methylsulfinylpropyl, benzyl, and *p*-hydroxybenzyl glucosinolates.

Diamondback moth, *Plutella xylostella*, is stimulated to feed by many glucosinolates, but two of these (3-butenyl and 2-phenylethyl) are toxic to them at high concentration. The glucosides sinigrin, sinalbin, and glucocheriolin act as specific feeding stimulants for *P. xylostella*, and 40 plant species containing one or more of these chemicals serve as hosts. Nonhost plants may contain these stimulants but also contain feeding inhibitors or toxins (77).

The stimulatory effect of feeding sucrose and sinigrin was studied with fifth-instar nymphs of *Schistocerca gregaria*. The food consisted of filter paper impregnated with the phagostimulants dissolved at different concentrations in water. The intake of sucrose- or sinigrin-impregnated paper gradually increased up to a limit, depending on the concentration of the phagostimulant, then decreased. The results with the two phagostimulants were comparable; however, they were obtained with lower amounts of sinigrin than of saccharose. There was no dissuasive or toxic effect even at very high concentration (78).

7.5.5 Adult Beetle Response

The number of live cabbage stem flea beetles (*Psylliodes chrysocephala* L.) in oilseed rape was reduced by application of 2-phenethylisothiocyanate precursors, whereas application of but-3-enyl isothiocyanate precursors reduced feeding by both flea beetles and damage by adult seed weevils (79). The pent-4-enylisothiocyanate was shown to be repellent (the effect being concentration dependent), whereas no significant response was elicited by 2-phenethylisothiocyanate (80).

The influence of secondary plant substances (glucosinolates, alkaloids, and saponins) on the food choice by *Leptinotarsa decemlineata* was investigated. Choice and no-choice assays with potato leaf discs were carried out. Insect feeding was decreased on leaf discs treated with tested substances, such as glucosinolates, and depended on the concentration of used compounds (81).

High concentrations of *p*-hydroxybenzyl glucosinolate in young cotyledons (up to 20 mM) and leaves (up to 10 mM) of mustard seedlings (*S. alba*) have been identified as responsible for deterring feeding (82) by the flea beetle, *Phyllotreta cruciferae*, while lower concentrations (2–3 mM) offer little protection or may even act to stimulate feeding. Pivnick et al. (83), testing several isothiocyanates at the same concentration, showed allylisothiocyanate to be more attractive to *Phyllotreta cruciferae* and *Phyllotreta striolata* than benzyl-, ethyl-, phenyl-, and 2-phenethylisothiocyanates, with nitriles being the least attractive compounds. Larsen et al. (84) reported indol-3-ylmethyl glucosinolate to be one of the most stimulatory glucosinolates for flea beetles (*Phyllotreta* spp). Comparative studies have indicated but-3-enyl, pent-4-enyl, benzyl, and *p*-hydroxybenzyl glucosinolates to be more stimulatory than 2-propenyl glucosinolate to feeding by the cabbage seed weevil *Ceutorhynchus assimilis* Payk., whereas indol-3-ylmethyl glucosinolate has no effect. These results are in aggreement with the finding that antennae of *Ceutorhynchus assimilis* and *Psylliodes chrysocephala* L. possess a high proportion of olfactory receptors that respond strongly to but-3-enyl- and pent-4-enylisothiocyanates but not to allylisothiocyanate.

High concentrations of glucosinolates can protect plants against insects. For example, 2-propenyl glucosinolate deters the feeding of mustard aphid *Lipaphis erysimi* (85). The desert locust (*Schistocerca gregaria* Forsk.) is repelled by high levels of glucosinolates (214 µmol/g dry matter), while low concentrations (21 µmol/g dry matter) acted as phagostimulants. Thus, new cultivars with reduced glucosinolate levels may be more susceptible to particular insects (86). The feeding behavior of crucivorous beetles also shows differential response to R groups. Benzyl glucosinolate was the most highly stimulatory compound for several flea beetle species examined. The most stimulatory compound was indolylmethyl glucosinolate. In contrast to results from flea beetles, laboratory tests with the seed weevils *Ceutorhynchus assimilis* showed that the longer chain alkyl glucosinolates and benzyl and *p*-hydroxybenzyl glucosinolates stimulated greater feeding activity than did allyl glucosinolates (75,86). In general, beetle feeding and oviposition activities are poorly correlated with seed glucosinolate profiles. Some glucosinolates are not stimulatory to crucifer specialists, e.g., indolylmethyl glucosinolate to seed weevils. Similar negative results are reported for *Entomoscells americana* which does not respond to 2-hydroxy-3-butenyl glucosinolate (86).

7.5.6 Toxicity

Lichtenstein et al. (87) found root extracts of Brussels sprouts to be very toxic to *Drosophila melanogaster* as demonstrated by 50% mortality in 10 min. Macerated root tissue was also quite toxic; they observed 50% mortality in 3 h for *Drosophila* and 50% mortality in 24 h for the common house fly *Musca domestica*. Toxicity was in most cases strongly correlated with phenylethyl isothiocyanate content. Insecticidal activity of several isothiocyanates has been demonstrated, especially for aromatic compounds (88). In screening products to protect seedling corn from injury by Southern corn root worm (*Diabrotica undecimpunctata howardi*), allylisothiocyanate killed the worm but also prevented the corn seeds from germinating. Methyl-isothiocyanate is mainly used to control various fungi that cause damping off and root rot, but it is also used to control insects, such as wireworms and symphilids, and nematodes. Allylisothiocyanate was the most effective of 57 volatile compounds tested against the wireworms *Limonium californicus* and *Limonius canus*, with an LC_{50} of 2.33 $\mu g\,mL^{-1}$. Rapeseed meal amended (3% on a weight basis) to soil repelled wireworms (*L. californicus*) but in uncovered containers did not kill them (89). However, it is important to note that glucosinolate products like isothiocyanate are not always toxic to insects at concentrations found in plant tissues. In fact, isothiocyanate can act as a cue or even an attractant influencing the behavior of certain insects (90). Other glucosinolate hydrolysis products have insecticidal properties as well. Organic thiocyanates have been used as insecticides to control weevils in grain and to produce quick knockdown of flying insects such as flies while having a relatively low impact on mammals. Indolylacetonitrile, known more for its auxin-like activity on plants, also inhibits growth of insects. Thiocyanate is ineffective when used alone but is insecticidal in combination with other chemicals. Thiocyanate was toxic to wireworms only at concentrations much higher than observed in rapeseed meal–amended soil (91).

Relatively few reports of the insecticidal activity of glucosinolates and hydrolysis products have been produced, but methylisothiocyanate has been offered commercially as an insecticide and fumigant. Lichtenstein et al. (68) reported that 2-phenylethylisothiocyanate in the edible parts of turnips was insecticidal toward vinegar flies (*Drosophila melanogastor* Meig), house flies (*Musca domestica* L.), confused flour beetles (*Tribolium confusum* Duvae), spider mites (*Tetranychus atlantic*), and pea aphids (*Macrosiphum pisi*).

Crambe (*Crambe abyssinica*) seed meal in diets of house flies resulted in high mortalities and was repellant and/or deterred feeding of the beetles *Tribolium castaneum* and *Oryzeaphilus surinamensis* (92). Soil amended with allylisothiocyanate showed both lethal and sublethal effects on *Limonius californicus* Mann. depending on concentration. Isothiocyanates can also provide some protection against the Bertha armyworm, *Mamestra configurata* Walker (93). Phenylethylisothiocyanate, the myrosinase hydrolysis product of watercress, was shown to be a feeding deterrent to amphipods (*Gammarus pseudolimnaeus*), caddisflies (*Hesperophylax designatus* and *Limnephilus* spp) (94).

Allylisothiocyanate was shown to be acutely toxic to *Papilio polyxenes* L. larvae, which normally attack only plants of the Umbeliferae. Larval growth of *Spodoptera eridania*, a generalist feeder, was inhibited by high concentrations of the compound, while larval growth of *Pieris rapae* was not affected (70). Insecticidal activity of phenylethylisothiocyanate and phenylisothiocyanate to larvae of the soldier fly, *Inopus rubiceps* (Macq.), at high dosage levels was reported by Lowe et al. (95).

7.5.7 Fumigant

Naturally occurring aglucones of three glucosinolates (sinigrin, glucotropaeolin, and epiprogoitrin) were tested for fumigation activity against the house fly, *Musca domestica*, and the lesser grain borer, *Rhizopertha dominica*. A total of eight natural aglycones were evaluated in the bioassays (96). Two aglucones of sinigrin showed efficacy against both species, which was comparable with that of a commercial fumigant, chloropicrin. None of the aglucones tested was comparable in activity to dichlorvos. Aglucones of glucotropaeolin were also insecticidal but not to the same level as the sinigrin aglucones. The aglucones of epiprogoitrin were only slightly effective as fumigants. A quantitative structure–activity relationship (QSAR) was developed for synthetic analogues of the sinigrin and glucotropaeolin aglucones. An electronic parameter provided the best predictor of activity in *R. dominica*, whereas a hydrophobicity parameter best predicted activity in *M. domestica* (96).

7.5.8 Toxicity to Nematode

Winkler and Otto (97) found that rotational plantings of rape or mustard in strawberries checked the spread of some nematodes, particularly *Pratylenchus penetrans*. Leguminosae, potato, or grass rotations did not alter nematode populations, although green manure applications of leguminosans (peas, beans, and vetch), rape, or mustard reduced nematode numbers. Similarly, growing rapeseed in soil and incorporating the tissue as a green manure significantly reduced populations of *Meloidogyne chitwoodi* compared to fallow, although results were not as clear in a study involving *Meloidogyne incognita* and *Meloidogyne javanica* (98,99). In the presence of myrosinase, glucosinolates are toxic to *Heterodera schachtii* at concentrations of 0.5–5.0 mg mL^{-1} (100). Some varieties of *Brassica* crops are being offered commercially as "trap crops" for nematodes. Rather than killing the nematode, *Brassica* plants interfere with nematode reproductive cycles.

7.5.9   Parasitoid Attractant

Parasitoids are known to be attracted to hostplant volatiles. Thus, manipulation of the hostplant chemistry may provide a means of enhancing the attraction of parasitoids to their prey. In a study, Bradburn and Mithen (101) describe the different attraction of the braconid wasp *D. rapae* to two near-isogenic lines of *Brassica oleracea* that differ in a gene that alters the chemical structure of the isothiocyanate, which are emitted following tissue damage. They demonstrated that by enhancing the production of but-2-enyl isothiocyanates in *Brassica oleracea* and *Brassica napus* they can increase the attraction of *D. rapae* to these plants under standard field conditions.

7.5.10   Induction of Glucosinolates

Pests and disease infections can markedly alter glucosinolate levels. It has been demonstrated that attack by insects including the flea beetle *Psylliodes chrysocephala* (L.) in oilseed rape *Brassica napus* and the root fly *Delia floralis* Fall in cruciferae plants can change both the total concentration of glucosinolates in different plant tissues and the relative proportion of aliphatic and aromatic compounds (102). Infestation of oilseed rape by the cabbage stem flea beetle leads to a substantial increase of indole glucosinolates, glucobrassicin, and neoglucobrassicin. Damage to the roots of *Brassica napus* by *D. floralis* has led to two- to fourfold increases in indole-based glucosinolates. The largest increase from 4- to 17-fold for an individual compound after root fly attack was found for 1-methoxy-3-indolylmethyl glucosinolate. Induction of glucosinolates indicates a defensive role for these compounds.

7.6   MECHANISM OF ACTION

Glucosinolates and their degradation products function as part of the plant's defense against insect attack to act as phagostimulants, and host plant location and colonization by many phytophagous insects. They also play a role in the location of insects by their parasitoids. Antixenosis (nonpreference) and tolerance have been identified as two mechanisms in seedlings of *S. alba* that probably account for flea beetle resistance in this species. The insecticidal activity of glucosinolates is a result of changes in the metabolism of the insect, specifically the inhibition of the glycolysis, Krebs cycle by decreasing the total O_2 uptake and CO_2 expired, as demonstrated using yellow mealworm larvae (*Tenebrio molitor* L.). Glucosinolates serve as gustatory stimulants for insect pests of cruciferous plants and their cleavage products; isothiocyanates are feeding and oviposition attractants for many insect species.

Isothiocyanates are generally biocides whose activity results from interaction with proteins (103). They interact nonspecifically and irreversibly with proteins and amino acids to form stable products by reacting with sulfhydryl groups, disulfide bonds, and amines. Isothiocyanates are known to inactivate enzymes *in vitro*. Cyanide (salts or gas) inactivate certain enzyme systems, especially those involved in cellular respiration such as cytochrome oxidase (104).

7.7 MOLECULAR BIOLOGY

The presence of glucosinolates in seeds like canola is disadvantageous because their presence reduces the nutritional status of the seed meal, whereas reducing glucosinolate levels in rape and mustard may cause problems by inducing susceptibility to pest insects and pathogens. It was later realized that cultivars could be developed that have high levels of total glucosinolates in leaves but reduced total seed glucosinolate levels. Thus, genetic engineering allows developing plants that combine high glucosinolate levels in leaves that produce plant–pest, plant–pathogen, and plant–herbivore interactions advantageous to plants, with low seed glucosinolate levels to impart quality advantages (105).

Several genetic engineering approaches were made to reduce the glucosinolate levels in the plants. One approach is to create a new pathway that competes for the sulfate donor 3-phosphoadenosine-5-phosphosulfate (PAPS). This is possible by introduction of the flavonol-3-sulfotransferase (F3ST) gene to create novel but innocuous metabolite, flavonol-3-sulfate. This strategy is based on the assumption that the foreign F3ST enzyme will compete with the native desulfoglucosinolate sulfotransferase (DGST) enzyme involved in the sulfation of desulfoglucosinolate to glucosinolate. Since the molecular affinity of the F3ST enzyme for its substrate and PAPS is one to two orders of magnitude higher that of the DGST, the sulfation reaction of native flavones is expected to predominate over that of native desulfoglucosinolates. The chimeric gene promoter–F3ST coding region from *Flaveria* spp has been used to transform several plant species including canola (106).

Another approach is to limit the supply of the substrates, namely, amino acids. The substrates are redirected to the synthesis of another metabolite. For example, tryptophan, the biosynthetic precursor of indolyl glucosinolates, is redirected to form tryptamine by tryptophan decarboxylase (TDC) chemeric gene with promoter–TDC coding region from *Catherantus roseus* and marker in vector has been used to create transgenic tobacco, potato, and canola. Transgenic plants that expressed the decarboxylase activity accumulated tryptamine and lower levels of tryptophan-derived glucosinolates were produced in all plant parts in comparison with nontransformed control plants. The indole glucosinolate content of mature seeds from transgenic plants was 3% of that found in nontransformed seeds. Thus, the undesirable indole glucosinolates, which decrease the nutritional quality of canola oilseed meal, have been reduced (107).

A cDNA encoding CYP79B1 has been isolated from *Sinapis alba* (108). CYP79B1 from *S. alba* shows 54% sequence identity and 73% similarity to sorghum CYP79A1 and 95% sequence identity to the *Arabidopsis* T42902, assigned CYP79B2. The high identity and similarity to sorghum CYP79A1 and 95% sequence identity to the *Arabidopsis* T42902, assigned CYP79B2. The high identity and similarity to sorghum CYP79A1, which catalyzes the conversion of tyrosine to *p*-hydroxyphenylacetaldoxime in the biosynthesis of the cyanogenic glucoside dhurrin, suggests that CYP79B1 similarly catalyzes the conversion of amino acid(s) to aldoxime(s) in the biosynthesis of glucosinolates. Within the highly conserved "PIRF" and the heme binding region of A-type cytochromes, the CYP79 family has unique substitutions that define the family-specific consensus sequences of FXP(E/D)RH and SFSTG(K/R)RGC(A/I)A, respectively. Sequence analysis of PCR products generated with CYP79B subfamily-specific primers identified CYP79B homologues in *Tropaeolum majus, Carica papaya, Arabidopsis, Brassica napus, and Sinapis alba*. The five glucosinolate-producing plants identified a CYP79B amino acid consensus sequence KPERHHLNECE-SEVTL TENDERFISFSTGKRGC. The unique substitutions in the "PERF" and the heme binding domain and the high sequence identity and similarity of CYP79B1, CYP79B2, and CYP79A1, together with the isolation of CYP79B homologues in the distantly related Tropaeolaceae, Caricaceae, and Brassicaceae within the Capparales order, show that the initial part of the biosynthetic pathway of glucosinolates and cyanogenic glucosides is catalyzed by evolutionarily conserved cytochromes P_{450}. This confirms that the appearance of glucosinolates in Capparales is based on a cyanogen "predisposition." Identification of CYP79 homologues in glucosinolate-producing plants provides an important tool for tissue-especific regulation of the level of glucosinolates to improve nutritional value and pest resistance. Nucleotide sequence data appear under accession numbers CYP79B1 and CYP79B2.

Studies of the biosynthetic pathway of the tyrosine-derived cyanogenic glucoside dhurrin in *Sorghum bicolor* have shown that tyrosine is converted to *p*-hydroxyphenylacetaldoxime by the multifunctional cytochrome P_{450}. CYP79A1 catalyzes two consecutive N-hydroxylation reactions followed by a dehydration and decarboxylation reaction. The oxime is then converted by another cytochrome P_{450}, CYP71E1, to the aglycone *p*-hydroxymandeloni-trile. The presence of highly conserved CYP79 homologues in the Capparales suggests that the enzymes involved in the conversion of amino acids to oximes in the biosynthesis of glucosinolates and cyanogenic glucosides are evolutionarily conserved. Introduction of CYP79A1 into *Arabidopsis thaliana* resulted in the production of high levels of *p*-hydroxybenzyl glucosinolate without altering the level and profile of the major endogenous glucosinolates (109). The fourfold increase in total level of glucosinolates in the selected line 79.1 shows that the postoxime enzymes have the capacity to produce substantially more glucosinolates which, combined with the low substrate specificity, shows significant flexibility in the glucosinolate biosynthetic pathway. By the expression of CYP79A1 they have obtained levels of *p*-hydroxybenzyl glucosinolate in the leaves of *Sinapis alba*, which has *p*-hydroxybenzyl glucosinolate as its major glucosinolate. The transgenic *A. thaliana* lines accumulate up to 4.7 nmol *p*-hydroxybenzylglucosinolate mg-1 FW compared to *S. alba* which contain 2 nmol mg^{-1} FW in young leaves. Production of *p*-hydroxybenzyl glucosi-nolate in transgenic *Arabidopsis thaliana* expressing sorghum CYP79A1 demonstrates that it is possible to use genetic engineering for production of *Brassica* crops with new glucosinolate profiles. The availability of glucosinolate-producing plants with altered glucosinolate profiles is important for studies of the biological role of individual glucosinolates with respect to insect interactions and nutritional value.

The enzyme involved in the degradation of glucosinolate is myrosi-nase. The first cDNA gene encoding myrosinase was isolated from *Sinapis alba*. Five functional genomic myrosinase genes have so far been sequenced, two from *Brassica napus*, two from *A. thaliana*, and one from *Brassica campestris*. A sequence comparison of the five functional myrosinase genomic genes has been presented (110).

Kliebenstein et al. (111) studied the role of the glucosinolate-myrosinase chemical defense system in protecting *A. thaliana* from specialist and generalist insect herbivory. Two *Arabidopsis* recombinant inbred population were used in which one had previously mapped QTL controlling variation in the glucosinolae-myrosinase system. QTL controlling resistance to specialist (*Plutella xylostella*) and generalist (*Trichoplusia ni*) herbivores was also mapped. Kliebenstein et al. (111) identified a number of QTL that are specific to one herbivore or the other, as well as a single QTL that controls resistance to both insects. Comparision of QTL for herbivory, glucosinolates, and myrosinase showed that *T. ni* herbivory is strongly deterred by higher glucosinolate levels, faster breakdown rates, and specific chemical structures. In contrast, *P. xylostella* herbivory is uncorrelated with variation in the glucosinolate-myrosinase system. This agrees with evolutionary theory stating that specialist insects may overcome host plant chemical defenses, whereas generalists will be sensitive to these same defenses.

7.8 PERSPECTIVES

Significant progress has been made in recent years toward understanding the biochemistry and molecular biology of the biosynthetic pathway of glucosinolates. Based on this knowledge, genetic manipulation of glucosinolates may be used to produce a glucosinolate profile more suited to a particular purpose, such as improved resistance to pests or diseases or enhancement of flavor or nutritional quality. CYP79 homologues with different substrate specificities from other glucosinolate and cyanogenic glucoside–producing plants can be used to introduce novel glucosinolates with desired biological function in crop plants. Introduction of the oxime-metabolizing cytochrome P_{450} from cyanogenic plants into glucosinolate-producing plants may channel the many oximes produced into hydroxynitriles, which may become glucosylated into new cyanogenic glucosides. The biological function of the different glucosinolates and their degradation products is not well understood. Design of transgenic plants with altered glucosinolate profiles provides an important tool with which to elucidate the biological function of these compounds, including their role in insect resistance. Cultivars could be developed that have high levels of glucosinolates in leaves but reduced levels of seed glucosinolates. High glucosinolate levels in leaves produce plants with enhanced and profitable interaction with pests, pathogens, and herbivores. On the other hand, plants containing low seed glucosinolate levels impart quality advantages.

REFERENCES

1. Kjaer, A. Naturally derived isothiocyanates and their parent glucosides. In *Progress in the Chemistry of Organic Natural Products*; Zechmeister, L., Ed.; Springer-Verlag: Berlin, 1960; 22–176.

2. Challenger, F. The natural mustard oil glucosides and the related isothiocyanates and nitriles. In *Aspects of the Organic Chemistry of Sulphur*; Butterworths: London, 1959; 115–161.

3. Ettlinger, M.G.; Kjaer, A. Sulfur compounds in plants. In *Rec. Adv. Phytochem*; Mabry, T.J., Alston, R.E., Runeckles, V.C., Eds.; 1968; 1, 89–144.

4. Larsen, P.O. Glucosinolates. In *The Biochemistry of Plants*; Conn, E.E., Ed.; Academic Press: New York, 1981; Vol. 7, 501–525.

5. Rosa, E.A.S.; Heaney, R.K.; Fenwick, G.R.; Portas, C.A.M. Glucosinolates in crop plants; Horticult. Rev. **1997**, *19*, 99–215.

6. Underhill, E.W. Glucosinolates. In *Encyclopedia of Plant Physiology*; Bell, E.A., Charlwood, B.V., Eds.; Springer-Verlag: New York, 1980; Vol. 8, 493–511.

7. Louda, S.; Mole, S. Glucosinolates: Chemistry and Ecology. In *Herbivores*. Rosenthal, G.A., Berenbaum, M.R., Eds.; Academic Press: San Diego, 1991; 123–164.

8. Tookey, H.L.; VanEtten, C.H.; Daxenbichler, M.E. Glucosinolates. In *Toxic Constituents in Food Crops*; Liener, I.E., Ed.; Academic Press: New York, 1980; 103–142.

9. Benn, M. Glucosinolates. Pure Appl. Chem. **1977**, *49*, 197–210.

10. Poulton, J.E.; Moller, B.L. Glucosinolates. In *Methods in Plant Biochemistry*; Lea, P.J., Ed.; Academic Press: New York, 1993; Vol. 9, 209–237.

11. Duncan, A.L. Glucosinolates. In *Toxic Substances in Crop Plants*; D'Mello, J.P.F., Duffus, C.M., Duffus, J.H., Eds.; Royal Society of Chemistry; London, 1992; 126–147.

12. Chew, F.S. Biological effects of glucosinolates. In *Biologically Active Natural Products: Potential Use in Agriculture*; Cutler, H.G., Ed.; American Chemical Society Symposium, 1988; 155–181.

13. Sorensen, H. Glucosinolates: structure–properties function. In *Canola and Rapeseed: Production, Chemistry, Nutrition and Processing Technology*; Shahidi, F., Ed.; Van Nostrand Reinhold: New York, 1991; 149–172.

14. Berneys, E.A. Nitrogen in defense against insects. In *Nitrogen as an Ecological Factor*; Lee, J.A., McNeil, S., Rorison, I.H., Eds.; Blackwell Scientific: Oxford, 1983; 321–344.

15. Halkier, B.A. Glucosinolates. In *Naturally Occurring Glycosides*; Ikan, R., Ed.; Wiley: Chichester, 1999; 193–223.

16. Rodman, J.E. A taxonomic analysis of glucosinolate-producing plants. Part 1: Phenet. Syst. Bot. **1991**, *16*, 598–618.

17. Clossais-Besnard, N.; Larher, F. Physiological role of glucosinolates in *Brassica napus*. Concentration and distribution pattern of glucosinolates among plant organs during a complete life cycle. J. Sci. Food Agric. **1991**, *56*, 25–38.

18. Kjaer, A. Glucosinolates and related compounds. Food Chem. **1980**, *6*, 223–234.

19. VanEtten, C.H.; Daxenbichler, M.E.; Williams, P.H.; Kwolek, F. Glucosinolates and derived products in cruciferous vegetables: analysis of the edible part from twenty two varieties of cabbage. J. Agric. Food Chem. **1976**, *24*, 452–455.

20. MacLeod, A.J.; Nussbaum, M.L. The effects of different horticultural practices on the chemical flavour composition of some cabbage cultivars. Phytochemistry **1977**, *16*, 861–865.
21. MacLeod, A.J.; Pikk, A.E. A comparison of the chemical flavour composition of some Brussels sprouts cultivars grown at different crop spacings. Phytochemistry **1978**, *17*, 1029–1032.
22. Fenwick, R.G.; Heaney, R.K.; Mullin, W.J. Glucosinolates and their breakdown products in foods and food plants. CRC Crit. Rev. Food Sci. Nutr. **1983**, *18*, 123–201.
23. Bjorkman, R. Plant myrosinases. In *The Biology and Chemistry of the Cruciferae*; Vaughan, J.G., MacLeod, A.J., Jones, B.M.G., Eds.; Academic Press: London, 1976; 191–205.
24. Halkier, B.A., Du, L. The biosynthesis of glucosinolates. Trends Plant Sci. **1997**, *2*, 425–431.
25. Rask, L.; Anderson, E.; Ekborn, B.; Eriksson, S.; Pontoppidan, B.; Mercer, J. Myrosinase: gene family evolution and herbivore defense in Brassicaceae. Plant Mol. Biol. **2000**, *42*, 93–113.
26. Bones, A.M.; Rossiter, J.T. The myrosinase-glucosinolate system, its organisation and biochemistry. Physiol. Plant. **1996**, *97*, 194–208.
27. Brown, P.D.; Morra, M.J. Control of soil-borne plant pests using glucosinolate-containing plants. Adv. Agron. **1997**, *61*, 167–231.
28. Kutacek, M.; Prochazka, Z.; Veres, K. Biogenesis of glucobrassicin, the *in vitro* precursor of ascorbigen. Nature **1962**, *194*, 393–394.
29. Underhill, E.W.; Chisholm, M.D.; Wetter, L.R. Biosynthesis of mustard oil glucosides. Can. J. Biochem. Physiol. **1962**, *40*, 1505–1514.
30. Underhill, E.W.; Chisholm, M.D. Biosynthesis of mustard oil glycosides. III. Formation of glucotropaeolin from L-Phenylalanine-C^{14}-N^{15}. Biochem. Biophys. Res. Commun. **1964**, *14*, 425–430.
31. Benn, M.H. Biosynthesis of mustard oils. Chem. Ind (London) **1962**, *12*, 1907.
32. Bennett, R.N.; Kiddle, G.; Wallsgrooe, R.M. Biosynthesis of benzylglucosinolate, cyanogenic glycosides and phenylpropanoids in *Carica papaya*. Phytochemistry **1997**, *45*, 59–66.
33. Underhill, E.W.; Wetter, L.R.; Chisholm, M.D. Biosynthesis of glucosinolates. In *Nitrogen Metabolism in Plants*; Goodwin, T.W., Smellie, R.M.S., Eds.; Biochem. Soc. Symp. 1973; Vol. 38, 303–326.
34. Du. L.; Halkier, B.A. Biosynthesis of glucosinolates in the developing silique walls and seeds of *Sinapis alba* Phytochemistry **1998**, *48*, 1145–1150.
35. Graser, G.; Schneider, B.; Oldham, N.J.; Gershenzon, J. The methionine chain elongation pathway in the biosynthesis of glucosinolates in *Eruca sativa* (Brassicaceae). Arch. Biochem. Biophys. **2000**, *378*, 411–419.
36. Dornemann, D.; Loffelhardt, W.; Kindl, H. Chain elongation of aromatic amino acids: the role of 2-benzylmalic acid in the biosynthesis of a C_6C_4 amino acid and a C_6 C_3 mustard oil glucoside. Can. J. Biochem. **1974**, *52*, 916–921.

37. deQuiros, H.C.; Magrath, R.; McCallum, D.; Kroymann, J.; Scnabelrauch, D.; Mitchellolds, T.; Mithen, R. α-Keto acid elongation and glucosinolate biosynthesis in *Arabidopsis thaliana*. Theor. Appl. Genet. **2000**, *101*, 429–437.
38. Kindl, H.; Schiefer, S. Aldoximes as intermediates in the biosynthesis of tyrosol and tyrosol derivatives. Phytochemistry **1971**, *10*, 1795–1802.
39. Du, L.; Lykkesfeldt, J.; Olsen, C.E.; Halkier, B.A. Involvement of cytochrome P450 in oxime production in glucosinolate biosynthesis as demonstrated by an in vitro microsomal enzyme system isolated from jasmonic acid–induced seedlings of *Sinapis alba* L. Proc. Natl. Acad. Sci. USA. **1995**, *92*, 12505–12509.
40. Du, L.; Halkier, B.A. Isolation of a microsomal enzyme system involved in glucosinolate biosynthesis from seedlings of *Tropaeolum majus* L. Plant Physiol. **1996**, *111*, 831–837.
41. Muller, J.L.; Hilgenberg, W. A plasma membrane bound enzyme oxidises L-tryptophan to indole 3-acetaldoxime. Physiol. Plant. **1988**, *74*, 240–250.
42. Dawson, G.W.; Hick, A.J.; Bennett, R.N.; Donald, A.M.; Pickett, J.A.; Wallsgrove, R.M. Synthesis of glucosinolate precursors and investigations into the biosynthesis of phenylalkyl and methylthioalkylglucosinolates. J. Biol. Chem. **1993**, *268*, 27154–27159.
43. Bennett, R.N.; Donald, A.M.; Dawson, G.W.; Hick, A.J.; Wallsgrove, R.M. Aldoxime-forming microsomal enzyme systems involved in the biosynthesis of glucosinolates in oilseed rape leaves. Plant Physiol. **1993**, *102*, 1307–1312.
44. Matsuo, M.; Kirkland, D.F.; Underhill, E.W.; 1-Nitro-2-phenyl ethane, a possible intermediate in the biosynthesis of benzylglucosinolate. Phytochemistry **1972**, *11*, 697–701.
45. Reed, D.W.; Davin, L.; Jain, J.C.; Deluca, V.; Nelson, L.; Underhill, E.W. Purification and properties of UDP-glucose: thiohydroximate glucosyltransferase from *Brassica napus* L. seedlings. Arch. Biochem. Biophys. **1993**, *305*, 526–532.
46. Glendening, T.M.; Poulton, J.E. Glucosinolate biosynthesis. Sulfation of desulfo glucosinolate by cell-free extracts of cress (*Lepidium sativum* L.) seedlings. Plant Physiol. **1988**, *86*, 319–321.
47. Giamoustaris, A.; Mithen, R. Genetics of aliphatic glucosinolates. IV. Side chain modification in *Brassica oleracea*. Theor. Appl. Genet. **1996**, *93*, 1006–1020.
48. Ma, W.C.; Schoonhoven, L.M. Tarcel contact chemosensory hairs of the large white butterfly *Pieris brassicae* and their possible role in oviposition behaviour. Entomol. Exp. Appl. **1973**, *16*, 343–357.
49. Stadler, E. Chemoreception of host plant chemicals by ovipositing females of *Delia brassicae*. Entomol. Exp. Appl. **1978**, *24*, 711–720.
50. Renwick, J.A.A.; Radke, C.D. Chemical recognition of host plants for oviposition by the cabbage butterfly, *Pieris rapae* (Lepidoptera: Pieridae). Environ. Entomol. **1983**, *12*, 446–450.
51. Chew, F.S. Searching for defensive chemistry in the Cruciferae or, do glucosinolates always control interactions of Cruciferae with their potential herbivores and symbionts? No! In *Chemical Mediation or Coevolution*; Spencer, K.C., Ed.; Academic Press: San Diego, 1988; 81–112.

52. Huang, X.; Renwick, J.A.A. Differential selection of host plants by two *Pieris* species: the role of oviposition stimulants and deterrents. Entomol. Exp. Appl. **1993**, *68*, 59–69.

53. Huang, X.; Renwick, J.A.A. Relative activities of glucosinolates as oviposition stimulants for *Pieris rapae* and *P. napi oleracea*. J. Chem. Ecol. **1994**, *20*, 1025–1037.

54. Huang, X.; Renwick, J.A.A.; Gupta, K.S. A chemical basis for differential acceptance of *Erysimum cheiranthoides* by two *Pieris* species. J. Chem. Ecol. **1993**, *19*, 195–210.

55. Huang, X.; Renwick, J.A.A.; Gupta, K.S. Oviposition stimulants and deterrents regulating differential acceptance of *Iberis amara* by *Pieris rapae* and *P. napi oleracea*. J. Chem. Ecol. **1993**, *19*, 1645–1663.

56. Renwick, J.A.A.; Radke, C.D.; Gupta, K.S.; Stadler, E. Leaf surface chemicals stimulating oviposition by *Pieris rapae* (Lepidoptera: Pieridae) on cabbage. Chemoecology **1992**, *3*, 33–38.

57. Finch, S. Volatile plant chemicals and their effect on host plant finding by the cabbage root fly (*Delia brassicae*). Entomol. Exp. Appl. **1978**, *24*, 150–159.

58. Finch, S.; Skinner, G. Trapping cabbage root flies in traps baited with plant extracts and with natural and synthetic isothiocyanates. Entomol. Exp. Appl. **1982**, *31*, 133–139.

59. Roessingh, P.; Stadler, E.; Fenwick, G.R.; Lewis, J.A.; Nielsen, J.K.; Hurter, J.; Ramp, T. Oviposition and tarsal chemoreceptors of the cabbage root fly are stimulated by glucosinolates and host plant extracts. Entomol. Exp. Appl. **1992**, *65*, 267–282.

60. Simmonds, M.S.J.; Blaney, W.M.; Mithen, R.; Birch, A.N.E.; Lewis, J. Behavioural and chemosensory responses of the turnip root fly (*Delia floralis*) to glucosinolates. Entomol. Exp. Appl. **1994**, *71*, 41–57.

61. Birch, A.N.E.; Stadler, R.J.; Hopkins, R.J.; Simmonds, M.S.J.; Baur, R.; Griffiths, D.W.; Ramp, T.; Hurter, J.; McKinlay, R.G. Mechanisms of resistance to the cabbage and turnip root flies: collaborative field, behavioural and electrophysiological studies. Bull OILB/SROP. **1993**, *16*, 1–5.

62. Reed, D.W.; Pivnick, K.A.; Underhill, E.W. Identification of chemical oviposition stimulants for the diamondback moth, *Plutella xylostella*, present in three species of Brassicaceae. Entomol. Exp. Appl. **1989**, *53*, 227–236.

63. Wensler, R.J.D. Mode of host selection by an aphid. Nature **1962**, *195*, 830–831.

64. Traynier, R.M.N. Chemostimulation of oviposition by the cabbage root fly, *Hylemia brassicae* (Bouche). Nature **1965**, *210*, 218–219.

65. Nair, K.S.S.; McEwen, F.L. Host selection by the adult cabbage maggot, *Hylemia brassicae* (Diptera: Anthomyiidae): effect of glucosinolates and common nutrients on oviposition. Can. Entomol. **1976**, *108*, 1021–1030.

66. Seo, S.T.; Tang, C.S. Hawaiian fruit flies (Diptera: Tephritidae): Toxicity of benzyl isothiocyanate against eggs of 1st instars of three species. J. Econ. Entomol. **1982**, *75*, 1132–1135.

67. Ahman, I. Toxicities of host secondary compounds to eggs of the *Brassica* specialist *Dasineara brassicae*. J. Chem. Ecol. **1986**, *12*, 1481–1488.
68. Lichtenstein, E.P.; Strong, F.M.; Morgan, D.G. Identification of 2-phenylethyl isothiocyanate as an insecticide occurring in the edible part of turnips. J. Agric. Food Chem. **1962**, *10*, 30–33.
69. Borek, V.; Elberson, L.R.; McCaffrey, J.P.; Moora, M.J. Toxicity of isothiocyanates produced by glucosinolate in Brassicaceae species to black vine weevil eggs. J. Agric. Food Chem. **1998**, *46*, 5318–5323.
70. Thorsteinson, A.J. The chemotactic responses that determine host specificity in an oligophagous insect [*Plutella maculipennis* (curt) Lepidoptera]. Can. J. Zool. **1953**, *31*, 52–72.
71. Blau, P.A.; Feeny, P.; Contardo, L. Allylglucosinolate and herbivorous caterpillars: a contrast in toxicity and tolerance. Science **1978**, *200*, 1296–1298.
72. Beck, S.D.; Resec, J.C. Insect–plant interactions: nutrition and metabolism. Rec. Adv. Phytochem. **1976**, *10*, 41–92.
73. Traynier, R.M.M.; Truscott, R.J.W. Potent natural egg laying stimulant for cabbage butterfly, *Pieris rapae*. J. Chem. Ecol. **1991**, *17*, 1371–1380.
74. Pawlowski, S.H.; Riegert, P.W.; Krzymanski, J. Use of grasshoppers in bioassay of thioglucosides in rapeseed (*Brassica napus*). Nature **1968**, *220*, 174–175.
75. Nielson, J.K. Host plant discrimination within cruciferae: feeding responses of four leaf beetles (Coleoptera: Chrysomelidae) to glucosinolates, cucurbitacins and cardinolides. Entomol. Exp. Appl. **1978**, *24*, 41–54.
76. Larsen, L.M.; Nielsen, J.K.; Sorensen, H. Host plant recognition in monophagous weevils: specialization of *Ceutorhynchus inaffectatus* to glucosinolates from its host plant *Hesperis matronalis*. Entomol. Exp. Appl. **1992**, *64*, 49–55.
77. Bodnaryk, R.P. Will low-glucosinolate cultivars of the mustards *Brassica juncea* and *Sinapis alba* be vulnerable to insect pests? Can. J. Plant Sci. **1997**, *77*, 283–287.
78. Aspirot, J.; Lauge, G. Phagostimulatory action of saccharose and sinigrin and proof of their regulation in the larvae of crickets, *Schistocerca gregaria*. Reprod. Nutr. Dev. **1981**, *21*, 695–704.
79. Griffiths, D.C.; Hick, J.A.; Pye, B.J.; Smart, L.E. The effects on insect pests of applying isothiocyanate precursors to oilseed rape. Asp. Appl. Biol. **1989**, *23*, 359–364.
80. Garraway, R.; Leake, L.D.; Ford, M.G.; Henderson, I.F.; Hick, A.J.; Wadhams, L.J. The action of oilseed rape metabolites on olfactory nerve activity and behaviour of *Deroceras reliculatum*. Brighton Crop Protection Conf. Pests and Diseases **1992**, *6C-10*, 593–596.
81. Waligora, D. The influence of different secondary plant substances, glucosinolates, alkaloids and saponins on the food choice of larvae and beetles of Colorado Potato beetle (*Leptinotarsa decemlineata* Say). J. Plant Prot. Res. **1998**, *38*, 93–104.

82. Bodnaryk, R.P. Developmental profile of sinalbin (p-hydroxybenzyl glucosinolate) in mustard seedlings, *Sinapis alba* L., and its relationship to insect resistance. J. Chem. Ecol. **1991**, *17*, 1543–1556.
83. Pivnick, K.A.R.; Lamb, J.; Reed, D. Response of flea beetles *Phyllotreta* spp. to mustard oils and nitriles in field trapping experiments. J. Chem. Ecol. **1992**, *18*, 863–873.
84. Larsen, L.M.; Nielsen, J.K.; Ploger, A.; Sorensen, H. Responses of some beetle species to varieties of oilseed rape and to pure glucosinolates. In *Advances in the Production and Utilization of Cruciferous Crops*; Sorensen, H., Ed.; Martinus Nijhoff: Dordrecht, 1985; 230–244.
85. Rohilla, H.R.; Singh, H.; Kumar, P.R. Stratagies for the identification of the source of resistance in oilseeds Brassicae against *Lipaphis erysimi* (Kalt.). Ann. Biol. **1993**, *9*, 174–183.
86. Nielson, J.K.; Dalgaard, L.; Larsen, M.; Sorensen, H. Host plant selection of the horse radish flea beetle *Phyllotreta armoraciae* (Coleoptera: Chrysomelidae): feeding responses to glucosinolates from several crucifers. Entomol. Exp. Appl. **1979**, *25*, 227–239.
87. Lichtenstein, E.P.; Morgan, D.G.; Mueller, C.H. Insecticides in nature: naturally occurring insecticides in cruciferous crops. J. Agric. Food Chem. **1964**, *12*, 158–161.
88. Borek, V.; Elberson, L.R.; McCaffrey, J.P.; Morra, M.J. Toxicity of aliphatic and aromatic isothiocyanates to eggs of the black vine weevil (Coleoptera: Curculionidae). J. Econ. Entomol. **1995**, *88*, 1192–1196.
89. Brown, P.D.; Morra, M.J.; McCaffrey, J.P.; Auld, D.L.; Williams, L. Allelochemicals reduced during glucosinolate degradation in soil. J. Chem. Ecol. **1991**, *17*, 2021–2034.
90. Pivnick, K.A.; Reed, D.W.; Millar, J.G.; Underhill, E.W. Attraction of northern false chinch bug *Nysius niger* (Heteroptera: Lygaeidae) to mustard oils. J. Chem. Ecol. **1991**, *17*, 931–941.
91. McCaffrey, J.P.; Williams, L.; Borek, V.; Brown, P.D.; Morra, M.J. Toxicity of ionic thiocyanate-amended soil to the wireworm, *Limonius californicus* (Coleoptera: Elateridae). J. Econ. Entomol. **1995**, *88*, 793–797.
92. Tsao, R.; Coats, J.R.; Johnson, L.A. Insecticidal glucosinolates in *Crambe* (*Crambe abyssinica*) seed extracts. ABS papers Am. Chem. Soc. **1993**, *206*, 49.
93. McCloskey, C.; Isman, M.B. Influence of foliar glucosinolates in oilseed rape and mustard on feeding and growth of the Bertha armyworm, *Mamestra configurata* Walker. J. Chem. Ecol. **1993**, *19*, 249–266.
94. Newman, R.M.; Hanscom, Z.; Kerfoot, W.C. The watercress glucosinolate myrosinase system: a feeding deterrent to caddisflies, snails and amphipods. Oecologia **1992**, *92*, 1–7.
95. Lowe, M.D.; Henzell, R.F.; Taylor, H.J. Insecticidal activity to soldier fly larvae, *Inopus rubriceps* (Macq.) of isothiocyanates occurring in "Choumoellier" (*Brassica oleracea* cv.). N. Z. J. Sci. **1971**, *14*, 322–326.
96. Peterson, C.J.; Tsao, R.; Coats, J.R. Glucosinolate aglucones and analogues: insecticidal properties and a QSAR. Pes. Sci. **1998**, *54*, 35–42.

97. Winkler, H.; Otto, G. Replant losses with strawberries and suggestions for their reduction. Hort. Abst. **1980**, *50*, 344.
98. Mojtahedi, H.; Santo, G.S.; Hang, A.N.; Wilson, J.H.; Suppression of root-knot nematode populations with selected rapeseed cultivars as green manure. J. Nematol. **1991**, *23*, 170–174.
99. Johnson, A.W.; Golden, A.M.; Auld, D.L.; Sumner, D.R. Effects of rapeseed and vetch as green manure crops and fallow on nematodes and soil-borne pathogens. J. Nematol. **1992**, *24*, 117–126.
100. Lazzeri, L.; Tacconi, R.; Palmieri, S. In vitro activity of some glucosinolates and their reaction products toward a population of the nematode *Heterodera schachtii*. J. Agric. Food Chem. **1993**, *41*, 825–829.
101. Bradburn, R.P.; Mithen, R. Glucosinolate genetics and the attraction of the aphid parasitoid *Diaretiella rapae* to Brassicae. Proc. R. Soc. Lond. B. Biol. Sci. **2000**, *267*, 89–95.
102. Koritsas, V.M.; Lewis, J.A.; Fenvik, G.R. Glucosinolate responses of oilseed rape, mustard and kale to mechanical wounding and infestation by cabbage stem flea beetle. Ann. Appl. Biol. **1991**, *118*, 209–221.
103. Kawakishi, S.; Kaneko, T. Interaction of proteins with allyl isothiocyanate. J. Agric. Food. Chem. **1987**, *35*, 85–88.
104. Johnson, P.J.; Cyanide. In *Encyclopedia of Science and Technology*; McGraw-Hill: New York, 1987.
105. Vageeshbabu, H.S.; Chopra, V.L. Genetic and biotechnological approaches for reducing glucosinolates from rapeseed-mustard meal. J. Plant Biochem. Biotechnol. **1997**, *6*, 53–62.
106. Ibrahim, P.K.; Chavadej, S.; deLuca, V. In Rec. Adv. Phytochem; Ellis, B.E., Kuroki, G.W., Stafford, H.A., Eds.; 1994; Vol. 8, 125–138.
107. Chavadej, S.; Brisson, N.; McNeil, J.N.; deLuca, V. Redirection of tryptophan leads to production of low indole glucosinolate canola. Proc. Natl. Acad. Sci. USA. **1994**, *91*, 2166–2170.
108. Bak, S.; Nielsen, H.L.; Halkier, B.A. The presence of CYP79 homologues in glucosinolate-producing plants shows evolutionary conservation of the enzymes in the conversion of amino acid to aldoxime in the biosynthesis of cyanogenic glycosides and glucosinolates. Plant Mol. Biol. **1998**, *38*, 725–734.
109. Bak, S.; Olsen, C.E.; Peterson, B.L.; Halkier, B.A. Metabolic engineering of p-hydroxybenzylglucosinolate in *Arabidopsis* by expression of the cyanogenic CYP79A1 from *Sorghum bicolor*. Plant J. **1999**, *20*, 663–671.
110. Xue, J.; Jorgensen, M.; Philgren, U.; Rask, L. The myrosinase gene family in *Arabidopsis thaliana*, gene organization, expression and evolution. Plant Mol. Biol. **1995**, *27*, 911–922.
111. Kliebenstein, D.; Pedersen, D.; Barker, B.; Mitchell-Olds, T. Comparative analysis of quantitative trait loci controlling glucosinolates, myrosinase and insect resistance in *Arabidopsis thaliana*. Genetics **2002**, *161*, 325–332.

8

Alkaloids

8.1 INTRODUCTION

Alkaloids have been known to mankind for several centuries. The alkaloids represent one of the most diverse and complicated group of secondary metabolites found in living organisms, especially in plants. More than 12,000 alkaloids have been described. The term *alkaloid*, meaning "alkali-like" or compounds possessing some basicity, was coined by W. Meibner, a German pharmacist. F. W. Serturner first isolated the alkaloid morphine in 1806 from latex of opium poppy. It is difficult to give a brief but adequate definition of alkaloids because they encompass a wide variety of structures. The definition has been revised several times to accommodate the ever-increasing number of alkaloids.

Pelletier (1) suggested a simple and general definition for an alkaloid. According to him, an alkaloid is a heterocyclic compound containing nitrogen. The term alkaloid normally covers a vast array of unrelated structures. Compounds with exocyclic nitrogen bases were termed *pseudo-alkaloids*, which are usually not derived from amino acids, whereas the term *protoalkaloids* includes compounds that are derived from amino acids without any heterocyclic ring system. Currently, the definition of alkaloid is much more practical and includes all nitrogen-containing natural products that are not otherwise classified as nonprotein amino acids, amines, cyanogenic glycosides, glucosinolates, peptides, cofactors, phytohormones, or primary metabolites such as purine and pyrimidines. Alkaloids are the important secondary metabolites, which impart resistance against many insect pests.

There are few comprehensive and authoritative publications about alkaloids. The *Dictionary of Alkaloids*, compiled by Southon and Buckingham (2), describes 10,000 structures. The series *The Alkaloids*, edited by Brossi (3), provides information on the chemistry, methodology, and pharmacology of the various classes of alkaloids. Several volumes were published in this series (4). Other books on alkaloids are also available (5–8).

8.2 CLASSIFICATION OF ALKALOIDS

Alkaloids have been classified based on taxonomy (as solanaceous, papilionaceous, etc.), or based on pharmacological properties (as analgesic, cardioactive, etc.), or based on chemical structure. A classification based on chemical structure is now universally accepted and adopted (9–12).

1. *Pyridine and piperidine alkaloids.* Nicotine, anabasine, nicotinic acid and its derivatives are the major pyridine alkaloids from *Anabasis aphyla, Nicotiana tabacum, N. sylvestris, and N. rusticas.* Trigonelline, a methyl derivative of nicotinic acid, is present in *Trigonella foenum-graceum.*

2. *Tropane alkaloids.* The alkaloids atropine and hyoscine found in the Solanaceae family are anticholinergics. The main source of hyoscyamine and hyoscine are *Datura stramonium* and *D. innoxia.* Cocaine is found in coca (*Erythroxylon coca*).

3. *Quinolizidine alkaloids.* These group of alkaloids are common in Leguminosae family and in all species of the genus *Lupinus.* Hence, these alkaloids are also known as lupin alkaloids.

4. *Pyrrolizidine alkaloids.* These alkaloids are common in several genera of the Asteraceae and Boraginaceae. More than 200 pyrrolizidine alkaloids have been isolated from plants. The common alkaloid is senecionine.

5. *Quinoline alkaloids.* The alkaloids quinine and quinidine isolated from *Cinchona* species belong to quinoline type. The antimalarial activity of these alkaloids has been known for many centuries.

6. *Isoquinoline or benzylquinoline alkaloids.* This group of alkaloids is widely distributed in Papaveraceae, Berbidaceae, Ranunculaceae, and Menispermaceae. Papaver alkaloids from *P. somniferum* belong to this group of alkaloids. Morphine, codeine, berberine, and coculine are some important isoquinoline alkaloids.

7. *Indole alkaloids.* These alkaloids represent a large and diverse group of plant products mainly present in Loganiaceae, Apocynaceae, and Rubiaceae. Monoterpene molecules are usually attached to the indole nucleus. Reserpine, ajmaline, cinchonamine, tabersonine, vincristine, and vinblastine are important alkaloids in this group.

8. *Purine alkaloids.* Purine alkaloids are widely distributed in the plant kingdom in at least 90 species belonging to 30 genera. Caffeine and theobromine are two important alkaloids of this group.

9. *Phenyl alkylamines.* Phenylalkylamine alkaloids are distributed in *Ephedra* (Gentaceae), *Catha edulis*, and *Taxus baccata.* Ephedrine, pseudoephedrine, taxin, and hordenine are important phenylalkylamine alkaloids.

10. *Isoprenoid (terpenoid) alkaloids.* These alkaloids possess monoterpene (gentianine, actinidine), sesquiterpene (nupharidine, nupharamine, dentrobine), diterpene (delpheline, garryfoline), triterpene (daphniphylline), or steroid units (tomatidine, germine).

11. *Polyhydroxy alkaloids.* An important and fascinating group of plant alkaloids that resemble sugars structurally and stereochemically are polyhydroxy alkaloids. Such alkaloids interfere with enzymes such as glycosidases and receptors of carbohydrate metabolism. Polyhydroxy derivatives of piperidine, pyrrolidine, indolizine, and pyrrolizidine are found in Moraceae, Leguminosae, Polygonaceae, Aspidiceae, and Euphorbiaceae.

12. *Benzoxazinoids (hydroxamic acids).* Benzoxazinoids were recently introduced into the alkaloids based on the revised definition. They are derived from anthranilic acid. The predominant benzoxazinoids are 2,4-dihydroxy-7-methoxy-2H-1,4-benzoxazin-3(4H)-one (DIMBOA) and its desmethoxy derivative (DIBOA) isolated from corn seedlings. Benzoxazinoids have been found in members of the following genera in the Graminae: *Aegilops, Arundo, Chusquea, Coix, Elymus, Secale, Sorghum, Tripsacum, Triticale, Triticum,* and *Zea.* They are not detected in *Avena, Hordeum,* and *Oriza* (13). The structures of alkaloids representative of each group are shown in Fig. 8.1.

8.3 OCCURRENCE

Alkaloids have been known to occur in 20% of all plants and in more than 150 families (9–14). They also occur in bacteria, fungi, and animals. Within the plant kingdom, the alkaloids occur in primitive groups such as *Lycopodium*

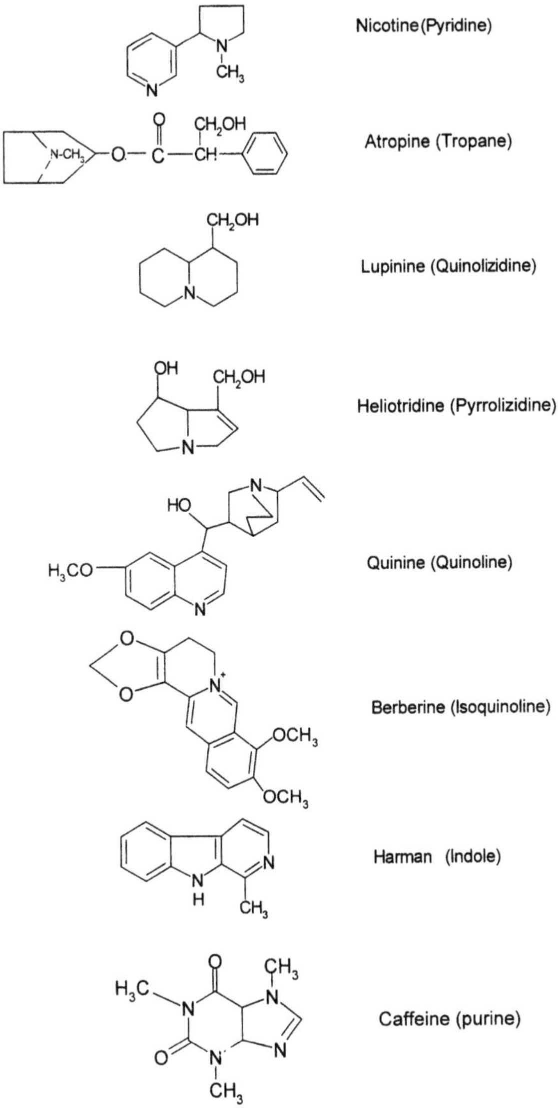

Nicotine (Pyridine)

Atropine (Tropane)

Lupinine (Quinolizidine)

Heliotridine (Pyrrolizidine)

Quinine (Quinoline)

Berberine (Isoquinoline)

Harman (Indole)

Caffeine (purine)

FIGURE 8.1 Structures of alkaloids.

or *Equisetum* in gymnosperms. In higher plants, some families contain more alkaloidal taxa than others. The important alkaloid-rich plant families are Apocynaceae, Papaveraceae, Asclepiadaceae, Liliaceae, Gnetaceae, Rubiaceae, Rutaceae, and Solanaceae. Monocotyledons are generally poor in alkaloids, but exceptions are the Liliaceae and Graminae. Specific

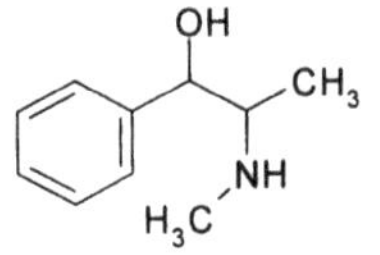

Ephedrine (Phenyl alkylamine)

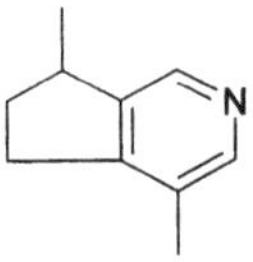

Actinidine (Monoterpenoid)

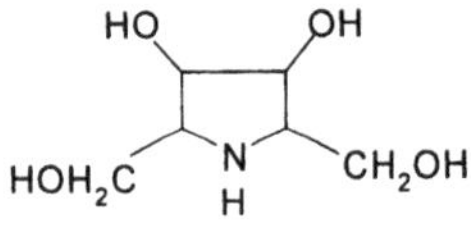

DMDP (Polyhydroxy)

(2R, 5R-Dihydroxymethyl -3R, 4R-dihydroxy pyrrolidine)

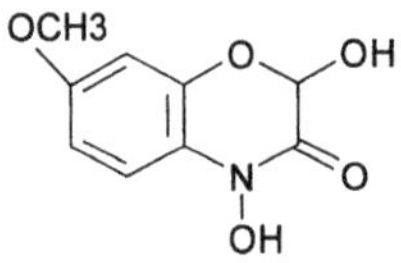

DIMBOA (Benzoxazinoid)

FIGURE 8.1   Continued

alkaloid types are restricted to certain specific plants only. For example, benzylisoquinoline alkaloids are typical for Papaveraceae, Berberidaceae, and Ranunculaceae, which are phylogenetically related (10). Alkaloids tend to accumulate mainly in four types of plant tissue: actively growing tissues, epidermal and hypodermal cells, vascular sheaths, and latex vessels. The alkaloids accumulate in the vacuoles and therefore do not appear in young cells until they are vacuolated.

8.4   BIOSYNTHESIS

The basic skeleton of most alkaloids is derived from amino acids, although moieties derived from other pathways, such as terpenoids, are attached to the skeleton. In a few alkaloids, such as steroid alkaloids, the nitrogen is added in the final steps of a biosynthetic pathway. In these alkaloids the basic skeleton is not formed from amino acids. The amino acids used for alkaloid synthesis are lysine, aspartic acid, histidine, tryptophan, phenylalanine, tyrosine, and ornithine. Biosynthetic pathways have been worked out in detail for a few alkaloids (14–24). There is no general or common pathway for the synthesis of alkaloids like that of phenolics or terpenoids.

The pathways for many other alkaloids are based on intelligent guesses and preliminary precursor experiments. Precursor experiments usually involve the feeding of radioactively labeled compounds (^{3}H or ^{14}C) to intact plant or to cell and tissue cultures. Nonradioactive ^{13}C-labeled precursors have also been used recently. Isolation and characterization of the enzymes catalyzing the particular reactions of a biosynthetic sequence is also employed to identify the biosynthetic pathways. The important amino acids used for the biosynthesis of few important alkaloids are shown in Table 8.1. A review of all the biosynthetic pathways reported for alkaloids is beyond the scope of this book. Moreover, pathways for many alkaloids have not been fully worked out. However, four well-established pathways are discussed here.

8.4.1 Nicotine and Tropane Alkaloids

The early steps of tropane alkaloid and nicotine biosynthesis are common. Although nicotine is not a member of the tropane class, the N-methyl-Δ^1-pyrrolinium cation involved in tropane alkaloid biosynthesis is also an intermediate in the nicotine pathway. The biosynthetic pathway for nicotine is shown in Scheme 8.1. Nicotine is derived from diamine putrescine (17) produced from ornithine by ornithine decarboxylase (ODC). Arginine also

TABLE 8.1 Biosynthetic Precursor for a Few Important Alkaloids

Precursor amino acid	Alkaloid type	Example	Source
a. Ornithine	Pyridine	Nicotine	*Nicotiana* spp
	Tropane	Cocaine	Solanaceae
		Hyoscyamine	*Datura, Atropa*
	Pyrrolizidine	Heliotrine	Boraginaceae
	Pyrrolidine	Stachydrine	*Medicago sativa*
b. Lysine	Quinolizidine	Lupanine	Gentisteae, *Lupinus*
c. Aspartic acid	Piperidine	Arecoline	*Areca catechu*
d. Histidine	Imidazole	Pilocarpine	*Pilocarcarpus* spp
e. Tryptophan	Indole	Ajmalicine,	Rubiaceae
		Nindoline	Loganiaceae
	Quinoline	Quinine	Cinchona
f. Phenylalanine/	Isoquinoline	Emetine	*Lyphophora*
tyrosine		Berberine	Berberidaceae
		Morphine	Papaveraceae
g. Anthranilic acid	Benzoxazinoids	DIMBOA	Graminae

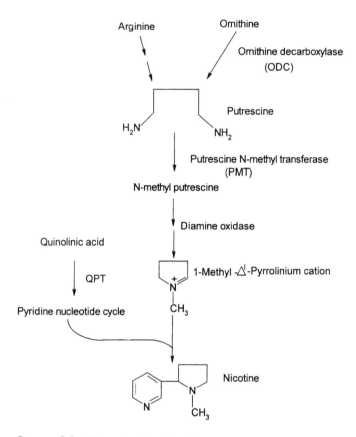

SCHEME 8.1 Biosynthesis of nicotine.

serves as a precursor for putrescine. Putrescine is N-methylated by putrescine *N*-methyltransferase (PMT) and then oxidatively deaminated by diamine oxidase to form 1-methyl-Δ^1-pyrrolinium cation. This cation is condensed with a derivative of nicotinic acid derived through pyridine nucleotide cycle to form nicotine. The *N*-methyl-Δ^1-pyrrolinium cation condenses with acetoacetic acid to yield hygrine as a precursor of the tropane ring. Tropionine is located at a branch point in the tropane alkaloids pathway (17).

8.4.2 Quinolizidine Alkaloids

It has been shown using tracer studies that lysine and its decarboxylation product, cadaverine; serve as the precursors for bi- and tetracyclic

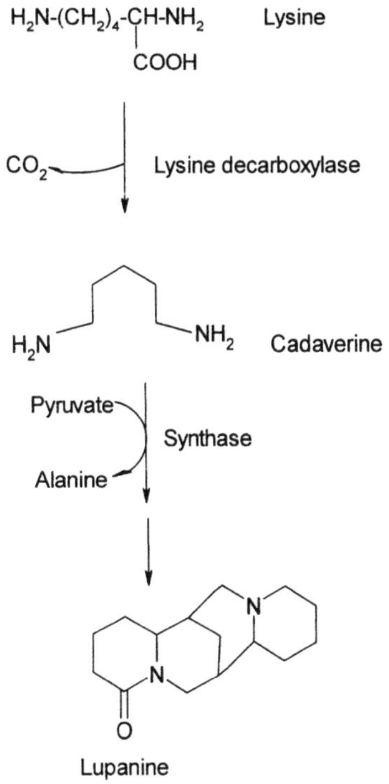

SCHEME 8.2   Biosynthesis of lupanine.

quinolizidine alkaloids (QAs). The tricyclic quinolizidine alkaloids are derived from tetracyclic precursors. Lupanine seems to be the common precursor for the tri- and tetracyclic α-pyridone quinolizidine alkaloids (20). The possible biosynthetic pathway for lupanine from lysine is shown in Scheme 8.2.

8.4.3   Benzylisoquinoline Alkaloids

The biosynthesis of berberine from tyrosine is shown in Scheme 8.3. Berberine synthesis in plant cells has been well investigated at the enzyme level (15,19,20). The biosynthetic pathway leading from tyrosine to berberine has 13 different enzymatic reactions (Scheme 8.3). The pathway involves several methyltransferases and oxidases that accept more than one potential intermediate (17).

8.4.4 Terpenoid Indole Alkaloids

Terpenoid indole alkaloids (TIAs) consist of an indole moiety provided by tryptamine and a terpenoid component derived from the iridoid glucoside secologanin. Tryptophan is converted to tryptamine by tryptophan decarboxylase (TDC). The first committed step in secologanin biosynthesis is the hydroxylation of geraniol to 10-hydroxygeraniol (Scheme 8.4). The production of terpenoid precursors might play a regulatory role in TIA

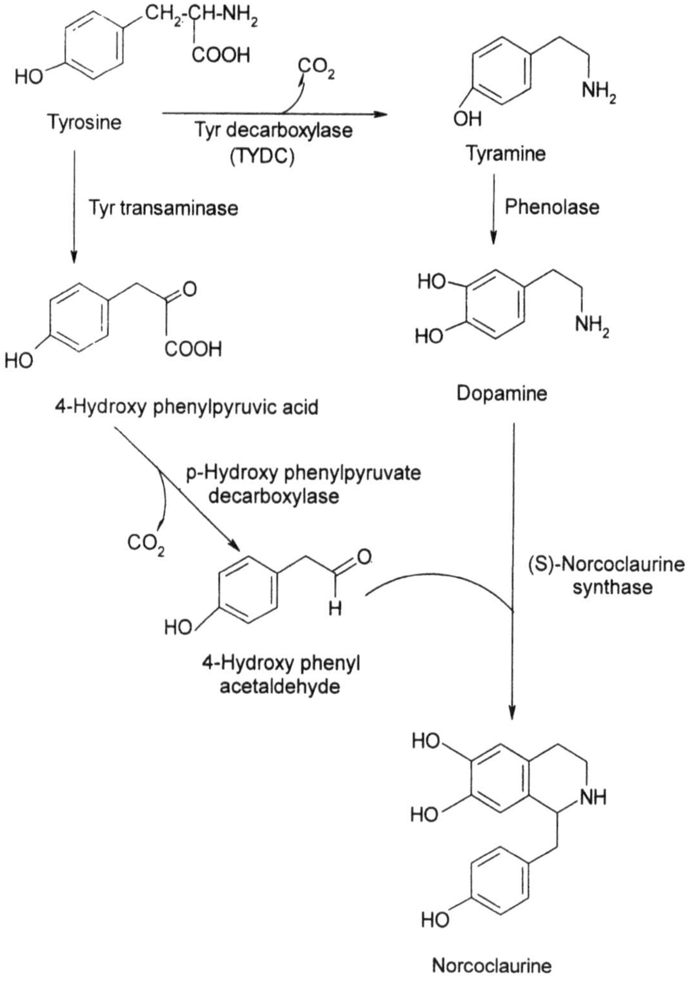

SCHEME 8.3 Biosynthesis of berberine (BIA).

Alkaloids   **205**

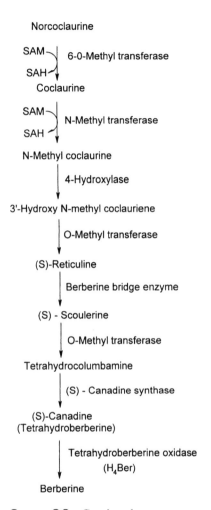

SCHEME 8.3 Continued.

biosynthesis. Tryptamine and secologanin are condensed by strictosidine synthase (STR) to form strictosidine, the common precursor for all TIAs. The strictosidine is then deglucosylated by strictosidine β-D-glucosidase (SGD). Deglucosylated strictosidine is converted via several unstable intermediates to 4,21-dehydrogeissoschizine followed by the formation of several TIAs.

Enzymes involved in alkaloid biosynthesis are associated with diverse subcellular compartments, including cytosol, vacuole, tonoplast membrane,

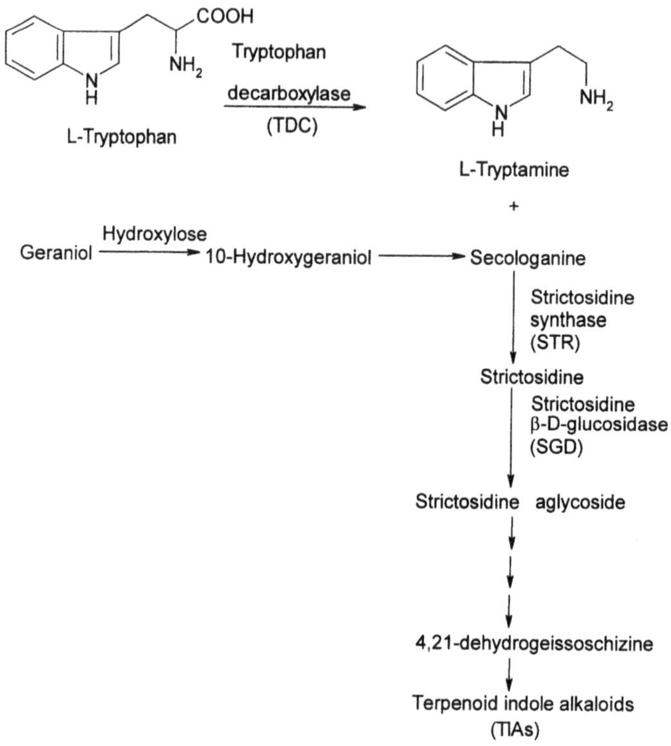

SCHEME 8.4   Biosynthesis of terpenoid indole alkaloids.

endoplasmic reticulum, chloroplast stroma, and thylakoid membranes. Localization studies have shown that sequential alkaloid biosynthetic enzymes can also occur in distinct cell types, suggesting the intercellular transport of pathway intermediates.

8.5 BIOACTIVITY

Many alkaloids found in plants have been recognized for their ability to prevent or reduce predation by insects. The repellency to insect feeding by a number of well-known alkaloids, including the plant sources from which they come and the insect species deterred, have been reviewed (9,11–13).

A great number of structurally unrealed alkaloids, such as pyrrolizidines, quinolizidines, indole alkaloids, benzylisoquinolines, steroid alkaloids, and methylxanthines, are feeding deterrents to many insects. The isoquinoline alkaloids and monoterpenoid indole alkaloids are synthesized and stored at specific sites in the producing plants. When herbivores attack these plants, the alkaloids prove to be insecticidal to the herbivores. Thus, alkaloids are regarded as part of the plant's constitutive chemical defense. Since alkaloids are often synthesized and stored at strategically important sites or vulnerable plant parts, plants have a greater chance for remaining unaffected by herbivory, microbial attack, or mechanical damage or stress.

8.5.1 Pyrrolizidine Alkaloids

Pyrrolizidine alkaloids (PAs) are assumed to serve as protective chemicals for the producing plant. However, direct evidence for this assumption is sparse. Boppre (25) reported that tests with cockroaches, locusts, and various lepidopteran larvae have demonstrated that PA-contaminated food is rejected due to taste. A number of isolated PAs were tested for their deterrent activity against larvae of spruce budworm (*Choristoneura fumiferana*), and, except for senkirkine and lasiocarpine, most were inactive (26).

More convincing evidence favoring a defensive role of plant PAs comes from a number of insects that have evolved adaptations not only to cope with these compounds but also to use them for their own benefit (27). Important adaptations include (a) insect herbivores that feed on PA plants and are able to cope with PAs; (b) insects that not only feed on PA plants but also store the PAs for their own protection; and (c) insects that use PAs as essential precursors for the production of pheromones. There are polyphagous insects that use PA-bearing plants as food sources but do not sequester the alkaloids. Examples are the noctuids *Spodoptera littoralis* and *Melanchra persicariae* and the migratory grasshopper *Melanoplus sanguinipes*, which readily feed on *Syringa vulgaris*. Pharmacokinetic studies with *Melanoplus* showed that these insects possess efficient mechanisms to eliminate ^{14}C-labeled senecionine and its *N*-oxide, fed orally or adminstered by injection, within 24 h.

A great number of lepidopterans and other insects use PA-bearing plants and are capable of storing PAs in their bodies (28). The majority of these species advertise their unpalatability by a conspicuous aposematic warning coloration. The arctiid *Tyria jacobaeae* (European cinnabar moth) is one of the classical examples. *Tyria* larvae feed almost exclusively on *Senecio jacobaea* and sequester all plant PAs, which are also found in pupae and adults. A recent reinvestigation revealed that PAs offered as free base or *N*-oxide were taken up without preference and were stored in the insect body exclusively as *N*-oxide. Larvae and pupae are able to N-oxidize tertiary PA. In larvae, about 75% of total PAs were located in the integument. Similar results were reported for the arctiid *Creatonotos transiens*, in which a large proportion of PA N-oxide was found in the integument during all developmental stages. Evidence has been presented that a specific carrier in the midgut of *Creatonotos* larvae is responsible for PA resorption.

The larva of the green lacewing (*Ceraeochrysa cubana*) (Neuroptera: Chrysopidae) is a natural predator of eggs of *Utetheisa ornatrix* (Lepidoptera: Arctiidae), a moth that sequesters pyrrolizidine alkaloids from its larval food plant (*Crotalaria* spp). *Utethesia* eggs are ordinarily endowed with the alkaloid (29). Alkaloid-free *Utetheisa* eggs, produced experimentally, are pierced by the larva with its sharp tubular jaws and sucked out. Alkaloid-laden eggs, in contrast, are rejected. When attacking an *Utethesia* egg cluster (numbering on average 20 eggs), the larva subjects it to an inspection process. It prods and/or pierces a small number of eggs (on average two to three) and, if these contain alkaloid, it passes "negative judgment" on the remainder of the cluster and turns away.

Larvae of *Creatonotos transiens* (Lepidoptera: Arctiidae) and *Zonocerus variegatus* (Orthoptera: Pyrgomorphidae) were ingested in the ^{14}C-labeled senecionine and its N-oxide with the same efficiency but sequester the two tracers exclusively as N-oxide (30). Larvae of the non-sequestering *Spodoptera littoralis* efficiently eliminate the ingested alkaloids. During feeding on the two alkaloidal forms transient levels of senecionine (but not of the N-oxide) are built up in the hemolymph of *S. littoralis* larvae. Based on these results, senecionine ^{18}O,N-oxide was fed to *C. transiens* larvae and *Z. variegatus* adults. The senecionine N-oxide recovered from the hemolymph of the two insects shows an almost complete loss of ^{18}O label, indicating reduction of the orally fed N-oxide in the guts, uptake of the tertiary alkaloid, and its re-N-oxidation in the hemolymph. The enzyme responsible for N-oxidation is a soluble mixed-function monooxygenase.

The mechanism by which the moth *Creatonotos transiens* produces its male pheromone, (7R)-hydroxydanaidal, from heliotrine, an alkaloidal precursor of opposite (7S) stereochemistry, was investigated (31). Specifically deuterated samples of heliotrine and epiheliotrine were prepared and fed to *C. transiens* larvae, and the steps in the biosynthetic process were monitored by gas chromatography/mass spectrometry. These analyses indicate that heliotrine is initially epimerized to (7S)-epiheliotrine by oxidation to the corresponding ketone followed by stereospecific reduction. The order of the subsequent steps is (a) aromatization of the dihydropyrrole ring, (b) ester hydrolysis, and (c) oxidation of the resulting primary alcohol to the final aldehyde. The ecological implications of this insect's ability (and the inability of another moth, *Utetheisa ornatrix*) to use representatives of two stereochemial families of alkaloids as pheromone precursors are discussed.

Males of the arctiid moth *Cosmosoma myrodora* acquire pyrrolizidine alkaloid by feeding on the excrescent fluids of certain plants (e.g., *Eupatorium capillifolium*). They incorporate the alkaloid systemically and as a result are protected against spiders (32). The males have a pair of abdominal pouches, densely packed with fine cuticular filaments, which in alkaloid-fed males are alkaloid laden. The males discharge the filaments on the female in bursts during courtship, embellishing her with alkaloid as a result. The topical investiture protects the female against spiders. Alkaloid-free filaments from alkaloid-deprived males convey no such protection. The males also transmit alkaloid to the female by seminal infusion. The systemic alkaloid thus received, which itself may contribute to the female's defense against spiders, is bestowed in part by the female on the eggs. Although paternal contribution to egg defense had previously been demonstrated for several arctiid moths, protective nuptial festooning a female by its mate, such as is practiced by *C. myrodora*, appears to be without parallel among insects.

8.5.2   Quinolizidine Alkaloids

The observation that "sweet" lupines with low QA content are much more susceptible to attack by plant pests and herbivores strongly supports a defensive role for QAs. QAs have been shown to deter pea aphids (33). In the latter case, 13-hydroxylupanine esters were highly active, but other QAs, including α-lupanine, were not. Stermitz and colleagues (34) studied the influence of QAs acquired by *Pedicularis semibarbata* from *Lupinus fulcatus* by root parasitism on the specilist herbivore *Euphydryas editha* (Lepidoptera). The authors were not able to detect significant difference in the behavior of *Euphydryas* larvae feeding on QA-free and QA-bearing *Pedicularis* leaves.

Acquisition of QAs by specialized insects for defensive purposes appears to be rare. Two specialist phloem feeders, the broom aphid *Aphis cytisorum* and *Macrosiphon albifrons*, accumulate QAs in concentrations similar to those of their host plants (35). *Macrosiphon* is attracted by QAs, preferably lupanine. When offered the choice of QA-rich and QA-low lupine varieties, it only heavily infested the former. The alkaloids may protect the aphid against carnivorous insects; the carabid beetle, *Carabus problematicus*, became narcotized after an "aphid meal" for up to 48 h.

Although alkaloids are known feeding deterrents for lepidopteran larvae, including the spruce budworm *Choristoneura fumiferana* (Clemens), few studies have investigated their effects on liepidopteran oviposition. Quinolizidine containing alkaloid extracts and isolated quinolizidine alkaloids were obtained from two Chinese plants, *Sophora alopecuroides* L. and *Thermopsis lanceolata* Robert Brown, in order to evaluate their effects on spruce budworm oviposition (36). Application of extracts from either plant on Parafilm substrate significantly deterred oviposition at dosages as low as 4.7 and 7.9 μg/cm^2, respectively, in dual-choice bioassays, of 9 QAs produced by these plants. Six deterred oviposition on Parafilm substrate treated at 7.9 μg/cm^2 or less. The most effective compounds were aloperine (0.3 μg/cm^2–1.3 nmol/cm^2), sparteine (1.6 μg/cm^2–6.8 nmol/cm^2), and cytisine (1.6 μg/cm^2–8.4 nmol/cm^2). Oviposition was deterred also on fresh host foliage, 8-cm twigs of balsam fir, *Abies balsamea* (L.) Miller, which had been sprayed with either a 1-mL aliquot of alkaloid extract (10 mg/mL) of *S. alopecurosides* or 1 mL of an aloperine solution (1 mg/mL). There was no effect on longevity of males or females continuously exposed to artificial substrate treated with aloperine or alkaloid extract from *S. alopecuroides* in no-choice bioassays, nor was the ability of males to inseminate females affected. However, female production of egg masses was inhibited, although clutch size (eggs per egg mass) was not affected. These results are the first demonstration of the behaviorally deterring and physiologically inhibiting effects of QAs on lepidopteran oviposition.

Adult *Helix aspersa* snails were maintained individually for one week in plastic cages with nine living *Lupinus albus* plants as their only food (37). Among these nine plants, three chemotypes—bitter, intermediate, and sweet—which differed in their alkaloid content, were equally represented. Each day, the leaf surface grazed and the number of leaves attacked by the snails were recorded for each chemotype and each snail. A consumption/attack (C/A) ratio was calculated by dividing the surface grazed (C) by the number of attacks (A). The number of attacks and the grazed area were positively correlated for each chemotype during the whole experiment, and the snails ate similar quantities of lupin each day. After 4 and 6 days of experiment, they noticed a rejection of the bitter chemotype in favor of the intermediate and sweet ones, respectively. After the sixth day, the surface grazed per attack was significantly higher on the sweet chemotype than on the bitter plants. The authors hypothesized that rejection of the bitter chemotype might be related to (a) an alkaloid reaction threshold associated with an increase in the amount of alkaloids in the wounded plants and/or (b) aversive ingestive conditioning.

8.5.3   Benzylisoquinoline Alkaloids

A methanol extract of tubers of *Corydalis bulbosa* had insecticidal activity against larvae of *Drosophila melanogaster* Meigen. Four protoberberine alkaloids—(−)-tetrahydroberberine (a), (−)-tetrahydrocopticine (b), (+)-corydaline (c), (±)-tetrahydropalmatine (d), and (±)-dehydrocorydaline (e) as its iodide—were isolated from the extract (38). Compounds e, b, and d exhibited LC$_{50}$ values toward larvae of *D. melanogaster* of 0.23 μmol/mL,

0.9 µmol/mL, and 1.70 µmol/mL diet concentration, respectively. Against adults, compound a showed the most potent activity with an LD_{50} value of 2.5 mg/adult. Compounds a, b, c, and d inhibited acetylcholinesterase by 78.7, 71.8, 68.2, and 64.6%, respectively, at 1.0 mM. Furthermore, compound e showed the most potent activity with inhibition of 61.3% at 0.40 mM. Compounds e, a, and b were identified as insecticidal compounds from *C. bulbosa*. Investigation of the structure–activity relationship indicated the importance of the methylenedioxyl group and double bonds in the isoquinoline moiety for enhanced activity of the protoberberine alkaloids

Antifeedants of some alkaloids isolated from *Thalictrum minus* L. (Ranunculaceae) and *Corydalis solida* L. (Fumariaceae) were investigated (39). The insect *Spodoptera littoralis* was treated with these alkaloids; α-allocryptopine, protopine, thalicarpine, and tetrahydropalmatine expressed an antifeedant activity, whereas thalfoetidine and tetrahydropalmatine had a phagostimulant effect on *Spodoptera littoralis*. Alkaloids from the first group belong to protopine, aporphinbenzylisoquinoline, and tetrahydroprotoberberine type of structure, while the remaining ones belong to bisbenzylisoquinoline and phthalidisoquinoline types. The above alkaloids led to development of different growth inhibition zones of *Corydalis herbarum*. Thalicarpine inhibited the growth completely, thalfoetidine and capnoidine partially, whereas growth of the cultures treated with protopine, α-allocryptopine, and tetrahydropalmatine was on par with the control.

The antifeeding activity of three isoquinoline alkaloids identified from roots of *Coptis japonica* Makino toward fourth-instar larvae of *Hyphantria cunea* Drury and adults of *Agelastica coerulea* Baly was examined using the leaf-dipping bioassay (40). The biologically active constituents of the *Coptis* roots were characterized as the isoquinoline alkaloids berberine, palmatine, and coptisine by spectroscopic analysis. In a test with *H. cunea* larvae, the antifeeding activity was much more pronounced in an application of a mixture of palmatine iodide and berberine chloride (1:1, wt/wt) at 250 ppm (82.3%) and 500 ppm (100%) compared with palmatine iodide (76.0%) and berberine chloride (75.4%) alone at 500 ppm. These results indicate a synergistic effect. With *A. coerulea* adults, berberine chloride showed 57.5% and 91.1% antifeeding activity at 125 and 250 ppm, respectively; whereas weak activity was obtained in application of 500 ppm of palmatine iodide (41.4%) and coptisine chloride (52.4%) alone. The *Coptis* root–derived compounds merit further study as potential insect control agents.

8.5.4 Nicotine and Tropane Alkaloids

Wild species of *Nicotiana* in the section Repandae were found to be highly toxic to the tabacco hornworm, *Manduca sexta*, a tobacco-adapted insect that is not poisoned by nicotine. The causative agent was found in crude leaf extract and was identified as a mixure of N-acyl analogues of nornicotine (41,42). The acyl moiety is always β-hydroxylated, with branched or unbranched carbon chains ranging from C_{12} to C_{15}. The major fatty acid was identified to be a C_{14}-acid, 3-hydroxy-12-methyltridecanoic acid. Topical application of 0.2 mg of the mixture of the N-acylnornicotines causes 100% mortality in 24–48 h; under the same conditions, 0.5 mg nicotine was almost inactive (7% mortality). The amides are exclusively localized in the trichomes and the trichome exudate of the aerial parts of *Nicotiana stocktonii*, and they are synthesized in leaves from nicotine via nornicotine. It would be of interest to relate the toxicity of the nornicotine amides to the insecticidal isobutylamides found in plants of the Asteraceae and Rutaceae. The tobacco budworm (*Heliothis virescens*), another "nicotine specialist," is not affected by the N-acylnornicotine (41).

The correlation between nicotine accumulation and its defensive role in *Nicotiana sylvestris* has convincingly been demonstrated in a series of carefully performed studies (43,44). Tobacco plants subjected to leaf damage show a fourfold increase in the alkaloid content of their undamaged leaves. The increase begins 19 h after the end of the damage regime, reaches a maximum at 9 days, and wanes to the control level after 14 days. The increase of the nicotine level was found, because of enhanced alkaloid synthesis, to cause a 10-fold increase in the alkaloid concentration of the xylem fluid. Experimental evidence indicates that alkaloid induction may be triggered by a phloem-translocated signal. The increased alkaloid synthesis could also be demonstrated by true herbivory. Furthermore, freshly hatched larvae of *M. sexta* rearing on damaged plants gained, because of the higher alkaloid content, about 50% less weight than larva fed on undamaged leaves.

Shell scientists prepared and tested a broad array of heterocyclic compounds incorporating nitromethylene substituent. Among the best of these was tetrahydro-2-(nitromethylene)-2H-1,3-thiazine. This compound proved highly active against a variety of insects and was particularly effective against lepidopterous larvae (45). Acute mammalian toxicity was also quite acceptable, being far better than that of nicotine. The mode of action was shown to be an agonist at the nicotinic acetylcholine receptor (46).

The major drawback to the further development of these new nitro-methylene compounds was an extreme photochemical instability. Extensive synthesis efforts by the Shell group, and later by Bayer scientists in colla-boration with a Japanese group, were undertaken in an attempt to find a solution. Most of this work is documented in the patent literature. Eventually from these efforts the new structurally related N-nitroimino analogue imidacloprid evolved for commercialization (47). Imidacloprid is effective against a wide range of insects and is particularly active as a systemic insecticide against sucking insects on many crops. Thus, imidacloprid appears to be the new improved "nicotine" long sought by many people. As with tetrahydro-2-(nitromethylene)-2H-1,3-thiazine, a representative of these newer compounds has been shown to function as a nicotinic acetylcholine receptor agonist.

The extracts of *Delphinium* seeds contain the major active toxin methyllycaconitine (48). Nicotine was 10,000-fold less potent in this competition for the nicotinic acetylcholine receptors. Unlike nicotine, methyllycaconitine showed activity against lepidopterous larvae. The LC_{50} for the Southern armyworm (*Spodoptera eridania*) fed treated lima bean leaves was 308 ppm; the number for nicotine was much greater than 1000 ppm. Lycaconitine, which lacks the aromatic ester function, was inactive against the armyworm and 1000-fold less effective in blocking binding of α-bungarotoxin to the insect cholinergic receptor.

Although cocaine has a fascinating and complex medicinal history in man, its natural function in plants is unknown. Studies demonstrated that cocaine exerts insecticidal effects at concentrations, which occur naturally in coca leaves (49). Unlike its known action on dopamine reuptake in mammals, cocaine's pesticidal effects are shown to result from a potentiation of insect octopaminergic neurotransmission. Amine reuptake blockers of other structural classes also exert pesticidal activity with a rank order of potency distinct from that known to affect vertebrate amino transporters. These findings suggest that cocaine functions in plants as a natural insecticide and that octopamine transporters may be useful sites for targeting pesticides with selectivity toward invertebrates.

A number of atropine analogues were synthesized and their effects on larval growth of the silkworm, *Bombyx mori*, were investigated by both topical applicaion and dietary administration (50). Among the tested compounds, 8-methyl-8-azabicyclo(3.2.1)octan-3-α-ol-2,2-diphenylpropionate, an antagonist of the muscarinic acetylcholine receptor in mammals, significantly prolonged the duration of the instar. When fed on the above compound at 30 ppm, some of the larvae failed to molt. A 2,2-diphenylpropionate moiety was indispensable for this activity. Synthesized compound had more potent activity than atropine, which is known to inhibit prothoracicotrophic hormone (PTTH) release in vitro.

8.5.5   Purine Alkaloids

It has been shown that caffeine and other methylxanthines are effective insect toxins (51). Caffeine at a dietary concentration of 0.3%, which is well below the levels of caffeine in fresh tea leaves or coffee beans ($\sim$0.7–2%), kills nearly all larvae of the tobacco hornworm within 24 h; similar results were obtained with other insects, including mealworm larvae, butterfly larvae (*Vanessa cardui*), milkweed bug nymph, and mosquito larvae. It was conclusively shown that the toxicity is primarily caused by inhibition of phosphodiesterase activity (40).

8.5.6   Indole Alkaloid

The plant-produced indole alkaloid tryptamine is one of a large arrary of neuroactive substances that may affect insect behavior, development, and physiology. Thomas et al. (52) tested the role of tryptamine on insect reproduction using the fruitfly, *Drosophila melanogaster* (Meigen), as a model system. Measurements were made on reproductive success, oviposition rate, and preadult survival of insects on artificial diets containing tryptamine, its precursor tryptophan, as well as glycine and serotonin (5-hydroxytryptamine). *Drosophila* reproduction was reduced to 15% of controls when adult insects mated and the young were allowed to develop on medium containing 75 mM tryptamine. Tryptamine-induced depression in reproductive success was due to decreased oviposition rate and preadult survival. Serotonin, but not tryptophan or glycine, also reduced oviposition rate. Preference tests indicated that tryptamine may act as an antiattractant or antifeedant in this species. The accumulation of the indole alkaloid tryptamine in plants may provide a mechanism for reducing insect reproduction, which is potentially useful in protecting crop plants.

8.5.7   Piperidine Alkaloids

Both the dichloromethane and methanol extracts of *Microcos paniculata* stem bark showed moribund/toxic and growth inhibitory effects on the second-instar larvae of the mosquito *Aedes aegypti* (53). Moribund larvae appeared to be dead but made occasional zig-zag movements when exposed to a light source. They did not develop into the next instar and their survival time depended on the concentration of extract in the medium. Larvae in media containing the extract in low concentrations (<20 ppm) remained moribund for up to 5 or 6 days and died. The bodies of affected larvae became lean, long, and transparent, suggesting that sublethal concentrations of the extract have effects on growth and development. The MC_{50} (concentration at which 50% of larvae were moribund) of the dichloromethane and methanol extracts of stem bark against the second instar were found to be 17.5 and 16.0 ppm, and the LC_{50} of the extract 24 h after treatment were found to be 47.5 and 47.0 ppm, respectively, indicating that both extracts showed comparable activity. The constituent responsible for the larvicidal activity was identified. Isolation of the piperidine alkaloid

N-Methyl-6-α-(deca-1′,3′,5′-trienyl)-3β-methoxy-2α-hydroxymethylpiperidine (micropine) from *M. philippinensis* and the isolation of an insecticidal alkaloid N-methyl-6β-(deca-1′,3′,5′-trienyl)-3β-methoxy-2β-methylpiperidine from *M. paniculata* stem bark have been achieved.

A crude methanol extract of *Piper longum* fruits was found to be active against the fourth-instar larvae of *Aedes aegypti* and the hexane fraction of the methanol extract showed a strong larvicidal activity of 100% mortality (54). The biologically active component of *P. longum* fruits was characterized as pipernonaline by spectroscopic analyses. The LC_{50} value of pipernonaline was 0.25 mg/L. The toxicity of pipernonaline is comparable to that of pirimiphos-methyl as a mosquito larvicide.

The insecticidal activity of materials derived from the fruits of *Piper nigrum* against third-instar larvae of *Culex pipiens pallens*, *Aedes aegypti*, and *A. togoi* was examined and compared with that of commercially available piperine, a known insecticidal compound from *Piper* species (55). The biologically active constituents of *P. nigrum* fruits were characterized as the isobutylamide alkaloids pellitorine, guineensine, pipercide, and retrofractamide A by spectroscopic analysis. Retrofractamide A was isolated from *P. nigrum* fruits as new insecticidal principle. In the basis of 48 h LC_{50} values, the compound most toxic to *C. pipiens pallens* larvae was pipercide (0.004 ppm) followed by retrofractamide A (0.028 ppm), guineensine (0.17 ppm), and pellitorine (0.86 ppm). Piperine (3.21 ppm) was least toxic. Against *A. aegypti* larvae, larvicidal activity was more pronounced in retrofractamide A (0.039 ppm) than in pipercide (0.1 ppm), guineensine (0.89 ppm), and pellitorine (0.92 ppm). Piperine (5.1 ppm) was relatively ineffective. Against *A. togoi* larvae, retrofractamide A (0.01 ppm) was much more effective, compared with pipercide (0.26 ppm), pellitorine (0.71 ppm), and guineensine (0.75 ppm). Again, very low activity was observed with piperine (4.6 ppm). Structure–activity relationships indicate that the N-isobutylamine moiety might play a crucial role in the larvicidal activity, but the methylenedioxyphenyl moiety does not appear essential for toxicity.

8.5.8 Piperizine Alkaloids

In the course of screening for novel naturally occurring insecticides from Chinese crude drugs, methanol extract of rhizomes of *Nuphar japonicum* DC was found to give an insecticidal activity against larvae of *D. melanogaster* Meigen. Four alkaloids—(−)-7-epideoxynupharidine (A), (−)-castoramine (B), (−)-nupharolutine (C), and (−)-nupharmine (D)—were isolated by bioassay-guided fractionation from the extract (56). Insecticidal activity against larvae of *D. melanogaster* was demonstrated; B and A showed

LC_{50} values of 1.00 and 4.33 μmol/mL of diet concentration, respectively. Acute toxicity against adults of *D. melanogaster* was also found. Compound A had the most potent activity, with an LD_{50} value of 0.86 μg/adult. Compounds A, B, and C caused acetylcholinesterase inhibitions of 99.6%, 65.0%, and 59.7%, at 0.5 mmol/L concentration, respectively; however, D had only slight activity in this study. Therefore, A and B were identified as insecticidal compounds from rhizomes of *N. japonicum* DC. A structure–bioactivity relationship study indicated the importance of the position of the hydroxyl group and the piperizine skeleton for enhanced activity.

8.5.9   Diterpene Alkaloids

The insect antifeedant and toxic activity of the *Delphinium* diterpene alkaloid 15-acetylcardiopetamine, cardiopetamine along with its amino alcohol, the β-unsaturated ketone, and the acetylated ketone derivatives were studied in *Spodoptera littoralis* and *Leptinotarsa decemlineata* (57). Cardiopetamine and 15-acetylcardiopetamine strongly inhibited the feeding activity of *S. littoralis* and *L. decemlineata*, respectively. Structure–activity studies with *S. littoralis* showed that the C_{13} and C_{15} hydroxy substituents are essential features of the active molecule, while a C_{13} hydroxy and/or a C_{15} acetate determined their effect on *L. decemlineata*. The C_{11} benzoate group enhanced the biological effect on both insect species. These alkaloids were not toxic to *S. littoralis*, while their toxicity on *L. decemlineata* was inversely correlated with their antifeedant effects, the β-γ-unsaturated ketone derivative being the most toxic.

8.5.10   Steroid Alkaloids

The inhibitory effect of a series of secondary plant compounds including steroidal alkaloids and glycoalkaloids on larvae of the red flour beetle *Tribolium castaneum* was investigated (58). Larval growth was inhibited on artificial diets containing 1 μmol g^{-1} diet of the glycoalkaloids solamargine, solasonine and tomatine, whereas the corresponing aglycones solasodine and tomatidine, as well as tomatidenol, were inactive. The inhibitory effect of solamargine and tomatine, but not of solasonine, was completely abolished by addition of 1 μmol g^{-1} dietary cholesterol and/or sitosterol. Nonetheless, synthetic cholesteryl tomatide displayed significant activity at 2 μmol g^{-1} diet. Parallel studies with the tobacco hornworm, *Manduca sexta*, showed marked inhibitory activity of tomatine at a dietary concentration of 1 μmol g^{-1}, whereas the other compounds did not affect sterol metabolism or larval development. An appraisal of the factors influencing the mode of action of the active steroidal glycoalkaloids was also attempted (58).

8.5.11   Benzoxazinoids (Hydroxamic Acid)

Hydroxamic acid isolated from the root tissue of four maize lines retarded the western corn rootworm, *Diabrotica virgifera virgifera* LeConte larval development and survivorship (59). European corn borer ovipositional nonpreference for resistant line B96 was explained by the profile of leaf tissue defensive compounds (60). 4-Hydroxy-7-methoxy-2H-1,4-benzoxazin-3(4H)-one (HMBOA) was isolated and characterized as a primary leaf compound that deters European corn borer (ECB) oviposition. HMBOA was synthesized to authenticate its presence in the leaf and verify it as an active oviposition mediator. HMBOA and related compounds had a significant effect on ECB oviposition and female adult mortality. The oviposition index (OI) for the compounds ranged from 0.22 for 6-methoxybenzoxazilin-2(3H)-one (MBOA), a decomposition product of 2,4-dihydroxy-7-methoxy-2H-1,4-benzoxazin-3(4H)-one (DIMBOA), to 0.66 for (4R)-(+)-4-isopropyl-2-oxazolidinone. ECB females distributed egg masses almost evenly between substrates coated with DIMBOA and the solvent control as indicated by a neutral OI of 0.02. Elimination of the cyclic carbamate structure of MBOA or BOA as in benzoxazolidinone, 2-pyrrolidinone, and 2-pyrrolidinone-5-carboxylic acid did not alter the activity from that of DIMBOA. However, HMBOA significantly deterred females from depositing egg masses on the substrate (OI 0.30). When the benzyl group of HMBOA was replaced with isopropyl functionality on the cyclic carbamate, as illustrated by (4R)-(+)-4-isopropyl-2-oxazolidinone, OI was −0.66, which showed that oviposition activity was significantly reduced in comparison with that of all other compounds. Mortality of females ranged from 1.33% for 2-pyrrolidinone to 13.00% for HMBOA. By using adult-resistant lines such as B96, manipulation of European corn adult host selection and oviposition behavior seem feasible as a pest management strategy.

The biological activities of DIMBOA were tested on the Asian corn borer (ACB) *Ostrinia furnscalis* (Guenee). The antifeedant property of DIMBOA on ACB was confirmed by the methods leaf disc and feeding-weighing (61). Antifeeding ratio in unselected experiments was lower than that in selected ones. The results showed that DIMBOA had some effect on embryonic development of ACB and significantly affected the growth and development of ACB's larvae. The larvae's growth and development were delayed, mortality increased, weight of both the larvae and the pupae decreased, sex ratio (f/m) decreased, and deformed pupae appeared after the treatment with DIMBOA. Most ACB holes in plants are at the lower parts of internodes, where DIMBOA concentrations are higher.

Aphids are among the most important pests of cereals on account of direct feeding damage and transmission of viral diseases. Inverse relationships were obtained between hydroxamic acid (Hx) levels in a maize plant and infestation numbers of the corn leaf aphid *Rhopalosiphum maidis*, under both field and greenhouse conditions (60). Similar relationships were obtained for Hx levels in a wheat plant and infestation numbers of the rose-grain aphid *Metopolophium dirhodum*, the greenbug *Schizaphis graminum*, and the grain aphid *Sitobion avenae* under greenhouse conditions. When severed barley leaves lacking Hx were immersed in DIMBOA-containing solution, the levels of DIMBOA incorporated also correlated with infestation numbers of *M. dirhodum*. Aphid infestation numbers were lower in those tissues with lower Hx levels resulting from older plant. DIMBOA exhibited antifeedant effect on cereal aphids when incorporated into holidic diets. The aphid most sensitive to DIMBOA levels was *S. graminum*, the least sensitive *R. maidis*. The glucoside of DIMBOA was less active than DIMBOA itself against *S. graminum*. Inverse relationships were obtained between DIMBOA levels in wheat plants and DIMBOA levels in aphids feeding on them. This result, coupled with a tendency to decreased honeydew production and weight gain, indicated feeding deterrency by DIMBOA also under natural conditions.

8.5.12   Polyhydroxy Alkaloids

Alkaloids that are inhibitors of glycosidases are found in microorganisms and plants. Four structural types can be distinguished, namely, polyhydroxy derivatives of piperidine, pyrrolidine, indolizine, and pyrrolizidine. Such compounds have so far been isolated from species out of five plant families: Moraceae, Fabaceae, Polygonaceae, Aspidiaceae, and Euphorbiaceae. It seems reasonable to accept the function of polyhydroxy alkaloids in plants as defense agents that deter most opportunistic herbivores from feeding (62,63). Although extensive studies are still hindered by a limited supply of purified compounds, experimental evidence clearly supports a defensive role of the alkaloids. Dihydroxymethyldihydroxypyrrolidine (DMDP) is a potent inhibitor of insect digestive α-glucosidase. Larvae of the seed beetle, *Callosobruchus maculatus*, a pest of stored grain legumes, were killed at concentrations of only 0.03% (64). Castanospermine inhibits cellobiose, lactose, maltose, sucrose, and trehalose-hydrolyzing enzymes from a broad spectrum of insects (19 species from 12 families). The extent of inhibition is highly dependent on the insect taxa and the particular disaccharide sub-strate. For the pea aphid (*Acyrthosiphon pisum*), an ED_{50} of 20 ppm was estimated when the alkaloid was incorporated into an artificial diet. Phloem-feeding mealybugs (*Pseudococcus longispinus*) that opportunistically colonized *Castanospermum* were found to excrete castanospermine with the honeydew. They are obviously able to tolerate the alkaloid, although their glycosidases did not show any tolerance adaption to the inhibitor. However, castanospermine had little effect on maltase and sucrase of *Spodoptera*, despite being a strong inhibitor of the mammalian (mouse) enzymes (65).

Furthermore, in addition to inhibiting carbohydrate digestion, DMDP deters feeding young nymphs of *Locusta migratoria* at 0.001% in artificial diet (66). The antifeedant behavior was investigated by electrophysiological studies. With *Spodoptera littoralis*, evidence has been presented that indicates that DMDP, as a structural analogue of fructose, may interact or block the neuronal receptor site for fructose. Thus, if fructose acts as a phagostimulant, interference of the fructose receptor with DMDP would render the insect "blind" to the phagostimulatory sugar.

8.5.13   Other Alkaloids

Sabadilla, long known for its insecticidal properties, is the dried powdered ripe seeds of *Schoencaulon officinale* A. Gray, also known as *Sabadilla officinarum* Brant and *Veratrum sabadilla* Retz (67). The major insecticidal components of sabadilla are veratridine, cevadine, and the 3-angenoyl esters of veracevine. Of a series of benzoyl esters, the 3,5-dimethoxybenzyl derivative was most active. The LD_{50} by topical application was 1.5 and 0.51 µg/g to house files (pretreated with piperonyl butoxide) and milkweed bugs (*Oncopeltus fasciatus*), respectively. Veratridine gave values of 6.6 and 2.6 µg/g, respectively, in the same test. In the aliphatic ester series, cevadine remained the most active with an LD_{50} of 10 µg/g against house files and 0.51 µg/g for milkweed bugs. Veracevine itself was nontoxic to the insects and mice at the upper levels used in the study. The closely related protovaratrines A and B, which are components of "hellebore," the dried rhyzome of *Varatrum album*, were also examined in the veracevine ester study. Only protovaratrine A proved highly insecticidal with an LD_{50} of 1.6 µg/g for the housefly (with PB) and 2.2 µg/g for the milkweed bug.

3-*O*-Vanilloylveracevine has been synthesized for the first time in 70% overall yield by conversion of veracevine ino its 3-*O*-(4-benzyloxy-3-methoxybenzoate) followed by catalytic hydrogenation. The insecticidal activity of the semisynthetic substance against three pest species is inferior to that of cevadine and veratridine, the major components of the insecticidal sabadilla alkaloid mixture (68).

Since the earlier review by Crosby (69), considerable work has been accomplished in isolating, identifying, and establishing the relative insecticidal activities of the compounds of ryania, the powder obtained by grinding the dried plant *Ryania speciosia* Vahl (Flacourtiaceae). In a study of ryania, the Casida group, utilizing radial thin-layer chromatography

(TLC) and preparative high-performance liquid chromatography (HPLC), isolated 10 of 11 compounds previously identified and provided an HPLC and ^1H nuclear magnetic resonance (NMR) procedure for monitoring different lots of powders (70). Ryanodine and dehydroryanodine composed over 80% of two ryania preparations analyzed and were among the most active of the components by injection into adult house flies, pretreated with piperonyl butoxide, and more active in diet using first-instar flour beetles (*Tribolium castaneum*). Of interest, the naturally occurring pyridine-3-carboxylate analogue of ryanodine showed little activity. The toxic action against the flour beetle for the 10 compounds paralleled their ability to bind to the calcium release channel protein of muscle sarcoplasmic reticulum, possibly preventing channel closing, and a mode of action which defines ryanodine's functions as a muscle poison. In other studies, which included a number of ryanodine derivatives and degradation products, ryanodol and didehydroryanodol were shown to have good knockdown activity when injected into houseflies and cockroaches, even though toxicities to mice were low and there was little binding to the ryanodine receptor.

Gonzalex-Coloma et al. (71) have studied the antifeedant and insecticidal effects of several natural ryanoid diterpenes. These compounds can be classified in two groups according to their chemical structures: ryanodol/isoryanodol-type (nonalkaloidal type) and ryanodine-type (alkaloidal type) ryanoids. The nonalkaloidal ryanoids were isolated from *Persea indica* (Lauraceae) while the alkaloidal ryanoids (ryanodines and spiganthines) were isolated from *Spigelia anthelmia* (Loganiaceae). The effects of these compounds on the feeding behavior and performance (with and without piperonyl butoxide pretreatment) of *Spodoptera littoralis* larvae and *Leptinotarsa decemlineata* adults indicate that some strongly deterred these insects, *L. decemlineata* being less sensitive than *S. littoralis*. Their antifeedant effects did not parallel their toxic action. In addition, more than 60% of the nonalkaloidal ryanoids were antifeedant and/or toxic in contrast to 30% of active alkaloidal ones, supporting the hypothesis of a ryanodol-specific mode of action in insects. Two pyrrolizidine alkaloids, megalanthonine (a) and lycopsamine (b), have been isolated from *Heliotropium megalanthum*. The structure of the novel compound a was determined by spectroscopic methods. The insecticidal, antifeedant, and antifungal effects of compounds a and b have been evaluated (72).

Extracts of the roots of Stemonaceae plants have been used for the control of agricultural insect pests in China. Stemofoline, an alkaloid isolated from the leaves and stems of *Stemona japonica* Miq., has been found to be the major insecticidal component (73). When fed an artificial diet at 100 ppb to fourth-instar silkworm larvae (*Bombyx mori* L.), there were immediate toxic symptoms and death within 24 h. Conversely, stemofoline was inactive at 100 ppm in artificial diet fed to cabbage armyworms (*Mamestra brassicae*).

In assessing the pesticidal activity of soosan, *Pancratium maritimum*, the bulbs and leaves were extracted using acetone/ethanol and ethanol as solvents and mosquito larvae *Culex pipiens*, as a test organism (74). The acetone/ethanol extract of bulbs was more toxic (LC_{50} 25 ppm) than that of the leaves (LC_{50} 75 ppm). Crude alkaloids, lycorine, terpenes, sterols, and fixed oil were isolated from soosan bulb; their percentages were 0.193%, 0.02%, 0.13%, and 3.3%, respectively. The acetone/ethanol extract showed a strong aphicidal activity to *Aphis gossypii* with LC_{50} of 0.07%, 0.28%, and 0.3%, respectively. Also, the acetone/ethanol extract showed high toxicity to *Spodoptera littoralis* fourth-instar larvae, with LD_{50} value of 2 mg per larva. In addition, ethanol extract containing alkaloids and the oil of soosan bulbs showed acaricidal activity against the two-spotted spider mite, *Tetranychus urticae*, with LC_{50} values of 0.2%, 0.36%, and 1.5%, respectively. Synergism studies on *A. gossypii* indicated that lycorine, the principal alkaloid of soosan bulbs, strongly synergized the organophosphorous insecticide cyanophos and reduced its LC_{50} values from 120 to 48 ppm. On the other hand, the aqueous extract of soosan bulbs synergized the toxicity of actellic and permethrin in *Tribolium castaneum*, which reduced their LC_{50} values from 80 to 46 ppm and from 1000 to 550 ppm, respectively. Also, ethanol and petroleum-ether extracts synergized the toxicity of reldan and permethrin, respectively, in the same insect.

A resistant factor from a strain (TC1966) of wild mung bean (*Vigna radiata* var. *sublobata*) active against the azuki bean weevil (*Callosobruchus chinensis*) was genetically transferred to a susceptible cultivar. The resulting strain (BC20F4) tolerated an infestation by *C. chinensis*. The ethanol extract of BC20F4 inhibited larval growth (75). Vignatic acid A was isolated and evaluated as one of the inhibitors present in BC20F4. The structure of vignatic acid was determined to be a cyclopeptide alkaloid composed of L-tyrosine, 3(S)-hydroxyl-L-leucine, L-phenylalanine, and 2-hydroxyisocaproic acid by multidimensional nuclear magnetic spectroscopy. This compound is the first example of an insecticidal cyclopeptide alkaloid base of plant origin.

The structure–activity relationship was investigated for avenanthramide alkaloids isolated from the eggs of *Pieris brassicae*, the large white cabbage butterfly, and eight synthesized related compounds as oviposition deterrents for this insect (76). The activity of all compounds was tested in a dual-choice bioassay. The two most active oviposition deterrents for *P. brassicae* were *trans*-2-[3-(4-hydroxyphenylpropenoyl)amino]-3,5-dihydroxybenzoic acid and *trans*-2-[3-(3,4-dihydroxyphenylpropenoyl)-amino]-3,5-dihydroxybenzoic acid. Among members of this compound class, alteration of the substituents of the cinnamic acid part of the molecule affected the oviposition deterrent activity more profoundly than other structural changes. Modification of the anthranilic acid part of the molecule resulted in lower activity.

8.6   MECHANISM OF ACTION

Several alkaloids are toxic to insects and vertebrates, and, in addition, can inhibit the growth of bacteria (Table 8.2). In vitro assays were established to elucidate their modes of action and to clarify their allelochemical properties (77). Basic molecular targets studied, present in all cells, included DNA intercalation, protein biosynthesis, and membrane stability. The degree of DNA intercalation was positively correlated with inhibition of DNA polymerase I, reverse transcriptase, and translation at the molecular level and with toxicity against insects and vertebrates at an organismic level. Inhibition of protein biosynthesis was positively correlated with animal toxicity. Molecular targets studied, present only in animals, included neuroreceptors (α_1, α_2, serotonin, muscarinic, and nicotinic acetylcholine receptors) and enzymes related to acetylcholine (acetylcholinesterase and choline acetyltransferase). The degree of binding of

TABLE 8.2   Mechanism of Action of Alkaloids at the Molecular Level

Alkaloids	Mechanism
Berberine, quinine, β-carbolines	DNA/RNA intercalation
Pyrrolizidines	Alkylation of DNA/RNA
Coniine	Mutations
Lycorine, vincristine	Inhibition of DNA/RNA polymerases
Vinblastine	Microtubules/cytoskeleton
Quinolizidines	Inhibition of protein biosynthesis
Steroidal alkaloids	Membrane stability
Quinine, aconitine	Inhibition of ion channels, carrier
Nicotine, heliotrine	Acetylcholine receptors
Cocaine	Dopamine receptors
β-Carbolines	Serotonine receptors, GABA receptors
Cocaine, β-carbolines	Transport or degradation of neurotransmitters
Tetrahydroberberine	Adenyl cyclase inhibition
Polyhydroxyalkaloids	Inhibitors of hydrolases

alkaloids to adrenergic, serotonin, and muscarinic acetylcholine receptors was positively correlated in six-protein-coupled receptors. Receptor binding and toxicity was correlated in insects. The biochemical properties of alkaloids were discussed. It is postulated that their structures were shaped in a process termed "evolutionary molecular modeling" to interact with a single molecular target and, more often, with several molecular targets at the same time. Many alkaloids are compounds with a broad activity spectrum that apparently have evolved as "multipurpose" defense compounds. The evolution of allelochemicals affecting more than one target could be a strategy to counteract adaptations by specialists and to help fight off different groups of enemies.

Alkaloids are multipurpose compounds which, depending on the situation, may be active in more than one environmental interaction. Quinolizidine alkaloids are the most important defense chemicals in Leguminosae against insects and other herbivores. Alkaloids repel or deter the feeding of many organisms due to their bitter or pungent taste. Many alkaloids are also toxic.

8.6.1   Nicotinoids

Nicotinoids and neonicotinoids are characterized by the presence of the 3-pyridylmethylamine moiety in their structure. In the former, the amino nitrogen atom is ionized, while in the latter the corresponding nitrogen atom is not ionized but bears a partial positive charge. Both types of insecticides interact with nicotinic acetylcholine receptor (nAChR) of insect origin (78). The poor interaction of neonicotinoids with vertebrate (nAChR) was shown by its poor binding affinity to the nAChR from Torpedo electric organ and rat brain and poor activation with nAChR expressed in *Xenopus oocytes*. The full positive charge was essential to interact with the vertebrae nAChR while the 3-pyridylmethylamine moiety with a partial positive charge was enough to interact with the insect nAChR. For penetration into the insect central nervous system, hydrophobicity seemed to play an important role, as indicated by the binding of the injected compounds to the housefly head nAChR. The ionization reduced hydrophobicity and limited the penetration of nicotinoids, resulting in less insecticidal activity. Among neonicotinoids, nitromethylene-type compounds, though far higher in binding affinity, were less hydrophobic than the corresponding nitroimine-type, and the net result was better or inferior insecticidal activity. A chlorine atom at the 6-position of the 3-pyridyl group found in commercialized neonicotinoids contributes to increased binding affinity and, more importantly, to hydrophobicity, thus increasing insecticidal activity. N-Me-imidacloprid was found to be a propesticide of imidacloprid.

Gaertner et al. (79) examined the accumulative transport properties of the Malpighian (excretory) tubules of the tobacco hornworm *Manduca sexta* to test the hypothesis that a P-glycoprotein-like multidrug transporter is active and is responsible for the excretion of dietary nicotine in this tissue. Isolated tubules were cannulated and exposed to radiolabeled forms of either nicotine (5 min exposure) or the P-glycoprotein substrate vinblastine (60 min exposure) in the bathing (basal surface) fluid. The luminal (apical) contents were then flushed and lumen-to-bath ratios measured. Although these ratios provide conservative estimates of the physiological ability of Malpighian tubules to move compounds from blood to lumen, tubules conentrated nicotine 10-fold from an initial bath concentration of $0.5 \, \text{mmol} \, \text{L}^{-1}$ and vinblastine threefold (from an initial concentration of $1 \, \mu\text{mol} \, \text{L}^{-1}$). Vectorial transport of vinblastine and nicotine was eliminated by $25 \, \mu\text{mol} \, \text{L}^{-1}$ verapamil (a P-glycoprotein inhibitor) and was not dependent on the presence of a transepithelial electrical potential. Nicotine transport was inhibited by atropine ($3 \, \text{mmol} \, \text{L}^{-1}$), while nicotine ($\geq 50 \, \mu\text{mol} \, \text{L}^{-1}$) significantly reduced vinblastine transport. Verapamil was effective at reducing vinblastine transport when applied to the basal side alone, but not when applied to the apical side alone. Taken together, these results are consistent with the idea that the active excretion of nicotine and other alkaloids by the tobacco hornworm is mediated by a P-glycoprotein-like mechanism.

8.6.2 Ryanodine

Ryanodine receptors/Ca^{2+} release channels play an important role in regulating the intracellular free calcium concentrations in both muscle and nonmuscle cells. Ryanodine, a neutral plant alkaloid, specifically binds to and modulates these Ca^{2+} release channels (80). The interaction of a tritium-labeled, photoactivable derivative of ryanodine [³H-labeled 10-*O*-(3-(4-azidobenzamido)propionyl)ryanodine] [(³H)ABRy)] with the ryanodine receptor of skeletal, cardiac, and brain membranes was characterized. Scatchard analysis demonstrates that this ligand binds to a single class of high affinity sites in skeletal muscle triads. Furthermore, competition binding assays of (³H)ryanodine with skeletal, cardiac, and brain membranes in the presence of increasing concentrations of unlabeled ABRy illustrate that this azido derivative of ryanodine can specifically displace (³H)ryanodine from its binding site(s). Analysis of the effects of Ca^{2+}, ATP, and KCl on (³H)ABRy binding in triad membranes shows a similar modulation of binding to that seen in these membranes with (³H)ryanodine. Photoaffinity labeling of triads with (³H)ABRy resulted in specific and covalent incorporation of (³H)ABRy into a 565-kDa protein that was shown to be the skeletal muscle ryanodine receptor. Digestion of the labeled ryanodine receptor revealed a (^{3}H)ABRy-labeled 76-kDa tryptic fragment that was identified with an antibody directed against the COOH terminal to the receptor. These results demonstrate that the 76-kDa COOH terminal tryptic fragment contains the high-affinity binding site for ryanodine.

Kinetic and equilibrium measurements of (^{3}H)ryanodine binding to the Ca^{2+} release channel of rabbit skeletal and rat cardiac sarcoplasmic reticulum (SR) were examined to ascertain the nature of cooperative interactions among high- and low-affinity binding sites and to quantitate their distribution (81). Equilibrium studies reveal affinities of 1–4 nM for the highest affinity binding site and of 3–50 nM and 2–4 µM for the lower affinity sites in both preparations, with Hill coefficients of significantly less than 1. Initial rates of association and dissociation increase with increasing concentrations of ryanodine. SR vesicles are actively loaded in the presence of pyrophosphate, and fluctuations in extravesicular Ca^{2+} are measured by the absorbance change of antipyrylazo III. The data demonstrate a biphasic, time- and concentration-dependent action of ryanodine on the release of Ca^{2+} with an initial activation and a subsequent inactivation phase. Kinetic analysis of the activation of Ca^{2+} release by ryanodine, in consonance with the binding data, demonstates the existence of multiple binding sites for the alkaloid on the channel complex with nanomolar to micromolar affinities. Based on the present findings obained by receptor binding analysis and Ca^{2+} transport measurements, they suggest a model that describes four, most plausibly negatively cooperative binding sites on the Ca^{2+} release channel. Occupation of ryanodine binding sites produces sequential activation followed by inactivation of the SR channel, revealing the strong possibility of an irreversible uncoupling of the native function of the receptor–channel complex by high concentrations of ryanodine. A model relating ryanodine receptor occupancy with SR Ca^{2+} release stresses two important new findings regarding the interaction of ryanodine with its receptor. First, ryanodine binds to four sites on the oligomeric channel complex with decreasing affinities, which can be best described by allosteric negative cooperativity. Second, binding of ryanodine to its receptor activates the Ca^{2+} release channel in a concentration-dependent and saturable manner in the range of 20 nM to 1 mM and produces a kinetically limited and sequential inactivaion of the Ca^{2+} channel, with the concomitant attainment of full negative cooperativity. The result presented suggests that driving of the complex toward full negative cooperativity with high concentration of ryanodine promotes a long-lived conformational state in which ryanodine is physically occluded and hindered from free diffusion from its binding site.

8.6.3   Pyrrolizidine Alkaloids

Pyrrolizidine alkaloids (PAs) are toxic constituents of hundreds of plant species, some of which people are exposed to in herbal products and traditional remedies. The bioactivities of PAs are related, at least in part, to their ability to form DNA–protein complexes (DPCs). Previous studies indicated a possible role for actin in PA-induced DPCs. Nuclei prepared from Madin-Darby bovine kidney (MDBK) and human breast carcinoma (MCF-7) cells were treated with the pyrrolic PAs dehydrosenecionine (DHSN) and dehydromonocrotaline (DHMO) (82). DPCs were purified and then analyzed by Western immunoblotting. Actin was found in DPCs induced by both DHSN and DHMO, but not in those from control nuclei. Actin was also present in DPCs induced by cisplatinum and mitomycin C, two bifunctional cross-linkers. In separate experiments, DHSN and DHMO were cross-linked to a mixture of Hind III digested phage with varying amounts of glutathione (GSH), cysteine, or methionine to identify the stoichiometry of competition between DNA and alternate nucleophiles for cross-link formation with pyrroles. GSH and cysteine, but not methionine, competed with phage for DNA cross-linking, indicating that reduced thiols may have a role in nucleophilic reactions with pyrroles in the cell. While actin involvement in cisplatinum induced DPCs is documented, the discovery of actin cross-linking in PA or mitomycin C-treated cells or nuclei is novel. Pyrrole-induced DPC formation with actin, a protein with structural and/or regulatory importance, may be a significant mechanism for PA toxicity and bioactivity.

8.6.4   Veratridine

Veratridine causes Na^+ channels to stay open during a sustained membrane depolarization by abolishing inactivation (83). The consequential Na^+ influx, either by itself or by causing a maintained depolarization, leads to many secondary effects such as increasing pump activity, Ca^{2+} influx, and in turn exocytosis. If the membrane is voltage clamped in the presence of the alkaloid, a lasting depolarizing impulse induces, following the "normal" transient current, another much more slowly developing Na^+ current that reaches a constant level after a few seconds. An inward tail current that slowly subsides then follows repolarization. Development of these slow currents is enhanced by additional treatment with agents that inhibit inactivation. Most of these phenomena can be satisfactorily explained by assuming that Na^+ channels must open before veratridine binds to them and that the slow current changes reflect the kinetics of binding and unbinding. It is unclear where the alkaloid stays when it is not bound. Although the effect sets in promptly, once this pool is filled, access to it from outside must be impeded since in most preparations veratridine can only be partially washed out.

8.7   MOLECULAR BIOLOGY

Several biosynthetic genes involved in the formation of benzylisoquinoline, terpenoid indole, tropane, and nicotine as well as purine alkaloids have been isolated. The functions of gene promoters have been studied in relation to the regulation of alkaloid metabolism. Isolated genes have been used to genetically alter the accumulation of specific alkaloids and other plant secondary metabolites.

8.7.1   Benzylisoquinoline Alkaloids

The biosynthetic pathway leading from L-tyrosine to berberine has 13 different enzymatic reactions that involve a norcoclaurine synthase, an N-methyltransferase, three O-methyltransferases (OMTs), a hydroxylase, a berberine bridge enzyme (BBE), a methylenedioxy ring–forming enzyme, and a tetrahydroprotoberberine oxidase. cDNAs of several enzymes in this pathway have been isolated and characterized.

Reticularine is a branch point intermediate in the biosynthesis of many benzylisoquinoline alkaloids (BIAs). The first committed step in benzo-phenanthridine (sangunarine), protoberberine (berberine, palmatine), and morphinan (morphine) alkaloid biosynthesis involves conversion of the N-methyl group of (S)-reticuline to the methylene bridge moiety of (S)-scoulerine by the berberine bridge enzymes. The berberine bridge enzyme (84) (S)-reticuline:oxygen oxidoreductase (methylene bridge forming), EC1.5.3.9, is a vesicular plant enzyme that catalyzes the formation of the berberine bridgehead carbon of (S)-scoulerine from the N-methyl carbon of (S)-reticuline in a specific, unparalleled reaction along the biosynthetic pathway that leads to benzophenanthridine alkaloids. Cytotoxic benzophenanthridine alkaloids are accumulated in certain species of Papaveraceae and Fumariaceae in response to pathogenic attack and, therefore, function as phytoalexins. The berberine bridge enzyme has been purified to homogeneity from elicited cell suspension cultures of *Eschscholtzia californica*, and partial amino acid sequence has been determined. A cDNA, isolated from an Agt/11 cDNA bank of elicited *E. californica* cell suspension cultures, coded for an open reading frame of 538 amino acids. The first 22 amino acids constitute the putative signal peptide. The mature protein has a molecular weight (MW) of 57,352. Daltons excluding carbohydrate. The berberine bridge enzyme was heterologously expressed in a catalytically active form in *Saccharomyces cerevisiae.* Southern hybridization with genomic DNA suggests that there is only one gene for the enzyme in the *E. californica* genome. Hybridized RNA blots from elicited *E. californica* cell suspension cultures revealed a rapid and transient increase in ploy $(A)^+$ RNA levels that preceded both the increase in enzyme activity and the accumulation of benzophenanthridine alkaloids, emphasizing the integral role of the berberine bridge enzyme in the plant response to pathogens.

cDNA clones for the (*S*)-tetrahydroberberine (H4Ber) oxidase of cultured berberine-producing *Coptis japonica* cells were isolated by screening a *C. japonica* cDNA library with synthetic nucleotides that can encode the NH_2-terminal sequence of this enzyme (85). Analysis of the nucleotide sequences of the cloned cDNA inserts revealed a 759-bp open reading frame that encoded a 253-amino-acid polypetide with an *MW* of 27,089 and NH_2 terminal and internal sequences identical with those of the (*S*)-H4Ber oxidase, as determined by microsequencing methods. *Escherichia coli* cells were transformed with an expression vector carrying (*S*)-H4Ber oxidase cDNA. The transformed bacteria were induced to overproduce a 28-kDa protein that reacted with *Coptis* (*S*)-H4Ber oxidase-specific antibody. A comparision of the derived amino acid sequence of (*S*)-H4Ber oxidase with sequences in the protein database of the Protein Research Foundation showed a marked similarity between (*S*)-H4Ber oxidase and the NH_2 terminal protein of mouse P1-450, which is encoded by a single exon of the mouse P1-450 gene. The availability of cloned cDNA for (*S*)-H4Ber oxidase allows use of the methods of molecular biology to study the regulation of (*S*)-H4Ber oxidase gene expression in cultured *C. japonica* cells in relation to berberine biosynthesis.

S-Adenoslyl-L-methionine:3'-hydroxy-*N*-methylcoclaurine4'-*O*-methyltransferase (4'-OMT) catalyzes the conversion of 3'-hydroxy-*N*-methylcoclaurine to reticuline, an important intermediate in the synthesis of isoquinoline alkaloids (86). In an earlier step in the biosynthetic pathway to reticuline, another *O*-methyltransferase, *S*-adenosyl-L-methionine norcoclaurine: 6-*O*-methyltransferase (6-OMT), catalyzes methylation of the 6-hydroxyl group of norcoclaurine. Two kinds of cDNA clones were isolated that correspond to the internal amino acid sequences of a 6-OMT/4'-OMT or a 4'-OMT preparation from cultured *C. japonica* cells. Heterologously expressed proteins have 6-OMT activtities, indicating that each cDNA encodes a different enzyme. 4'-OMT was purified using recombinant protein, and its enzymological properties were characterized. It had enzymological characteristics similar to those of 6-OMT; the active enzyme was the dimer of the subunit, no divalent cations were required for activity, and there was inhibition by Fe^{2+}, Cu^{2+}, Co^{2+}, Zn^{2+}, or Ni^{2+} , but none by the SH reagent. 4'-OMT clearly had different substrate specificity.

It methylated (R,S)-6-O-methylnorlaudanosoline, as well as (R,S)-laudanosoline and (R,S)-norlaudanosoline. Laudanosoline, and N-methylated substrate, was a much better substrate for 4′-OMT than norlaudanosoline. 6-OMT methylated norlaudanosoline and laudanosoline equally. The molecular evolution of these two related O-methyltransferases is discussed (86).

Coptis SMT cDNA was introduced into California poppy (E. californica) cells that produce the benzophenanthridine alkaloid sanguinarine rather than berberine-type alkaloids to modify the metabolite profile (87). Transformation of E. californica was confirmed by RNA and protein blot analyses of Coptis SMT. All of the transformants selected on hygromycin-containing media had an accumulation of the SMT transcript as high as that of the Coptis cells, whereas wild-type E. californica cells had no hybridization signal for the SMT transcript. Protein blot analysis findings further support the expression of Coptis SMT in transgenic E. californica cells, but the accumulation of SMT protein was less than that in cultured Coptis cells. Transgenic E. californica cells also had evident but lower SMT activities than Coptis cells, whereas the wild-type E. californica cells had none. These findings indicate that the introduced SMT gene(s) of Coptis was successfully expressed in E. californica cells.

Opium poppy (Papaver somniferum) contains a large family of tyrosine/dihydroxyphenylalanine decarboxylase (tydc) genes involved in the biosynthesis of benzylisoquinoline alkaloids and cell wall–bound hydroxycinnamic acid amides. Eight members from two distinct gene subfamilies have been isolated, tydc1, tydc4, tydc6, tydc8, and tydc9 in one group and tydc2, tydc3, and tydc7 in the other (88). The tydc8 and tydc9 genes were located 3.2 kb apart on one genomic clone, suggesting that the family is clustered. Transcripts for most tydc genes were detected only in roots. Only tydc2 and tydc7 revealed expression in both roots and shoots, and TYDC3 mRNAs were the only specific transcripts detected in seedlings. TYDC1, TYDC8, and TYDC9 mRNAs, which occurred in roots, were not detected in elicitor-treated opium poppy cultures. Expression of tydc4, which contains a premature termination codon, was not detected under any conditions. Five tydc promoters were fused to the β-glucuronidase (GUS) reporter gene in a binary vector. All constructs produced transient GUS activity in microprojectile-bombarded opium poppy and tobacco (Nicotiana tabacum) cell cultures. The organ- and tissue-specific expression pattern of tydc promoter-GUS fusion in transgenic tobacco was generally parallel to that of corresponding tydc genes in opium poppy. GUS expression was most abundant in the internal phloem of shoot organs and in the stele of roots. Select tydc promoter–GUS fusions were also wound induced in transgenic tobacco, suggesting that the basic mechanisms of developmental and inducible tydc regulation are conserved across plant species.

The metabolic engineering of BIA pathways was lacking due to the nonavailability of transformation protocols for BIA-producing plants. The recently developed transformation procedures for opium poppy plants, root and cell cultures, and *E. californica* plants provide the opportunity to alter the activity of individual enzymes of BIA biosynthesis.

8.7.2 Monoterpenoid Indole Alkaloids

Isolated *tdc* (tryptophan decarboxylase gene) and, more recently, *tydc* (see Schemes 8.3 and 8.4) genes have been used to genetically alter the regulation of secondary metabolic pathways derived from aromatic amino acids in several plant species (89). The biotechnological objectives of these modifications include (a) increasing the accumulation of valuable pharmaceuticals in medicinal plants, or introducing novel phytochemical pathways into plants of agronomic importance; (b) reducing the accumulation of undesirable compounds in plant-derived products; and (c) improving the resistance of crop species against pests and disease-causing pathogens. However, these efforts to alter plant metabolic pathways using *tdc* and *tydc* genes have often produced unpredictable results, primarily due to our limited understanding of the network architecture of metabolic pathways. Most current models of metabolic regulation in plants are still based on individual reactions and do not consider the integration of several pathways sharing common branch points. The use of transgenic plants expressing a heterologous *tdc* or *tydc* gene has proven to be a powerful technique to study metabolic regulation and in improving our understanding of the physiological roles for specific secondary metabolic pathways.

Chimeric *tdc* and *tydc* gene constructs, consisting of the cauliflower mosaic virus (CaMV) 35S promoter, followed by the *tdc* or *tydc* coding region and the nopaline synthase (NOS) transcription terminator, have been introduced into various plant species. Tobacco plants expressing *Catharanthus roseus tdc* also showed a 97% reduction in the reproduction rate of sweet potato whitefly (*Bemisia tabaci*) larvae, relative to wild-type tobacco, suggesting that tryptamine production in transgenic crops might be a useful strategy to confer protection against insect damage without the application of agrochemicals (89). The mechanism by which tryptamine production in transgenic tobacco affects insect reproduction is not known. However, it has been suggested that tryptamine disrupts whitefly larval and pupal development, and the process of adult leaf selection for feeding and oviposition, by blocking neuromuscular transmission.

Introduction of the *tdc* gene into *Brassica napus* (canola) resulted in the redirection of tryptophan into tryptamine, rather than indole glucosinolates, in all parts of the plant (90). The indole glucosinolate content of mature seed from transgenic plants was only 3% of that found in wild-type seeds. In oil seed crops, such as canola, the presence of indole glucosinolates in seeds decreases the meal palatability and, consequently, its value as animal feed. This study is an elegant example of how the introduction of a heterologous aromatic amino acid decarboxylase can be used to create an artificial sink to divert metabolic flow and reduce the levels of undesirable aromatic amino acid–derived products. Detailed knowledge of the biochemistry and molecular regulation of aromatic amine biosynthesis in plants has the potential to create new agricultural biotechnology opportunities for pharmaceutical production and pest/disease resistance.

A *tdc* transgene was first introduced into *C. roseus* cells by infecting seedlings with an oncogenic strain of *Agrobacterium tumefaciens* (91). Tumori- genic calli expressing the *tdc* transgene showed increased TDC activity and tryptamine content, but alkaloid levels were not affected compared to wild-type controls.

Strictosidine β-D-glucosidase (SGD) is an enzyme that contributes to the biosynthesis of terpenoid indole alkaloids (TIAs) by converting strictosidine to cathenamine (92). The biosynthetic pathway toward strictosidine is thought to be similar in all TIA-producing plants. Somewhere downstream of strictosidine formation, however, the biosynthesis diverges to give rise to the different TIAs found. SGD may play a role in creating this biosynthetic diversity. They have studied SGD at both the molecular and enzymatic levels. Based on the homology between different plant β-glucosidases, degenerate polymerase chain reaction primers were designed and used to isolate a cDNA clone from a *Catharanthus roseus* cDNA library. A full-length clone gave rise to SGD activity when expressed in *Saccharomyces cerevisiae*. SGD shows 60% homology at the amino acid level to other β-glucosidases from plants and is encoded by a single copy gene. SGD expression is induced by methyl jasmonate with kinetics similar to those of two other genes acting prior to SGD in TIA biosynthesis. These results show that coordinate induction of the biosynthetic genes forms at least part of the mechanism for the methyl jasmonate-induced increase in TIA production. Using a novel in vivo staining method, subcellular localization studies of SGD were performed. This showed that SGD is most likely associated with the endoplasmic reticulum, which is in accordance with the presence of a putative signal sequence, in contrast to previous localization studies. This new insight in SGD localization has significant implications for our understanding of the complex intracellular trafficking of metabolic intermediates during TIA biosynthesis.

In contrast to other groups of plant products, which produce many glycosides; indole alkaloids rarely occur as glucosides. Plants of *Rauwolfia serpentina* accumulate ajmaline as a major alkaloid, whereas cell suspension cultures of *Rauwolfia* mainly accumulate the glycoalkaloid raucaffricine at levels of 0.6 g/L. Cell cultures contain a specific glucosidase, known as raucaffricine-O-β-glucosidase (RG), which catalyzes the in vitro formation of vomilenine (a direct intermediate in ajmaline biosynthesis). The authors described the molecular cloning and functional expression of this enzyme in *E. coli* (93). RG shows up to 60% amino acid idendity with other glucosidases of plant origin and it shares several sequence motifs with family 1 glucosidases that have been characterized.

Nontumorigenic *C. roseus* cell cultures transformed with an STRI transgene showed 10-fold higher STR activity and accumulated higher levels of strictosidine and other TIAs than wild-type cultures; however, TDC activity was not affected (94). In contrast, high TDC activity conferred by a TDC transgene introduced alone or in combination with the STRI transgene did not affect alkaloid accumulation. These results further suggest that STR catalyzes a rate-limiting step of alkaloid biosynthesis in *C. roseus* cell cultures. The influence of precursor availability on TIA accumulation was investigated by feeding various concentrations and combinations of tryptamine and loganin to a trasgenic *C. roseus* cell line overexpressing STRI. High rates of tryptamine synthesis were found to occur even when TDC activity was low. Moreover, efficient STR activity was possible even when the tryptamine pool was small. However, the overall formation of strictosidine was shown to require a sufficient supply of both secologanin and tryptamine such that the efficient utilization of one depends on the availability of the other. Transgenic tobacco expressing *C. roseus* TDC and STRI was used to establish a cell culture with high constitutive TDC and STR activity (95). This transgenic tobacco cell line accumulated tryptamine and produced strictosidine when secologanin was added to the culture medium. These results demonstrate that two consective steps in the TIA pathway can be cooperatively expressed in a foreign plant species that does not normally produce these metabolites. Examination of the integration frequencies and expression levels of TDC and STRI in tobacco showed that both transgenes were expressed in only 33% of the plants. Thus, the extensive phenotypic variation in alkaloid production in transgenic tissues is partly caused by gene-silencing phenomena affecting TDC and STRI.

Recent studies have begun to reveal the location of cis elements and the identity of transcription factors involved in the development and inducible regulation of TDC, STRI, and CPR. The activity of the TDC promoter linked to the GUS reporter gene was initially examined in transgenic tobacco plants and transfected protoplasts (96). Progressive 5'-truncations gradually reduced GUS activity levels until deletion to −112 essentially eliminated TDC promoter activity. Three functional regions involved in basal or elicitor-induced expression were identified in the TDC promoter from −160 to −37 by a loss-of-function assay. The −160 to −99 region was shown to act as the main transcriptional enhancer for basal expression, and two separate elicitor-responsive elements were found between −99 and −87, and between −87 and −37. In vitro binding of nuclear factors to the −572 to −37 region of the TDC promoter has also been described. Two binding sites that interact with multiple TDC promoter regions were identified as GT-1 and 3AF1 in tobacco and *C. roseus* nuclear protein extracts. Mutagenesis of the GT-1 binding sites did not affect basal or elicitor-induced expression but did reduce TDC promoter activation by UV light.

Transcription factors are undoubtedly involved in the basal expression of the TDC, STRI, and CPR genes. A G-box motif at −105 was shown to bind G-box binding factors (GBFs) in vitro but was not essential for the elicitor-induced expression of STRI in vivo. This G-box element also interacts with tobacco nuclear factors and the G-box–binding factor TAF-1 (97). Mutation of the G-box motif prevented binding of these factors and reduced the functional activity of constructs containing tetramers of the STRI G box sequence. A G-box tetramer fused only to a TATA box conferred seed-specific expression in transgenic tobacco but required the enhancer region from the CaMV promoter for expression in leaves. These results suggest that sequences flanking the G-box motif determine STRI promoter activity in different tissues.

8.7.3 Nicotine and Tropane Alkaloids

The Solanaceae produce a range of biologically active alkaloids that include nicotine and the tropane alkaloids. Tropane alkaloids, such as hyoscyamine (atropine) and scopolamine (hyoscine), which are found mainly in *Hyoscyamus*, *Duboisia*, *Atropa*, and *Scopolia* species, together with their semisynthetic derivatives, are used as parasympatholytics that competitively antagonize acetylcholine. Both the tropane ring moiety of the tropane alkaloids and the pyrrolidine ring of nicotine are derived from putrescine by way of *N*-methylputrescine (MP). Because putrescine is metabolized to polyamines such as spermidine and spermine, the *N*-methylation of putrescine catalyzed by putrescine *N*-methyltransferase (PMT) is the first committed step in the biosynthesis of these alkaloids.

Conversion of hyoscyamine to its epoxide scopolamine proceeds in two oxidative steps: hydroxylation at the 6β position of the tropane ring followed by intramolecular epoxide formation by removal of the 7β hydrogen. Expression of a hyscyamine 6β-hydroxylase (H6H) cDNA clone in *E. coli* has demonstrated that H6H catalyzes both reactions and that the hydroxylase activity is about 40-fold stronger than the epoxidase activity (98).

This bifunctional enzyme requires 2-oxoglutarate, ferrous ion, ascorbate, and molecular oxygen for catalysis and therefore belongs to the 2-oxoacid-dependent oxygenase family. The amino acid sequence of *H. niger* H6H is homologous, among others, to ethylene-forming enzymes (EFE; also called ACC oxidase) and the positions of three introns found in the *H6H* gene and the tomato *EFE* gene are conserved strictly. Because the *H6H* gene has been found only in scopolamine-producing solanaceous species, while the *EFE* gene exists in most (probably all) higher plants, the *H6H* gene must have evolved from a ubiquitous plant 2-oxoacid-dependent oxygenase gene, such as the *EFE* gene, during diversificaton of the Solanaceae.

A surprisingly successful case of metabolic engineering has been reported for scopolamine production in *Atropa belladonna* (99). Wild-type belladonna accumulates primarily hyoscyamine, and the content of scopolamine, which is much preferred to hyoscyamine as an anticholinergic medicine, is low because of the low hyoscyanmine 6β-hydroxylase (H6H) activity in the root. The H6H cDNA of *H. niger* was introduced into belladonna, in which the enzyme was stongly and constitutively expressed in all plants. The primary transformant, and its progeny that inherited the transgene, contained scopolamine almost exclusively in the leaf and the stem, while the conversion of hyoscyamine to scopolamine in the root was not as efficient. Feeding of hyoscyamine or 6β-hydroxyhyoscyamine to the culture medium of a tobacco plantlet that had been transformed with the same *H6H* transgene suggested that hyoscyamine was converted efficiently to scopolamine during translocation from the root to the aerial parts. When transgenic belladonna hairy roots were used for the overexpression of *H6H*, the conversion efficiency of hyoscyamine to scopolamine was considerably lower than in the transgenic plants. The transgenic approach that involves fortifying rate-limiting enzyme activities can be applied successfully to only a few metabolic steps. The future success of biotechnological metabolic engineering will depend on the molecular cloning of elusive regulatory genes that control the expression of a series of alkaloid biosynthesis genes. Enhanced and constitutive expression of such regulatory genes in transgenic medicinal plants is expected to increase significantly the total amount of useful alkaloids produced per plant.

Leech et al. (100) reported that overexpression of hyoscyamine 6β-hydroxylase in *Atropa belladonna* efficiently converts this species' main alkaloid, hyoscyamine, to scopolamine. This successful metabolic engineering of a medicinal plant has raised prospects for biotechnological applications of secondary metabolite production, but fundamental difficulties remain in transforming the host plants (e.g., *Catharanthus*). Attempts to improve the production of putrescine-derived alkaloids and isoquinoline alkaloids by molecular engineering are reported.

Nicotine alkaloids are synthesized in the root of *Nicotiana* species, and their synthesis increases after insect attack, wounding, and jasmonate treatment of the leaf. Putrescine *N*-methyltransferase (PMT) catalysis the first committed step in nicotine biosynthesis. The expression patterns of the three *Nicotiana sylvestris PMT* genes (*NsPMT1, NsPMT2, NsPMT3*) are reported (101). Transcripts of the *NsPMT* genes are detected only in the root and were up-regulated by methyl jasmonate treatment. When the 5′ flanking regions of *NsPMT1, NsPMT2*, and *NsPMT3* were fused independently to β-glucuronidase reporter gene and introduced into *N. sylvestris* by *Agrobacterium*-mediated transformation, all introduced transgenes were expressed in the cortex, endodermis, and xylem in the root, as well as up-regulated by methyl jasmonate treatment.

Tobacco PMT cDNA was expressed under the same CaMV35S promoter in *N. sylvestris* plants to assess the impact of altered PMT expression in another, shorter, alkaloid pathway. Three transgenic lines that expressed four- to eight-higher PMT transcript levels in their roots were selected, as was a cosuppression line whose PMT expression was approximately 16% that of the wild type (20). These overexpression lines had an approximately 40% increase in leaf nicotine contents as compared with the wild type and accumulated *N*-methylputrescine in the leaf, whereas the spermidine and spermine contents were somewhat decreased. In contrast, the cosuppression line accumulated only a very small amount of nicotine (about 2% that of the wild type). Instead it had increased amounts of putrescine and spermidine, indicating that the efficient inhibition of PMT activity shifted the nitrogen flow from nicotine to polyamine synthesis. Interestingly, the reduction of PMT activity in the root caused the accumulation of polyamines in the leaf. One explanation for this correlation is that polyamines, like nicotine, may be transported from the root to the aerial parts of the plant.

Deoxyhypusine synthase from tobacco was cloned and expressed in active form in *E. coli* (102). It catalyzes the formation of a deoxyhypusine residue in the tobacco eIF 5A substrate as shown by gas chromatography coupled with mass spectrometry. The enzyme also accepts free putrescine as the aminobutyl acceptor, instead of lysine bound in the eIF 5A polypeptide chain, yielding homospermidine. Conversely, it accepts homospermidine instead of spermidine as the aminobutyl donor, whereby the reactions with putrescine and homospermidine proceed at the same rate as those involving the authentic substrates. The conversion of deoxyhypusine synthase–catalyzed eIF5A deoxyhypusinylation pinpoints a function for spermidine in plant metabolism. Furthermore, and quite unexpectedly, the substrate spectrum of deoxyhypusine synthase hints at a biochemical basis behind the sparse and skewed occurrence of both homospermidine and its pyrrolizidine derivatives across distantly related plant taxa.

Transgenic *Nicotiana rustica* root cultures expresing a yeast ODC gene produced higher levels of putrescine and nicotine (103). However, despite strong heterologous ODC expression, nicotine, putrescine, and *N*-methylputrescine levels increased only twofold, suggesting that ODC is not a rate-limiting step in nicotine biosynthesis. Overexpression of oat ADC in tobacco increased the accumulation of agmatine, the ADC reaction product. However, increased nicotine production was not detected despite suggestions that the putrescine required for nicotine biosynthesis is generated via ADC rather than ODC. It is possible that the additional agmatine was not accessible to the nicotine pathway. Tobacco root cultures transformed with a bacterial lysine decarboxylase (LDC) gene produced higher levels of cadaverine, the product of the LDC reaction, and the alkaloid anabasine, produced by the coupling of cadaverine and *N*-methylputrescine (104).

8.7.4   Purine Alkaloids

Caffeine is one of the important purine alkaloids distributed in tea and coffee plants. The major route begins with xanthosine and proceeds through three *N*-methylations. The pathway contains three S-adenosyl methionine dependent *N*-methyltransferase activities found in young tea leaves. An *N*-methyltransferase known as caffeine synthase (CS) catalyzing two consecutive methylations involved in the conversion of *N*-methyl-xanthine to caffeine was found in tea leaves. A CS cDNA has been isolated and expressed in *E. coli* (105). The predicted amino acid sequence shows that CS shares homology with salicylic acid *O*-methyltransferase.

8.7.5   Vignatic Acid

Bruchid resistance, controlled by a single dominant gene (*Br*) in a wild mung bean accession (TC 1966), has been incorporated into cultivated mung bean (*Vigna radiata*). The resistance gene simultaneously confers inhibitory activity against the bean bug, *Riptortus clavatus* Thunberg (Hemiptera: Alydidae). The resultant isogenic line (BC20 generation) was characterized by the presence of a group of novel cyclopeptide alkaloids called vignatic acids. A linkage map was constructed for Br and the vignatic acid gene (*Va*) using restriction fragment length polymorphism (RFLP) markers and a segregating BC20F2 population (106). By screening resisant and susceptible parental lines with 479 primers, eight randomly amplified polymorphic DNA (RAPD) markers linked to Br were identified and cloned for use as RFLP probes. All eight RAPD based markers, one mung bean, and four common bean genomic clones were effectively integrated around *Br* within a 3.7-cM interval. *Br* was mapped to a 0.7-cM segment between a cluster consisting of six markers and a common bean RFLP marker, Bng 110. The six markers are closest to the bruchid resistance gene, approximtely 0.2 cM away. The vignatic acid gene, *Va*, cosegregated with bruchid resistance. However, one individual was identified in the BC20F2 population that retained vignatic acids in spite of its bruchid susceptibility. Consequently, *Va* was mapped to a single locus at the same position as the cluster of markers and 0.2 cM away from *Br*. These results suggest that the vignatic acids are not the principal factors responsible for bruchid resistance in *V. radiata* but will facilitate the use of map-based cloning strategies to isolate the *Br* gene.

8.7.6 Ryanodine

Ryanodine is a plant alkaloid that was orginally used as an insecticide. To study the function and regulation of the ryanodine receptor (RyR) from insect cells, the entire cDNA sequence of RyR from the fruitfly *Drosophila melanogaster* was cloned (107). The primary sequence of the *Drosophila* RyR contains 5134 amino acids, sharing 45% identity with RyRs from mammalian cells, with a large cytoplasmic domain at the amino terminal end and a small transmembrane domain at the carboxyl terminal end. To characterize the Ca^{2+} release channel activity of the cloned *Drosophila* RyR, both full-length and a deletion mutant of *Drosophila* RyR lacking amino acids 277–3650 (*Drosophila* RyR-C) were expressed in Chinese hamster ovary cells. For subcellular localization of the expressed *Drosophila* RyR and *Drosophila* RyR-C proteins, green fluorescent protein (GFP)–*Drosophila* RyR and GFP–*Drosophila* RyR-C fusion constructs were generated. Confocal microscopic imaging identified GFP-*Drosophila* RyR and GFP-*Drosophila* RyR-C on the endoplasmic reticulum membranes of transfected cells. Upon reconstitution into the lipid bilayer membrane, *Drosophila* RyR-C formed a large-conductance cation-selective channel, which was sensitive to modulation by ryanodine. Opening of the *Drosophila* RyR-C channel required the presence of micromolar concentrations of Ca^{2+} in the cytosolic solution, but the channel was insensitive to inhibition by Ca^{2+} at concentrations as high as 20 mM. Their data are consistent with the previous observation with the mammalian RyR that the conduction pore of the calcium release channel resides within the carboxyl terminal end of the protein and further demonstrate that structural and functional features are essentially shared by mammalian and insect RyRs.

Maize and a variety of other plant species release volatile compounds to herbivore attack that serve as chemical cues to signal natural enemies of the feeding herbivore. N-(17-Hydroxylinolenoyl)-1-glutamine is an elicitor component that has been isolated and chemically characterized from the regurgitant of the herbivore-pest beet armyworm. This fatty acid derivative, referred to as volicitin, triggers the synthesis and release of volatile components, including terpenoids and indole in maize. There have been reports on a previously unidentified enzyme, indole-3-glycerol phosphate lyase (IGL), that catalyzes the formation of free indole and is selectively activated by volicitin. IGL's enzymatic properties are similar to BX1, a maize enzyme that serves as the entry point to the secondary defense metabolites DIBOA and DIMBOA. Gene sequence analysis indicates that Igl and Bx1 are evolutionarily related to the tryptophan synthase α subunit (108).

8.8 PERSPECTIVES

Even though a great array of plant alkaloids are known, very little work has been done to determine the exact role of alkaloids in insect resistance. A detailed study must be undertaken to study the activity of individual alkaloids against insect pests. Several interesting aspects of alkaloid biogenesis and its control are now emerging. Several alkaloid biosynthetic genes have been cloned. The genes that regulate alkaloid metabolism must be elucidated. There is reason to believe that control of multistep pathways by specific regulatory genes is not unique to phenol or terpenoid pathways. Molecular cloning of putative alkaloid-specific master genes may require some ingenuity and time, but it should be rewarding considering the biotechnological applications. Many aspects of alkaloid biosynthesis, such as elaborate subcellular compartmentation of enzymes and the intercellular translocation of pathway intermediates, need more attention. Emerging knowledge of the biochemistry, molecular biology, and cell biology of alkaloid biosynthesis will also lead to exciting opportunities to engineer alkaloid metabolism in transgenic plants. Transformation protocals have to be evolved for those plants for which hitherto no workable method has been available.

Engineering alkaloid metabolism by introducing cloned genes into plants or cell cultures promises to be a great challenge for commercial exploitation of plant biotehnology. However, the rate at which an alkaloid is synthesized is generally controlled at multiple steps by several enzymes, subsets of which are regulated both spatially and temporally. Thus, over-expression of only one enzyme in a multistep pathway does not usually result in a significant increase in the amount of end product of the pathway. For example, constitutive enhanced expression of one decarboxylase (ornithine decarboxylase, tryptophan decarboxylase, or lysine decarboxylase) in tobacco, *Catharanthus roseus*, or *Peganum harmala* increased the concentration of the enzyme's reaction product or its immediate metabolite, but only marginally (if at all) affected the accumulation of the final, desired alkaloids (109).

Metabolic pathway inhibition may prove to be much more straightforward. Several efficient techniques are available for the suppression of endogenous gene expression: cosuppression, antisense suppression, RNA interference, and gene disruption by reverse genetics. These techniques should open up the field for the metabolic engineering of medicinal plants. The intracellular localization of target products is also an important target for metabolic engineering. The high toxicity of berberine in nonproducing plant species as compared with the low toxicity in high berberine–producing *Coptis* cells suggests that metabolite accumulation is crucial in the creation of a novel metabolite producer(s). Our current knowledge of metabolite accumulation is very limited; much research is needed. What must be carefully evaluated is the undesirable accumulations of pathway precursors and their metabolic effects on related pathways, such as Okada et al. (110) have found in PMT-suppressd *N. sylvestris* plants. Primary metabolism enzymes that supply amino acids or other precursors to particular secondary pathway often are regulated in concert with secondary metabolism enzymes and therefore may be genetically engineered for the optimal manipulation of secondary metabolism.

REFERENCES

1. Pelletier, S.W. Alkaloids. Chem. Biol. Persp. **1983**, *1*, 1–31.
2. Southon, I.W.; Buckingham, J., Eds. In *Dictionary of Alkaloids*; Chapman and Hall: London, 1989.
3. Brossi, A., Ed. *The Alkaloids: Chemistry and Pharmacology*; Academic Press: London, Vol. 5, 1–48.
4. Kutchan, T.M. In *The Alkaloids*; Cardell, G., Ed.; Academic Press: San Diego, 1998; Vol. 50, 257–316.
5. Pelletier, S.W., Ed. *Chemistry of the Alkaloids*. Van Nostrand Reinhold: New York; 1970.
6. Roberts, M.F.; Wink, M., Eds. *Alkaloids: Biochemistry, Ecology and Medicinal Applications*; Plenum Press: New York, 1998.
7. Wiesner, K. *The Alkaloids*. Butterworths: London, 1973.
8. Bentley, K.W. *The Alkaloids*. Interscience: New York, 1957.
9. Hartman, T. Alkaloids. In *Herbivores: Their Interaction with Secondary Plant Metabolites*; Rosenthal, G.A., Berenbaum, M.R., Eds.; The Chemical Participants. Academic Press: New york, 1991; Vol. 1, 79–121.
10. Wink, M. Special nitrogen metabolism. In *Plant Biochemistry*; Dey, P.M., Horborne, J.B., Eds.; Academic Press: New York, 1997; 439–486.

11. Wink, M. Plant secondary metabolites from higher plants: biochemistry, function and biotechnology. In *Biochemistry of Plant Secondary Metabolism. Annual Plant Reviews*; Wink, M., Ed.; Academic Press: Sheffield, 1999; Vol. 2, 1–16.

12. Caporale, L.H. Chemical ecology: a view from the pharmaceutical industry. Proc. Natl. Acad. Sci. USA **1995**, *92*, 75–82.

13. Hashimoto, Y.; Shudo, K. Chemistry of biologically active benzoxazinoids. Phytochemistry **1996**, *43*, 551–559.

14. Mothes, K.; Schutte, H.R.; Luckner, M. *Biochemistry of Alkaloids*. Verlag Chemie: Weinheim, 1985.

15. Facchini, P.J. Alkaloid biosynthesis in plants: biochemistry, cell biology, molecular regulation and metabolic engineering applications. Annu. Rev. Plant Physiol. Plant Mol. Biol. **2001**, *52*, 29–66.

16. DeLuca, V.; St Pierre, B. The cell and developmental biology of alkaloid biosynthesis. Trends Plant Sci. **2000**, *4*, 168–173.

17. Hashimoto, T.; Yamada, Y. Alkaloid biogenesis: molecular aspects. Annu. Rev. Plant Physiol. Plant Mol. Biol. **1994**, *45*, 257–285.

18. Herbert, R.B. The biosynthesis of plant alkaloids and nitrogenous microbial metabolites. Nat. Prod. Rep. **2001**, *18*, 50–65.

19. DeLuca, V.; Laflamme, P. The expanding universe of alkaloid biosynthesis. Curr. Opin. Plant Biol. **2001**, *4*, 225–233.

20. Sato, F.; Hashimoto, T.; Hachiya, A.; Tamura, K.; Choi, K.; Morishige, T.; Fujimoto, H.; Yamada, Y. Metabolic engineering of plant alkaloid biosynthesis. Proc. Natl. Acad. Sci. USA **2001**, *98*, 367–372.

21. DeLuca, V. Enzymology of indole alkaloid biosynthesis. In *Methods in Plant Biochemistry*; Dey, P.M., Horborne, J.B., Eds.; Academic Press: London, 1993; Vol. 9, 345–368.

22. Hashimoto, T.; Yamada, Y. Tropane alkaloid biosynthesis: regulation and application. In *Proc. Annu. Penn. State Symp. Plant Physiol.* Am. Soc. Plant Physiol. Press Rockville, MD, 1992; 122–134.

23. Kutchan, T.M.; Zenk, M.H. Enzymology and molecular biology of benzophenanthridine alkaloid biosynthesis. J. Plant Res. **1993**, *3*, 165–173.

24. Meijer, A.H.; Verpoorte, R.; Hoge, J.H.C. Regulation of enzymes and genes involved in terpenoid indole alkaloid biosynthesis in *Catharanthus roseus*. J. Plant Res. **1993**, *3*, 145–164.

25. Boppre, M. Insects pharmacophagously utilizing plant chemicals (pyrrolizidine alkaloids). Naturwissenschaften **1986**, *73*, 17–26.

26. Bentley, M.D.; Leonard, D.E.; Reynolds, E.K.; Leach, S.; Beck, A.B.; Murakoshi, I. Lupine alkaloids as larval feeding deterrents for spruce budworm, *Choristoneura fumiferana* (Lepidoptera Tortricidae). Entomol. News **1984**, *77*, 989–400.

27. Ehmke, A.; Von Borstel, K.; Hartmann, T. Specific uptake of the N-oxides of pyrrolizidine alkaloids by cells, protoplasts and vacuoles from *Senecio* cell cultures. NATO AVV. Sci. Inst. Ser. A Life Sci. **1987**, *134*, 301–304.

28. Wink, M. Chemical ecology of quinolizidine alkaloids. In *Allelochemicals: Role in Agriculture and Forestry*; Waller, G.R., Ed.; ACS Symp. Ser. Am. Chem. Soc. 1987; Vol. 330, 524–533.

29. Eisner, T.; Eisner, M.; Rossini, C.; Iyengar, V.K.; Roach, B.L.; Benedikt, E.; Meinwald, J. Chemical defense against predation in an insect egg. Proc. Natl. Acad. Sci. USA **2000**, *97*, 1634–1639.

30. Lindigkeit, R.; Biller, A.; Buch, M.; Schiebel, H.M.; Boppre, M.; Hartmann, T. The two facies of pyrrolizidine alkaloids: the role of the tertiary amine and its N-oxide in chemical defense of insects with acquired plant alkaloids. Eur. J. Biochem. **1997**, *245*, 626–636.

31. Schulz, S.; Francke, W.; Boppre, M.; Eisner, T.; Meinwald, J. Insect pheromone biosynthesis: stereochemical pathway of hydroxydanaidal production from alkaloidal precursors in *Creatonotos transiens* (Lepidoptera, Arctiidae). Proc. Natl. Acad. Sci. USA **1993**, *90*, 6834–6838.

32. Conner, W.E.; Boada, R.; Schroeder, F.C.; Gonzalez, A.; Meinwald, J.; Eisner, T. Chemical defense: bestowal of a nuptial alkaloidal garment by a male moth on its mate. Proc. Natl. Acad. Sci. USA **2000**, *97*, 14406–14411.

33. Dreyer, D.L.; Jones, K.C.; Molyneux, R.J. Feeding deterrency of some pyrrolizidine, indolizidine, and quinolizidine alkaloids towards pea aphid (*Acyrthosiphon pisum*) and evidence for phloem transport of indolizidine alkaloids swainsonine. J. Chem. Ecol. **1985**, *8*, 1045–1051.

34. Stermitz, F.R.; Belofsky, G.N.; Ng, D.; Singer, M.C. Quinolizidine alkaloids obtained by *Pedicularis semibarbata* from *Lupinus fulcratus* (Leguminosae) fail to influence the specialist herbivore *Euphydryas editha* (Lepidoptera). J. Chem. Ecol. **1989**, *15*, 2521–2530.

35. Wink, M.; Hartmann, T.; Witte, L.; Rheinheimer, J. Interrelationship between quinolizidine alkaloid producing legumes and infesting insects: exploitation of the alkaloid-containing phloem sap *Cytisus scoperius* by broom aphid *Aphis cytisorum*. Z. Naturforsch. **1982**, *37C*, 1081–1086.

36. Zhao, B.G.; Grant, C.G.; Langevin, D.; MacDonald, L. Deterring and inhibiting of quinolizidine alkaloids on spruce budworm (Lepidoptera: Tortricidae) oviposition. Environ. Entomol. **1998**, *27*, 984–992.

37. Chevalier, L.; Desbuquois, C.; Papineau, J.; Charrier, M. Influence of the quinolizidine alkaloid content of *Lupinus albus* (Fabaceae) on the feeding choice of *Helix aspersa* (Gastropoda: Pulmonata). J. Moll. Stud. **2000**, *66*, 61–68.

38. Miyazawa, M.; Yoshio, K.; Ishikawa, Y.; Kameoka, H. Insecticidal alkaloids from *Corydalis bulbosa* against *Drosophila melanogaster*. J. Agric. Food Chem. **1998**, *46*, 1914–1919.

39. Popovic, M.; Gasic, O.; Malencic, D.J.; Simmonds, M. Antifeedant and antifungal activity of some alkaloid isolated from *Thalictrum minus* (Ranunculaceae) and *Corydalis solida* L. (Fumariaceae). Arch. Biol. Sci. (Yugoslavia) **1997**, *49*, 101–104.

40. Park, I.K.; Lee, H.S.; Lee, S.G.; Park, J.D.; Ahn, Y.J. Antifeeding activity of isoquinoline alkaloids identified in *Coptis japonica* roots against *Hyphantria cunea* (Lepidoptera: Arctiidae) and *Agelastica coerulea* (Coleoptera: Galerucinae). J. Econ. Entomol. **2000**, *93*, 331–335.

41. Huesing, J.E.; Jones, D. A new form of antibiosis in *Nicotiana*. Phytochemistry **1987**, *26*, 1381–1384.
42. Severson, R.F.; Arrendale, R.F.; Cutler, H.G.; Jones, D.; Sisson, V.A.; Stephenson, M.G. Chemistry and biological activity of acylnornicotine from *Nicotiana repande*. In *Biologically Active natural Products, Potential Use in Agriculture*; Cutler, H.G., Ed.; ACS Sympo. Ser. 1988; Vol. 380, 335–362.
43. Baldwin, I.T. Damage induced alkaloids in tobacco: pot-bound plants are not inducible. J. Chem. Ecol. **1988**, *14*, 1113–1119.
44. Baldwin, I.T.; Sims, C.L.; Kean, S.E. The reproductive consequence associated with inducable alkaloidal responses in wild tobacco. Ecology. **1990**, *71*, 252–262.
45. Sattelle, D.B.; Buckingham, S.D.; Wafford, K.A.; Sherby, S.M.; Bakry, N.M.; Eledefrawi, T.A.; Eldefrawi, M.E.; May, T.E. Vertebrate nicotinic acetylcholine receptors. Proc. R. Soc. London B **1989**, *237*, 501–507.
46. Schroeder, M.E.; Flattum, R.F. The mode of action and neurotoxic properties of the nitromethylene heterocycle insecticides. Pest. Biochem. Physiol. **1984**, *22*, 148–160.
47. Cheung, H.; Clarke, B.S.; Beadle, D.J. A patch-clamp study of the action of nitromethylene heterocycle insecticide on cockroach neurons growing in vitro. Pest. Sci. **1992**, *34*, 187–193.
48. Jennings, K.D.; Brown, D.G.; Wright, D.P.; Chalmers, A.E. Site of action of neurotoxic pesticides. In *ACS Symp. Ser.*; Hollingworth, R.M., Green, M.B., Eds.; 1987; Vol. 356, 274–296.
49. Nathanson, J.A.; Hunnicutt, E.J.; Kantham, L.; Scavone, C. Cocaine as a naturally occurring insecticide. Proc. Natl. Acad. Sci. USA **1993**, *90*, 9645–9648.
50. Toyomura, N.; Kuwano, E. Effect of tropine derivatives, antimuscarinic agents on the growth of *Bombyx mori* larvae. Biosci. Biotechnol. Biochem. **1998**, *62*, 2046–2048.
51. Nathanson, J.A. Caffeine and related methylxanthines: possible naturally occurring pesticides. Science. **1984**, *226*, 184–186.
52. Thomas, J.C.; Saleh, E.F.; Alammar, N.; Akroush, A.M. The indole alkaloid tryptamine impairs reproduction in *Drosophila melanogaster*. J. Econ. Entomol. **1998**, *91*, 841–846.
53. Bandara, K.A.N.P.; Kumar, V.; Jacobsson, U.; Molleyres, L.P. Insecticidal piperidine alkaloid from *Microcos paniculata* stem bark. Phytochemistry **2000**, *54*, 29–32.
54. Yang, Y.C.; Lee, S.G.; Lee, H.K.; Kim, M.K.; Lee, S.H.; Lee, H.S. A piperidine amide extracted from *Piper longum* L. fruit shows activity against *Aedes aegypti* mosquito larvae. J. Agric. Food Chem. **2002**, *50*, 3765–3767.
55. Park, I.K.; Lee, S.G.; Shin, S.C.; Park, J.D.; Ahn, Y.J. Larvicidal activity of isobutylamides identified in *Piper nigrum* fruits against three mosquito species. J. Agric. Food Chem. **2002**, *50*, 1866–1870.
56. Miyazawa, M.; Yoshio, K.; Ishikawa, Y.; Kmeoka, H. Insecticidal alkaloids against *Drosophila melanogaster* from *Nuphar japonicum* DC. J. Agric. Food Chem. **1998**, *46*, 1059–1063.

57. Gonzalez-Coloma, A.; Guadano, A.; Gutierrez, C.; Cabrera, R.; Pena-de-la, E.; de la Fuenta, G.; Reina, M. Antifeedant *Delphinium* diterpenoid alkaloids. Structure activity relationship. J. Agric. Food Chem. **1998**, *46*, 286–290.

58. Weissenberg, M.; Levy, A.; Svoboda, J.A.; Ishaaya, I. The effect of some *Solanum* steroidal and glycoalkaloids on larvae of the red flour beetle, *Tribolium castaneum*, and the tobacco hornworm, *Manduca sexta*. Phytochemistry **1998**, *47*, 203–209.

59. Davis, C.S.; Ni, X.; Quisenberry, S.S.; Foster, J.E. Identification and quantification of hydroxamic acids in maize seedling root tissue and impact on western corn rootworm (Coleoptera: Chrysomelidae) larval development. J. Econ. Entomol. **2000**, *93*, 989–992.

60. Niemeyer, H.M. Hydroxamic acids (4-hydroxy-1,4-benzoxazin-3-ones) defence chemicals in the Gramineae. Phytochemistry **1998**, *27*, 3349–3358.

61. Yan, F.; Xu, C.; Li, S.; Lin, C.; Li, J. Effects of DIMBOA on several enzymatic systems in Asian corn borer *Ostrinia furnacalis* (Gunee). J. Chem. Ecol. **1995**, *21*, 2047–2056.

62. Fellows, L.E.; Evans, S.V.; Nash, R.J.; Bell, E.A. Polyhydroxy plant alkaloids as glucosidase inhibitors and their possible ecological role. ACS. Symp. Ser. **1986**, Vol. *296*, 72–78.

63. Fellows, L.E.; Kite, G.C.; Nash, R.J.; Simmonds, M.S.J.; Scofield, A.M.; Castanospermine, swainsonine and related polyhydroxyalkaloids: structure, distribution and biological activity. Rec. Adv. Phytochem. **1989**, *23*, 395–427.

64. Evans, S.V.; Fellows, L.E.; Shing, T.K.M.; Fleet, G.W.J. Glycosidase inhibition by plant alkaloids which are structural analogues of monosaccharides. Phytochemistry. **1985**, *24*, 1953–1955.

65. Simmonds, M.S.J. Chemoecology: the legacy left by Tony Swain. Phytochemistry **1998**, *49*, 1183–1190.

66. Fellows, E. The biological activity of polyhydroxyalkaloid from plants. Pes. Sci. **1986**, *17*, 602–606.

67. Ujvary, I.; Eya, B.K.; Grendell, R.L.; Toia, R.F.; Casida, J.E. Insecticidal activity of various 3-acyl and other derivatives of veracevine relative to the veratrum alkaloids, veratridine and cevadine. J. Agric. Food Chem. **1991**, *39*, 1875–1881.

68. Ujvary, I.; Casida, J.E.; Partial synthesis of 3-*O*-vanilloylveracevine, an insecticidal alkaloid from *Schoenocaulon officinale*. Phytochemistry **1997**, *47*, 1257–1260.

69. Crosby, D.G. Alkaloids. In *Naturally Occurring Insecticides*; Jacobson, M., Crosby, D.G., Eds.; Marcel Dekker: New York, 1971; 198–216.

70. Jeffries, P.R.; Toia, R.F.; Brannigan, B.; Pessah, I.; Casida, J.E. *Ryania* insecticide: analysis and biological activity of 10 natural ryanoids. J. Agric. Food Chem. **1992**, *40*, 142–146.

71. A Gonzalez-Coloma, Gutierrez, C.; Hubner, H.; Achenbach, H.; Terrero, D.; Fraga, B.M. Selective insect antifeedant and toxic action of ryanoid diterpenes. J. Agric. Food Chem. **1999**, *47*, 4419–4424.

72. Reina, M.; Gonzalez-Coloma, A.; Gutierrez, C.; Cabrera, R.; Henriquez, J.; Villarroel, L. Pyrrolizidine alkaloids from *Heliotropium megalanthum*. J. Nat. Prod. **1998**, *61*, 1418–1420.

73. Jiwajinda, S.; Hiral, N.; Watanabe, K.; Santisopasri, V.; Chuengsamamyart, N.; Koshimizu, K.; Ohigashi, H. Occurrence of the insecticidal 16,17-dihydro-16(E) stemofoline in stemona collinase. Phytochemistry **2001**, *56*, 693–695.

74. Abbassy, M.A.; el-Gougary, O.A.; el-Hamady, S.; Sholo, M.A. Insecticidal, acaricidal and synergistic effects of soosan, *Pancratium maritimum* extracts and constituents. J. Egypt. Soc. Parasitol. **1998**, *28*, 197–205.

75. Sugawara, F.; Ishimoto, M.; Le-Van, N.; Koshino, H.; Uzawa, J.; Yoshida, S.; Kitamura, K. Insecticidal peptide from mungbean: a resistant factor against infestation with azuki bean weevil. J. Agric. Food Chem. **1996**, *44*, 3360–3364.

76. Blaakmeer, A.; van der Wal, D.; Stork, A.; van Beek, T.A.; de Groot, A.; van Loon, J.J. Structure–activity relationship of isolated avenanthramide alkaloids and synthesized related compounds as oviposition deterrents for *Pieris brassicae*. J. Nat. Prod. **1994**, *57*, 1145–1151.

77. Wink, M.; Schmeller, T.; Bruning, B.L. Modes of action of allelochemical alkaloids: interaction with neuroreceptors, DNA and other molecular targets. J. Chem. Ecol. **1998**, *24*, 1881–1937.

78. Yamamoto, I.; Tomizawa, M.; Saito, T.; Miyamoto, T.; Walcott, E.C.; Sumikawa, K. Structural factors contributing to insecticidal and selective action of neonicotinoids. Arch. Insect. Biochem. Physiol. **1998**, *37*, 24–32.

79. Gaertner, L.S.; Murray, C.L.; Morris, C.E. Transepithelial transport of nicotine and vinblastine in isolated malpighian tubules of the tobacco hornworm (*Manduca sexta*) suggests a P-glycoprotein like mechanism. J. Exp. Biol. **1998**, *201*, 2637–2645.

80. Witcher, D.R.; McPherson, P.S.; Kahl, S.D.; Lewis, T.; Bentley, P.; Mullinnix, M.J.; Windass, J.D.; Campbell, K.P. Photoaffinity labeling of the ryanodine receptor/Ca^{2+} release channel with an azido derivative of ryanodine. J. Biol. Chem. **1994**, *269*, 13076–13079.

81. Pessah, I.N.; Zimanyi, I. Characterization of multiple (3H) ryanodine binding sites on the Ca^{2+} release channel of sarcoplasmic reticulum from skeletal and cardiac muscle: evidence for a sequential mechanism in ryanodine action. Mol. Pharmacol. **1991**, *39*, 679–689.

82. Coulombe, R.A. jr.; Drew, G.L.; Stermitz, F.R. Pyrrolizidine alkaloids crosslink DNA with actin. Toxicol. Appl. Pharmacol. **1999**, *154*, 198–202.

83. Ulbricht, W. Effects of veratridine on sodium currents and fluxes. Rev. Physiol. Biochem. Pharmacol. **1998**, *133*, 1–54.

84. Ditirich, H.; Kutchan, T.M. Molecular cloning, expression, and induction of berberine bridge enzyme, an enzyme essential to the formation of benzophenanthridine alkaloids in the response of plants to pathogenic attack. Proc. Natl. Acad. Sci. USA **1991**, *88*, 9969–9973.

85. Okada, N.; Koizumi, N.; Tanaka, T.; Ohkubo, H.; Nakanishi, S.; Yamada, Y. Isolation, sequence and bacterial expression of a cDNA for (*S*)-tetrahydroberberine oxidase from cultured berberine-producing *Coptis japonica* cells. Proc. Natl. Acad. Sci. USA. **1990**, *87*, 6928–6934.

86. Morishige, T.; Tsujita, T.; Yamada, Y.; Sato, F. Molecular characterization of the *S*-adenosyl-L-methionine: 3'-hydroxy-*N*-methylcoclaurine4'-*O*-

methyltransferase involved in isoquinoline alkaloid biosynthesis in *Coptis japonica*. J. Biol. Chem. **2000**, *275*, 23398–23405.

87. Sato, F.; Takeshita, N.; Fujiwara, H.; Katagiri, Y.; Huan, L.; Yamada, Y. Characterization of *Coptis japonica* cells with different alkaloid productivities. Plant Cell Tissue Organ Cult. **1994**, *38*, 249–256.

88. Facchini, P.J.; Yost, C.P.; Samanani, N.; Kowalchuk, B. Expression patterns conferred by tyrosine dihydroxy phenylalanine decarboxylase promoters from opium poppy are conserved in transgenic tobacco. Plant Physiol. **1998**, *118*, 69–81.

89. Thomas, J.C.; Adams, D.G.; Nessler, C.L.;Brown J.K.; Bohnert, H.J. Tryptophan decarboxylase, tryptamine, and reproduction of the whitefly. Plant Physiol. **1995**, *109*, 717–720.

90. Chavadej, S.; Brisson, N.; McNeil, J.N.; DeLuca, V. Redirection of tryptophan leads to production of low indole glucosinolate canola. Proc. Natl. Acad. Sci. USA **1994**, *91*, 2166–2170.

91. Goddijn, O.J.M.; Pennings, E.J.; van der Helm, P.; Schilperoort, R.A.; Verpoorte, R.; Hoge, J.H.C. Overexpression of a tryptophan decarboxylase cDNA in *Catharanthus roseus* crown gall calluses results in increased tryptamine levels but not in increased terpenoid indole alkaloid production. Transgenic Res. **1995**, *4*, 315–323.

92. Geerlings, A.; Ibanez, M.M.L.; Memelink, J.; Van der Heijden, R.; Verpoorte, R. Molecular cloning and analysis of strictosidine β-D-glucosidase, an enzyme in terpenoid indole alkaloid biosynthesis in *Catharanthus roseus*. J. Biol. Chem. **2000**, *275*, 3051–3056.

93. Warzecha, H.; Gerasimenko, I.; Kutchan, T.M.; Stockigt, J. Molecular cloning and functional bacterial expression of a plant glucosidase specifically involved in alkloids biosynthesis. Phytochemistry. **2000**, *54*, 657–666.

94. Canel, C.; Lopes-Cardoso, M.I.; Whitmer, S.; van der Fits, L.; Pasquali, G. Effects of overexpression of strictosidine synthase and tryptophan decarboxylase on alkaloid production by cell cultues of *Catharanthus roseus*. Planta. **1998**, *208*, 414–419.

95. Hallard, D.; Van der Heijden, R.; Verpoorte, R.; Lopes Cardoso, M.I.; Memelink, J.; Hoge J.H.C. Suspension cultured transgenic cells of *Nicotiana tabacum* expressing the tryptophan decarboxylase and strictosidine synthase cDNAs from *Catharanthus roseus* produce strictosidine upon feeding of secologanin. Plant Cell Rep. **1997**, *17*, 50–54.

96. Goddijn, O.J.M.; Lohman, F.P.; de Kam, R.J.; Schilperoort, R.A.; Hoge, J.H.C. Nucleotide sequence of the tryptophan decarboxylase gene of *Catharanthus roseus* and expression of tdc-gus A gene fusions in *Nicotiana tabacum*. Mol. Gen. Genet. **1994**, *242*, 217–225.

97. Ouwerkerk, P.B.; Memelink, J. A G-box element from the *Catharanthus roseus* strictosidine synthase (str) gene promoter confers seed specific expression in transgenic tobacco plants. Mol. Gen. Genet. **1999**, *261*, 635–643.

98. Hashimoto, T.; Matsuda, J.; Yamada, Y. Two-step epoxidaion of hyoscyamine to scopolamine is catalysed by functional hyoscyamine 6β-hydroxylase. FEBS. Lett. **1993**, *329*, 35–39.

99. Yun, D.J.; Hashimoto, T.; Yamada, Y. Metabolic engineering of medicinal plants: transgenic *Atropa belladonna* with an improved alkaloid composition. Proc. Natl. Acad. Sci. USA **1992**, *89*, 11799–11803.

100. Leech, M.J.; May, K.; Hallard, D.; Verpoorte, R.; De Luca, V.Z.; Christou, P. Expression of two consecutive genes of a secondary metabolic pathway in transgenic tobacco: molecular diversity influences levels of expression and product accumulation. Plant Mol. Biol. **1998**, *38*, 765–774.

101. Shoji, T.; Yamada, Y.; Hashimota, T. Jasmonate induction of putrescine *N*-methyltransferase genes in the root of *Nicotiana sylvestris*. Plant Cell Physiol. **2000**, *41*, 831–839.

102. Ober, D.; Hartmann, T. Deoxyhypusine synthase from tobacco. cDNA isolation, characterization, and bacterial expression of an enzyme with extended substrate specificity. J. Biol. Chem. **1999**, *274*, 32040–32047.

103. Hamill, J.D.; Robins, R.J.; Parr, A.J.; Evans, D.M.; Furze, J.M.; Rhodes, M.J.C. Overexpression of a yeast ornithine decarboxylase gene in transgenic roots of *Nicotiana rustica* can lead to enhanced nicotine accumulation. Plant Mol. Biol. **1990**, *15*, 27–38.

104. Fecker, L.F.; Rugenhagen, C.; Berlin, J. Inereased production of cadaverine and anabasine in hairy root cultures of *Nicotiana tabacum* expressing a bacterial lysine decarboxylase gene. Plant Mol. Biol. **1993**, *23*, 11–21.

105. Ashihara, H.; Gillies, F.M.; Crozier, A. Metabolism of caffeine and related purine alkaloids in leaves of tea (*Camillia sinensis*). Plant Cell Physiol. **1997**, *38*, 413–419.

106. Kaga, A.; Ishimoto, M. Genetic localization of a bruchid resistance gene and its relationship to insecticide cyclopeptide alkaloids, the vignatic acids in mungbean (*Vigna radiata* L-Wilczek). Mol. Gen. Genet. **1998**, *258*, 378–384.

107. Xu, X.; Bhat, M.B.; Nishi, M.; Takeshima, H.; Ma, J. Molecular cloning of cDNA encoding a *Drosophila* ryanodine receptor and functional studies of the carboxyl terminal calcium release channel. Biophys. J. **2000**, *78*, 1270–1281.

108. Fray, M.; Stettner, C.; Pare, P.W.; Schmelz, E.A.; Tumlinson, J.H.; Gierl, A. An herbivore elicitor activates the gene for indole emission in maize. Proc. Natl. Acad. Sci. USA **2000**, *97*, 14801–14806.

109. Herminghaus, S.; Schreier, P.H.; Mc Carthy, J.E.G.; Landsmann, J.; Botterman, J.; Berlin, J. Expression of a bacterial lysine decarboxylase gene and transport of the protein into chloroplasts of transgenic tobacco. Plant Mol. Biol. **1991**, *17*, 475–486.

110. Okada, N.; Koizumi, N.; Tanaka, T.; Ohkubo, H.; Nakanishi, S.; Yamada, Y. Isolation, sequence, and bacterial expression of a cDNA for (*S*)-tetrahydroberberine oxidase from cultured berberine producing *Coptis japonica* cells. Proc. Natl. Acad. Sci. USA **1989**, *86*, 534–538.

9

Phenolics

9.1 INTRODUCTION

Aromatic rings bearing a hydroxyl group, including functional derivatives such as esters, methyl ethers, and glycosides, are called *phenolics*. The Plant Phenolics Group, constituted in England in 1957, recognized phenolics as a discrete group of biogenetically related plant metabolites. In 1960 the Plant Phenolics Group was renamed the Phytochemical Society of Europe. Another organization called Le Group Polyphenols, was constituted by French scientists in 1970. The first monograph on phenolic compounds, written by Harborne (1), was published in 1964.

A number of derivatives and polymers of phenols occur in plants and perform different functions. Phenolics are important cellular support materials. The polymeric compounds, such as lignins, cutins, and suberins, form an integral part of cell wall structures. They provide mechanical support and act as barriers against pathogenic invasion. Lignins are the second most abundant organic compounds on earth. The formation of woody stems and conducting cell elements for water transport are possible with the help of lignins. The important plant pigments anthocyanins along with flavones and flavonols as copigments contribute to flower and fruit colors. They also play a role in attracting animals and insects to the plant for pollination and seed dispersal. The dermal tissues of the plant contain

UV light–absorbing flavonoids and other phenolics. Phenolic acids, tannins, and phenolic resins at the plant surface function as effective deterrents. The microbial attack on the plants induce low molecular weight phenolics known as phytoalexins. The hydroxycoumarins and hydroxycinnamate conjugates are important phenolic phytoalexins and contribute to disease resistance mechanisms in plants.

Phenolics are gaining importance in applied science. Flavonoids are attracting the attention of medical scientists because of their anticarcinogenic, antiallergic, and anti-inflammatory properties. Flavonoids are involved in the process of nitrogen fixation in plants and open the way for agricultural applications of these substances. Lignins and quinones are being studied as part of modern plant biotechnology.

9.2 OCCURRENCE

9.2.1 General

Phenolic compounds are found throughout the plant kingdom, but the type of compound present varies according to the phylum under consideration (2). The presence of phenolics in bacteria, fungi, and algae is not common and only very few classes of phenolics are recorded. The chlorinated gossypetin derivative chloroflavonin is found in the fungus *Aspergillus candidus*. The symbiotic associates of fungi and algae lichens produce special colored phenolic compounds. Notable compounds among these are the depsides and depsidones. They also produce xanthones and anthraquinones.

Bryophytes like mosses and liverworts produce polyphenols and simple flavonoids. Vascular plants contain a full range of polyphenols. All ferns, gymnosperms, and angiosperms have lignin in the cell wall. Some classes of phenol have a more discrete distribution. For example, the isoflavonoids are confined to the Leguminosae, anthraquinones occur in about six families, and xanthones are recorded in Gentianaceae, Guttiferae, Moraceae, and Polygalaceae. Hydroxybenzoic acids, hydroxycinnamic acids, and flavonoids are universly distributed.

The pattern of occurrence of phenolics within the plant vary. Compounds that occur in leaves may be different from those present in roots, stems, flowers, and fruits. The concentration of different phenolics in different parts may also vary significantly.

9.2.2 Phenols and Phenolic Acids

The simple monohydric phenol occurs in heartwood of *Populus tremuloides* Michx, as well as in essential oils and seeds of *Pinus sylvestris* (3).

Alkylphenols, such as 3-ethylphenol and 3,4-dimethylphenol, were found in cocoa beans. *p*-Cresol occurs in several essential oils. Thymol has been detected in the leaves and essential oils of Labiatae. Salicyl alcohol and its glucoside salicin occurs in *Populus* leaves and barks of six varieties.

The dihydric phenol catechol occurs in *Populus* leaves, grapefruit, and avacodo. The catechol derivative urushiol is also found in plants. Guaicol occurs in beechwood tar and in several plant oils and soups. Resorcinol occurs in *Pinus rigida* needles. Orcinol has been found in *Eric arborea*. The seeds of Graminae contain 5-alkyl- and 5-alkenylresorcinol. Hydroquinone is commonly found simple phenol in Ericaceae, Rosaceae, and Saxifragaceae. The hydroquinone derivative sesamol is found in sesame oil.

The trihydric phenol pyrogallol and phloroglucinol occur in *Allium*, *Sequoia*, and *Cerotonia siliqua*. The glycoside of phloroglucinol occurs in citrus peel and *Medinilla magnifica*.

Salicylaldehyde is found in essential oils and *p*-hydroxybenzaldehyde occurs in the essential oils of fennel, vanilla, and *Mimosa*. Potato tubers contain protocatechualdehyde and vanillin. Vanilla pod, *Dahlia* tuber, and several other essential oils contain vanillin.

Gallic acid occurs in the form of quinic acid ester or as hydrolyzable tannins in woody plants. Methyl and ethyl esters of gallic acids and 4-methylgallic acid occur in the flowers of *Tamarix nilotica*. *p*-Hydroxybenzoic, protocatechuic, and vanillic acids are widely distributed in gymnosperms and ferns. Hydroxyphenylacetic acids such as *p*-hydroxyphenylacetic acid occur free and as a glucoside in bamboo shoots.

9.2.3 Stilbenes and Phenanthrenes

The bibenzyls widely occur in higher plants (4). They occur together with related stilbenes in *Pinus* and *Morus* and with dihydrophenanthrenes and phenanthrenes in the Orchidaceae, Dioscoreaceae, and Combretaceae as well as a few other higher plants. Stilbenes are more widely distributed in both gymnosperms and angiosperms. Prenylated stilbenes, stilbene glycosides, and polymeric stilbenes occur in higher plants.

Phenanthrene and dihydrophenanthrene derivatives have been found in liverworts, Orchidaceae, and *Clusia paralycola*.

9.2.4 Flavonoids

Flavones, flavonols, and their glycosides occur widely in the plant kingdom with the exception of the algae, the fungi, and the hornworts. The occurrence of flavone C-glycosides in green algae and chloroflavonin in the fungus *Aspergillus candidus* are some exceptions.

Flavonones, dihydroflavonols, and dihydrochalcones are widespread in the Leguminosae and the Compositae (5). In these families *C*-alkylation is a characteristic feature of the flavonoids. Dihydrochalcones and flavanone are present in the genera *Uvaria* and *Unona* of Annonaceae family. Flavanones are rich in the genus *Citrus* belonging to Rutaceae family. Flavonones accumulate in approximately 60 other plant families. Dihydroflavonols occur in over 50 plant families and are characteristic of the Anacardiaceae, Compositae, Coniferae, Ericaceae, and Leguminosae.

Dihydrochalcones are found in fungus, liverwort, ferns, conifer, and in 17 angiosperm families. Leguminosae, Compositae, and Annonaceae contain the greatest variety of structures. In the remaining families dihydrochalcones occur sporadically.

9.2.5   Anthocyanins

Anthocyanins are universal plant colorants responsible for the orange, pink, scarlet, red, mauve, violet, and blue colors of flowers and fruits of higher plants. They also occur in other plant organs such as roots and leaves, accumulating in the vacuoles of epidermal and subepidermal cells (2).

9.2.6   Biflavonoids

Biflavonoids are the characteristic constituents of gymnosperms, the Psilotales, and the selaginellates, and they have a limited distribution in the angiosperms (6). Bisflavonoids have been reported only in just six moss species. Biflavonoids with oxygenation patterns in different positions are identified in these lower plants. The major biflavonoid in Psilotales and selaginellates is amentoflavone. Among the gymnosperms, all the Cycadales except the monotypic Stangeriaceae contain biflavonoids. Cycadaceae is characterized by having both the amentoflavone and hinokiflavone series. Within the Coniferales, biflavonoids are major leaf constituents of all families except the Pinaceae where only 3 out of the 45 species examined were found to produce them.

Biflavonoids are comparatively uncommon in the angiosperms, having been recorded in only 34 genera from 16 families. Amentoflavone is widespread in the angiosperms, and hinokiflavone has been found in only three unrelated groups: Casuarinaceae, Anacardiaceae, and Iridaceae. Even though biflavonoids are widespread in dicotyledons, they have been found only in a limited number of monocotyledonous species.

9.2.7   Tannins

Tannins are widely distributed in plants and occur in especially high amounts in the bark of trees and in galls (7).

9.2.8 Isoflavonoids

The isoflavonoids show a very limited distribution in the plant kingdom, being largely confined to the subfamily Papilionoidae of the Leguminosae with only occasional occurrences in the subfamilies Caesalpinioidae and Mimosoidae (8). Isoflavonoids occur in the monocotyledons, the richest source being the rhizomes of *Iris* species. They have also been identified in leaves of *Patersonia*, another genus of the Iridaceae. Isoflavonoids occur in the two gymnosperm genera *Juniperus* and *Podocarpus* and the moss *Bryum capillare*.

9.2.9 Quinones and Xanthones

The majority of quinones found in plants are simple benzoquinones, naphthoquinones, or anthraquinones (9). In higher plants, anthraquinones are found in the Rubiaceae, Leguminosae, Rhamnaceae, Polygonaceae, and Scrophulariaceae. Naphthoquinones mostly occur in Bignoniaceae, Verbenaceae, Juglandaceae, Plumbaginaceae, Boraginaceae, Proteaceae, Lythraceae, Balsaminaceae, Sterculiaceae, Ulmaceae, Ebenaceae, and Droseraceae. Benzoquinones usually occur in Myrsinaceae, Boraginaceae, and Primulaceae.

Xanthones have been found in a limited number of families. As aglycones, they occur in Gentianaceae, Guttiferae, Polygalaceae, Leguminosae, Lythraceae, Moraceae, Loganiaceae, and Rhamnaceae. Aglycones and their corresponding *O*-glycosides are found in Gentianaceae and Polygalaceae.

9.3 CLASSIFICATION

The major classes of plant phenolic compounds are listed in Table 9.1. The phenolic compounds are classfied according to the number of carbon atoms of the basic skeleton. Phenolic compounds also include *O*-methylated or other substituents.

9.4 BIOSYNTHESIS

9.4.1 General

Plant phenolics are biosynthesised through three different pathways. The shikimate pathway leads through the aromatic amino acids phenylalanine and tyrosine to the majority of phenolics. Some phenolic compounds are synthesized from intermediates of the shikimate pathway. The

TABLE 9.1   The Major Classes of Phenolics in Plants

Class	No. carbon atoms	Basic carbon skeleton	Examples
1. Simple phenols	6	C_6	Catechol, hydroquinone Phloroglucinol
2. Acetophenones Phenylacetic acid	8	$C_6\text{–}C_2$	4-hydroxyacetophenone P-Hydroxyphenylacetate
3. Hydroxycinnamates	9	$C_6\text{–}C_4$	Caffeic, ferulic
Coumarins	9	$C_6\text{–}C_3$	Umbelliferone, esculetin
Isocoumarins	9	$C_6\text{–}C_3$	Bergenin
Chromones	9	$C_6\text{–}C_3$	Eugenin
4. Hydroxybenzoates	7	$C_6\text{–}C_1$	Salicylic, gallic
5. Naphthoquinones	10	$C_6\text{–}C_3$	Juglone, plumbagin
6. Xanthones	13	$C_6\text{–}C_1\text{–}C_6$	Mangiferin
7. Stilbenes	14	$C_6\text{–}C_2\text{–}C_6$	Resveratrol
Anthraquinones	14	$C_6\text{–}C_2\text{–}C_6$	Emodin
8. Flavonoids	15	$C_6\text{–}C_3\text{–}C_6$	Cyanidin
Isoflavonoids	15	$C_6\text{–}C_3\text{–}C_6$	Genistein
9. Lignans	18	$(C_6\text{–}C_3)_2$	Pinoresinol
10. Biflavonoids	30	$(C_6\text{–}C_3\text{–}C_6)_2$	Amentoflavone
11. Hydrolyzable tannins	n	$(C_6\text{–}C_1)_n$: Glc	Gallotannins
12. Condensed tannins	n	$(C_6\text{–}C_3\text{–}C_6)_n$	Catechin polymers
13. Lignins	n	$(C_6\text{–}C_3)_n$	Guaiacyl lignins
14. Catechol melanins		$(C_6)_n$	

acetate-malonate pathway or the polyketide pathway is responsible for the formation of some plant quinones and various side chain–elongated phenylpropanoids, e.g., flavonoids. The acetate-mevalonate pathway leads by dehydrogenation reactions to some aromatic terpenoids, mostly monoterpenoids. The shikimate and polyketide pathways are the most important ones in biosynthesis of the phenolic compounds.

9.4.1.1 Shikimate Pathway

The Japanese term *Shikimino-ki* refers to *Illicium anisatum* (Illiciaceae), the plant from which shikimate was first described by Eykmann in 1885. The sequence of reactions, leading to the aromatic amino acids phenylalanine and tyrosine, require 11 enzymes (10–12). The precursor compounds phosphoenolpyruvate and erythrose 4-phosphate, required for this pathway, are obtained from glycolytic and pentose phosphate pathway, respectively

(Scheme 9.1). The first condensation reaction is between erythrose-4-phosphate and phosphoenolpyruvate and is catalyzed by the enzyme 2-dehydro-3-deoxyarabinoheptulosonate-7-phosphate synthase (DAHP synthase). The other names are 2-dehydro-3-deoxy-D-arabinoheptonate-7-phosphate-D-erythrose-4-phosphate lyase and 2-dehydro-3-deoxyphospho-heptonate aldolase. The product formed is an open-chain C_7 sugar 2-dehydro-3-deoxyarabinoheptulosonate-7-phosphate (DAHP). The second step is the conversion of DAHP to 3-dehydroquinate. This involves a sequence of complex reactions of oxidation, a β elimination, a reduction, and an intramolecular aldol condensation to give a cyclic structure and finally NADH-dependent reduction to 3-dehydroquinate which is catalyzed by 3-dehydroquinate synthase. The 3-dehydroquinate is then cis-dehydrated by the enzyme 3-dehydroquinate dehydratase to 3-dehydroshikimate. The 3-dehydroshikimate is reduced by a NADPH-dependent dehydrogenase, namely, shikimate 3-dehydrogenase, to shikimate. The shikimate is then phosphorylated at the C-3 position by shikimate kinase in presence of ATP. The shikimate 3-phosphate then condenses with phosphoenolpyruvate to produce the enol ether 5-enolpyruvylshikimate-3-phosphate (EPSP). This reaction is catalyzed by EPSP synthase. The next step in the pathway is elimination of phosphate from EPSP, catalyzed by chorismate synthase, which gives chorismate. A pericyclic Claisen rearrangement catalyzed by chorismate mutase produces a quinonoid prephenate. In this rearrangement reaction, the pyruvate side chain is transferred from C-5 to C-1, giving the basic phenylpropanoid skeleton.

A pyridoxal 5′-phosphate-dependent prephenate aminotransferase converts the prephenate into arogenate in higher plants. The arogenate dehydrogenase catalyzes the formation of tyrosine and arogenate dehydratase catalyzes the conversion arogenate to phenylalanine (Scheme 9.1). An overview of phenolic metabolism is shown in Scheme 9.2. The phenolic compounds derived from phenylalanine and tyrosine contain a phenyl ring with a C-3 side chain and these compounds are collectively termed phenylpropanoids. The flavonoids, such as flavones, isoflavones, and anthocyanins, contain a second aromatic ring formed from three molecules of malonyl-CoA. This also applies to the stilbenes, but here, after the introduction of the second aromatic ring, one C atom of phenylpropane is split off.

9.4.2 Phenylpropanoids

The interface between phenylalanine and the phenylpropanoid metabolism is controlled by the enzyme phenylalanine ammonia lyase (PAL). This enzyme catalyzes the nonoxidative deamination of phenylalanine to form

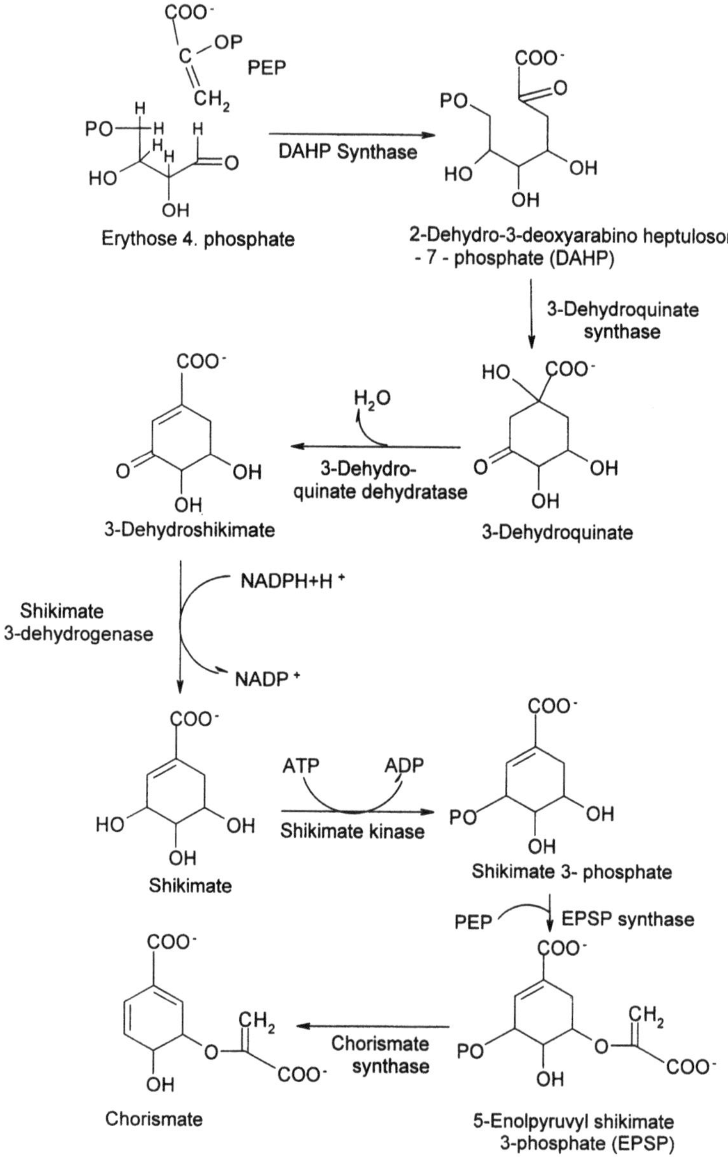

SCHEME 9.1   Shikimate pathway.

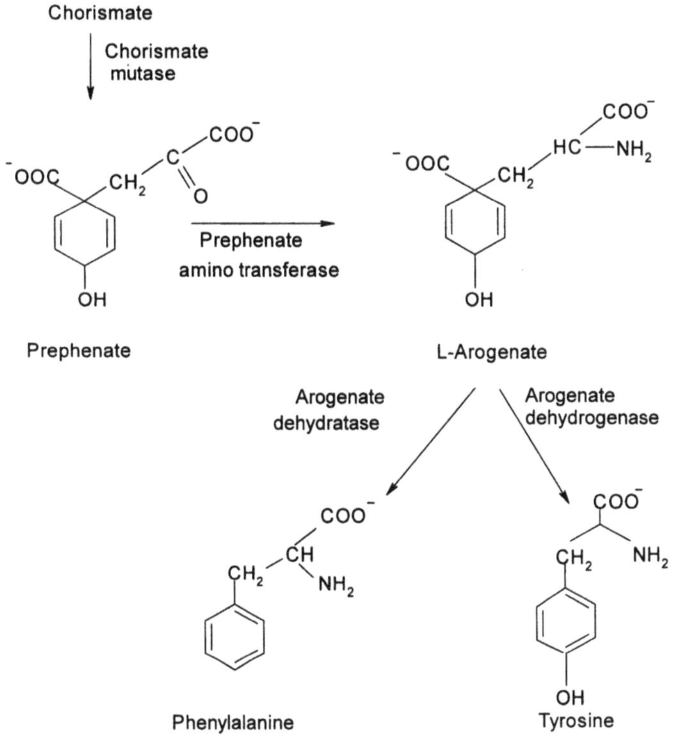

Scheme 9.1 (Continued).

the phenylpropane structure (13). PAL is one of the most intensely studied enzymes of plant secondary metabolism. The common hydroxycinnamates are formed by series of hydroxylation and methylation reactions.

The introduction of the hydroxyl group into the phenyl ring of cinnamic acid proceeds via monooxygenase or mixed-function oxidases, which introduce a single atom of oxygen into the substrate (14). The oxygen is split during this reaction and the second oxygen atom is reduced to H_2O. This reaction also requires a second oxidizable substrate, NADPH.

$$NADPH + H^+ + RH + O_2 \longrightarrow NADP^+ + R\text{-}OH + H_2O$$

The hydroxylases are of two types: (a) the microsomal (membrane bound) cytochrome P_{450}-dependent oxygenase and (b) the soluble phenolase

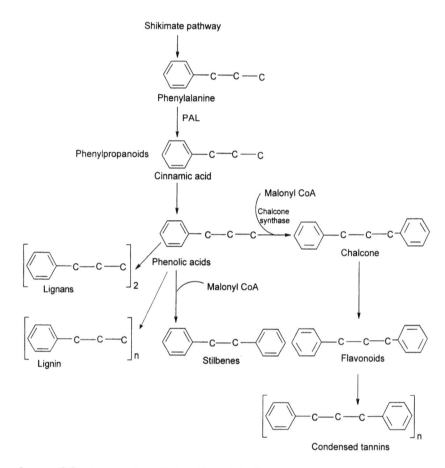

SCHEME 9.2   An overview of phenolic metabolism.

which catalyzes the introduction of a second hydroxyl group into a monophenol.

   Cinnamate 4-hydroxylase introduce a hydroxyl group in the 4-position of cinnamate (Scheme 9.3). The phenolase 4-coumarate hydroxylase introduces a second hydroxyl group in the 4-coumarate to form caffeate. The hydroxyl group in the 3-position is methylated via O-methyltransferase with S-adenosylmethionine as methyl donor. The product formed is ferulic acid. This is again hydroxylated employing ferulate hydroxylase, which is again a membrane-bound cytochrome P_{450}-dependent oxygenase. The hydroxyl group generated in the 5-position is methylated using O-methyltransferase to form sinapate.

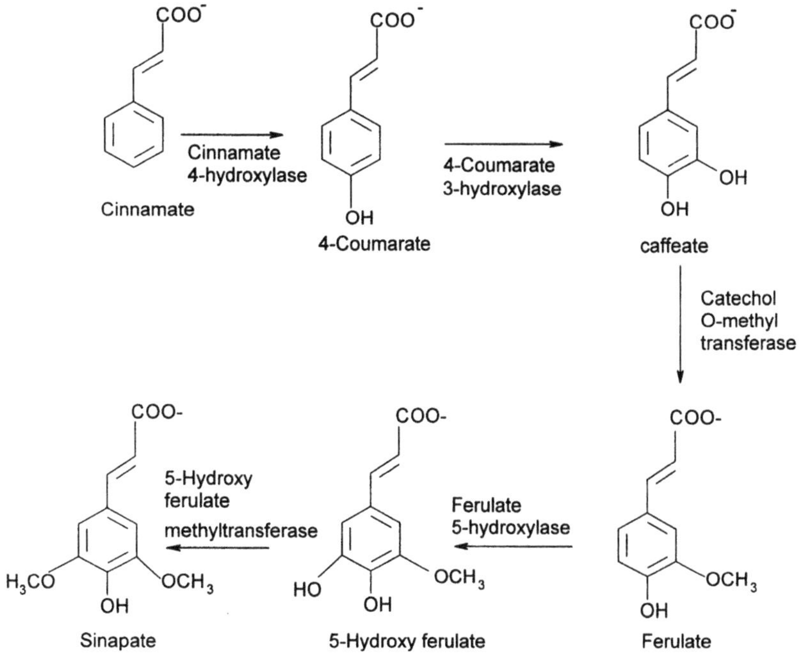

SCHEME 9.3 Biosynthesis of hydroxycinnamates.

The hydroxycinnamates are utilized in various pathways. There are four predominant types of side chain reactions in hydroxycinnamate.

1. Side chain elongation with malonyl CoAs. Sequential addition of three acetyl units from three malonyl CoAs leads to the formation of flavonoids.
2. Shortening of side chains by removal of an acetyl unit heading to formation of hydroxybenzoates.
3. Reduction of carboxyl group leading to formation of hydroxycinnamyl alcohols, the lignin precursors.
4. Conjugation with hydroxyl or amino group–bearing molecules leading to formation of esters or amides.

9.4.3 Hydroxybenzoates

The hydroxybenzoates (C_6–C_1) are universely distributed in plants like hydroxycinnamates. The important compounds include 4-hydroxybenzoate protocatechuate, vanillate, gallate, and syringate. The 2-hydroxybenzoate salicylate is widely distributed in Ericaceae and in essential oils as its methyl ester. There are several pathways leading to the individual hydroxybenzoates, depending on the plant. One major route is the side chain degradation of hydroxycinnamates by removal of acetate by a mechanism analogous to the β oxidation of fatty acids (Scheme 9.4).

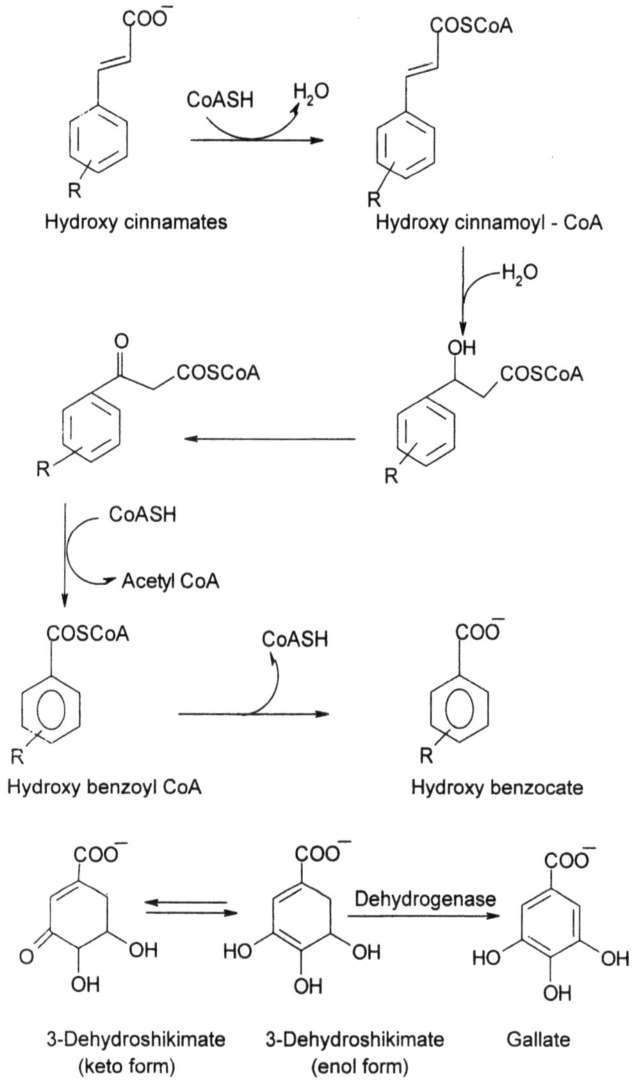

SCHEME 9.4   Biosynthesis of hydroxybenzoates.

Hydroxybenzoates with three hydroxyl groups in the phenyl ring (gallate) can also be formed by a pathway that deviates from the shikimate pathway at 3-dehydroshikimate. Direct aromatization of the enol form of 3-dehydroshikimate occurs by dehydrogenase leading to production of gallate in plants (15).

The substitution patterns of hydroxybenzoates may be determined by the hydroxycinnamate precursor. However, hydroxylation and methylation reactions may occur with the hydroxybenzoates analogous to the hydroxycinnamate pathway.

9.4.4 Hydroxycinnamate Conjugates

Esters and amides are the most common type of conjugates whereas glycosides occur rarely. Conjugating moieties can be carbohydrates, proteins, lipids, amino acids, amines, carboxylic acids, terpenoids, flavonoids, or alkaloids. Conjugate reactions may determine whether a compound is converted to a metabolically inactive end product for permanent storage or into a transiently accumulating intermediate. Hydroxycinnamate glucose esters or glucosides are formed with the help of UDP-glucose-dependent transferases. The other type of transferases involved in the conjugate formation are (a) hydroxycinnamoyl-CoA thioester, (b) hydroxycinnamate-1-O-acylglucoside, and (c) hydroxycinnamate O ester–dependent transferases (15).

9.4.5 Flavonoids

The chalcone synthase condenses 4-coumaryl-CoA with three molecules of malonyl-CoA to form naringenin chalcone. This enzyme is considered to be the rate-limiting enzyme in flavonoid synthesis. It catalyzes the formation of the basic C_{15} skeleton by channeling hydroxycinnamates into flavonoid biosynthesis. The enzyme from most sources is highly specific for 4-coumaryl-CoA, although there are examples where other hydroxycinnamoyl-CoAs are accepted. The proposed mechanism of the malonyl transfer is a stepwise addition of acetate from malonyl-CoA to 4-coumaryl-CoA. Malonyl-CoA is supplied for the chalcone synthase by the ATP-dependent action of acetyl-CoA carboxylase. It supplies malonyl-CoA as an important building block for the flavonoid aglycones and as the acyl donor for malonylation of the glycosides. The cyclization of chalcone to the flavanone naringenin is catalyzed by chalcone isomerase, which is found in a tight complex with the chalcone synthase during flavanone synthesis (Scheme 9.5). Aurones are derived directly from the chalcone intermediate. The flavanone is the precursor for flavones, isoflavones, and dihydroflavonols (16).

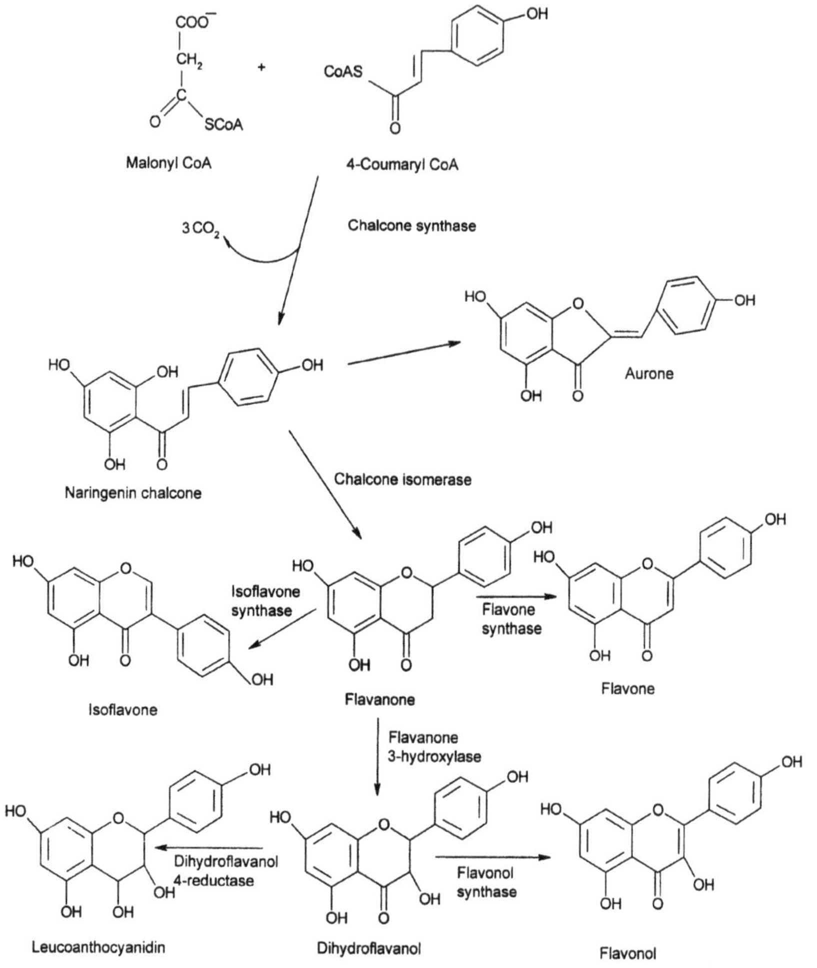

SCHEME 9.5   Biosynthesis of flavonoid aglycones.

9.4.5.1   Isoflavonoids

Oxidative rearrangement of naringenin with a 2,3-aryl shift yields the isoflavone genistein. The initial step in isoflavone formation may be an epoxidation reaction catalyzed by a cytochrome P_{450}-dependent monooxygenase. After structural rearrangement, aryl shift, and addition of a hydroxyl ion to C-2, elimination of water by a dehydratase gives the isoflavone structure (Scheme 9.6).

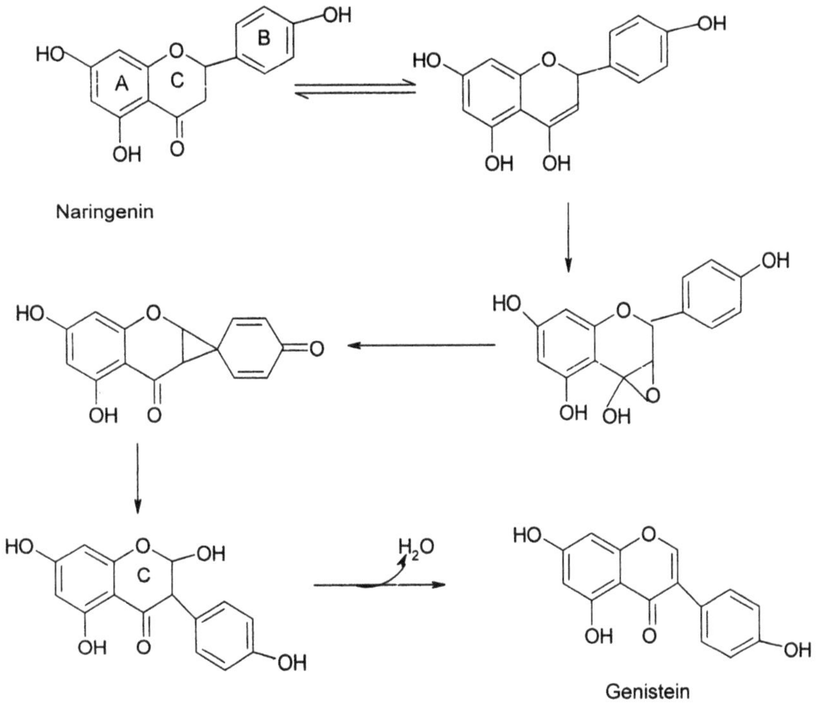

Scheme 9.6 Biosynthesis of isoflavonoid-genistein from naringenin.

9.4.5.2 Flavones and Flavonols

A double bond introduced between C-2 and C-3 into naringenin leads to the flavones. Two reactions are involved in this conversion. In the first step 2-hydroxyflavanone is formed and in the second step water is removed by dehydratase. An analogous reaction sequence leads to the most widespread flavonols. A flavanone 3-hydroxylase (dioxygenase) hydroxylates flavanone to dihydroflavonol. Then a 2-hydroxylase and a dehydratase convert the dihydroflavanol to flavonol. The products formed are the flavone apigenin and the flavonol kaempferol (Scheme 9.7).

9.4.5.3 Flavanols and Anthocyanins

Flavan-3,4-*cis*-diol, the common intermediate compound for both flavonols and anthocyanins, is formed from dihydroflavonol catalyzed by an NADPH-dependent dihydroflavonol 4-reductase (Scheme 9.8). A hypothetical three-step pathway includes a hydroxylation and two dehydrations.

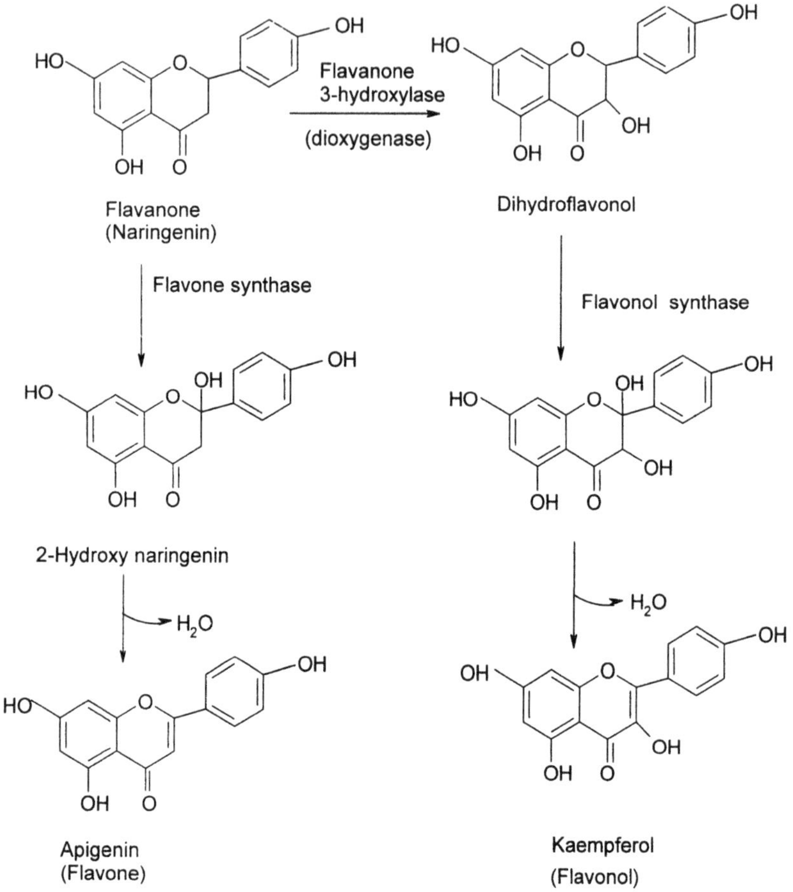

SCHEME 9.7   Biosynthesis of flavone and flavonol.

The produced labile flavylium structure is stabilized by a glucosylation of the 3-OH group giving pelargonidin 3-O-glucoside. This is catalyzed by a 3-O-glucosyltransferase in the presence of the glucose donor, UDP-glucose.

9.4.5.4   Conjugation of Flavonoids and Anthocyanidins

Flavonoids usually occur as water-soluble glycosides in vacuoles. These glycosides are synthesized at the endoplasmic reticulum and then stored in the vacuoles. The important sugar conjugation products include 3-O-glycosides, 3-O-diglycosides, and C_3, C_5, and C_7 glycosides. The occurrence of flavonoids with sugars attached to the B ring is rare. There are several hundred different

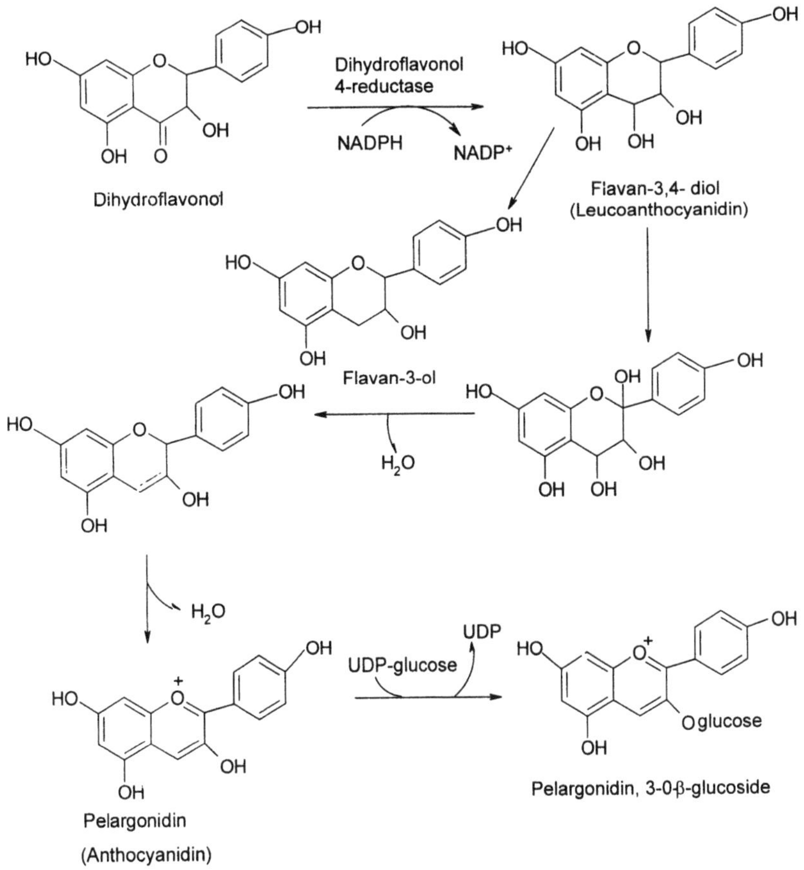

SCHEME 9.8 Biosynthesis of anthocyanidin and flavan-3-ol.

glycosides known with glucose, galactose, rhamnose, xylose, mannose- and arabinose. The two major types of linkages are *O*- and C-glycosides. All known flavonoid glycosyltransferases use UDP sugars as glycosyl donors.

Many flavonoids also possess acylated sugars. The acyl groups are either hydroxycinnamates or aliphatic acids such as malonate. The transferases use CoA activated acids as acyl donors. Hydroxycinnamate 1-*O*-acylglucosides serve as donors for the formation of hydroxycinnamate-acylated flavonoids. Glycosylation of anthocyanidins at C-3 to produce anthocyanins has no marked effect on plant colors *in situ* because all anthocyanidins are glycosylated at C-3. Glycosylation at other positions has only minor effects on color (17).

9.4.6 Quinones

Benzo-, naphtho-, and anthraquinones occur in nature. Their structures contain phenolic moieties but are derived from different biosynthetic routes. Secondary benzoquinones are rare in higher plants. Benzoquinone conjugates and hydroquinone glucosides occur in higher plants (Rosaceae and Ericaceae). Hydroquinones are formed by oxidative decarboxylation of 4-hydroxybenzoate derivatives (Scheme 9.9) (18). Naphthoquinones occur in 20 families of higher plants. Shikonin, lawsone, plumbagin, chimaphilin, alkannin, and juglone are important naphthoquinones. The biosynthetic routes are indicated below:

i. Chorismate → succinoylbenzoate → juglone
                                    → lawsone
ii. Polyketide pathway → plumbagin
iii. Tryosine → homogentisic → chimaphilin
iv. p-Hydroxybenzoate + GPP → alkannin (shikonin)

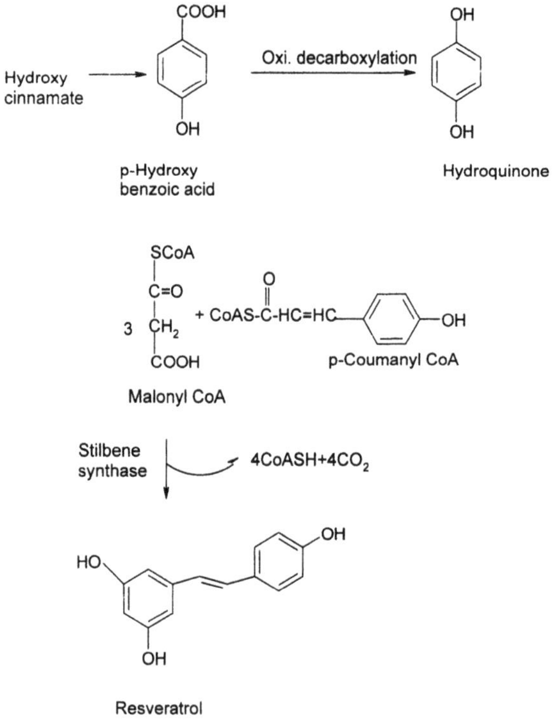

SCHEME 9.9 Biosynthesis of hydroquinone and resveratrol.

Important anthraquinone-bearing plant families are Caesalpiniaceae, Polygonaceae, Rhamnaceae, and Rubiaceae. The important anthraquinones are alizarin and emodin. Alizarin is formed from succinoylbenzoate and mevalonate.

9.4.7 Stilbenes and Xanthones

Some plants, such as pine, grapevine, and peanuts, possess a stilbene synthase by which p-coumaryl-CoA reacts with three molecules of malonyl CoA. In contrast to chalcone synthase, the carboxyl group of the phenylpropane is released as CO_2. Resveratrol synthesized by this process is a stilbene (Scheme 9.9).

Xanthones of higher plants are formed from shikimate and acetate as precursors. Xanthone nucleus is formed from a C_6–C_3 precursor coupled with two malonate units. Glycosylation seems to occur at the benzophenone stage and is followed by oxidative cyclization.

9.4.8 Tannins

The biosynthesis of hydrolyzable tannins, gallo, and ellagitannins starts by the formation of 1-O-galloyl-β-D-glucose from gallate and UDP-glucose catalyzed by O-glucosyltransferase. Addition of further glucose units leading to pentagallonylglucose is catalyzed by gallonyltransferases and acyltransferases. Ellagitannin arises from secondary C–C linkage between adjacent galloyl groups (19).

Gallate + UDP-glucose

$\downarrow$ O-glucosyltransferase

1-Galloyl glucose

Galloyltransferase $\downarrow$ 1-Galloyl glucose

1,6-Digalloyl glucose

Galloyltransferase $\downarrow$ 1-Galloyl glucose

1,2,6-Tetragalloyl glucose

Galloyltransferase $\downarrow$ 1-Galloyl glucose

1,2,3,6-Tetragalloyl glucose

Galloyltransferase $\downarrow$ 1-Galloyl glucose

1,2,3,4,6-Pentagalloyl glucose

$\swarrow$ $\searrow$

Ellagitannins   Gallotannins

The key enzymes involved in condensed tannin (proanthocyanidin) are unknown. The reaction leading to flavan-3-ol from flavan-3,4-*cis*-diol has already been described (Sec. 9.4.5). The proanthocyanidins may be formed as by-products of the reactions leading to the flavan-3-ols. The stereospecific capture of the intermediate carbocations or their quinone methides by the end product flavan-3-ol results in the oligomeric structure. Nothing is known about the enzymes involved in this process.

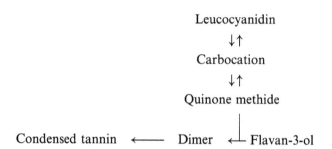

9.4.9   Lignins, Lignans, and Neolignan

The basic components required for lignin synthesis, i.e., *p*-coumaryl, sinapyl, and coniferyl alcohols, are collectively called *monolignols*. Synthesis of monolignols requires reduction of the carboxylic group of the corresponding acids to an alcohol. The carboxyl group is reduced by NADPH to an aldehyde via CoA derivative. In the subsequent reduction to an alcohol NADPH is again used as reductant (Scheme 9.10). The cleavage of the energy-rich thioester bond drives the reduction of the carboxylate to the aldehyde. The same enzymes that catalyze the conversion of *p*-coumarate to *p*-coumaryl alcohol also catalyze the formation of sinapyl and coniferyl alcohols from sinapic and ferulic acids, respectively.

Dimerization of monolignols leads to the formation of lignans. Free radicals are probably involved in the formation of lignans (20–22). Plant lignans are widely distributed as defensive substances. Lignin is formed by polymerization of a mixture of the three monolignols. Peroxidases are involved in linking of the monolignols. Oxidation of a phenol results in the formation of a resonance-stabilized phenol radical. These phenol radicals can dimerize and finally polymerize to form C–C or C–O linkages (Scheme 9.11). Neolignans are phenylpropane dimers that are linked head to tail.

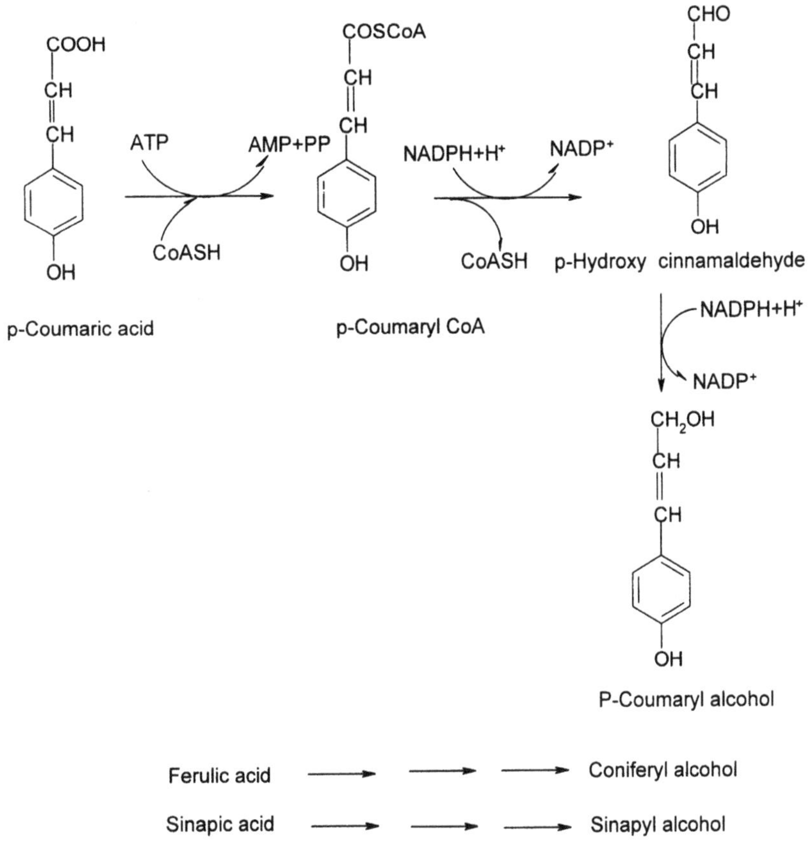

SCHEME 9.10   Biosynthesis of monolignols.

9.5   BIOACTIVITY

9.5.1   Phenols and Phenolic Acids

Choice and no-choice feeding assays were conducted to determine the level of resistance among 10 taxa of *Malus* spp Mill (23). Under no-choice conditions, *M. baccata* (L.) Borkh. 'Jackii', *M.* x 'Hargozam' Harvest Gold and *M. transitoria* (Balatin) Schneider 'Schmitcutleaf' Golden Raindrops were highly resistant, with less than $2\,cm^2$ leaf area consumed in $24\,h$. *M.* x 'Radiant' was highly susceptible, with $7.6\,cm^2$ consumed, and the remaining six cultivars were intermediate. Under choice conditions, eight taxa were resistant with less than 10% defoliation, *M.* x 'Red Splendor' was

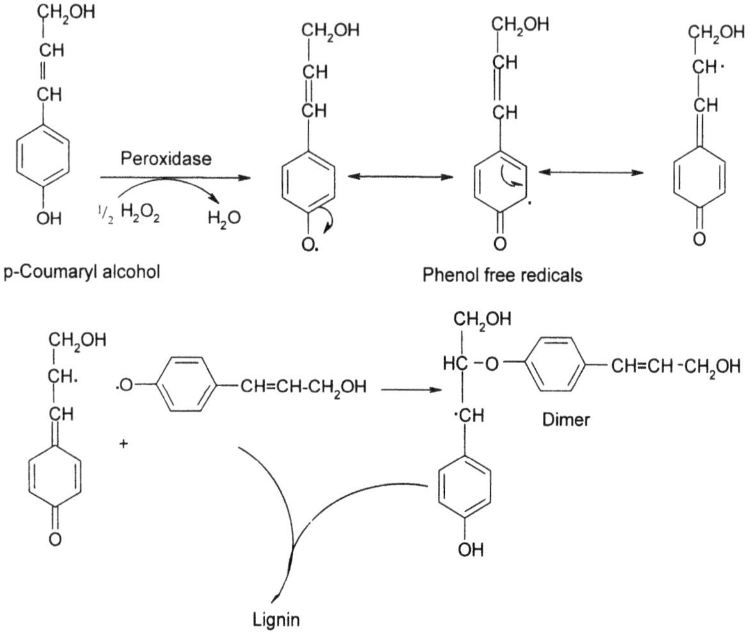

SCHEME 9.11   Biosynthesis of lignin.

intermediate with 26%, and *M*. x 'Radiant' was susceptible with 73% defoliation. Feeding responses to eight individual phenolics were tested in artificial diets over a range from 0 to 100 mM. Phloridzin, phloretin, naringenin, and catechin were all feeding deterrents, whereas quercetin and rutin were feeding stimulants. Chlorogenic acid stimulated feeding at low concentrations and deterred feeding at higher concentrations (i.e., a peak response). Kaempferol had no effect. Analysis of endogenous foliar phenolics showed considerable variation in concentrations among taxa. Stepwise multiple regression analysis identified phloridzin as the only endogenous phenolic that was significantly related to resistance under both choice and no-choice feeding conditions to Japanese beetles, *Popilla japonica* Newman.

The impact of phenolics on a specialist herbivore, *Manduca sexta*, and a generalist herbivore, *Heliothis virescens*, was investigated using transgenic tobacco with differential expression of PAL (24). Foliar phenolics such as chlorogenic acid, rutin, and total flavonoids differentially accumulated in the respective transgenic tobacco lines; the amount of chlorogenic acid ranged from 201 to 2202 $\mu g\, g^{-1}$ of fresh leaf, that of total flavonoids from 211 to 500 $\mu g\, g^{-1}$ of fresh leaf, and that of rutin from 73 to 172 $\mu g\, g^{-1}$

of fresh leaf. However, the levels of the phenolics and larval growth of *M. sexta* or *H. virescens* were not significantly correlated. Likewise, phenolic levels were not correlated with larval survival of *M. sexta*.

Leaf beetle larvae *Galerucella lineola* feeding on *Salix dasyclados* showed lower survival, took longer to develop, and were smaller than those feeding on *Salix viminalis* (25). Larval performance was not correlated with the total concentration of phenolics. Nor did any individual phenolic compounds appear to influence the insect. However, in a multivariate analysis, larval performance was correlated with phenolic composition. Thus performance seems to be affected by interactions between several compounds rather than by single compound. Hybridization affected insect performance; *S. viminalis* x *dasyclados* hybrids were of intermediate suitability as food for *G. lineola* larvae compared with the parental plants.

The attack of phytophagous insects can change the secondary chemistry of a host plant and increase its natural resistance (26). In a preliminary experiment, different patterns of phenolic compounds were detected by high-performance liquid chromatography (HPLC) in uninfested and *Psylla*-infested pear trees. In both young and mature uninfested pear leaves (variety conference), a group of phenolic compounds eluted between 10 and 20 min from a Lichrosorb RP-18 column. In *Psylla*-infested young leaves, different and more prominent number of phenolics appeared between 30 and 35 min whereas the first group diminished.

Chemical defense of the tomato plant against noctuid larvae was argued to result from suites of interactive chemical traits that simultaneously impair the acquisition of nutrients and toxify the insect. Defense results from tomatime, catecholic phenolics and phenol oxidases, proteinase inhibitors, and lipoxygenase (27).

A methylene chloride extract of the spice *Cinnamomum aromaticum* seeds was shown to be insecticidal to *T. castaneum* and *S. zeamais* (28). The contact, fumigant, and antifeedant effects of cinnamaldehyde were tested against *T. castaneum* adults and larvae, and *S. zeamais* adults. Both adults showed similar susceptibility to the contact toxicity of cinnamaldehyde, both having an LC_{50} of $0.7\,mg/cm^2$ and an LC_{95} of $0.9\,mg/cm^2$. However, cinnamaldehyde had a higher level of fumigant toxicity to *T. castaneum* than *S. zeamais* with LC_{50} values of 0.28 and $1.78\,mg/cm^2$, respectively. *T. castaneum* adults were more susceptible than larvae to the contact and fumigant actions of cinnamaldehyde. The larvae became less susceptible to both contact and fumigant toxicity of cinnamaldehyde with age. A flour disc bioassay using no-choice tests was employed to study the antifeedant activity of cinnamaldehyde against the insects and effects on consumption and utilization of food by the insects. Cinnamaldehyde had no significant effects on diet consumption and growth. *T. castaneum* adults had no antifeedant action against them at concentrations of up to $3.6 \, \text{mg g}^{-1}$ food. The combined contact, fumigant, and antifeedant properties of cinnamaldehyde make it potentially useful grain protectant.

Feeding performance of herbivorous insects is influenced by host plant nutritional quality, which can be improved for insect resistance by artificial selection (29). This study was conducted to determine which biochemical constituents in maize (*Zea mays* L.) changes during recurrent selection for resistance to first and second generations of European corn borer (ECB) [*Ostrinia nubilalis* (Hubner)]. Four cycles of selection (C_0, C_2, C_4, and C_5) from the BS9 population were field grown, artificially and naturally infested with ECB and the following tissues sampled for biochemical analysis: immature and mature leaf blade, leaf sheath, rind, node, and pith. Tissue was analyzed for percent protein, DIMBOA [2,4-dihydroxy-7-methoxy-2H-1,4-benzoaxazin-3-(4H)-one], fiber, and cell wall–bound phenolics, which included *p*-coumaric acid (pCA), ferulic acid (FA), cyclobutane dimers (CBDs), and diferulic acid (DFA). Leaf and stalk toughness were also determined and showed significant increases over cycles of selection. Protein content was lowest in stalk tissues with advanced cycles having lower levels, but leaf protein content did not differ significantly. DFA reached high levels in the rind ($0.85 \, \text{mg/g}$) and leaf sheath ($1.35 \, \text{mg/g}$) tissues, increased significantly in immature leaf tissue (0.55–$1.02 \, \text{mg/g}$) over cycles of selection, and may serve to fortify tender whorl tissue. Number of tunnels per stalk was negatively correlated with DFA content in the pith ($r = -0.77$, $P = 0.02$). Microspectrophotometric determinations of epidermal cell wall absorbance for leaf blade and rind tissue showed increased absorbances (23% and 27%, respectively) in the spectral region characteristic of phenolic acids over cycles of selection. Phenolic acids, in particular DFA, have increased over cycles of selection to render maize tissue more resistant through fortification of cell walls, especially in leaf and rind epidermal tissue.

Leaf tissue of multiple borer resistance (MBR) genotypes of maize is tough, which may restrict feeding by early-instar larvae (30). MBR genotypes also tend to have reduced nutritional value (lower nitrogen content), and elevated levels of fiber and cell wall phenolics, which may account for the elevated leaf toughness. Cell wall phenolics can cross-link the hemicellulose of the cell wall by the action of peroxidase to produce diferulic acid. Approximately 80% of the variation in field leaf ratings for *Ostrinia nubilalis* could be accounted for by protein, fiber, and diferulic acid content in leaf tissue at the midwhorl stage in plant development.

The mechanism of resistance in maize to the stored product insects such as the maize weevil, *Sitophilus zeamais* Motsch, and the larger grain borer, *Prostephanus truncatus* Horn, has been investigated in relation to

secondary plant chemicals and other biochemical and physical characteristics of maize genotypes (31). Performance parameters of weevils (number
of eggs laid, number of progeny, Dobie index, grain consumption) were
negatively and significantly correlated ($r = -0.8$, $P = 0.5$) to the most
abundant phenolic of grain, ferulic acid. With *P. truncatus*, the weight loss
of grain also showed a negative correlation with ferulic acid whereas percent
damage of kernels by insects was negatively correlated to *p*-coumaric acid.
These phenolic acids were found in highest concentration in the pericarp
and cell walls of the endosperm by fluorescence microscopy. Phenolic acid
content was also found to correlate strongly with hardness of the grain,
which may be related to the mechanical contributions of phenolic dimers to
cereal cell wall strength. In the aleurone layer, phenolic acid amines have
been detected that have toxic effects on insects.

The genetics of maize grain resistance to the maize weevil, *Sitophilus
zeamais* Motsch., infestation was analyzed by means of additive linear models
that considered genetic contributions of maize caryopsis through embryo,
endosperm, and pericarp (32). Specific traits associated with these grain
tissues were phenolic acids (pericarp, embryo), proteinase inhibitors
(endosperm, embryo), and hardness of grain (pericarp, endosperm,
embryo). The susceptibility of the grains to weevil infestation was measured
by consumption and reproductive activities of insect populations. Inbred
lines of quality protein maize contrasting in resistance to maize weevil
infestation were used for the genetic analysis of resistance. Concentrations of
phenolic acids in grain have a highly negative and significant correlation with
indices of susceptibility of maize to the maize weevil. However, the
correlation between susceptibility of grain and contents of proteinase
inhibitors in the endosperm is low, though negative and significant.
Resistance of pericarp testa to compression forces was the only rheological
trait of grain inversely correlated with susceptibility of maize to colonization
by maize weevils, but neither the correlation coefficient nor the significance
was high. The negative relationship of biochemical and biophysical traits of
maize grain with feeding and reproductive activities of insects on the grain
suggests detrimental effects of these grain characteristics on the colonization
success of insect populations. The estimated genetic parameters for additivity
of endosperm and dominance of pericarp associated with the expression of
phenolic acid concentration in the grain were highly significant and inversely
correlated to estimated susceptibility parameters of genetic action.

Steryl esters of ferulic and *p*-coumaric acids were tested for their effect
on corn fungal pathogens and insects (33). The steryl ester fraction from
corn bran had no effect on corn earworm larvae or dried fruit beetle adults,
but dried fruit beetle larvae showed a significant increase in weight and the
triglyceride fraction was active.

Phenolics in maize grain were assessed as a possible factor or indicator of resistance to two stored product insect pests (34). Weight loss of grain of seven maize genotypes due to *Prostephanus truncatus* (Horn) (Coleoptera: Bostrichidae) was found to be strongly and negatively correlated to the phenolic content of maize grain determined by HPLC.

9.5.2 Phenolic Glycosides

Snook et al. (35) surveyed the maysin content of the silks from over 600 corn inbreds, populations, plant introductions, and various unassigned collections. Several lines were identified that contained high levels of related flavones (Fig. 9.1). The other compounds identified includes apimaysin, 3′-methoxymaysin, dihydromaysin, and dihydro-3′-methoxymaysin. When corn earworms were allowed to feed on the diet containing maysin, concentration of more than 0.2% maysin reduced larval growth to over 50% than that of the control, while higher levels of maysin (up to 1%) reduced weights of the larvae to greater than 80%.

Laboratory bioassays on corn earworm were performed with maysin, luteolin, rutin, and chlorogenic acid. Maysin reduced the weight of corn earworm by about 84% of controls at 12.6 mM concentration. Rutin possessing the requisite *ortho*-dihydroxy structure was found to be just as active as maysin. The presence of the sugar moiety is not needed, because the aglycone luteolin was found to be just as active as maysin. Chlorogenic acid having an *ortho*-dihydroxyphenyl structure was found to be active against the corn earworm resulting in 80% reduction of growth at 20.5 mM concentration. Galactosyl luteolin was almost as active as maysin in reducing corn earworm growth. The natural resistance of corn silks to the corn earworm, *Helicoverpa zea* (Boddie), has been attributed to the presence of a single flavone called maysin. Wiseman et al. (36) showed that maysin was active against the fall armyworm *Spodoptera frugiperda*.

Pieris napi oleracea, an indigenous butterfly to North America, lays eggs on *A. petiolata* an invasive weed introduced from Europe, but the larvae generally do not survive. A new apigenin glycoside, isovitexin 6″-*O*-glucoside, has been isolated from the leaves of *A. petiolata* and identified as a feeding deterrent for *P. napi* fourth-instar larvae (37).

Several plant phenol glycosides have been identified as attractants or stimulants for herbivores. Mixtures of quercetin, quercetin-7-*O*-glucoside and quercetin 3-*O*-glucoside, are stimulants for the boll weevil (38). Synergistic effects between phenol glycosides and some unidentified plant metabolites serve as oviposition stimulants for papilionid butterflies (39). Similarly, a mixture of chlorogenic acid and luteolin 7-*O*-(6′-*O*-malonyl)-β-D-glucopyranoside isolated from carrot foliage serves as an oviposition stimulant

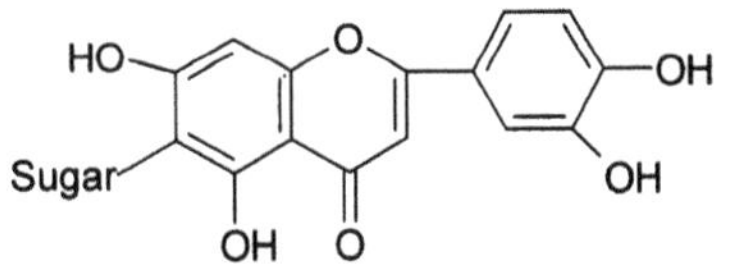

Maysin (Luteolin - C - glycosides)

Sugar = 2′ - O -α -L - rhamnosyl -
6-c-(6-deoxy - xylo - hexos - 4 -ulosyl)

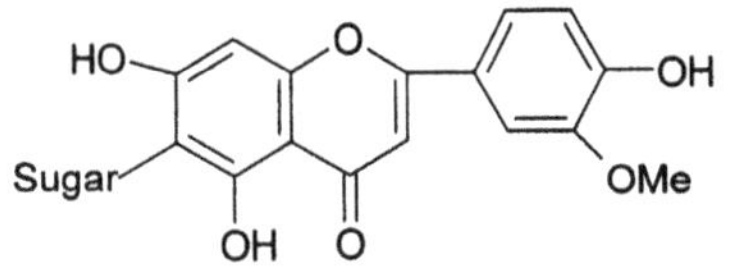

3 - Methoxymaysin

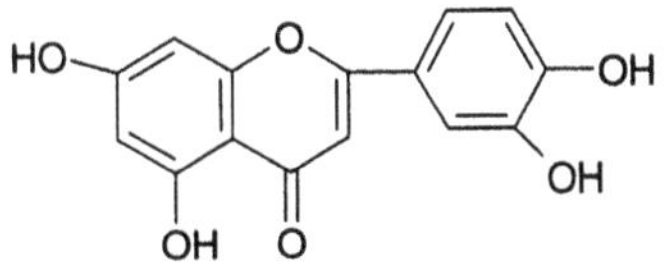

Luteolin

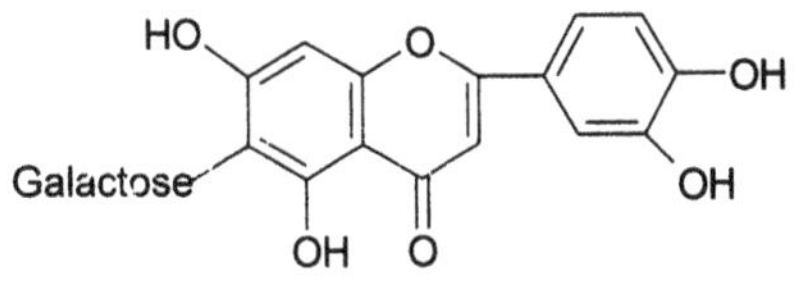

6-C-β-D-Galactopyranosyl luteolin

FIGURE 9.1   Structure of maysin and maysin analogues.

for the black swallowtail butterfly, even though neither pure compound was active (40). An example of the specificity of a response elicited by a unique mixture of phenol glycosides has been provided by Tahavanainen et al. (41) who demonstrated that four leaf beetle species select their favored host plants (*Salix* spp.) based on phenol glycoside content. Furthermore, when forced to shift host plant, the insects selected the alternate host with phenol glycoside content close to that of the preferred host.

Another method for herbivores' exploitation of plant phenol glycosides has also been described. Several species of oligophagous chrysomelid beetles prefer willow species as hosts that have high levels of salicin, and one beetle utilizes a salicortin-containing willow (42). The insects do not use the phenol glycosides as feeding cues; rather, they metabolize them to salicaldehyde, a defensive chemical used by the insects, and glucose, which is a significant source of energy. Furthermore, the protection afforded by this defense can be passed on to eggs.

Other studies have demonstrated that plant phenol glycosides can have adverse effects on insect herbivores. Rutin and several other flavonoids adversely affect lavae of the tobacco budworm, the cotton bollworm, the pink bollworm, and the greenbug (43). However, effects of these substances are limited to certain life stages of the larvae and do not extend to all insects (44). One component of cotton's defense against insects is the ability of some of its phenol glycosides to inhibit larval growth (45).

Perhaps the most thoroughly investigated case of plant chemical defense against insects is based on phenol glycosides in quaking aspen (*Populus tremuloides*). In one set of investigations, Lindroth (46) showed four phenol glycosides in aspen foliage (salicin, tremuloiden, salicortin, and tremulacin), in which the latter two adversely affect the performance of larvae of the swallowtail butterfly. Two subspecies (*Papilio glaucus canadensis* and *P. glaucus glaucus*) show significant differences in their abilities to tolerate these chemicals. The former uses aspen as a normal host being less sensitive to the effects of salicortin and tremulacin. He proposed that *P. g. canadensis* has adapted to salicortin and tremulacin by decreasing the hydrolysis of the β-glucoside to the presumably more toxic aglycone and by rapidly hydrolyzing the carboxylic acid ester to innocuous metabolites. A separate study has revealed that the phenol glycosides of aspen have similar effects on larvae of the great aspen tortrix [*Choristoneura conflictana* (Walker)]. The investigation has focused on metabolic changes in the phenol glycoside content of the plant during herbivory (47). In crushed aspen leaves, salicortin and tremulacin were transformed into salicin and tremuloidin (respectively), and 6-hydroxycyclohexenone (6-HCH), a reactive substance that adversely affects the larvae. Furthermore, damaging foliage on intact plants causes increases in concentrations of salicortin and tremulacin, a situation that could lead to even higher levels of 6-HCH in damaged foliage. Taken together, these studies have identified aspen phenol glycosides that were harmful to insect herbivores and have pointed out the importance of metabolic transformations of these compounds in the plant–insect interaction.

Roth et al. (48) examined the effects of dietary phenolic glycosides and tannic acid on performance of the gypsy moth, *Lymantria dispar* (L.), and its suitability as a host for the parasitoid *Cotesia melanoscela* (Ratzeburg). Gypsy moth growth was reduced on diets containing allelochemicals, and the magnitude of reduction was greater on diets supplemented with phenolic glycosides than on diets supplemented with tannic acid. Reductions in gypsy moth performance translated into prolonged development, reduced cocoon weight, and a higher incidence of mortality for *C. melanoscela*. Overall, phenolic glycosides reduced *C. melanoscela* performance slightly more than did tannic acid. These results show that interactions of *C. melanoscela* with its host were influenced by the diet of the host, demonstrating that plant allelochemicals can have profound effects not only on plant–herbivore interactions but on herbivore–natural enemy interactions as well.

Although the studies reviewed above represent only a limited beginning of attempts to understand the role of phenol glycosides in plant–insect interactions, two general points emerge. The first is that the chemicals or mixtures mediating a given interaction are quite specific. That is, there appears to be no general role that can be ascribed to phenol glycosides as a class, but there are individual phenol glycosides that play important roles in specific cases. Second, metabolic transformation is a key aspect of the way in which phenol glycosides influence plant–insect interactions.

9.5.3 Coumarins

Simple coumarin is ovicidal to Colorado potato beetles and toxic to mustard beetles and a variety of insects. Furanocoumarins are capable of photosensitizing insects. When xanthotoxin, a furanocoumarin, was incorporated in an artificial diet at 0.1% and UV irradiated, it caused 100% mortality to the Southern armyworm (49). α-Terthienyl and related furanocoumarin compounds are among the most promising chemicals for insect control under field conditions (50). Coumarins exhibit a tremendous range of effects on insects (49). They are plant allomones for many taxa. The simple coumarin bergammotin is ovicidal to *Leptinotarsa decemlineata*, and mammein is toxic to mustard beetles, houseflies, and mosquitoes. Coumarin is also ovicidal to *Drosophila melanogaster* (49). Furanocoumarins are deterrents as well as toxin to a variety of insects such as *Helicoverpa zea*, *H. virescens*, *Spodoptera litura*, *Battus philenor*, *Depressaria pastinacella*, *Aedes aegypti*, and *Tetranychus urticae* (49). Toxicity is greatly enhanced in the presence of UV light. Coumarins also appear to act as host recognition compounds (kairomones) for certain insects that feed on coumarin-containing plants. Coumarin function as arrestant to the spring migratory flight of *Sitonia cylindricolla*, a weevil that feeds extensively on sweet clover. Furanocoumarins stimulate oviposition in the carrot rust fly, *Psila rosae*. It also enhances feeding in the black swallowtail caterpillar, *Papilio polyxenes*, which feeds on furanocoumarin containing species of umbelliferae.

9.5.4 Tannin

Leaf tannin extracts from trees on which gypsy moths grew and reproduced poorly precipitated lysolecithin more effectively than did extracts from trees on which gypsy moths performed well (51). Adding tannic acid to midgut fluid elevated surface tension, and about 25% of larvae feeding on oak leaves exhibited elevated midgut surface tension, suggesting a loss of surfactants. Larvae appear able to replace lost surfactants to a limited degree. An important effect on leaf tannins, and perhaps other phenolics, may be to reduce concentrations of surface active phospholipids in the midgut and produce lipid or other dietary deficiencies in insects.

The seeds of the tested genotypes of *Vicia faba* L. differed by the presence or absence of tannins in the seed coat and in the level of vicine and convicine in the cotyledons (52). In vertebrates, these two glycosides can be transformed into divicine and isouramil which influence the enzymatic activity of glucose-6-phosphate dehydrogenase. *Callosobruchus chinensis* (L.) develops in *V. faba* seeds and causes high losses during storage. *C. maculatus* (F.) is less able to develop in the seeds of *V. faba*. For both bruchids the seed coat represents a barrier that only 45–60% of larvae overcome. The presence of tannins did not affect the perforation rate.

The effect of tannic acid on gypsy moth, *Lymantria dispar* (L.), larvae was stadium specific. First- to third-instar gypsy moths were larger and developed faster on tannin-supplemented artificial diets than on control diets (53). However, after the start of the fourth stadium, larvae on tannin diets grew more slowly than larvae on control diets. The longer the insect's feeding period after the fourth stadium, the greater the negative effect of the tannin diets on eventual pupal weights. Female gypsy moths were most affected by tannin content of the diet because their extra instar results in a longer feeding period. This results in a proportionally greater influence of the negative effects of tannins.

Tannins have been implicated in the resistance of some cotton varieties to insect attack (54). The relationship between tannin concentration and susceptibility of some Egyptian cotton varieties to sucking insects was undertaken. Eight Egyptian cotton varieties derived from *Gossypium barbadense* and one variety from *G. hirsutum* were evaluated. There was a significant negative linear relationship between the concentration of tannins in two Egyptian varieties Giza 70 and 76 and the population density of sucking insects. On the contrary, a positive correlation existed between sucking insect infestation and tannin concentrations in the two cotton varieties Giza 80 and Giza 81. This may be attributed to the low level of tannins in these two varieties.

Tannin extracts of two sorghum [*Sorghum bicolor* (L.) Moench.] cultivars, BKS 5 and BTX 623, were assessed for their bioactivity against the rice weevil, *Sitophilus oryzae* (L.), and the Angoumois grain moth, *Sitotroga cerealella* (Oliver), in whole wheat flour pellets and against the red flour beetle, *Tribolium castaneum* (Herbst), in whole wheat flour at three treatment levels. The number of adult progeny was reduced and number of days to emergence increased by tannins extracted from both sorghum cultivars at all treatment levels. In a short-term (25 days) feeding test using *S. oryzae* adults, the presence of sorghum tannins from cv BKS 5 in the pellets significantly reduced feeding by the insects (55).

9.5.5 Flavonoids

Larvae of the lycaenid butterfly *Polyommatus icarus* were reared on inflorescences of *Coronilla varia* and *Medicago sativa*, which are rich in flavonoids. Twelve different flavonoids (five compounds from the former and nine from the latter), including aglycones and *O*-glycosides of kaempferol, quercetin, and myricetin, were isolated and identified by spectroscopic means (56). Nuclear magnetic resonance (NMR) and mass spectroscopy (MS) data for the new acylated glycoside kaempferol 3-*O*-6″-(3-hydroxy-3-methylglutaroyl)-β-D-glucopyranoside are reported. Comparative HPLC analysis of the respective host plants and of larvae, pupae, and imagines of *P. icarus* indicated selective uptake and accumulation of kasmpferol vs. quercetin and myricetin derivatives. The latter were excreted largely unchanged through the feces. Irrespective of the larval host plant kaempferol 3-*O*-glucoside was found as the major flavonoid in larvae, pupae, and imagines of *P. icarus*, accounting for approximately 83–92% of all soluble flavonoids in adult butterflies. Within the imagines, approximately 80% of all flavonoids are stored in the wings (especially in the orange submarginal lunules), whereas the remaining 20% reside in the bodies. Feeding experiments with artificial diet demonstrated that the insects are able to form kaempferol 3-*O*-glucoside by glucosylation of dietary kaempferol. Possible functions of the sequestered flavonoids, especially for mate recognition of *P. icarus*, are discussed (56).

Thin-layer chromatography resolved nine major compounds from the 60% methanol extractables from PI 227687 soya bean [*Glycine max* (L.) Merrill] leaves. The flavonoids daidzein, an unidentified flavonoid X2 (Rf 0.19), glyceollin, sojagol, and coumestrol exhibited antifeedant and/or antibiotic effects against the larvae of cabbage looper, *Trichoplusia ni* Hb. The results indicate that several compounds in PI 227687 soya beans contribute to its antifeedant and/or antibiotic effects against *T. ni* (57).

Quercetin and three of its glycosides (3-rhamnoside, 3-glucoside, and 3-rutinoside) repelled feeding by the tobacco budworm, *Heliothis virescens*, tobacco bollworm, *Helicoverpa zea*, and cotton pink bollworm, *Pectinophora gossypiella* when fed on low concentrations. These common flavonoids killed the larvae at 0.2% concentration or greater when applied in the diet, whereas the silkworm, *Bombyx mori*, uses the flavonoids present in mulberry leaf as dietary feeding stimulents (58). The flavonoids morin and isoquercitrin present in the leaves of mulberry (*Morus alba*) attract the silkworm, *Bombyx mori*. Other flavonoids that act as feeding attractant include 6-methoxyluteolin 7-rhamnoside in alligator weed (*Alternanthera phylloxeroides*), catechin-7-xyloside in elm bark, and various flavonones in *Prunus* bark which stimulate the feeding of different beetles (59).

The cotton boll weevil, *Anthonomus grandis*, is repelled by rutin but stimulated to feed by both the 3-glucoside and the 3-rhamnoside. Larvae of *Pectinophora gossypiella* are killed if fed diets containing 0.2% of these three flavonol glycosides. Cabbage white butterfly (*Pieris brassicae*) larvae are highly sensitive to flavonoids when fed with artificial diets (60).

Laboratory studies were conducted to assess the toxicity and growth inhibitory activity against *Aedes aegypti* larvae of the crude acetone extracts of *Polygonum senegalense* (Meissn.) leaves containing flavonoid component 2',6'-dihydroxy-4'-methoxydihydrochalcone and the internal tissue flavonol quercetin. The first showed significantly high insecticidal and growth inhibitory activity even at low concentrations, whereas quercetin was toxic only at concentration above 7 µg/mL (61).

The only group of flavonoids that are known to be highly toxic to many insects are the isoflavonoid-based rotenoids. Rotenone is the active principle obtained from the roots of *Derris elliptica* (Leguminosae). Rotenoids have low mammalian toxicity but are highly toxic to insects and fish. Rotenone is potent against wide range of pests. It controls many leaf-chewing beetles, caterpillars, flex beetles, and aphids. Coumesterol found in mung bean and soybean seedlings have antifeedant and antibiotic activities against cabbage semilooper, *Trichoplusia ni*, feeding (62). Glyceollin in soybean has strong antifeedant activities against the Mexican bean beetle. High concentrations of rutin and chlorogenic acids have been found in the trichomes of tomato leaves. These compounds are toxic to corn earworm (44).

9.5.6 Lignin

In laboratory assays with intact foliage, both *Callosamia angulifera* and the polyphagous *C. promethea* fed readily on sweetbay but were unable to survive past the third instar. Two neolignan compounds, magnolol and a biphenyl ether, were found to reduce neonate growth and survival of unadapted herbivore species when painted on acceptable host leaves at concentrations similar to those found in sweetbay foliage. Both compounds significantly reduced neonate growth of *C. angulifera* and *C. promethea* but had no effect on the sweetbay specialist, *C. securifera*, indicating that the latter species possesses the unique ability to tolerate, metabolize, or otherwise circumvent the phytochemical defenses of this host (63).

9.6   MECHANISM OF ACTION

Chemically, phenols are very reactive substances. All phenols are capable of taking part in hydrogen bonding. This may be intramolecular or may be intermolecular, which bring about interactions between plant phenols and the peptide links of proteins and enzymes. Another important property of many phenols with an *O*-dihydroxy grouping is their ability to chelate metals.

The protective effects of phenolic acids are thought to be due to their ability to form complexes with proteins via both covalent and non-covalent interactions (64), which may result in impaired enzyme functions, reduced protein digestion, and reduced bioavailability of amino acids (65). The *O*-dihydroxyphenolic acids, such as caffeic and chlorogenic acids, have the potential effect on lipid peroxidation and protein oxidation in the midgut tissues of *Helicoverpa zea* and hence the greatest potential for having a deleterious effect on the growth and development of the larvae. Phenolic acids that exert the largest effect on lipid peroxidation and protein oxidation have two ionizable hydroxyl groups on the aromatic ring (or in very close proximity). The importance of this structural feature may reside in its ability to redox cycle in vivo acting both to deplete important biological antioxidants and to generate direct oxidants such as reactive oxygen species, as well as form highly electrophilic quinones capable of alkylating proteins. In the midgut tissue, the chemical reduction of the resulting quinones may deplete ascorbic acid and glutathione. The depletion of these antioxidants may compromise the ability of larvae to protect themselves effectively from endogenous oxidative stress as well as challenges by exogenous oxidants. Appel (66) hypothesized that the toxicological effect of the *O*-hydroxyphenolics, caffeic and chlorogenic acids, is due primarily to their ability to act as prooxidants. Evidence in support of this hypothesis includes

1.   Previous studies indicating caffeic and chlorogenic acids producing reactive oxygen species.
2.   Electron spin resonance studies showing that these compounds form free radicals.

3. The inhibition of superoxide dismutase by diethylthiocarbamate, which dramatically increases the toxicity of quercetin in lepidopteran larvae.

4. The larval ingestion of caffeic and chlorogenic acid increasing midgut lipid peroxidation and protein oxidation and results in the loss of ascorbic acid.

5. The addition of high levels of ascorbic acid to diets containing caffeic and chlorogenic acid, thus ameliorating the toxicological effects of these phenolics to larvae.

Coumarins are phototoxic when exposed to UV light. They bind pyrimidine bases of DNA. Photosensitization of furanocoumarins is due to the formation of an excited triplet state on absorption of a photon. The excited triplet state furanocoumarin can react directly with the molecules of the pyrimidine base (67). Lignin provides a mechanical barrier of unpalatability and reduces the digestibility of cell wall carbohydrate. Tannins reduce the postabsorptive inhibition that is responsible for poor insect growth. Rotenone specifically inhibits the NADH-dependent dehydrogenase step of the mitochondrial respiratory chain.

9.7 MOLECULAR BIOLOGY

Genes encoding many enzymes active in the phenylpropanoid pathway have been isolated from various species. Phenylalanine ammonia lyase (PAL) catalyzes the first step in the phenylpropanoid metabolism and plays a central role in the biosynthesis of phenylpropanoid compounds. Two PAL genes, *PAL1* and *PAL2*, from *Populus trichocarpa* x *P. deltoides* F_1 hybrid have been cloned. The properties of PAL1 and PAL2 promoters and their expression patterns in transgenic tobacco and poplar are described by Gray-Mitsumune et al. (68) The promoters were 75% identical in the regions sequenced and each contained two copies of AC-rich putative cis-acting elements that matched a consensus plant myb transcription factor binding site sequence.

Gui et al. (69) quantified the levels of mRNA transcripts coding for chalcone synthase, an enzyme required for flavonoid biosynthesis and for caffeic-O-methyl transferase, an enzyme used in synthesis of lignin. These mRNA levels were greatly elevated in plants inoculated with *Verticillium dehliae* as compared to water-treated controls.

The flavonoid and isoflavonoid pathways are probably the best characterized natural product pathways in plants and are therefore excellent targets for metabolic engineering (70). Manipulation of flavonoid biosynthesis can be approached via several strategies, including sense or antisense manipulation of pathway genes, modification of the expression of regulatory genes, or generation of novel enzymatic specificities by rational approaches based on emerging protein structure data. In addition, activation tagging provides a novel approach for the discovery of uncharacterized structural and regulatory genes of flavonoid biosynthesis.

Isoflavones are synthesised by a branch of the phenylpropanoid pathway. In plants, isoflavones play major defensive roles. Using a soybean EST collection, the cDNA encoding isoflavone synthase has been expressed in *Arabidopsis thaliana* plant that does not normally make isoflavones and shown that this transgenic plant is now able to produce genistein (71).

Transgenic birdsfoot trefoil (*Lotus corniculatus*) plants harboring antisense dihydroflavonol reductase (dihydrokaempferol 4-reductase) (AS-DFR) sequences were produced and analyzed by Robbins et al. (72). The effect of introducing three different antisense *Antirrhinus majus* DFR constructs into a single recipient genotype (S50) was assesed. There were no obvious effects on plant biomass, but levels of condensed tannins showed a statistical reduction in leaf, stem, and root tissues of some of the antisense lines. Transformation events were also found that resulted in increased levels of condensed tannins. The powerful molecular techniques and reproducible transformation protocols would enable the manipulation of flavonoid pathway (73).

9.8   PERSPECTIVES

The only group of phenolics that are known to be highly toxic to many insects are the isoflavonoid-based rotenoids. Even though many other phenolic compounds have been tested for the bioactivity against insects, no compounds other than rotenoids proved to possess effective insecticidal or insect repellent activity. On the other hand, some phenolic glycosides have been identified as attractants or stimulants to insects. More definite tests have to be conducted to prove unequivocally the role of phenolics in insect control. Even where there is a control, the molecular mechanism has to be worked out. This will be starting point for further molecular work. Though the potential of metabolic engineering of phenylpropanoids in plants has been recognized, manipulation of the phenylpropanoid pathway genes has been done only in some plants. But most of the reported molecular biology and cloning work have been confined to the role of phenolics in disease resistance in plants. Hence, more work has to be carried out on the molecular biology as well as metabolic engineering of the phenolic pathway with respect to insect resistance in plants.

REFERENCES

1. Harborne, J.B. Biochemistry of Phenolic Compounds. Academic Press, London, 1964.
2. Harborne, J.B. General procedures and measurement of total phenolics. In *Methods in Plant Biochemistry, Plant Phenolics*; Harborne, J.B., Ed.; Academic Press: London, 1989; Vol. 1, 1–28.
3. Van Sumere, C.F. Phenols and phenolic acids: In *Methods in Plant Biochemistry, Plant Phenolics*; Harborne, J.B., Ed.; Academic Press: London, 1989; Vol. 1, 29–73.
4. Gorham, J. Stilbenes and phenanthrenes: In *Methods in Plant Biochemistry, Plant Phenolics*; Harborne, J.B., Ed.; Academic Press: London, 1989; Vol. 1, 159–196.
5. Wollenweber, E. In *The flavonoids: Advances in Research*; Harborne, J.B., Ed.; Chapman and Hall: London, 1988; Vol. 1, 233–296.
6. Williams, C.A.; Harbone, J.B. Biflavonoids. In *Methods in Plant Biochemistry, Plant Phenolics*; Harborne, J.B., Ed. Academic Press: London, 1989; Vol. 1, 357–388.
7. Haslam, E. Plant Polyphenols, Vegetable Tannins Revised. Cambridge University Press, Cambridge, 1989.
8. Dewick, P.M. In *The Flavonoids: Advances in Research Since 1980*; Harborne, J.B., Ed.; Chapman and Hall: London, 1988; 125–209.
9. Thomson, R.H. Naturally Occurring Quinones III. Chapman and Hall, London, 1987.
10. Harborne, J.B. Plant phenolics. In *Encyclopedia of Plant Physiology*; Bell, E.A., Charlhood, B.V., Eds.; New Series Springer: Berlin, 1980; Vol. 8, 329–402.
11. Schmid, J.; Amrhein, N. Molecular organization of the shikimate pathway in higher plants. Phytochemistry **1995**, *39*, 737–750.
12. Lea, P.J. Enzymes in secondary metabolism. In *Methods in Plant Biochemistry*; Dey, P.M., Harborne, J.B., Eds.; Academic Press: London, 1992; Vol. 9.
13. Jensen, R.A. The shikimate/arogenate pathway: link between carbohydrate metabolism and secondary metabolism. Physiol. Plant **1985**, *66*, 164–168.
14. Conn, E.E. In *The Biochemistry of Plants*; Stumpf, P.K., Conn, E.E., Eds.; A Comprehensive Treatise, Secondary Plant Products; Academic Press: London, 1981; Vol. 7.
15. Strack, D. Phenolic metabolism. In *Plant Biochemistry*; Dey, P.M., Harborne, J.B., Eds.; Academic Press: London, 1997; 387–416.
16. Shirley, B.W. Flavonoid biosynthesis: new function for an old pathway. Trends Plant Sci. **1996**, *1*, 377–382.
17. Heller, W. Flavonoid biosynthesis: an overview. In *Plant Flavonoids in Biology and Medicine*; Cody, V., Middlestone, E., Harborne, J.B., Eds.; Biochemical, Pharmacological and Structure Activity Relationships; Alan R Liss: New York, 1986; Vol. 1, 25–42.
18. Hiroyuki, I. Leistner, E. Biochemistry of quinones. In *The Chemistry of Quinonoid Compounds*; Patai, S., Rappoport, Z., Eds.; Wiley: Chichester, 1988; Vol. 2, 1293–1349.

19. Haslam E. Vegetable tannins. In *The Biochemistry of Plants: A Comprehensive Treatise*; Stumpf, P.K., Conn, E.E., Eds.; Secondary Plant Products. Academic Press: London, 1981; Vol, 7, 527–556.

20. Lewis, N.G.; Yamamoto, E. Lignin: Occurrence, biogenesis and biodegradation. Annu. Rev. Plant Physiol. Mol. Biol. **1990**, *41*, 455–496.

21. Davin, L.B.; Lewis, N.G. Phenylpropanoid metabolism. In *Biosynthesis of Monolignols, Lignans and Neolignans, Lignins and Suberins*. Stafford, H.A., Ibrahim, R.K., Eds.; Plenum Press: New York, 1992; 325–375.

22. Whetten, R.W.; Mackay, J.J.; Sederoff R.R. Recent advances in understanding lignin biosynthesis. Annu. Rev. Plant Physiol. Mol. Biol. **1998**, *49*, 585–610.

23. Fulcher, A.F.; Ranney, T.G.; Burton, J.D.; Wallgenbach, J.F.; Danehower, D.A. Role of foliar phenolics in host plant resistance of *Malus taxa* to adult Japanese beetles. Hort. Sci. **1998**, *33*, 862–865.

24. Bi, J.I.; Felton, G.W.; Murphy, J.B.; Howles, P.A.; Dixon, R.A.; Lamb, C.J. Do plant phenolics confer resistance to specialist and generalist insect herbivores? J. Agric. Food Chem. **1997**, *45*, 4500–4504.

25. Haggstrom, H.E. Variable plant quality and performance of the willow-feeding leaf beetle. *Galerucella lineola* Silvestria. Swedish University of Agricultural Sciences, Uppsala: Sweden; 1997.

26. Scutareanu, P.; Schoffelmeer, E. A preliminary experiment to detect induced phenolic compounds in *Psylla* infested pear trees. 46th Int Symp Crop Prot, Gent, Belgium, 3 May 1994.

27. Duffey, S.S.; Stout, M.J. Antinutritive and toxic components of plant defense against insects. Arch. Insect. Biochem. Physiol. **1996**, *32*, 3–37.

28. Huang, Y.; Ho, S.H. Toxicity and antifeedant activities of cinnamaldehyde against the grain storage insects, *Tribolium castaneum* (Herbst) and *Sitophilus zeamais* Motsch. J. Stored Prod. Res. **1998**, *34*, 11–17.

29. Bergvinson, D.J.; Arnason, J.T.; Hamilton R.T. Phytochemical changes during recurrent selection for resistance to the European corn borer. Crop. Sci. **1997**, *37*, 1567–1572.

30. Bergvinson, D.J.; Arnason, J.T.; Mihm, J.A.; Jewell, D.C. Phytochemical basis for multiple borer resistance to maize. Proc Int Symp CIMMYT, Mexico, 27 Nov–3 Dec, 1994.

31. Arnason, J.T.; Conilh-de-Beyssac, B.; Philogene, B.J.R.; Bergvinson, D.J.; Serratos, J.A.; Mihm J.A. Mechanisms of resistance in maize grain to the maize weevil and the larger grain borer. Proc. Int. Symp. CIMMYT, Mexico, 27 Nov–3 Dec, 1994.

32. Serratos, J.A.; Blanco-Labra, A.; Arnason, J.T.; Mihm J.A. Genetics of maize grain resistance to maize weevil. Proc. Int. Symp. CIMMYT, Mexico, 27 Nov–3 Dec, 1994.

33. Norton, R.A.; Dowd P.F. Effect of steryl cinnamic acid derivatives from corn bran on *Aspergillus flavus*, corn earworm larvae and dried fruit beetle larvae and adults, J. Agric. Food Chem. **1996**, *44*, 2412–2416.

34. Arnason, J.T.; Gale, J.; Conilh-de-Beyssac, B.; Sen, A.; Miller, S.S.; Philogene, B.J.R.; Lambert, J.D.H.; Fulcher, R.G.; Serratos, A.; Mihm J. Role of phenolics in resistance of maize to the stored grain insects,

Prostephanus truncatus (Horn) and *Sitophilus zeamais* (Motsch). J. Stored Prod. Res. **1992**, *28*, 119–126.

35. Snook, M.E.; Widstrom, N.W.; Wiseman, B.R.; Gueldner, R.C.; Wilson, R.L.; Himmelsbach, D.S.; Harwood, J.S.; Costell C.E. In *Bioregulators for Crop Protection and Pest Control*; Hedin, P.A., Ed.; ACS Symposium Series 1994; *557*, 122–134.

36. Wiseman, B.R.; Snook, M.E.; Isenhour, D.J.; Mihm, J.A.; Widstron N.W. J. Econ. Entomol. **1992**, *85*, 2473–2479.

37. Haribal, M.; Renwick, J.A.A. Isovitexin 6-*O*-β-D-glucopyranoside. A feeding deterrent to *Pieris napi oleracea* from *Alliaria petiolata*. Phytochemistry **1998**, *47*, 1237–1240.

38. Hedin, P.A.; Miles, L.R.; Thompson, A.C.; Minyard J.P. Constituents of a cotton bud: formulation of a boll weevil stimulant mixture. J. Agric. Food Chem. **1968**, *16*, 505–513.

39. Honda, K. Flavanone glycosides as oviposition stimulants in a papilionid butterfly, *Papilio protenor*. J. Chem. Ecol. **1986**, *12*, 1999–2010.

40. Feeny, P.; Sachdev, K.; Rosenberry, L.; Carter, M. Luteolin-7-*O*-(6''-*O*-malonyl)-β-D-glucoside and trans-chlorogenic acid: oviposition stimulants for the black wallowtail butterfly. Phytochemistry **1988**, *27*, 3439–3448.

41. Tahavanainen, J.; Julkunen-Tiitto, R.; Kettunen, J. Phenolic glycosides govern the food selection pattern of willow-feeding leaf beetles. Oecologia. **1985**, *67*, 52–56.

42. Smiley, J.T.; Horn, J.M.; Rank, N.R. Ecological effects of salicin at three trophic levels: new problems from old adaptations. Science **1985**, *229*, 649–651.

43. Elliger, C.A.; Chan, B.C.; Waiss, A.C. Flavonoids as larval growth inhibitors: structural factors governing toxicity. Naturwissenschaften. **1980**, *67*, 358–360.

44. Isman, M.B.; Duffey S.S. Toxicity of tomato phenolic compounds to the fruitworm, *Heliothis zea*. Entomol. Exp. Appl. **1982**, *31*, 370–376.

45. Hedin, P.A.; Jenkins, J.N.; Collum, D.H.; White, W.H.; Parrott, W.L.; MacGown, M.W. Cyanidin-3-β-glucoside, a newly recognized basis for resistance in cotton to the tobacco budworm *Heliothis virescens* (Lepidoptera: Noctuidae). Experientia **1983**, *39*, 799–801.

46. Lindroth, R.L. Hydrolysis of phenolic glycosides by midgut β-glucosidases in *Papilio glaucus* subspecies. Insect Biochem. **1988**, *18*, 789–792.

47. Clausen, T.P.; Reichardt, P.B.; Bryant, J.P. A simple method for the isolation of salicortin, tremulacin and tremuloidin from quaking aspen (*Populus tremuloides*). J. Nat. Prod. **1989**, *52*, 207–209.

48. Roth, S.; Knorr, C.; Lindroth, R.L. Dietary phenolics affects performance of the gypsy moth (Lepidoptera: Lymantriidae) and its parasitoid *Cotesia melanoscela* (Hymenoptera: Braconidae). Environ. Entomol. **1997**, *26*, 668–671.

49. Berenbaum, M.R. Coumarins. In *Herbivores: Their Interactions with Secondary Plant Metabolites*, Rosenthal, G.A., Berenbaum, M.R., Eds.; Academic Press: New York, 1991; Vol. 1, 221–249.

50. Arnason, J.T.; Philogene, B.J.R.; Towers, G.H.N. Phototoxins in plant–insect interactions. In *Herbivores: Their Interactions with Secondary Plant*

Metabolites; Rosenthal, G.A., Berenbaum, M.R., Eds.; Academic Press: New York, 1992; 317–341.

51. Ian-de-Veau, E.J., Schultsz, J.C. Reassessment of interaction between gut detergents and tannins in Lepidoptera and significance for gypsy moth larvae. J. Chem. Ecol. **1992**, *18*, 1437–1453.

52. Desroches, P.; El-Shazly, E., Mandon, N.; Due, G.; Huignard, J. Development of *Callosobruchus chinensis* (L.) and *C. maculatus* (F.) (Coleoptera: Bruchidae) in seeds of *Vicia faba* L. differing in their tannin, vicine and convicine contents. J. Stored Prod. Res. **1995**, *31*, 83–89.

53. Bourchier, R.S.; Nealis, V.G. Development and growth of early and late instar gypsy moth (Lepidoptera: Lymantriidae) feeding on tannin supplemented diets. Annu. Entom. Soc. Am. **1993**, *22*, 642–646.

54. Mansour, M.H.; Zohdy, N.M.; El-Gengaihi, S.F.; Amr, A.E. The relationship between tannins concentration in some cotton varieties and susceptibility to piercing sucking insects. J. Appl. Entomol. **1997**, *121*, 321–325.

55. Wongo, L.E. Biological activity of sorghum tannin extracts on the stored grain pests *Sitophilus oryzae* (L.), *Sitotroga cerealella* (Oliver) and *Tribolium castaneum* (Herbst). Insect. Sci. Appl. **1998**, *18*, 17–23.

56. Wiesen, B.; Krug, E.; Fieldler, K.; Wray, V.; Proksch, P. Sequestration of host–plant–derived flavonoids by lycaenid butterfly, *Polyommatus icarus*. J. Chem. Ecol. **1994**, *20*, 2523–2538.

57. Sharma, H.C.; Norris, D.M. Chemical basis of resistance in soya bean to cabbage looper, *Trichoplusia ni*. J. Sci. Food Agric. **1991**, *55*, 353–364.

58. Shaver, T.N.; Lukefahr, M.J. Effect of flavonoid pigments and gossypol on growth and development of bollworm, tobacco budworm and pink bollworm. J. Econ. Entomol. **1969**, *62*, 643–646.

59. Harborne, J.B. Flavonoid pigments. In *Herbivores: The Chemical Participants*; Rosenthal, G.A., Berenbaum, M.R., Eds.; Academic Press: New York, 1991; Vol. 1, 389–429.

60. Schoonhoven, L.M. Chemical mediators between plants and phytopha-gous insects. In *Semiochemicals: Their Role in Pest Control*; Nordlund, D.A., Jones, R.L., Lewis, W.J., Eds.; John Wiley and Sons: New York, 1981; 31–50.

61. Gikonyo, N.K.; Mwangi, R.W.; Midiwo, J.O. Toxicity and growth inhibitory activity of *Polygonum senagalense* (Meissn.) surface exudate against *Aedes aegypti* larvae. Insect. Sci. Appl. **1998**, *18*, 229–234.

62. Neupane, F.P.; Norris, D.M. Iodoacetic acid alteration of soybean resistance to the cabbage looper. Environ. Entomol. **1990**, *19*, 215–221.

63. Johnson, K.S.; Soriber, J.M.; Nair, M. Phenylpropanoid phenolics in Sweetbay mangolia as chemical determinants of host use in saturniid silkmoths (Callosamia). J. Chem. Ecol. **1996**, *22*, 1955–1969.

64. Pierpoint, W.S. Relations of phenolic compounds with proteins and their relevance in the production of leaf protein. In *Leaf Protein Concentrates*; Felek, I., Graham, H.D., Eds.; Avi. Greenwich, C.T, 1983, 235–267.

65. Felton, G.W.; Donato, K.K.; Broadway, R.M.; Duffey, S.S. Impact of oxidized plant phenolics on the nutritional quality of dietary protein to a noctuid herbivore. J. Insect Physiol. **1992**, *38*, 277–285.

66. Appel, H.M. Phenolics in ecological interaction: the importance of oxidation. J. Chem. Ecol. **1993**, *19*, 1521–1552.

67. Arnason, J.T.; Towers, G.H.N.; Philogene, B.J.R.; Lambert, J.D.H. The role of natural photosensitizers in plant resistance to insects. In *Plant Resistance to Insects*; Hedin, P.A., Ed.; ACS Symposium Series. American Chemical Society, Washington, 1983, Vol. 208, 139–151.

68. Gray-Mitsumune, M.; Molitor, E.K.; Cukovic, D.; Carlson, J.L.; Douglas, C.J. Developmentally regulated patterns of expression directed by poplar PAL promoters in transgenic tobacco and poplar. Plant Mol. Biol. **1999**, *39*, 657–669.

69. Gui, Y.; Bell, A.A.; Joost, O.; Magill, C. Expression of potential defense response genes in cotton. Physiol. Mol. Plant Pathol. **2000**, *56*, 25–31.

70. Dixon, R.A.; Steele, C.L. Flavonoids and isoflavonoids—a gold mine for metabolic engineering. Trends Plant Sci. **1999**, *4*, 394–400.

71. Jung, W.; Yu, O.; Lau, S.M.C.; O'Keefe, D.P.; Odell, J.; Fader, G.; McGonigle, B. Identification and expression of isoflavone synthase, the key enzyme for biosynthesis of isoflavones in legumes. Nature Biotechnol. **2000**, *18*, 208–212.

72. Robbins, M.P.; Bavage, A.D.; Strudwicke, C.; Morris, P. Genetic manipulation of condensed tannins in higher plants II. Analysis of birdsfoot trefoil plants harboring antisense dihydroflavonol reductase constructs. Plant Physiol. **1998**, *116*, 1133–1144.

73. Padmavati, M.; Reddy, A.R. Flavonoid biosynthetic pathway and cereal defense response: an emerging trend in crop biotechnology. J. Plant Biochem. Biotechnol. **1999**, *8*, 15–20.

10

Terpenoids

10.1 INTRODUCTION

According to chemical theories of the ancient Greeks, all matter was made up of four elements: earth, air, fire and water. The Pythagoreans and Aristotle added a fifth element, ether, of which the heavenly bodies are supposedly composed. In their search for a fifth element, alchemists made extensive investigations of the volatile oils of herbs and spices. These oils are still referred to as essential oils (1).

In the development of organic chemistry during the 1800s, plant secondary products played a central role. Among these were rubber and oil of turpentine. Dry distillation of rubber yielded a compound with an empirical formula C_5H_8. In 1835 this same compound was obtained by pyrolysis of turpentine by Himley, a student of Wohler. A number of compounds containing 10, 15, 20, or more carbon atoms have also been isolated from turpentine oil. Later such compounds have been found in many plants and hence the name terpenoids was given for these groups of compounds.

The name *isoprene* was coined for C_5H_8 by Williams in 1860. The structure of isoprene produced by pyrolysis of oil of turpentine was elucidated by Hlasiwetz and Hinterberger in 1868. In 1897, Ipatiew and Wittorf gave conclusive evidence for the correct structure of isoprene. An unambiguous synthesis of isoprene was finally carried out by Euler in 1898. Wallach, in 1887, proposed the "isoprene rule" that many natural products can be regarded as being composed of isoprene units. This rule proved extremely valuable in predicting the structures of unknown plant products. All of these products were shown to be synthesized from a biologically "active isoprene" (2). The terms *isoprenoid* and *terpenoid* are used interchangeably. Some authors have used *terpene* to refer only to hydrocarbons based on an integral number of C_5 units and *terpenoid* or *isoprenoid* to designate the whole class of compounds (hydrocarbons and oxygenated compounds) based on an integral number of C_5 units. Ruzicka et al. (2) suggested that these compounds should be called terpenes and that the term terpenoid should be reserved for compounds such as the steroids which have varying numbers of carbon atoms but are derivable from a $(C_5)_n$ structure. Nes and Mckean (3) used the term *isopentenoid* to describe the whole group. Since the term terpenoid is more popular than isoprenoid, it is used in this chapter to denote the whole group of compounds arising from isoprene units.

Terpenoids have been known since antiquity as ingredients of perfumes, soaps, flavorings, and food colorants. The development of chromatographic and spectroscopic techniques has led to the general understanding of the structure and biosynthesis of terpenoids. The series containing structural and spectroscopic details of individual monoterpenoids, diterpenoids, and triterpenoids were edited by Dev (4–6). The collation, a three-volume *Dictionary of Terpenoids* by Connolly and Hill (7), contains around 22,000 structures excluding steroids which are described separately in the *Dictionary of Steroids* (8). The structure and occurrence of terpenoids and steroids in plants have been reviewed by Bonner (9), Devon and Scott (10), Nicholas (11), and Robinson (12). The book on essential oil by Guenther (13) is a valuable reference book.

10.2 CLASSIFICATION

10.2.1 General

The terpenoids are composed of isoprene (C_5) units and the nomenclature of the major classes reflects the number of isoprene units present (Table 10.1). Various terpenoids are formed by condensation, cyclization, rearrangements, addition, or deletion of carbon atoms from linear arrangements of isoprene units.

Plant terpenoids comprise a structurally diverse group of compounds that can be divided into classes of primary and secondary metabolites. Terpenoids that are primary metabolites include sterols, carotenoids,

TABLE 10.1 Major Classes of Terpenoids Found in Plants

Major class	Number of carbon atoms	Parent compound	Abbreviations
Hemiterpenoids	5	Isopentenyl pyrophosphate	IPP
Monoterpenoids	10	Geranyl pyrophosphate	GPP
Sesquiterpenoids	15	Farnesyl pyrophosphate	FPP
Diterpenoids	20	Geranyl geranyl pyrophosphate	GGPP
Sesterterpenoids	25	Geranyl farnesyl pyrophosphate	GFPP
Triterpenoids	30	Squalene	
Tetraterpenoids	40	Phytoene	
Polyprenols	60–100	Geranyl pyrophosphate and IPPs	
Polyterpenes	>40	Geranyl pyrophosphate and IPPs	

growth regulators, and the polyprenol substituents of dolichols, quinones, and proteins. These compounds are essential for maintaining the integrity of membranes, photoprotection, and anchoring of essential biochemical functions to specific membrane systems. Terpenoids grouped as secondary metabolites include monoterpenoids, sesquiterpenoids, diterpenoids, sester-terpenoids, and phytosteroids. They mediate important interactions between plants and their environment. Specific terpenoids have been correlated with plant–plant, plant–insect, and plant–pathogen interactions (14). In this chapter, the terpenoids grouped as secondary metabolites alone are dealt for the reason that some are known to fight against insect attack in plants.

10.2.2 Monoterpenoids

Monoterpenoids have been classified according to their functional groups as well as based on their linear or cyclic nature. They can be divided into five broad structural categories (15–18):

1. Acyclic
2. Monocyclic
   a. Other than cyclohexanoid
   b. Cyclohexanoid
3. Bicyclic

4. Tricyclic
5. Irregular

The first four groups arise from a head-to-tail condensation of isopentenylpyrophosphate (IPP). The irregular monoterpenoids are formed by a head-to-middle joining of IPP units. Structure of some important monoterpenoids are shown in Fig. 10.1.

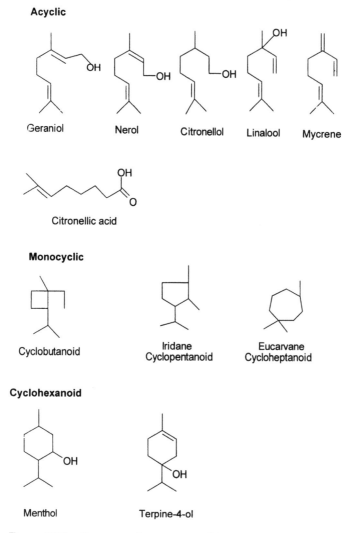

FIGURE 10.1   Structure of monoterpenoids.

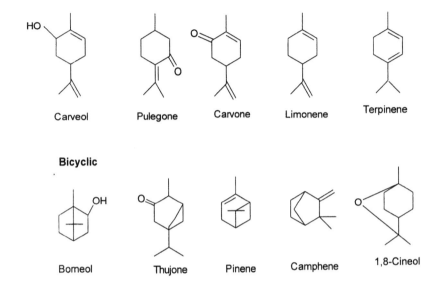

Carveol   Pulegone   Carvone   Limonene   Terpinene

Bicyclic

Borneol   Thujone   Pinene   Camphene   1,8-Cineol

Irregular monoterpenes

Artimisane   Fenchane

FIGURE 10.1   Continued.

10.2.2.1   Important Monoterpenoids

   i. Acyclic
      alcohol   : geraniol, linalool
      aldehydes   : citronellal, citral
      acids   : citronellic acid
      hydrocarbons : myrcene
  ii. Monocyclic
      a) other than cyclohexanoid
      cyclobutanoid : cyclobutanoid
      cyclopentanoid : iridane, loganin
      cycloheptanoid : eucarvane

b) Cyclohexanoid

alcohol	:	menthol, terpineol, 4-carvomenthenol, carvacrol, thymol, isopiperitenol
ketone	:	menthone, pulegone, carvone
aldehydes	:	perillaldehyde
hydrocarbons	:	*p*-cymene, limonene, α-terpinene, γ-terpenene, terpinolene, p-menthane

iii. Bicyclic

alcohol	:	verbenol, borneol, thujyl alcohol
ketone	:	thujone, fenchone, verbenone, camphor, *d*-sabinone
hydrocarbons	:	α-pinene, β-pinene, camphene
ether	:	1,8-cineole

iv. Tricyclic

alcohol	:	teresantalol

v. Irregular

Hydrocarbons	:	artimisane, santolinane, chrysanthemane, fenchane, isocamphane, tropane, lavandulane

10.2.3   Sesquiterpenoids

About 200 different carbon skeletons containing up to four carbocyclic rings are found in this terpenoid group and several thousand sesquiterpene compounds have been identified (18,19). They are the largest class of terpenoids with a wide structural diversity. A few important skeletal types are given in Fig. 10.2. Most of these are constructed by the typical head-to-tail condensation of isoprene units.

i. Acyclic

alcohol	:	farnesol, nerolidol
hydrocarbon	:	β-farnesene

ii. Monocyclic

alcohol	:	α-bisabolol
hydrocarbon	:	α-curcumene, lanceol, bisabolene
ketone	:	zerumbone, germacrone

iii. Bicyclic

hydrocarbon	:	sesquithujene, α-bergamotene, cuparene, α-murolene, caryophyllene
ketone	:	campherenone, β-vetivone, eremophilone
alcohol	:	eudesmol

Acyclic

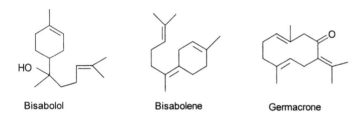

Farnesol   Nerolidol

Monocylic

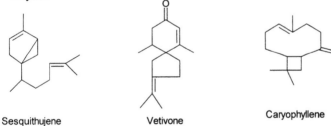

Bisabolol   Bisabolene   Germacrone

Bicyclic

Sesquithujene   Vetivone   Caryophyllene

FIGURE 10.2   Structure of sesquiterpenoids.

iv. Tricyclic
      hydrocarbon : sativene, longifolene
      alcohol   : cedrol, culmorin, trichothecin,
                      patchouli alcohol

10.2.4 Diterpenoids

Diterpenoids are classified on a biogenetic basis (1,14). There are a few open-chain diterpenoids. Phytol, a linear diterpene, is found as a side chain on the chlorophyll molecule. Most of the diterpenoids are cyclic compounds (Fig. 10.3).

Acyclic

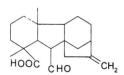

Phytol

Monocyclic

Vitamin A (Retinol)

Bicyclic   **Tricyclic**

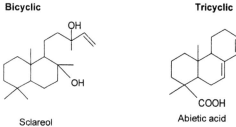

Sclareol   COOH
                                                Abietic acid

Tetracyclic

HOOC   CHO   CH₂

GA₁₂ - aldehyde (Gibberellin)

FIGURE 10.3   Structure of diterpenoids.

The important groups of diterpenes include

1.	Linear	: Phytol
2.	Monocyclic	: Vitamin A
3.	Bicyclic	: Sclareol
4.	Tricyclic	: Abietic acid
5.	Aromatic	: Carnosic acid, totarol
6.	Tetracyclic	: Gibberellins
7.	Macrocyclic	: Cembrene, sinulariolide

10.2.5 Sesterterpenoids

Six major classes are known (20). There are linear as well as compounds with five different ring structures. The structures of compounds for each class is shown in Fig. 10.4.

1. Linear: Eicosadien-1-ol, geranylnerolidol, geranyl farnesol, ircinin I and II, variablin

Linear

Geranyl farnesol

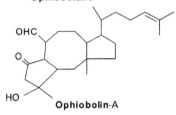

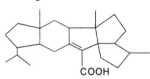

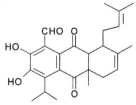

Heliocide H1

FIGURE 10.4 Structure of sesterterpenoids.

2. Ophiobolane type: Ophiobolin A, B, C, D, F, I, ceroplastol I, II, albolic acid,
3. Chelianthatriol type: Scalarin, chelianthatriol
4. Retigeranic acid type: Retigeranic acid
5. Gascardic acid type: Gascardic acid
6. Heliocides: Heliocides H1 to H4

10.2.6 Triterpenoids and Steroids

The triterpenoids are classified based on the linearity or number of cyclic ring systems present (22–24). The structures of compounds for each class is shown in Fig. 10.5.

1. Acyclic   : Squalene
2. Monocyclic: Presqualene alcohol
3. Bicyclic   : Lansic acid
4. Tricyclic   : Malabaricanediol
5. Tetracyclic : Azadirachtin
6. Pentacyclic : Cycloartanol
7. Hexacyclic : Phyllanthol

10.3 OCCURRENCE

The terpenoids in general are very widely distributed throughout the whole plant kingdom. All green plants have the ability to produce linear terpenoids such as phytol. Terpenoids containing more than five isoprene units appear in all plants and the less complex terpenoids (C_{10}–C_{25}) are mainly restricted to the Tracheophyta, although sesquiterpenoids have been found widely in the Bryophyta and the fungi.

10.3.1 Hemiterpenoids

There are relatively few naturally occurring hemiterpenoid compounds. Hemiterpenoids are only present in very small amounts and hence escapes detection. The development of sophisticated instruments such as GC-MS led to the identification of very minute quantities of hemiterpenoids. Isoprene occurs abundantly as a natural emission from the leaves of many plant species. Isopentenol and 3,3-dimethylallyl alcohol are also found in plants (14).

10.3.2   Monoterpenoids

Monoterpenoids are very widely distributed in the plant kingdom. The orders Rutales, Cornales, Lamiales, and Asterales are the most prolific in their production of monoterpenoids. The distribution of monoterpenoids is

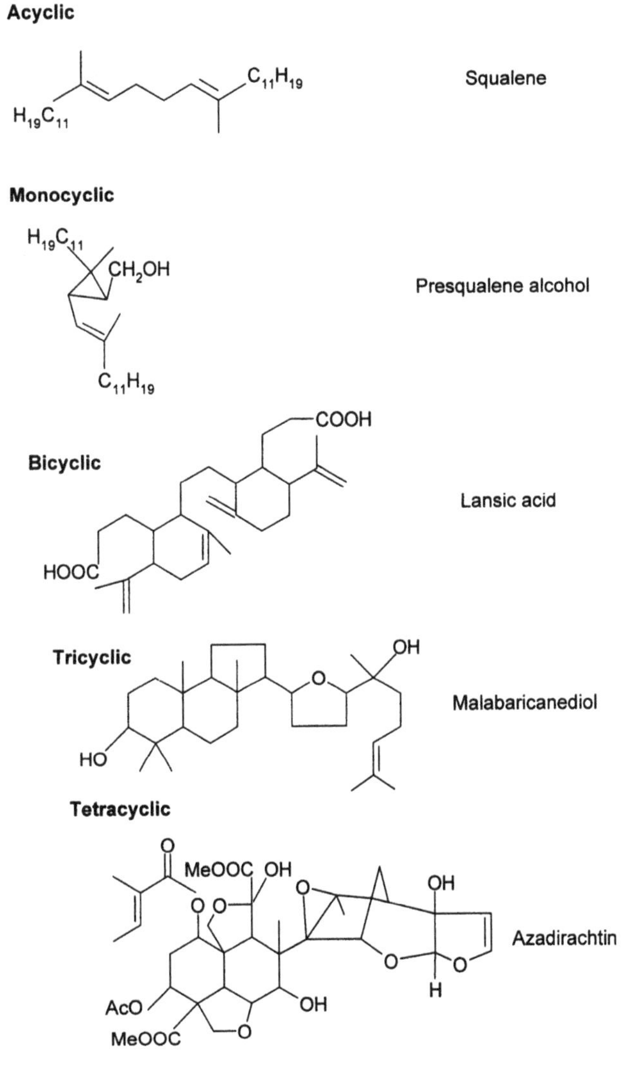

Acyclic

Squalene

Monocyclic

Presqualene alcohol

Bicyclic

Lansic acid

Tricyclic

Malabaricanediol

Tetracyclic

Azadirachtin

FIGURE 10.5   Structure of triterpenoids.

Pentacyclic

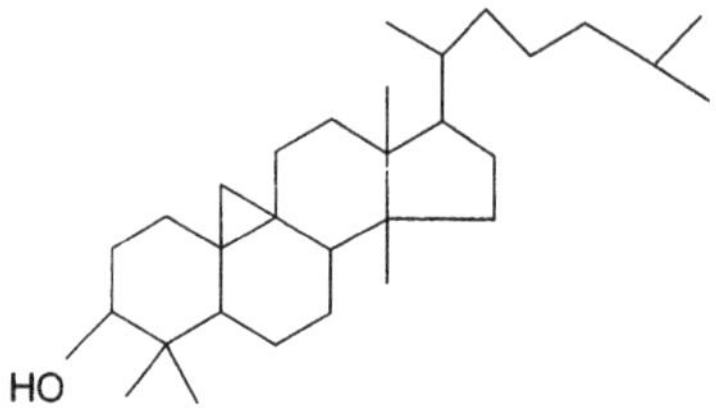

Cycloartanol

Hexacyclic

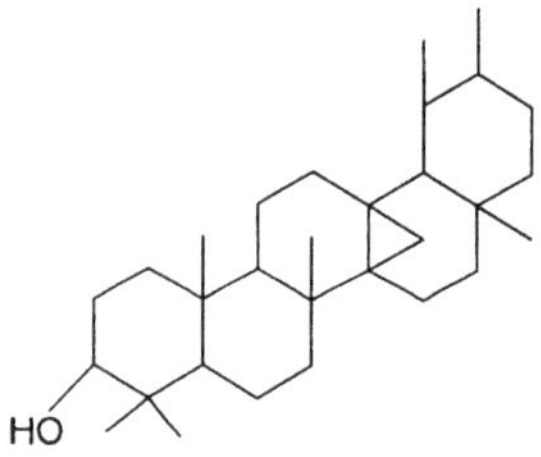

Phyllanthol

FIGURE 10.5   Continued.

not uniform throughout the orders. Four of the eight families of the order Rutales contain monoterpenoids, and only members of the Apiaceae (Umbelliferae) of the ten families in the order Cornales are abundant in monoterpenoids (14,21).

Monoterpenoids are synthesized and stored in globules within cells, in specialized ducts and glands, or in dead cells. Synthesis and distribution of the terpenoids are often restricted to tissues that are the most susceptible to attack by insects or pathogens. Typically young lateral leaves contain higher concentrations of monoterpenoids than main-stem leaves. Bottom main-stem leaves often contain the lowest concentration. The physical structures such as glandular hairs and leaf glands secrete a protective monoterpenoid that repels herbivores. The quantity of terpenoids produced is affected by climatic and edaphic conditions (21).

10.3.3   Sesquiterpenes

Hydrocarbons, alcohols, ketones, and other derivatives of sesquiterpenes occur along with monoterpenoids in essential oils. Nonvolatile glycosides, esters, polar polyhydroxylactones, and sesquiterpene alkaloids also occur widely in plants. Halogenated sesquiterpenes occur in marine algae. Sesquiterpene lactones are abundant in umbelliferae. The orders Magnoliales, Rutales, Cornales, and Asterales contain large number of sesquiterpenoids (10,21).

10.3.4   Diterpenoids

Diterpenoids are principally of higher plant origin. Their presence is also seen in fungi. The important compounds include gymnosperm resin acids and the plant hormones gibberellins. Diterpenoids are common in the order derived from the Gentianales, Lamiales, and the Asterales. Fabales and the Geraniales are also important sources (14,21). Macrocyclic diterpenes such as cembranes have been isolated from *Pinus* spp. Marine organisms are also good sources for a number of diterpenoids. Among the simpler derivatives are crinitol and oxicrinol obtained from the brown alga *Cystoseria crinita*. Bicyclic diterpenoids occur in the oleoresin of *Larix* species. The phenol sugiol and a number of bicyclic relatives of agathic acid have been obtained from *Araucaria angustifolia*.

10.3.5   Sesterterpenoids

Sesterterpenoids are a relatively recent addition to the terpenoid family. They have been isolated and identified from higher plants, fungi, insects, and sponges. Cotton contains a group of sesterterpenoids known as heliocides. Plant pathogenic fungi appear to be prime sources of sesterterpenoids (14,20,21).

10.3.6   Triterpenoids and Steroids

Traditionally C_{30} compounds were viewed as triterpenoids and compounds with cyclopentanoperhydrophenathrene ring structure were regarded as steroids (21,22). Free triterpenoids occur often as components of resins, latex, and cuticle. Sterols have been isolated from a large number of species and occur in all angiosperms and gymnosperms. Phytoecdysteroids were found in lower as well as in higher plants, in annuals or perennials. They have been found occasionally to accumulate in roots, leaves, flowers, fruits, or seeds. Ecdysteroids have been found in about 80 plant families.

10.4   BIOSYNTHESIS

10.4.1   General

The biosynthetic pathway of terpenoids was elucidated originally from study on sterols (chloesterol) in animals. In plants, the branching of main pathway occurs to yield the terpenoids characteristic of a particular species (14–19). The major discovery of the pathway was the identification of mevalonic acid (MVA), a C_6 compound. Then the "active isoprene" isopentenyl pyrophosphate, synthesized from mevalonic acid, was identified. This pathway is called the acetate-mevalonate pathway.

The key enzyme involved in this pathway is hydroxymethylglutaryl coenzyme A reductase (HMG-CoA reductase) which catalyzes the formation of MVA from hydroxymethylglutaryl coenzyme A (HMG-CoA). The reaction involves two molecules of NADPH per MVA produced (Scheme 10.1).

The conversion of MVA to IPP occurs in three steps. Each step requires one mole of ATP per mole of substrate. The IPP is then isomerized to dimethylallylpyrophosphate (DMAPP) by the IPP isomerase (Scheme 10.2). Formation of DMAPP is essential for the condensation of IPP with DMAPP. DMAPP acts as the prenyl donor to a molecule of IPP by a head-to-tail condensation reaction producing geranyl pyrophosphate (GPP). Further addition of IPP to geranylpyrophosphate occurs to form farnesyl pyrophosphate (FPP) and geranylgeranylpyrophosphate (GGPP) (Scheme 10.3). This condensation of allylic pyrophosphates with IPP produces the higher prenylpyrophosphates. The formation of all-*trans* prenylpyrophosphates is catalyzed by the prenyltransferases. The head-to-tail condensation occurs within the main pathway. However the formation of tri- and tetraterpenoids involves the head-to-head condensation of FPP and GGPP to form squalene and phytoene, respectively (Scheme 10.4).

10.4.2   Monoterpenoids

The role of GPP as the precursor of monoterpenoids was established in the 1960s. The classical essential oil–bearing plants are unique in possessing specialized secretory structures adapted to synthesize these compounds in large quantities.

The monoterpene cyclases or synthases are involved in the initial ionization of the allylic pyrophosphates with electrophilic attack of the resulting allylic carbocation on a double bond (17,18). More than 20 monoterpenoid cyclases have been identified and purified. The cyclic oxygenated monoterpenoids are formed from the cyclized products by oxidation and reduction, which usually involve cytochrome P450-dependent oxidases (23).

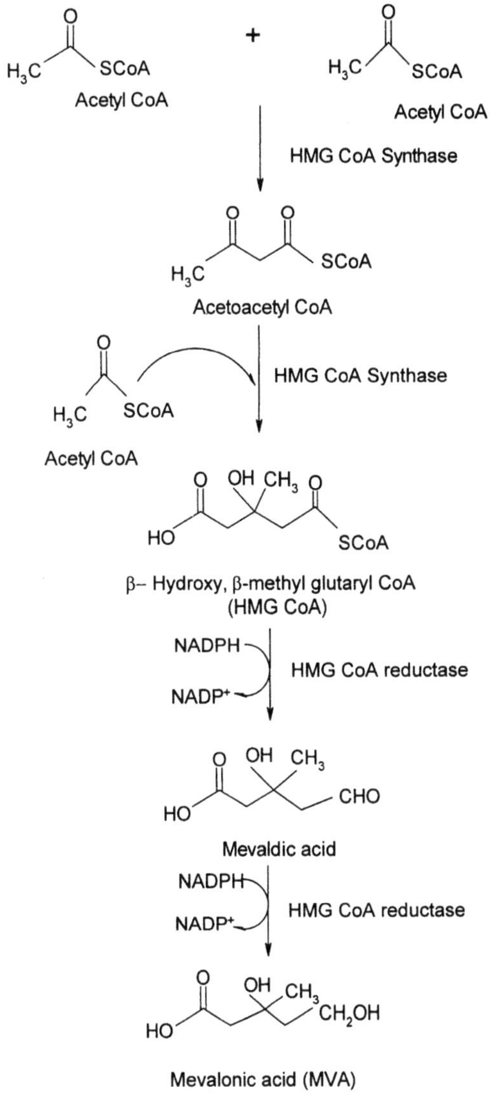

SCHEME 10.1 Synthesis of mevalonic acid.

Different monoterpene synthases catalyze the conversion of GPP to monoterpene olefins. For example, pinene cyclase I converts GPP to the bicyclic (+)α-pinene and (+) camphene; cyclase II forms (−)α-pinene and (−)β-camphene, whereas, cyclase III is involved in the formation of (+)α-pinene, (+)β-pinene, and monocyclic and acyclic olefins.

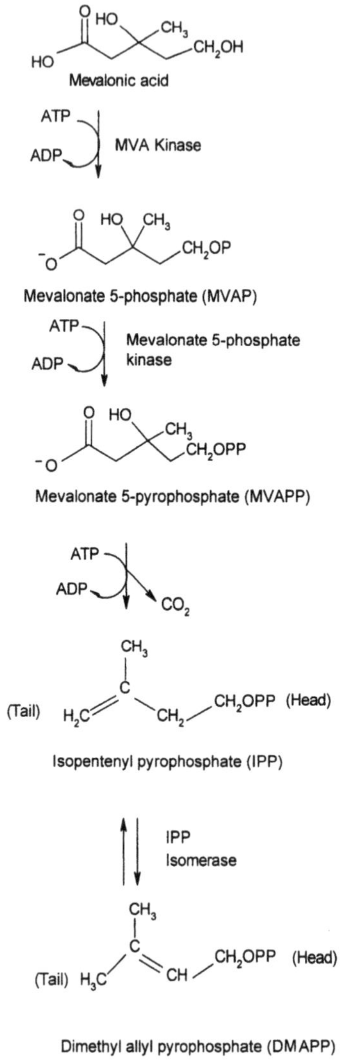

Mevalonic acid

ATP ⏋
          MVA Kinase
ADP ⏌

Mevalonate 5-phosphate (MVAP)

ATP ⏋ Mevalonate 5-phosphate
ADP ⏌ kinase

Mevalonate 5-pyrophosphate (MVAPP)

ATP
ADP
        CO_2

(Tail) H_2C CH_2 CH_2OPP (Head)

Isopentenyl pyrophosphate (IPP)

IPP
Isomerase

(Tail) H_3C CH CH_2OPP (Head)

Dimethyl allyl pyrophosphate (DMAPP)

SCHEME 10.2   Synthesis of IPP and DMAPP.

Evidence for a conjugate reduction was obtained from studies on the biosynthesis of oxygenated monoterpenes in *Mentha piperita* (15). Through labeled olefins (precursors), it was demonstrated that limonene was converted to piperitenone. Terpinolene was much less efficient precursor of the oxygenated products supporting the key role of limonene. The summary

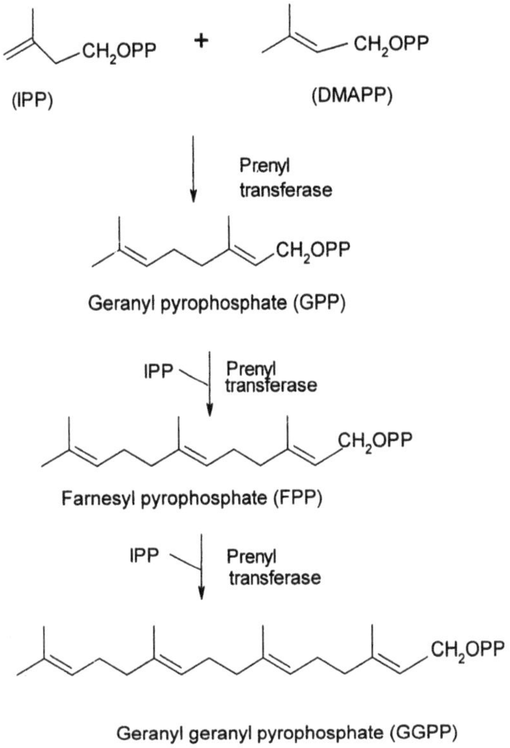

Geranyl pyrophosphate (GPP)

Farnesyl pyrophosphate (FPP)

Geranyl geranyl pyrophosphate (GGPP)

Scheme 10.3 Biosynthesis of GGPP.

of the evidence thus supports the direct synthesis of l-limonene from geranylpyrophosphate, the allylic oxidation of this olefin, the conversion of the product to a conjugated dienone, and the subsequent reduction of both conjugated double bonds as the major features of monoterpene metabolism in *M. piperita*.

Limonene can be hydroxylated at different positions employing different cytochrome P450-dependent hydroxylases (15,23). For example, peppermint (*Mentha piperita*) produces almost exclusively monoterpenoids with oxygen at C-3 (pulegone, menthone, and menthol) (Scheme 10.5), whereas spearmint (*Mentha spicata*) produce those with oxygen at C-6 (carvone, carveol). Limonene synthase is a rate-limiting enzyme involved in monoterpenoid biosynthesis in peppermint. The purified enzymes from peppermint and spearmint have similar basic properties, with a mechanism of action similar to that of other terpenoid cyclases of higher plants.

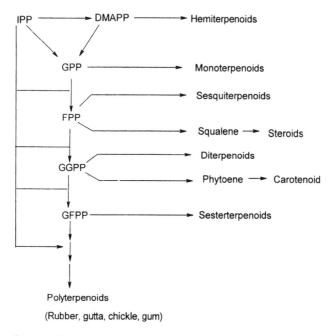

SCHEME 10.4 Pathway of terpenoid biosynthesis.

Nerol, the cis analogue of geraniol, is formed from nerylpyrophosphate (NPP). Nerol and geraniol appear to be necessary intermediates in the formation of cyclic monoterpenes. A common enzyme-bound cyclic intermediate is formed from NPP or GPP. This common intermediate when followed by a proton loss from carbon 5 and a 1,2-hydride shift with expulsion of the enzyme would yield γ-terpinene. A similar mechanism with the loss of a proton from carbon 6 and the formation of a cyclopropene ring could yield α-thujene. There is evidence that γ-terpinene is the precursor of the aromatic monoterpenes p-cymene and thymol (Scheme 10.6). Enzymes involved in secondary transformations of monoterpenes like monoterpenol dehydrogenases, acetyltransferases, and double-bond reductases have been isolated from several species and characterized (18,23).

A nonmevalonate pathway was recently established that was found to occur in plastids. In this pathway, glyceraldehyde-3-phosphate reacts with pyruvate to form isopentenylpyrophosphate with the consumption of three molecules of NADPH and one molecule of ATP with the release of CO_2 (25).

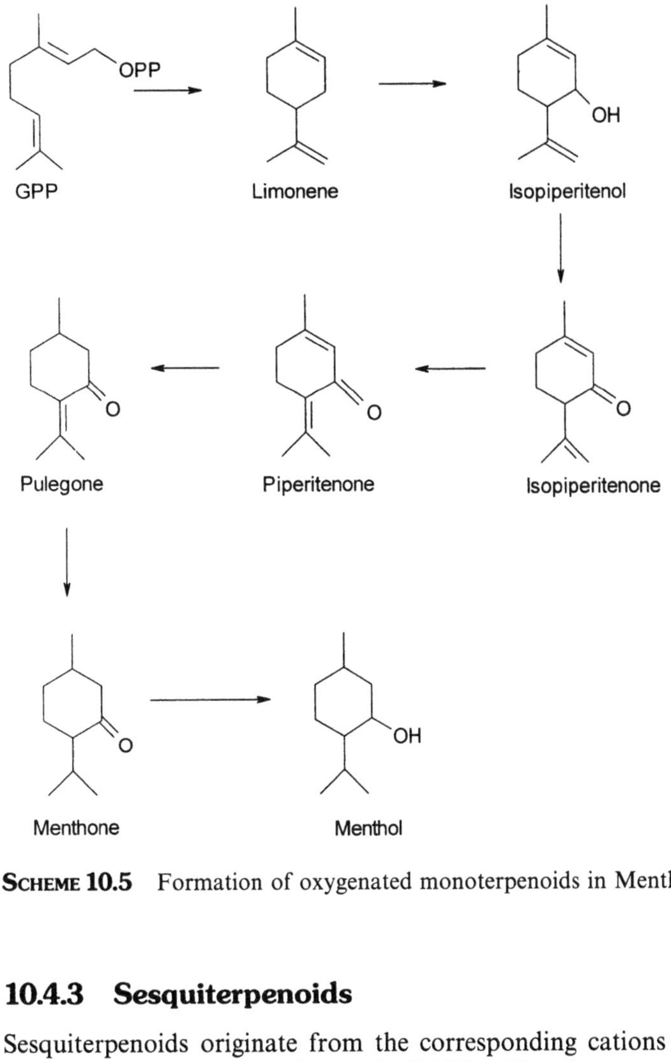

GPP   Limonene   Isopiperitenol

Pulegone   Piperitenone   Isopiperitenone

Menthone   Menthol

Scheme 10.5 Formation of oxygenated monoterpenoids in Mentha piperita.

10.4.3 Sesquiterpenoids

Sesquiterpenoids originate from the corresponding cations of *trans,trans*-farnesol, and *cis, trans*-farnesol (26,27). The former yields 10- and 11-membered ring intermediates (Scheme 10.7) and the latter yields 6- and 7-membered ring intermediates. Transformation of the monocyclic intermediates involve further cyclizations by addition of cationic centers to the remaining double bonds, hydride shifts, and rearrangements. The enzymes responsible for the cyclization of FPP, referred to as sesquiterpene synthases or cyclases, represent reactions committing carbon from the central terpenoid pathway to end products in the sesquiterpene classes. The prenyltransferases catalyze carbon–carbon bond formation between two

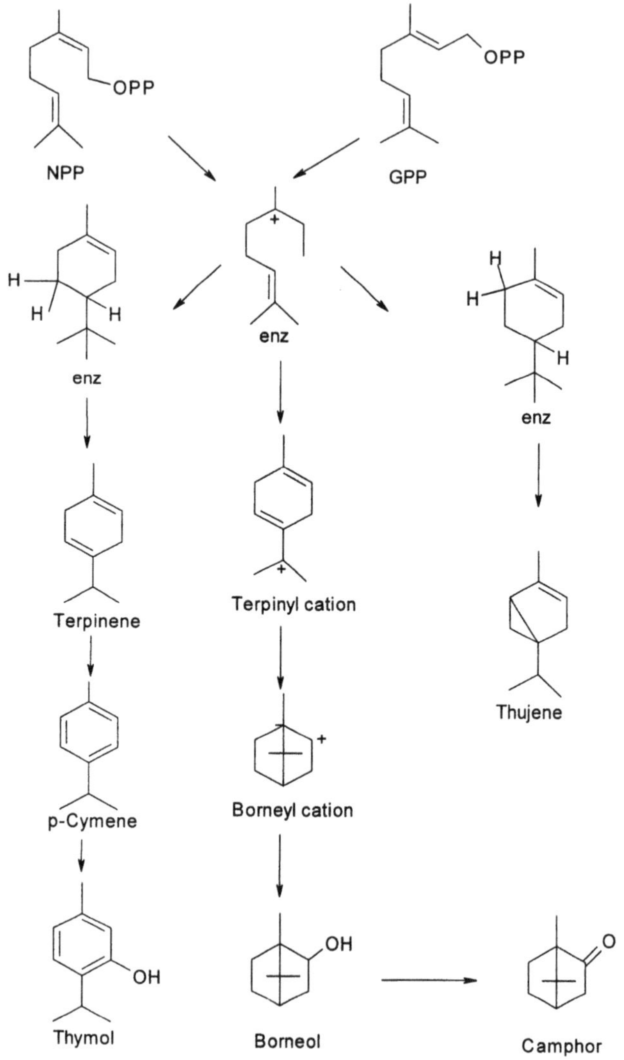

SCHEME 10.6   Biosynthetic pathway of monoterpenoids.

substrate molecules, whereas the cyclases catalyze an intramolecular cyclization. Different prenyltransferases vary only in the length of the allylic substrates they can accept in initiating these reactions. The cyclases are also highly substrate specific. However, different sesquiterpene cyclases can utilize the same substrate to produce dramatically different reaction

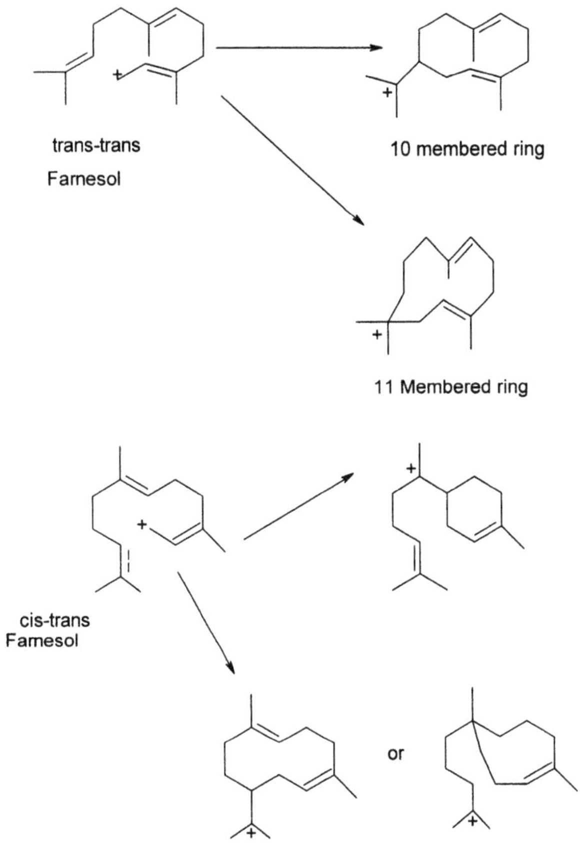

SCHEME 10.7 Hypothetical cyclization scheme for sesquiterpenoids.

products. The FPP synthases partially purified from pumpkin and castor bean endosperm were capable of using either DMAPP or GPP as the allylic diphosphate acceptor (15,16,27). The hypothetical cyclization scheme is summarized in Scheme 10.7. The synthesis of germacrane skeleton is illustrated in Scheme 10.8. The precursor for these group of sesquiterpenes is *trans,trans*-FPP. Stabilization of the intermediate ion by proton loss or addition of a hydroxyl can be noted. A 1,2-hydride shift and a proton loss yields germacrene C (similar to the formation of α-terpinene and γ-terpinene). The bicyclic sesquiterpenes, such as cadinene, originate from *trans,trans*-FPP with a loss of hydrogen from C-1 of the precursor.

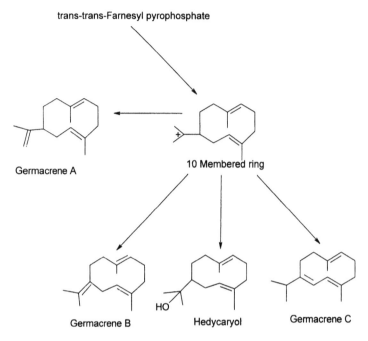

SCHEME 10.8 Biosynthesis of germacrene skeleton.

Acyclic sesquiterpenes are likely to be derived from *trans,trans*-FPP in a manner analogous to the generation of acyclic monoterpenes from GPP (15,27).

Sesquiterpenes are subjected to exhaustive oxidative metabolism. Oxygen may be introduced during cyclization (hedycaryol). Oxygen may also be introduced via allylic methylene or methyl hydroxylation or by epoxidation or hydration of double bonds. Dehydrogenases may be involved in the conversion of alcohols to corresponding ketones, aldehydes, and carboxylic acids. Such derivatives undergo changes to furans and lactones (14,19).

10.4.4 Diterpenoids

Geranylgeranylpyrophosphate is the precursor of the diterpenes (26,28). The biosynthesis of majority of diterpenes is initiated by an electrophilic attack on the terminal double bond. This triggers a series of cyclizations leading to mono-, di-, tri-, and tetracyclic derivatives. The cyclization normally progresses directly to the bicyclic decalin system, which is subsequently discharged by addition of water or proton elimination (Scheme 10.9).

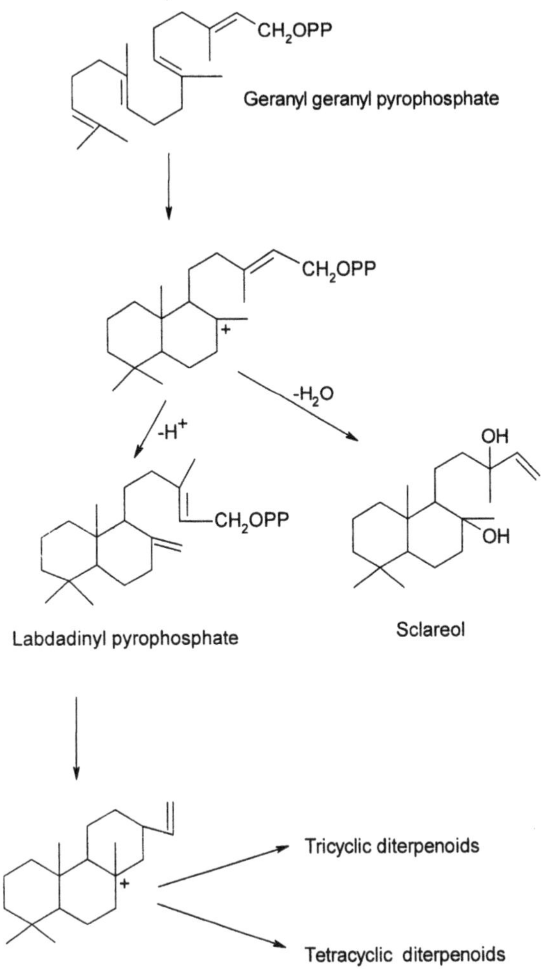

SCHEME 10.9   Biosynthesis of diterpenoids.

Another type of terpene cyclase relies on a proton addition mechanism to generate the first carbocation intermediate needed to initiate the reaction. Examples of cyclases utilizing this mechanism include diterpene synthases. This mechanism relies on proton addition at C-14 of the GGPP molecule to initiate a complex reaction cycle (15–17,28). The initial addition of a proton at C-14 creates a carbocation at C-15; this then attacks electron rich C-10 and results in a new carbocation at C-11. The C-11 carbocation attacks electron-rich C-6 followed by a proton elimination at C-7 to terminate the reaction (Scheme 10.9).

10.4.5   Triterpenoids and Steroids

The linear compounds are derived from the isoprene units linked head to tail or from phytyl residue by addition of a *cis*-isoprene unit. The former proposal is more likely. The squalene synthase reaction is complicated by proceeding in two distinct steps and catalyzing a head-to-head condensation. Prenyltransferases, in contrast, catalyze head-to-tail condensation (19,26,29).

In squalene synthase–catalyzed reaction, the strong electron withdrawing capacity of one diphosphate substituent sufficiently ionizes the C-1′ atom of one allylic substrate, which in turn attacks the electron-rich double bond between C-2 and C-3 of the second allylic diphosphate substrate to form an unusual cyclopropyl intermediate (Scheme 10.10). The reaction intermediate formed in this first step of the reaction is presqualene; it arises concomitant with the loss of one diphosphate group and one proton. Presqualene is converted into the corresponding C_{30} reaction product in

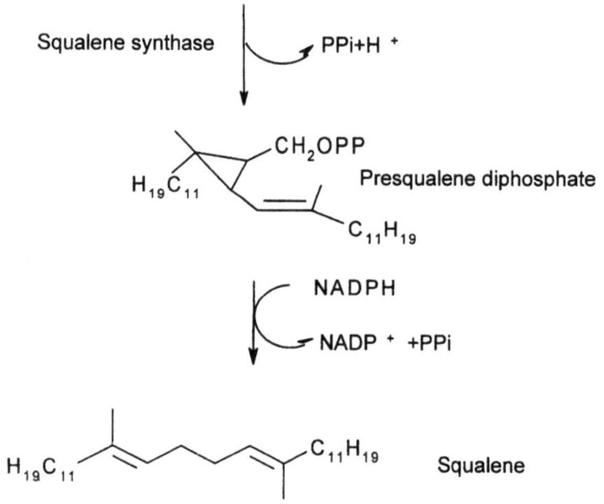

SCHEME 10.10   Biosynthesis of squalene.

a second rate-limiting step of the reaction. The second reaction step entails a complex rearrangement requiring the breakage of the cyclopropyl ring and formation of a new C-1 to C-1' bond. The squalene synthase step requires NADPH. A hypothetical model suggests that classes of terpenoids are synthesized by discrete metabolism channels (26). Entire metabolic channels for sterol and sesquiterpene biosynthesis are localized to the endoplasmic reticulum. Synthesis of other terpenoids, such as carotenoids and long-chain prenols of ubiquinone, require the coordinated activity of cytoplasmic and organellar biosynthetic channels.

For more details on the structure and biochemistry of terpenoids, earlier reviews may be consulted (19,26–29).

10.5   BIOACTIVITY

10.5.1   Monoterpenoids

Monoterpenoids are widely distributed in the essential oils of plants and more than 1000 naturally occurring monoterpenoids have been isolated from higher plants (30). Some of them have shown promise as insect pest control agents because they naturally provide plants with chemical defenses against phytophagous insects. Monoterpenoids may interfere with basic metabolic, biochemical, physiological, and behavioural functions of insects. Some have acute toxicity; others function as repellents, attractants, or antifeedants; and still others affect growth, development, and reproduction.

10.5.1.1   Repellents and Fumigants

Repellents in plants constitute the first barrier against insect attacks. A compound is considered repellent when there is an oriented movement away from the stimulus source.

Individual monoterpenoid or simple mixtures of these compounds function in repelling insect pests and browsing animals. Thus, volatile terpenoids provide the ability to influence another organism at some distance from the source plant, as well as uncommon structural features such as cyclopropane and cyclobutane rings; a lipophilic nature may permit a degree of persistence in the largely aqueous biosphere.

Oil of camphor and citronella were used commercially as insect repellents. Eucalyptus oil, which contains at least 70–85% 1,8-cineole and other high 1,8-cineole oils, are used as insect repellents (31). Pure 1,8-cineole repels many insects, including mosquitoes and American cockroach. Several monoterpenoids from *Artemisia* species are excellent mosquito repellents and terpen-4-ol is the most effective one (32).

The efficacy of 1,8-cineole a major constituent of essential oil of *Ocimum kenyense*, as repellent toxicant and grain protectant against *Sitophilus granarius*, *S. zeamais*, *Tribolium castaneum*, and *Prostephanus truncatus* was investigated in the laboratory using contact toxicity, grain treatment, and repellent assays (33). Whole wheat or maize grains when topically applied or impregnated on filter paper with 1,8-cineole were highly toxic to the tested four beetle species. However, 1,8-cineole was more toxic in grain than on filter paper since the lowest dosage of 0.5 µL/kg controlled all beetles exposed. It evoked strong repellent action against *S. granarius* and *S. zeamais* but was moderately repellent to the other two beetles.

The fumigant activity of some essential oils has been evaluated against a number of stored product insects. The essential oils of *Pogostemon heyneanus*, *Ocimum basilicum*, and *Eucalyptus* have shown insecticidal activity against *Sitophilus oryzae*, *Stegobium paniceum*, *Tribolium castaneum*, and *Callosobruchus chinensis* (34). Fumigant activity and reproductive inhibition induced by a number of essential oils and their monoterpenoids have also been evaluated against the bean weevil *Acanthoscelides obtectus* and the moth *Sitotroga cerealella* (35). The essential oil of *Cymbopogon citratus* has shown its in vitro fumigant activity in the management of storage fungi and insects of cereals without exhibiting mammalian toxicity (36).

Laboratory bioassays were conducted to characterize the acute toxicity of seven monoterpenoids to tracheal mites, *Acarapis woodi* (Rennie), and their honeybee, *Aphis mellifera* L. hosts (37). Citral, thymol, carvacrol, α-terpineol, pulegone, *d*-limonene, and menthol were applied as fumigants to mite-infested honeybees. Thymol and menthol were the most toxic compounds to honeybees and α-terpineol was the least toxic. Menthol, citral, thymol, and carvacrol were more toxic to tracheal mites than to honeybees. Pulegone, *d*-limonene, and α-terpineol were more toxic to honeybees than to tracheal mites. Menthol was 18.9 times more toxic to tracheal mites than to honey bees at the LC$_{50}$ concentrations. However, as the concentration increased, bee mortality increased more rapidly than mite mortality and menthol was only 5.7 times more toxic at the LC$_{50}$ concentrations. Citral and thymol were 2.9 and 2.0 times more toxic to tracheal mites, respectively, at all concentrations estimated.

10.5.1.2 Attractants

Many terpenoids also act as insect attractants. Attractants can be used to lure insects to insecticides or mask repellent properties of insecticides. They may also be used to attract beneficial insects to desired habitats. Essential oils of plants usually contain both attractants and repellents, and the effect of a constituent can vary considerably among insect species. The three terpenoid components of Scotch pine steam distillate were attractants to *Hylobius abietis* and other components were repellent to weevil (38). Geraniol repelled houseflies while attracting honeybees (39). Thus, certain terpenoids might have dual utility as attractants to beneficial insects and repellents to certain insect pests. Plant terpenoids are often metabolic precursors of pheromones synthesized by insects. Myrcene, a common constituent of conifer resin, is converted to the pheromones ipisidienol and ipsenol by beetles of the genus *Ips* that attack pine trees (40).

Botanical pesticides represent an important component of IPM systems in grain storage as they have a broad spectrum of activity based on local materials and are less expensive. Many are ecofriendly and harmless to man and other mammals. Six compounds attractive to the Western balsam bark beetle *Dryocoetes confusus* in laboratory bioassays were isolated by micropreparative GC from steam-distilled phloem oil from the subalpine fir, *Abies lasiocarpa* (Hook) Nutt (41). In the bioassays, α-pinene, *p*-cymene, terpinolene, longifolene, myrtenal, and *trans*-pinocarveol were attractive at 1 μg doses to both sexes of beetles. An expensive and stable compound, α-pinene could be used to improve the sensitivity of pheromone baited traps for monitoring *D. confusus*.

10.5.1.3 Feeding Deterrents

Sublethal or chronic effects of monoterpenoids appear to be more important in the plant's defense than any acute toxicity. The plants' first defense against polyphagous insects is their ability to repel insects by acting as feeding deterrent. The citronellal is a repellent present in yellow citronella candle. Cineole repels the American cockroach, *Periplanata americana*, verbenone is a deterrent to the spruce bark beetle, *Ips typographus*; and geraniol is repulsive to the red flour beetle, *Tribolium castaneum* (42). Linalool, isopulegol, pulegol, pulegone, carvone, and *d*-limonene repel the German cockroach, *Blattella germanica*. Feeding deterrents interfere with the insects' ability to ingest and utilize food, which ultimately results in reduced growth and delayed development. Limonene acts as a feeding deterrent to cat fleas, *Ctenocephalides felis* and pulegone acts as a strong feeding deterrent to the sixth-instar fall armyworm, *Spodoptera frugiperda*.

Exposure to verbenone vapors did not cause any acute toxicity symptoms in the pine weevil, *Hyloblus abietis* (L.), whereas weevils exposed to higher doses of both limonene and α-pinene exhibited signs of poisoning within a few hours. Verbenone consistently suppressed feeding, while symptom-free weevils exposed to limonene or α-pinene fed as much as controls. Feeding deterrence by verbenone was lost after 48 h and starved weevils were less deterred than fed weevils (43).

Four natural plant compounds [limonin, S(+) and R(−)-carvone, and cucurbitacin] and one insect pheromone (verbenone) were evaluated for antifeedant activity against the pales weevil, *Hyloblus pales* (Herbst), on *Pinus strobus* seedlings and for toxic activity against the pathogenic fungus. All compounds demonstrated significant antifeedant activity in a choice test on treated pine seedling but none completely eliminated feeding. Only cucurbitacin elicited a linear response relationship, with significant activity occurring at concentration as low as 0.10 μg/mL. The other compounds significantly reduced feeding at concentrations as low as 1 μg/mL (44).

Iridoid glycosides are present in 57 plant families of the Dicotyledoneae. Most of the iridoids are bitter substances. Iridoid glycosides act as feeding deterrents to a variety of generalist insects and have been shown to act as antifeedants to grasshoppers and lepidopteran larvae (45).

10.5.1.4 Neurotoxins

One of the best examples of monoterpene derivatives that are toxic to insects are pyrethroids found in leaves and flowers of some *Tanacetum* species. Pyrethroids are neurotoxins, causing hyperexcitation, uncoordinated movement, and paralysis of insects. Pyrethroids are a group of monoterpene esters. They are very active broad-spectrum insecticides and are generally not hazardous under normal circumstances due to their rapid metabolism and excretion in mammals (46). Pyrethrum is the common name for the dried flowers, and the active ingredients present in pyrethrum are known as pyrethrins. The commercially important pyrethrum plant species is *Tanacetum cinerariifolium* (which formerly belonged to the genus, *Chrysanthemum*); the other source is *T. coccineum*. These are dried either under the sun or with mechanical driers, and ground. The finely ground material contains 1.3% pyrethrins. Although all parts of the plant contain small amounts of pyrethrins, the highest concentration is found in fully opened flowers. More than 90% of the pyrethrins in the flower head are localized in the secretary ducts of the achenes.

A total of six closely related insecticidally active esters have been isolated from the pyrethrum extract. They are esters of two monoterpene carboxylic acids, chrysanthemic and pyrethric acids (Fig. 10.6).

The chrysanthemic or pyrethric acid forms esters with pyrethrolone, cinerolone, and jasmolone. The esters differ only in the terminal substituents on the side chains of the acid and alcohol moieties (R1 and R2). The pyrethrum extract usually contains about equal quantities of chrysanthemic and pyrethric acid esters, with pyrethrin I and II accounting for 65–75%.

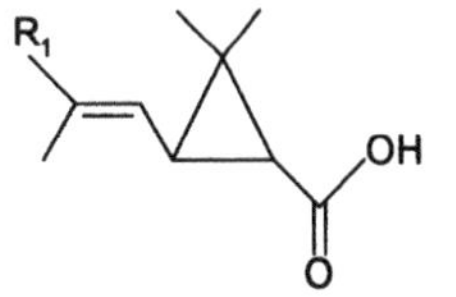

R₁=CH₃ : Chrysanthemic acid
R₁=COOCH₃ : Pyrethric acid

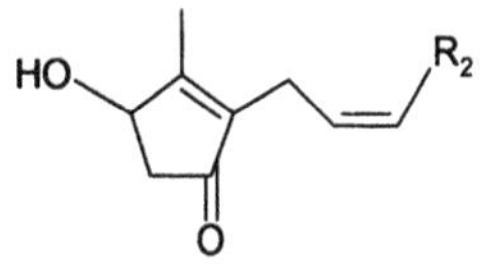

R₂ = CH = CH₂ Pyrethrolone
R₂ = CH₂ - CH₃ Jasmolone
R₂ = CH₃ Cinerolone

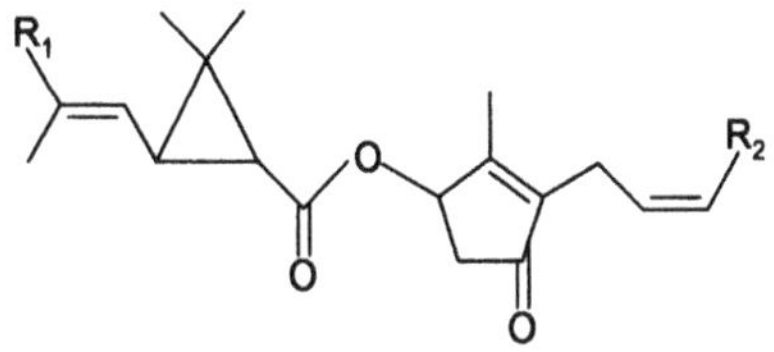

	R₁	R₂
Pyrethrin I	-CH₃	-CH=CH₂
Pyrethrin II	-COOCH₃	-CH=CH₂
Jasmolin I	-CH₃	-CH₂ - CH₃
Jasmolin II	-COOCH₃	-CH₂ - CH₃
Cinerin I	-CH₃	-CH₃
Cinerin II	-COOCH₃	-CH₃

FIGURE 10.6 Structures of natural pyrethroids.

The overall structural elucidation of the pyrethrins, including establishment of their complete absolute stereochemistry, was carried out in the early 1970s (47).

Some of the natural pyrethroids have been used against lice and fleas in homes as well as public buildings. It has also been successfully used against mosquitoes, hosueflies, and other insects that spread diseases in animals and human beings (46).

All of the natural pyrethrins are unstable in air and light. Hence, their principal use has been limited in the household sector. The pyrethrins are lethal against a wide range of insect species, but their relative toxicities vary with the insect species and the conditions of the treatment. Since these natural pyrethroids are unstable in light, derivatives that are light stable were prepared and are now available commercially. The first synthetic pyrethroid, named allethrin (active moiety chrysanthemic acid), was made in

TABLE 10.2   Relative Toxicities of Pyrethrins and Synthetic Pyrethroids

Compound	Heliothis virescens (topical) μg/ third-instar larvae, LC_{50}	Spodoptera exigua (topical) μg/third-instar larvae, LD_{50}	Musca domestica (topical) μg/adult, LD_{50}	Aphis fabae (foliar residue) adults, $LC_{50(ppm)}$
Pyrethrins	0.48	—	1.6	> 1000
Allethrin	10.0	—	1.0	310
Cypermethrin	0.0069	0.005	0.0065	—
Deltamethrin	0.00084	0.003	0.0019	—
Fenvalerate	0.012	0.018	0.042	8.5

1949, but it possessed a weaker knockdown effect than natural pyrethrins. Important synthetic pyrethroids widely used as insecticides are cypermethrin, deltamethrin, and fenvalerate. The relative toxicities of natural and synthetic pyrethroids are given in Table 10.2.

10.5.1.5   Growth and Development Inhibitors

Plant-derived terpenoid juvenile hormone mimics are used as effective insecticides. More than a dozen plant-derived mono- and sesquiterpenes have been described to have insect-sterilizing effects as juvenile hormone analogues (48).

Citronellol acts as an oviposition deterrent to the leafhopper, *Amrasca devastans* (Distant) (49). All four monoterpenoids—*d*-limonene, linalool, α-terpineol, and β-myrcene—significantly influenced the number of days required by nymphs to reach the adult stage of German cockroach (49). In general, higher monoterpenoid concentrations in the diet were associated with a reduction in the days required by nymphs to mature. Topical applications of linalool, β-myrcene, or α-terpineol at near-lethal rates to adult cockroaches before they mated had no significant influence on the reproductive parameters examined.

Sixteen natural monoterpenoids and six synthetic derivatives were selected for study of larvicidal activity and growth inhibitory effect against the European corn borer, *Ostrinia nubilalis* (Hubner) (50). For this study, two different exposure bioassays were used: compounds applied on the diet surface (on-diet) and compounds incorporated into the diet (in-diet). Most of the monoterpenoid compounds showed some degree of larvicidal activity in both bioassay procedures after a 6-day exposure period. Among the monoterpenoids, pulegone was the most active. Larvicidal toxicities were significantly enhanced for the structurally modified compounds; modified monoterpenoid derivatives MTEE-25 (2-fluoroethylthymyl ether) and MTEE-P (propargyl citronellate) were most toxic to borer larvae. In general, growth and development of the European corn borer were affected by monoterpenoid compounds, and some compounds such as l-menthol, pulegone, MTEE-25, and MTEE-P acted as insect growth inhibitors.

Monoterpenoids also interfere with the developmental processes of insects. Limonene inhibits embryonic development in the cat flea *C. felis* and pulegone decreases larval growth of the Southern armyworm, *Spodoptera eridania* (51). Plant allelochemicals also interfere with the reproduction process of insects by acting as oviposition repellent and by reducing mating success, egg production, and egg viability. Cineole is a feeding deterrent as well as an ovipositional repellent against adult yellow fever mosquitoes, *Aedes aegypti*. Limonene inhibits egg hatch of exposed Western corn rootworm eggs, *Diabrotica virgifera virgifera* (49).

The activity of the essential oils derived from 11 Greek aromatic plants belonging to the Lamiaceae family against egg hatching larval and adult growth of *Drosophila aurania* has been reported (52). Among the most active was that from *Mentha pulegium*, which includes pulegone as the major constituent.

10.5.1.6 Mixed Activities

Acute toxicities of 34 naturally occurring monoterpenoids were evaluated against three important arthropod pest species; the larva of the Western corn worm, *Diabrotica virgefera* LeConte; the adult of the two-spotted spider mite, *Tetranychus urticae* Koch; and the adult housefly, *Musca domestica* L (53). Potential larvicidal or acaricidal activity of each monoterpenoid was determined by topical application, leaf-dip method, soil bioassay, and greenhouse pot tests. Citronellic acid and thymol were the most toxic topically against the housefly, and citronellol and thujone were the most effective on the Western corn rootworm. Most of the monoterpenoids were lethal to the two-spotted spider mite at higher concentrations; carvomenthenol and terpinen-4-ol were especially effective. A wide range of monoterpenoids showed some larvicidal activity against the Western corn rootworm in the soil bioassay. Perillaldehyde, the most toxic ($LC_{50} = 3\,\mu g/g$) in soil, was one-third as toxic as carbofuran, a commercial soil insecticide ($LC_{50} = 1\,\mu g/g$). Selected monoterpenoids, especially α-terpineol, effectively protected corn roots from attack by the Western corn rootworm larvae under greenhouse conditions. In general, bioactivity of monoterpenoids was significantly less than that of conventional organic insecticides; however, they can be effective under conditions that allow use of high concentration of these safe chemicals.

House flies, *Musca domestica L.* and their eggs were treated with 22 monoterpenoids to determine the topical, fumigant, and ovicidal activity of each compound (54). Fumigant activity of 14 monoterpenoids was examined further using red flour beetles, *Tribolium castaneum* (Herbst). Third-instar Southern corn rootworm, *Diabrotica undecimpunctata* Howardi Barber, were treated with carvacrol, citral, citronellal, menthol, pulegone, verbenol, and verbenone to determine the activity of these compounds. Ketones were more effective than alcohols in the topical, fumigant, and ovicidal bioassays. Aldehydes were more toxic than alcohols in the topical and fumigant bioassays. In the topical and ovicidal bioassays, aromatic or acyclic alcohols, or both, were more effective than monocyclic and bicyclic alcohols. Vapors of bicyclic ketones were more toxic than monocyclic ketones to adult *M. domestica*. Monoterpenoid alcohols containing three carbon–carbon double bonds were more effective than saturated alcohols in the topical and larval bioassays. A monounsaturated ketone was more toxic than a structurally similar saturated ketone and two diunsaturated ketones when it was applied topically to adult *M. domestica*. A saturated monocyclic ketone inhibited egg hatch more effectively than unsaturated monocyclic ketones.

The structural characteristics of monoterpenoids can influence their insecticidal properties. Their shape, degree of saturation, and type of functional groups can influence penetration into the insect cuticle, affect the ability of the compound to move, and interact with the active site, and influence their degradation. The enhanced toxicity of ketones and phenols relative to monocyclic alcohols could be related to differential susceptibility to metabolism because applications of piperonyl butoxide, an inhibitor of mixed-function oxidases, resulted in greater toxicity of *d*-limonene to the German cockroach, *Blatella germanica* (*L.*), and to *M. domestica* (55). The ketone–monocyclic alcohol toxicity differences appears to be a factor in the peppermint *Mentha piperita* (*L.*) plant's defense against the variegated cutworm *Peridroma saucia* (Hubner). Pulegone and menthone predominate in young peppermint leaves, whereas monoterpenoid alcohols and aldehydes appear in the more mature leaves. The difference in the monoterpenoid profile renders the younger leaves less suitable for the diet of the variegated cutworm (56).

Commercial peppermint, *Mentha piperita* (*L.*), contains more than 30 monoterpenes (56). Larvae fed for 6 days from the beginning of the fifth stadium on semidefined artificial diet fortified with pulegone (0.0–0.1%) menthol (0.1% and 0.2%) or menthone (0.2%) weighed less than controls. Reduced growth was the result of feeding inhibition in larvae fed menthone and pulegone and to molting abnormalities in larvae receiving menthol. Menthol completely inhibited pupation at doses approximating those in pepperminet (0.05–0.2% wet wt). Limonene (0.2%), menthone (0.1%, 0.2%), and pulegone (0.1%) also inhibited pupation. Growth, feeding, and pupation were not affected, when limonene (0.05% or 0.1%) or α-pinene (0.05–0.2%) was ingested by the larvae.

The monocyclic terpenoid d-limonene is the main constituent of the terpenoid fractions of lemon and orange oils. In their undistilled form, citrus oils have exhibited toxicity against various insects including the cowpea weevil (*Callosobruchus maculatus*), the rice weevil (*Sitophilus oryzae*), the housefly (*Musca domestica*), and red imported fire ant (*Solenopis invicta*). This monocyclic monoterpenoid, d-limolene, was found to be acutely toxic to *Dendroctonus* bark beetles, eggs, and larvae of the fruit fly, *Anastrepha suspensa*, and all life stages of the cat flea (*Ctenocephalides felis*). Bioassays were performed to determine topical, fumigant, oral, repellent, residual, ovicidal, and larvicidal activities (57). The material was slightly toxic topically to German cockroaches and houseflies and was synergized with piperonyl butoxide. High concentrations of vapors caused mortality in German cockroaches and rice weevils. Repellent activity against German cockroaches was noted at high concentrations. No residual activity was observed on any of four surface types exposed to adult German cockroaches. Limonene inhibited Western corn rootworm egg hatch at high concentrations and showed moderate toxicity in soil against third-instar Western corn rootworm larvae. These findings indicated that the insecticidal properties of d-limonene are limited.

Monoterpenoid derivatives were synthesized and their insecticidal activities were evaluated against red flour beetles [*Tribolium castaneum* (iterbst)] in fumigant bioassays and against houseflies, *Musca domestica* (L.) in topical, fumigant, and ovicidal bioassays (58). Acetate derivatives were more active than the propionate derivatives of cyclic monoterpenoids in the topical, fumigant, and ovicidal bioassays. Pivalates were topically more insecticidal than acetates to adult houseflies, while the acetates had the greater ovicidal activity. Fluoroacetates of cyclic monoterpenoids were the most effective housefly fumigants followed by acetates.

Toxaphene, the polychlorinated product of camphene, is a mixture of almost 200 compounds many of which are polychloroboranes (59). Several specific forms of chlorinated camphene were found to be much more insecticidal than others. During 1970, it was the leading insecticide in the United States in terms of sales. However, it was removed from the market due to toxicological concerns.

Insecticidal and acaricidal components from sawdust of *Thujopsis dolabrata* var. *hondai* against eight species of arthropod pests were identified as the terpenoids carvacrol and β-thujaplicine. In tests using the filter paper diffusion method, carvacrol had broad insecticidal and acaricidal activity against agricultural, stored product, and medical arthropod pests. However, β-thujaplicine showed only weak termiticidal activity. The insecticidal activity of carvacrol was attributable to fumigant action (60).

10.5.2   Sesquiterpenoids

A review by Picman summarizes the effect of insect feeding, development, and survival for more than 50 sesquiterpene lactones (61). The structures of important sesquiterpene lactones are shown in Fig. 10.7. The feeding deterrency of these compounds confirms the general observation of the importance of the presence of α-cyclopentenone and/or α-methylene -γ-lactone functionalities. Sesquiterpene lactones are poisonous to several lepidopterans, flour beetles, and grasshoppers (62). A study of antifeedant activity of tenulin was reported (63). On the other hand, the sesquiterpene polyol ester angulatin A is claimed to be strongly antifeedant and insecticidal against a variety of insects (64). Angulatin A and its close

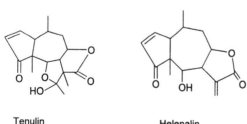

Alantolactone

Glaucolide A

Tenulin   Helenalin

Figure 10.7   Structures of important sesquiterpene lactones.

antifeedant analogues derived from the root bark of *Celastrus angulatus* Max have been used in China to protect plants from insects (64). Absinthin, a dimeric sesquiterpene obtained from *Artemisia absinthium*, has been reported to exhibit antifeedant activity against a number of insects (65). MeGovran and Meyer (66) tested helenalin, tenulin, and isotenulin as contact insecticides against nymphs and adults of the Mexican bean beetle, adult houseflies, and cockroaches. These three lactones topically applied were not significantly toxic. Although sesquiterpene lactones are not significantly toxic as contact insecticides, some are potent insect antifeedants, inhibitors of larval insect growth, and oviposition deterrents. Larvae of the fall armyworm, Southern armyworm, and saddleback caterpillar exhibited strong tendencies to avoid feeding on the fresh foliage of *Veronia* species (*V. gigantea*, *V. glauca*) that contain glaucolide A. But they readily consume *V. flaccidifolia* that does not contain sesquiterpene lactones. Incorporation of glaucolide A into larval diets resulted in growth inhibition of the fall, Southern, and yellow-striped armyworms and increased the number of days to pupation in most of the insect species studied (67–69). Tulip poplar, *Lirodendron tulipifera*, contains at least four sesquiterpene lactones: lipiferolide, epitulipinolide diepoxide, peroxyferolide, and tulirinol. These sesquiterpenes are antifeedants for larvae of the gypsy moth. When eggs of the fruitfly (*Drosophila melanogaster*) were exposed to xanthumin, 8-epixanthatin, and euponin (sesquiterpene lactones), reduced growth and delayed pupation were noted (68). Euponin was found to be an insect development inhibitor rather than an antifeedant. Euponin exhibited no ovicidal activity against *D. melanogaster*. Sesquiterpene lactones did not deter oviposition by lepidoptera. The presence of sesquiterpene lactones may deter oviposition and feeding by at least some generalist insect feeders like armyworm species. The sesquiterpene caryophyllene is an aphid repellent and the epoxide of this compound inhibits *Heliothis virescens* larval growth by 60%, when incorporated at a 5 mM concentration into the diet (70). Caryophyllene and several other terpenes had antifeedant activity against cabbage butterfly larvae. Sesquiterpenes from wild cotton have acted as antifeedants to several insect species (71). Henalin from sneezeweed (*Helenium microcephalum*) is one of the more acutely toxic sesquiterpenes with an oral median lethal dose of 150 mg/kg for mice. Alantolactone acts as a potent antifeedant against the confused flour beetle, *Tribolium confusum* (72). On the basis of data available, it is clear that at least some sesquiterpene lactones can function as insect oviposition deterrents, antifeedants, and/or growth inhibitory agents. However, the role of these chemicals in protecting plants from the damaging effects of insect feeding is far from clear.

Ethanol extract of *Bontia daphnoids* L. (Myoporaceae) leaves and stems led to the isolation of epingalone, a sesquiterpene furan. The compound showed growth regulatory activities on gravid adult female *Boophilus microplus canestrini* (the Southern cattle tick). The dose required for inhibiting the hatching of *B. microplus* eggs by 50% (Fid_{50}) was $0.4 + 0.06 \, mg/g$ of the tick's body weight (73).

The sesquiterpenes farnesol, caryophyllene oxide, and 2,10-bisabola-dien-1-one and the monoterpenes citronellol and geraniol were tested for settling inhibition and chronic effects on the aphid *Myzus persicae* by means of an improved leaf-disc assay (74). Of these compounds, geraniol, farnesol, and the natural bisabolene significantly inhibit settling in choice tests. Furthermore, application of the bisabolene to intact *Capsicum annum* leaves did not cause phytotoxicity but did affect the insects' probing behavior of intracellular punctures. Both compounds significantly reduced offspring production. Of those compounds tested, the natural product, bisabolene, could be a promising lead for future development of aphid control agents (74).

Gossypium species produce hemigossypolone with cadinene skeleton. The concentration of this compound appears to be directly related to disease and insect resistance. A new sesquiterpenoid aldehyde, raimondal, was isolated from *Gossypium raimondii* (75).

A novel sesquiterpene of the eudesman type, mutangin, with five hydroxyl functions has been isolated from unripe fruits of *Elacodendron buchanani*. Mutangin has demonstrated moderate antifeeding activity against the lepidopteron *Chilo partellus* (Swinhoe) (76).

The drimane-type sesquiterpenoids, such as polygodial from *Polygonum hydropiper* and warburganal from *Warburgia* sp are potent insect-feeding deterrents (77). These compounds inhibit the feeding of the armyworms *Spodoptera exempta*, *S. littoralis*, and *Leptinotarsa decemlineta* as well as the aphid *Myzus persicae*.

Monoterpenoids and sesquiterpenoids from the foliage of Douglas fir (*Pseudotsuga menziesii*) were bioassayed using agar diets to determine the effect of these compounds on natural and colony populations of Western spruce budworm (*Choristoneura occidentalis*). Larvae collected from two different populations were significantly different in their tolerance to terpenoids (78). For the Montana population, agar diet studies showed that camphene, myrcene, terpinolene, bornyl acetate, and tricyclene adversely affected larval growth rate and pupal weight. For the Idaho population, agar diet studies showed that only terpinolene and bornyl acetate were the most toxic sesquiterpenes; their combination with tricyclene, camphene, myrcene, limonene, terpinolene, or the acetate fraction appears to represent an effective mixture of defensive compounds against the budworm (78).

10.5.2.1   Juvenile Hormones and Their Analogues

There are four compounds known as JH 0, JH 1, JH II, and JH III, which represent the juvenile hormone (JH) activity of the majority of insects (79,80). These hormones are synthesized in the corpora allata, two tiny endocrine glands at the base of the insect brain. JH (Fig. 10.8), along with ecdysones, regulates molting and metamorphosis in insects. As JH is involved in the regulation of such physiological processes in insects as molting and metamorphosis, which do not have a counterpart in vertebrates, it was anticipated that juvanoids would disrupt such processes leading to the death of the insect without detrimental effects against most other classes of animals. JH is required in the adults of many species for several reproductive functions, such as ovarian development, yolk synthesis and maturation of eggs in females, phoremone function and accessory reproductive gland development in males. Regulation of diapause is also under JH control. It also appears to have a role in the regulation of molting by affecting the synthesis of α-ecdysone in the prothoracic gland (81).

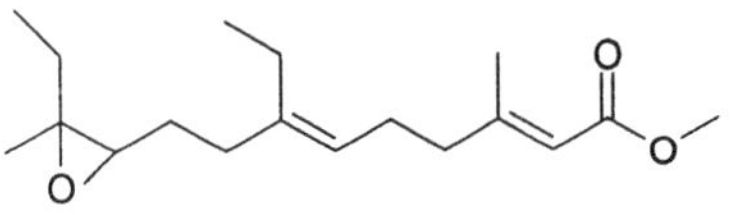

Juvenile hormone I

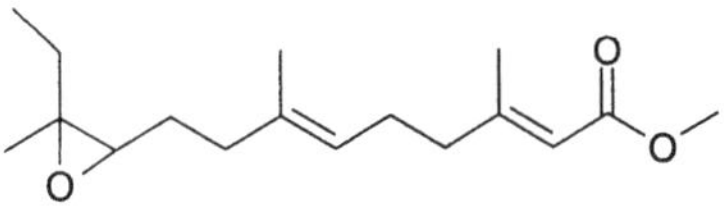

Juvenile hormone II

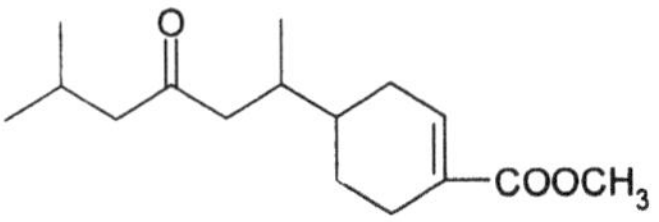

+ Juvabione

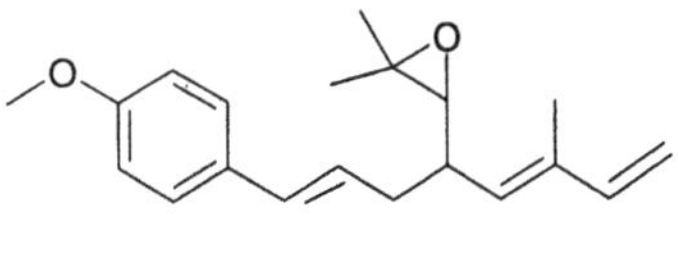

Juvocimene II

FIGURE 10.8   Structure of juvenile hormones and their analogues.

Slama and Williams in the mid-1960s observed JH activity in paper used to line the cages containing European linder bug, *Pyrrhocoris apterus* (82). Bowers and coworkers soon isolated (+)-juvabione from the wood of the balsam fir *Abies balsamea* and showed it to be one of the active materials of this paper factor (83). Later, several closely related juvenoids constributed to the JH activity of the paper factor. These compounds are also species specific and are highly active only against some hemopterous species of the family Pyrrhocoridae. Although this class of juvenoid has been extensively investigated, no commercial products of this type have been developed. Several other compounds with JH activity have been isolated that showed activity against a limited range of insect species. Juvocimene II (Fig. 10.8) was isolated from the oil of sweet basil (*Ocimum basilicum*) and showed good activity against the milkweed bug *Oncopeltus fasciatus* (84). JH analogues are also present in plants. One of the first observations was made by Williams who reported that the water of Rio Negro was rich in suspended organic materials from wood decay and free from mosquito larvae, unlike other rivers of the Amazon basin (85). JH analogues have been found in various plants and may interfere to some extent with the control of several species of insects. There have been numerous reports on the occurrence of plant compounds having JH activity (phytojuvenoids) such as juvabione (Fig. 10.8), farnesol, juvocimene I and II, sesamin, sesamolin, thujic acid, sterculic acid, tagetone, ostruthin, echinolone, bake-chiol, and juvadecene (86). Plant-derived terpenoid JH mimics are effective insecticides. More than a dozen plant-derived sesquiterpenoids have been described to have insect-sterilizing effects as JH analogues, along with phytochemical disruption of insect development and behavior (86).

Juvenile hormone analogues may keep the insects in an immature and potentially injurious stage longer than normal; a chemical that would shut off JH production could be more useful. Such a chemical should cause premature metamorphosis of larvae and sterilization of adult insects. Two such anti-JH substances were discovered in the garden plant *Ageratum houstanianum* and named precocene I and II (87). These anti-JH substances produce dwarf sterile insects that are unable to survive. The open-chain analogue of precocene, α,β-asarone, is a constituent of the rhizome of sweet flag (*Acorus calamus*) which exhibits antigonadal activity in insects. Such oils may be utilized as chemosterilants, and the problem of development of physiological races of the pests may be minimized.

10.5.3 Diterpenoid

Tigliane diterpene ester was isolated from the seeds of *Croton tiglium* L (88). Tigliane and two closely related esters showed growth inhibitory and insecticidal activities in diet against newly hatched larvae of *Pectinophora gossypiella*. Tigliane also gave 100% kill of *Culex pipiens* larvae at 0.6 ppm but was ineffective against *Oncopeltus fasciatus* and *Tribolium confusum*. Esters of this type are also known for their vesicant and tumor-producing properties. A novel diterpene was isolated from the medicinal plant *Croton linearis* (89). This compound showed a topical LD_{50} value of 0.32 µg/insect to the adult weevil, *Cylas formicarius elegantulus*, a serious pest on sweet potatoes. Typical symptoms of neurotoxic insecticides was noted. Three grayanoid diterpenes were isolated from the dried flowers of the Chinese insecticidal plant, *Rhododendron molle* (90). These diterpenes showed antifeedant activity against two insect species. Of these, rhodojuponin III was the most active antifeedant. It was also toxic to newly hatched *Spodoptera fruigiperda* organisms with an LD_{50} of 8.8 ppm, the same order of magnitude as malathion. It was nontoxic to *Heliothis virescens* and failed to retard growth. Various compounds isolated from different parts of *Podocarpus gracilior* and *P. nagi* were found to be responsible for resistance to many insects (91). These compounds belong to a series of norditerpene dilactones referred to as nagilactones. These compounds are strong feeding deterrents leading to reduction in larval growth and pupation. The most effective of these is nagilactone D, which prevented all larvae of *H. virescens* from reaching the third instar at 166 ppm in artificial diet. Twelve briaran diterpenes from *Briareum polyanthes* and *Ptilosarcus gurneyi* have been isolated and identified (92). Five of these diterpenes were tested against the tobacco hornworm *Manduca sexta* and found to cause weight loss.

The alkaloidal diterpene ryanodine (Fig. 10.9), was isolated from the stem wood of *Ryania speciosa* (93). The powdered stem of *Ryania* was used for about 40 years as a commercial insecticide. It is one of the most potent naturally occurring insecticides against lepidopteran larvae. It controls European corn borer at 40 g/ha, and improved preparations are effective against coddling moth larvae at 10 g/ha.

The diterpene acids 16-kauren-19-oic acid and trachyloban-19-oic acid present in sunflower (*Helianthus annus*) florets resist the attack by larvae of the sunflower moth *Homeosoma electellum* (94). Inflexin from *Isodon inflexus* and isodomedin from *I. shikokianus* exhibit feeding deterrent action towards lepidopteran larvae including the African armyworm, *Spodoptera exempta* (73).

The diterpene derivatives odoracin and odracin from the roots of sweet daphne (*Daphne odora*) have significant nematicidal activity against rice white-tip nematode. They gave 70–96% mortality at 1 ppm (95).

Twenty-four diterpenoids have been isolated from root or stem bark of neem. These chiefly belong to two groups podacarpanoids and abietanoids (96). Even though no screening of these compounds has been

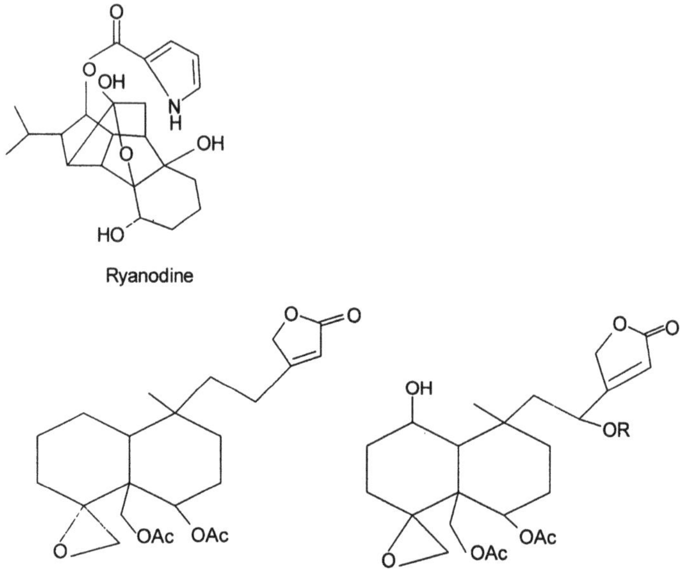

Ryanodine

Ajugarin I   Ajugamarin

FIGURE 10.9   Structure of ryanodine, ajugarin I, and ajugamarin.

reported, other tricyclic diterpenes of this type have been reported to possess a wide range of biological activities, including insecticidal activities. The important diterpenes include sugiol, nimbisonol, nimbinone, nimbiol, and nimosone.

Macrocyclic diterpenes, cembranes, isolated from *Pinus* spp and marine coelentrates such as sea fans and soft corals act as scent trail pheromones. The diterpene phorbol is present as an ester in the latex of plants of the spurge family. Phorbol acts as a toxin against herbivores. Skin contact with phorbol causes severe inflammation (92).

The diterpenoids ferruginol and manool are strong antifeedant compounds to subterranean termites, providing protection to bold crypress (97). Polygodial, a diterpene from water pepper (*Polygonium hydropiper*), increased yield of barley by 36% due to its antifeedant effects on the aphid, *Rhopalsoiphum padi*.

Ajugarins I–V (Fig. 10.9) and clerodin were the first insect antifeedants isolated from *Ajuga remota* (98,99). The antifeedant activities of ajugarins I–III were investigated by host-plant leaf disc method using *Zea mays* for the monophagous *Spodoptera exempta* and *Ricinus communis* for the polyphagous *S. littoralis*. Activity levels of 100 ppm against *S. exempta* and 300 ppm against *S. littoralis* were found. The other two ajugarins exhibited no antifeedant activities, and only moderate insecticidal and growth inhibitory activities were reported for ajugarin IV against different insects. The other compounds isolated from *Ajuga* spp include ajugamarins (Fig. 10.9) ajugapitin, ajugachins, ajugavensins, and ajugareptansin. Most of them exhibited good to excellent antifeedant activities at 30 ppm, with the remarkable exceptions of ajugareptansin and ajugareptansone A having activities at 300 ppm or more. The most active compounds were found in the ajugapitin series in which some products exhibited activity at 0.3 ppm dose.

Cinnzeylanine and cinnzeylanol, diterpenoids with a pentacyclic skeleton, were isolated from *Cinnamonum zeylanium* Nees (100). Ryanodine, an insecticidal constituent from *Ryania speciosa* Vahl, is a sole natural compound to have such a skeleton. Cinnzeylanine and cinnzeylanol killed the larvae of silkworm at a dose of 16 ppm and inhibited larval ecdysis at 2–14 ppm. The growth of larvae of long corn beetle, *Psacothea hilaris* Pasco, was strongly inhibited and larvae of the fall webworm, *Hyphantria cunea* Drury, died before pupation. The larvae of mulberry pyralid, *Glyphodes pyroalis* Walker, died at the stage of larvae at the dosage of 100 ppm.

Three new neoclerodane diterpenoids—14,15-dehydroajugareptansion, 3β-hydroxyajugavension β, and 3α-hydroxyajugamarin F4—have been isolated from aerial parts of *Ajuga reptans* cv catlins Giant, together with the known compound ajugareptansin. Insect antifeedant testing of all four compounds revealed that 14,15-dehydroajugareptansin had significant activity against sixth-stadium larvae of *Spodoptera littoralis* (101).

10.5.4 Sesterterpenoids

Even though a large number of sesterterpenoids have been isolated and identified from plants, only few compounds exhibit bioactivity against insects. Stipanovic et al. (102) have isolated a group of sesterterpenoids known as heliocides from glands of leaves and bolls of cotton plants. These compounds, as the name indicates, are toxic to *Helicoverpa* spp and appear to be involved in the resistance of certain varieties of cotton to this insect. The toxicity of heliocide H1 and H2 was tested against tobacco budworm. The ED_{50} values for heliocides H1 and H2 are 0.12 and 0.13, respectively. Hemigossypolone, gossypol, and heliocides H1 to H4 are toxic and deterrent to feeding of several generalist lepidopteran insects. HPLC analysis revealed greater amounts of these terpenoid aldehydes per gland in young foliage of damaged plants than control plants 1 day after initial injury (103). By 7 days after initial injury, greater quantities of hemigossypolone and all heliocides except H4 were detected in young foliage from damaged plants compared to control plants.

10.5.5   Triterpenoids and Steroids

The steroids show antifeedant as well as growth regulatory activities in insects. The important groups of steroids that show bioactivity against insects are limonoids and phytoecdysteroids.

10.5.5.1   Limonoids

Limonoids are modified triterpenes or derivatives of 4,4,8-trimethyl-17-furanylsteroid skeleton. More than 300 limonoids have been isolated to date (104). Their production is confined to the plants in the order Rutales and they are characteristics of the members of the family Meliaceae, where they are diverse and abundant.

   Azadirachta indica A. Juss, an indigenous tree of the Indo-Pakistan subcontinant, belongs to the family Meliaceae is widely distributed in Asia, Africa, and other tropical parts of the world. It has been an age-old practice in the Indo-Pakistan subcontinent to mix dried neem leaves with stored grains and then place them in folds of clothes to keep insect away. Siddiqui reported the isolation and characterization of bitter constituents, nimbin, nimbinin, and nimbidin (96). Subsequently, a host of compounds have been isolated from various parts of the tree, including meliantriol, salanin (Fig. 10.10) and azadirachtin, which are active feeding deterrents, toxicants, and/or disruptants of growth and development against a large variety of insect species and nematodes. Two insect growth regulatory triterpenoids, namely, nimocinolide, and isonimicinolide have been reported against *Aedes aegypti*. Nimocinolide and isonimocinolide have an LC_{50} value of 0.625 and 0.74 ppm, respectively, against *A. aegypti*. The fruit fractions are effective against *Callosobruchus analis* and other pests. Numerous triterpenoids such as epoxyazadiradione, azadiradione, kulactone, limocinol, limocinone, limocinin, limocin A and B, azadirol, azadironolide, and isoazadironolide have been isolated from neem and identified (105).

   Azadirachtin (Fig. 10.5) is considered the most important active ingredient in neem seed kernels. The high content reported was about 10 g/kg of seed kernels. Azadirachtin possesses feeding deterrent, antiovipositional, growth disrupting, fecundity, and fitness-reducing properties on insects. Azadirachtin is formed by a group of closely related isomers, the azadirachtins A–G. Azadirachtin is the most important compound in terms of its quantity in neem seed kernel extracts. Azadirachtin E is regarded as the most effective insect growth regulator (106).

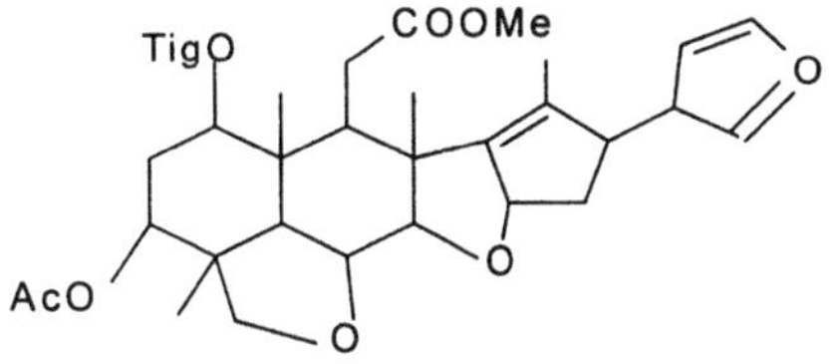

FIGURE 10.10   Structure of C-seco-limonoids.

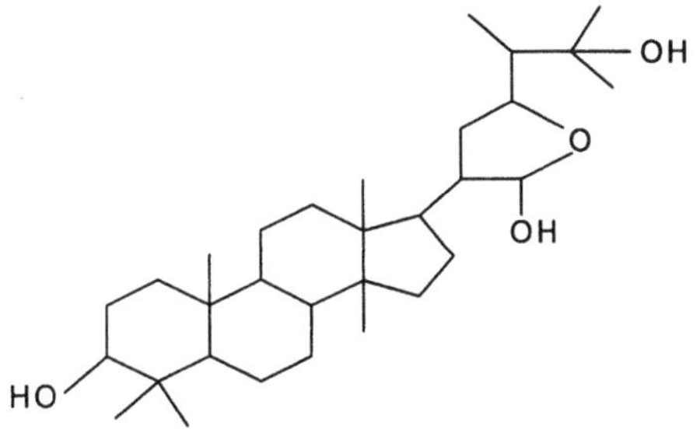

The antifeedant activity of azadirachtin, azadirachtin derivatives, and related limonoids was assessed in choice and no-choice bioassays against four species of *Lepidoptera*: *S. littoralis*, *S. frugiperda*, *H. virescens*, and *H. armigera*. Azadirachtin and dihydroazadirachtin were the most potent of the 40 compounds tested. The natures of substitutes at C-1, C-3, and C-11 are important (107).

At present, the compounds with the most activity against herbivorous insects appear to be the C-*seco* limonoids, restricted to one tribe of the Meliaceae (108,109). Laboratory investigations involving azadirachtin indicated that feeding deterrency contributed significantly to the control of aphid populations. Antifeedant activity to the strawberry aphid, *Chaetosiphon fragaefolii*, was lost within 24 h following application. Control of aphids result from the inhibition of adult reproduction and from the failure of nymphs to molt. Natural enemies of aphids are moderately susceptible to the insect growth–regulating effects of neem. Neem-based botanicals appear to be relatively benign to beneficial insects and are well suited in integrated pest management (IPM) programs (107).

Nymphs of the locust *Schistocerca gregaria* in antifeedancy tests were more sensitive to azadirachtin than to other compounds tested. To *S. littoralis*, *S. gregaria*, and the bug *O. fasciatus* azadirachtin was toxic, exerting severe growth and molt-disrupting effects. In contrast, compounds from lower structural levels than azadirachtin showed virtually no toxicity to these three insects. The combination of strong antifeedant and toxic properties of azadirachtin provides the more important chemical protection for the neem tree (108).

Four limonoids, humilinolides A–D from *Swietenia humilis* and cedrelanolide from *Cedrela salvadorensis*, were evaluated for their effect on the European corn borer; *Ostrinia nubilalis* in comparison with toosendandin, a commercial insecticide derived from *Melia azedarach* (110). When incorporated into the artificial diets of neonates at 50 ppm, all compounds caused larval mortality as well as growth reduction and increased the development time of survivors in a concentration-dependent manner. Humilinolide C also reduced growth and survivorship at 5 ppm. Additional effects observed in many of the limonoid-treated groups included a significant delay in time to pupation and adult emergence. The compounds showed comparable activity to toosendanin, a commercial insecticide.

Three limonoids—namely, limonin, nomilin, and obacunone—were isolated from the seeds of *Citrus reticulata* (Blanco coorg Mandarin). With fourth-instar larvae of mosquito, the EC_{50} for inhibition of adult emergence was 6.31, 26.61, and 59.57 ppm for obacunone, nomilin, and limonin, respectively (111).

Bitter gourd, *Momordica charantia*, was less palatable to two species of armyworms, *S. littura* and *Pseudaletia separata*, than two other cucurbitaceous plants (112). A methanolic extract of *M. charantia* leaves inhibited feeding of the armyworm larvae. The two most active fractions obtained by silica gel chromatography were purified by HPLC. Momordicine II, a triterpene monoglucoside, was identified as an antifeedant compound from the more active of these fractions. Momordicine II showed significant antifeedant effects on *P. separata* concentrations of 0.02%, 0.1%, and 0.5% in artificial diets. Momordicine II caused a significant feeding reduction in *S. littura* only at the highest concentration (0.5%) tested.

Four compounds—stigmesterol, stigmasterol-β-D-glucopyranoside, luteolin-7-O-glucopyranoside, and dotriacontanoic acid—isolated from *Verona cinerea* have shown significant antifeedant activity based on percent feeding deterrence and effective doses (ED_{50}) against two lepidopterous insects, *S. littura* and *Spilosoma obliqua* (113).

10.5.5.2 Phytoecdysteroids and Their Antagonists

The screening of plants by simple molting bioassays, such as the chilo dipping and fly pupation tests, revealed the presence of ecdysteroids. A variety of such compounds have been isolated and identified from *Ajuga* plants. Among them cyasterone and 20-hydroxyecdysone are the most abundant, but ajugalactone is the most characteristic with unique structural features (114). This compound exhibits dual bioactivity: molting hormone activity, when injected into diapausing pupae of *Manduca sexta*, and antiponasterone A activity, in the chilo dipping test. The structures of cyasterone, 20-hydroxyecdysone, and ajugalactone are shown in Fig. 10.11.

We know at present more than 200 ecdysteroids, most of which occur in plants. Several plant derived steroids are analogues of the insect molting

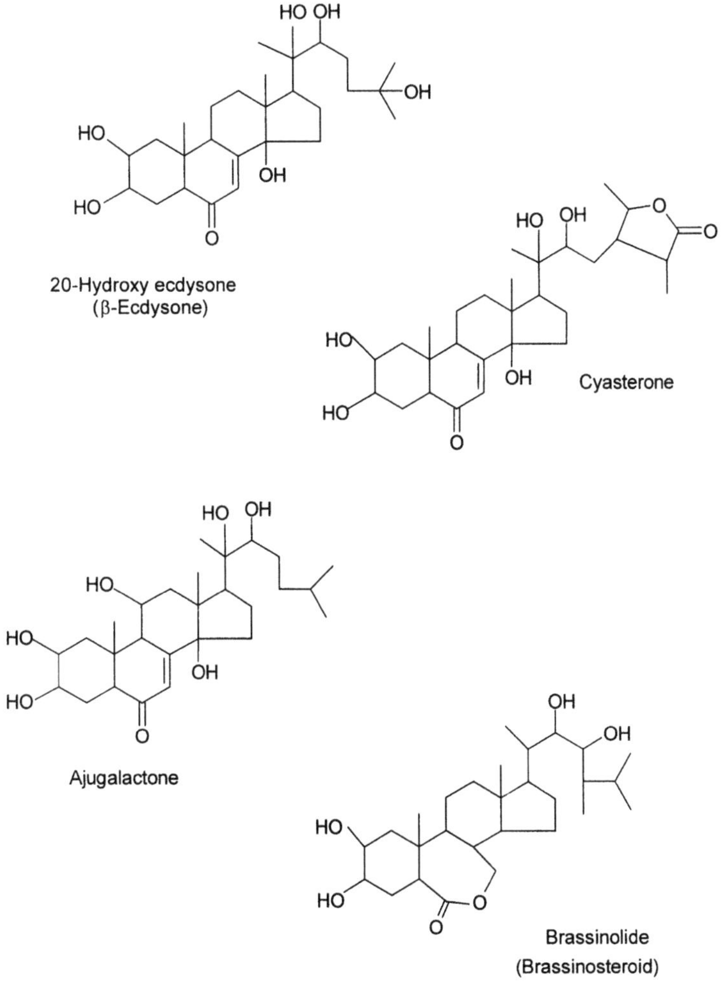

FIGURE 10.11 Structure of ecdysteroids.

hormone ecdysone. By preventing molting, these compounds act as potent insecticides. Ponasterone A, from the conifer *Podocarpus gracilior*, is lethal to a wide range of insect species (115).

Brassinosteroids (Fig. 10.11) represent the first true antiecdysteroids observed. They are a family of growth-promoting hormones found in plants and have striking structural similarities with the ecdysteroids (116). The effect of brassinosteroids on insects may be explained by their competition with molting hormones at the binding site of the hormone receptor which results in delayed molting. Two few triterpenoids isolated from seeds of cruciferous plant, cucurbitacin B and D, were demonstrated to be insect steroid hormones antagonists acting at the ecdysteroid receptor (117). Another ecdysis-inhibiting terpenoid is the azadirachtin from neem tree (109). The quantitative, concentration-dependent dose–response relationships suggest that ecdysteroids may be used by peripherals for homeostatic regulation and mutual synchronization of growth among different tissues and organs. An exogenous supply of ecdysteroid before the endogenous peak always accelerates ecdysis, whereas the same treatment after the peak always inhibits ecdysis in a dose-dependent manner.

Phytoecdysones are occasionally present in unbelievably large amounts. They are found in lower as well as in higher plants, in annuals or perennials. They occasionally accumulate in roots, but sometimes also in leaves, flowers, fruits, or seeds (118). Certain families of plants are a better source for ecdysteroids than others. The plant *Leuzea carthamoides* seems to be the richest source of ecdysteroids (119). These plants do not contain a variety of other polyhydroxylated sterols lacking the α,β-unsaturated 6-keto group. Certain plants in the course of their evolution acquired the ability to synthesize the α,β-unsaturated 6-ketoecdysteroid molecules with selective advantage to restrain most insect feeders. The ecdysteroid content in *Leuzea* surpasses the maximal tolerated doses in insect diet (25–50 ppm) by two orders of magnitude. Five-year-old plantations of this plant are absolutely free from any insect feeder. Ecdysteroid positively appears as a nontoxic, selectively acting, and ecologically safe defensive plant substance.

10.5.5.3   Miscellaneous Steroids and Triterpenoids

Gossypol is the predominant triterpenoid present in the lysigenous glands in seeds of most *Gossypium* spp. It also occurs in glands of foliar plant parts. Gossypol is toxic to *Helicoverpa* spp and appears to be involved in the resistance of certain varieties of cotton to this insect. The effects of gossypol on growth of the cotton bollworm, *Helicoverpa armigera* (Huebner), and development of its endoparasitoid *Campoletis chlorideae* (Uchida), have been studied (120). Growth of *H. armigera* larvae was accelerated by the addition of 0.1% gossypol in the artificial diet, causing 10.75% reduction of the vulnerable period to *C. chlorideae*, while the suppressive activity of 0.5% gossypol to *H. armigera* larvae prolonged the vulnerable period by 28.15%. Negative effects of gossypol on the development of the parasitoid were demonstrated by using the artificial diet and cotton varieties WD 151 (glandless) and HG-BR-8 (glanded).

The glycosylated steroid cardenolide inhibits the Na^+-K^+ pump present in animals. Larvae of certain butterflies can ingest cardenolides without being harmed. They store these substances, which renders them poisonous to birds. All of the cardenolides are toxic to vertebrates but apparently much less so to insects (121).

Saponins are of two types: steroidal saponins with C_{27} skeleton, such as diosgenin, and triterpene saponins with a C_{30} skeleton, such as β-amyrin. Both are glycosylated at C-3. Saponins have been shown to act as toxins and feeding deterrents to species of mites, lepidopterans, beetles, and many other insects (122). Digitonin, a steroidal saponin from foxglove (*Digitalis*), inhibits molting of the leek moth, *Acrolepiosis assectella*, when added to the insect's diet (123).

10.6 MECHANISM OF ACTION

The mode of action for all the bioactivities of monoterpenoids has not been elucidated, but there have been studies that provide some clues to their possible mechanism of action. Monoterpenoids can induce several different types of bioactivities. The monoterpenoid, limonene, induces neurotoxicity in the earthworm, *Eisenia foetida*. Neurotoxic activity has been recorded using electrophysiological recording techniques. The primary effects were reduced conduction velocity and blocking of all neural activity in the medial and lateral giant fibers. Several of the symptoms were similar to those demonstrated by dieldrin in earlier experiments. Pulegone, myrcene, α-terpineol, and linalool also exhibited the same neurotoxic effects (42,53,54). Polyhalogenated monoterpenoids from marine alga have been shown to elicit lindane (organochlorine pesticide) like GABA antagonist effects (124). Acetylcholinesterase inhibition has also been reported as a possible mode of action. The insects' sensory receptors could be affected by species-specific or compound-specific repellent activity of the monoterpenoids. A monoterpenoid in a plant may be repellent to many insects, yet be attractive to certain host-specific species. Specific studies are needed to elucidate their mechanisms of action as repellents. Although monoterpenoids have definite effects on insect growth, development, and reproduction, the mechanism is not well understood.

Sesquiterpene lactones function as repellent, antifeedant, and oviposition deterrents, and a few sesquiterpene lactones are inhibitors of insect development (68,72). The feeding deterrent activity of the sesquiterpenoids, polygodial, and warburganol (drimane dialdehyde) against insects appears to be a direct result of their action on taste receptors. These substances block the stimulatory effects of glucose and sucrose on chemosensory receptor cells located on the mouthparts (125). The sulfhydryl group at the receptor site binds covalently to the double bond of the α,β-unsaturated aldehyde moiety or the amino group at the receptor site interact with the aldehyde groups forming a pyrrole (126). The sesquiterpene lactones possess an exocyclic α-methylene group on the lactone ring. This group is involved in alkylating biological nucleophiles (sulfhydryl and amino groups of proteins). The sulfhydryl reagents, such as cysteine, prevent sesquiterpene lactone poisoning, indicating the role of alkylation reactions by sesquiterpene lactones (127). Juvenile hormone analogues cause premature metamorphosis of larvae and sterilization of adult insects (87).

Many diterpenoids function as feeding deterrents or antifeedants (101). The alkaloidal diterpene ryanodine poisons muscles by binding at the Ca^{2+}-activated open state of the channel involved in the release of contractile Ca^{2+} from sarcoplasmic reticulum. This results in uncoupling of the electrical signal of the transverse tubule from the Ca^{2+} release mechanism of the sarcoplasmic reticulum (93).

The toxicity of the terpenoid aldehyde compounds hemigossipolone, gossypol, and heliocides to insects is believed to result from interaction of the aldehyde group with the amino groups of proteins in the intestinal tract causing disturbances in the digestibility of protein or with the digestive enzymes themselves. Protein digestion will be hindered and insect growth reduced (128).

Saponins' toxicity to insects seems to be due to their binding with free sterols in the gut, thus reducing the rate of sterol uptake into the hemolymph (122). Reductions in sterol uptake from the gut could prove detrimental because insects cannot synthesize sterols. The steroids must be obtained from dietary sources only. Sterols serve as precursors for the synthesis of molting hormones. Hence, by reduced intake of sterol, saponins could interfere with the molting process of insects.

Ecdysteroids such as ecdysone controls the molting cycle of larvae. When insects eat plants containing phytoecdysone, the molting process is disturbed and the larvae die (129). Limonoids mostly function as antifeedant and growth retardant (104). Retention of azadirachtin in insects is extremely low and yet sufficient to initiate many endocrine and thus morphological and behavioral effects. A major action of azadirachtin is to modify insect hemolymph ecdysteroid titers due to a blockage of release of prothoraciocotropic hormone from the brain corpus (117).

10.7 MOLECULAR BIOLOGY

Advances in molecular biology require a more complete understanding of the structure–function relationship of terpenoid biosynthetic enzymes, along with a much better appreciation for metabolic channels and the regulation of terpenoid biosynthesis (130). Gray attempted to integrate the chemistry and biochemistry of all terpenoids produced in plants as a means of identifying the many mechanisms possibly regulating terpenoid biosynthesis in plants (131).

The recent review by Bohlmann et al. (132) on terpenoid synthesis deals with the synthesis of monoterpenoids, diterpenoids, and sesquiterpenoids. Sequence relatedness and phylogenetic reconstruction, based on 33 members of the *Tps* gene family, are delineated and comparison of important structural features of these enzymes is provided.

Considerable progress has been made on cloning genes of the terpenoid biosynthetic pathway including the enzymes of putative rate-limiting steps. They include genes of HMG-CoA reductase (26,133) phytoene synthase (134), and the cyclases of the mono-, sesqui-, and diterpenoids (135–137). Sequences of a monoterpene cyclase (135), a sesquiterpene cyclase (136), and a diterpene cyclase (137) have been cloned using conventional cloning strategies.

cDNA clones of sesquiterpene cyclase resembling cotton Δ-cadinene synthase with a deduced polypeptide of 548 amino acids have been studied in tomato (138). A family of HMG receptor genes has been isolated from higher plants (133), and expression of genes suggests that isoenzymes of HMGR are regulated independently from each other to form specific terpenoid classes. Potrykus et al. (139) activated the terpenoid pathway with an endosperm-specific transgenic phytoene synthase such that the endosperm synthesizes good quantities of phytoene for further conversion to tetraterpenoids.

The coding region of the yeast mevalonate kinase gene (*ERG12*) under the control of cauliflower mosaic virus (CaMV) 35S promoter has been inserted in tobacco using an *Agrobacterium tumefaciens* binary vector system. The specific mevalonate kinase activity in young plantlets was increased by about 60% on average (140).

The characterization of two pepper cDNAs, Cap TKT 1 and Cap TKT 2, that encode *trans*-ketolases having distinct and dedicated specificities is described (14). Cap TKT 2 initiates the synthesis of terpenoids in plastids via the nonmevalonate pathway. From pyruvate and glyceraldehyde-3-phosphate, Cap TKT 2 catalyzes the formation of 1-deoxyxylulose 5-phosphate, the IPP precursor. Cap TKT 2 is overexpressed during chloroplast to chromoplast transition (141). However, information on molecular biology to confer or increase resistance to herbivores is limited.

Grand fir (*Abies grandis* Lindl.) has been developed as a model system for the study of wound-induced oleoresinosis in conifers as a response to insect attack (142). Oleoresin is a roughly equal mixture of turpentine (85% monoterpenoids and 15% sesquiterpenoids) and rosin (diterpene resin acids) that acts to seal wounds and is toxic to both invading insects and their pathogenic fungal symbionts. The dynamic regulation of wound-induced oleoresin formation was studied over 29 days at the enzyme level by in vitro assay of the three classes of synthases directly responsible for the formation of monoterpenoids, sesquiterpenoids, and diterpenoids from the corresponding C_{10}, C_{15}, and C_{20} prenyl diphosphate precursors and at the gene level by RNA blot hybridization using terpene synthase class-directed DNA probes. In overall appearance, the shapes of the time–course curves for all classes of synthase activities are similar, suggesting coordinate formation of all of the terpenoid types. However, closer inspection indicates that the monoterpene synthases arise earlier, as shown by an abbreviated time course over 6–48 h. RNA blot analyses indicated that the genes for all three classes of enzymes are transcriptionally activated in response to wounding, with the monoterpene synthases up-regulated first (transcripts detectable 2 h after wounding), in agreement with the results of cell-free assays of monoterpene synthase activity, followed by the coordinately regulated sesquiterpene synthases and diterpene synthases (transcription beginning on 3–4 h). The differential timing in the production of oleoresin components of this defense response is consistent with the immediate formation of monoterpenoids to act as insect toxins and their later generation at solvent levels for the mobilization of resin acids responsible for wound sealing (142).

The field of isoprenoid biosynthesis has advanced to the point where several approaches can now be considered for the rational manipulation of essential oil formation. These include the mutagenesis of specific terpenoid synthases to alter product outcome, the manipulation of a biosynthetic pathway by either altering the activity of an existing enzyme or by introducing novel enzymes, and the use of genetic approaches to manipulate the formation of the specialized anatomical structures such as glandular trichomes. A yeast mutant that excretes large amounts of geraniol and linalool contains an altered FPP synthase that has a single Lys197 to Glu. The lysine at this position is involved in substrate binding. Mutation at this position aborts FPP production (143).

Studies on the site-directed mutagenesis of trichodiene synthase have demonstrated the ability of a fungal sesquiterpene cyclase to form aberrant products following minor modification of the active site. Replacement of Arg134 with Lys results in a drastically decreased rate of trichodiene formation with concomitant formation of several novel sesquiterpene hydrocarbons. Replacement of Tyr305 with Thr results in the formation of approximately 50% of one of these novel products (144).

An important approach to bioengineering of a metabolic pathway involves the introduction of new or modified enzymes not normally present in the plant. For example, the production of C_3 hydroxylated monoterpenoids by a mutant of Scotch spearmint was attempted. Replacement of the normal limonene-6-hydroxylase with a C_3 hydroxylase activity results in the formation of numerous C_3 hydroxylated monoterpenoids not naturally found in spearmint varieties (145). The availability of cDNA clones for limonene synthase raises the possibility of engineering limonene production into crop plants to improve resistance to insect pests (146). The availability of cDNA clones for other cyclases and related enzymes of monoterpenoid metabolism provides the opportunity for manipulating composition in essential oil–producing plants as well as introducing the formation of bioactive terpenoids in crop plants.

The biosynthetic pathways and genes for some terpenoids have been known for many years; new pathways have been found and known pathways further investigated. Barkovich and Liao (147) reviewed the recent advances in metabolic engineering of terpenoids, focusing on the molecular genetics that affects pathway engineering the most. Examples in mono-, sequi-, and diterpenoid synthesis as well as carotenoid production are discussed.

Following the recent discovery of the role of the 2-C-methyl-D-erythritol-4-phosphate (MEP) pathway in the biosynthesis of plastidial terpenoids, such as the carotenoids and monoterpenes and diterpenes, several genes of this pathway have been cloned (148–151). Modifying the MEP pathway is potentially useful for a wide range of applications. For example, a cosuppression and an antisense strategy used to knock out a cytochrome P450 enzyme in tobacco trichome glands conferred an increased resistance to aphids (152). There was a clear shift in the cembranoid spectrum, with a 19-fold increase of the diterpenoid cembratrieneol and a decrease in its oxidation product cembratrienediol. In another example, overexpression of a chimeric farnesyl diphosphate synthase gene in *Artemisia annua* was reported to increase the flux in the sesquiterpenoid biosynthetic pathway leading to a twofold to threefold increase in the antimalaria drug (153). In tomatoes, overexpression of an S-linalool synthase transgene increased many fold the production of the monoterpenoid flavor compound S-linalool, compared with control plants, although no changes in the levels of other terpenoids were observed (154). Possibilities for genetic engineering of essential oil production in mint have been reviewed (150). Overexpression of the gene encoding deoxyxylulose phosphate reductoisomerase (DXR) in mint resulted in plants that had twofold to fourfold higher DXR activities (149). The plants had a normal phenotype and an almost 50% increase in essential oil (monoterpenoid) production.

Increasing menthol levels in mint oil might be achieved by cutting one of the competitive branches in the monoterpenoid metabolic network leading to menthofuran (149). An attempt was made to block the branch that channels pulegone away from the menthol pathway by using the antisense gene of menthofuran synthase. Most transgenic plants did not show any effect, but a few had dereased levels of menthofuran (35–55%) and more menthol than wild-type plants. Surprisingly, these plants also had a lower pulegone level.

10.8 PERSPECTIVES

The discovery process for natural pesticides is more complicated than for synthetic pesticides because of the difficulties in isolation, purification, and identification. A large number of plant species are available for evaluation against the insect pests. A particular plant product has to be screened against a wide spectrum of insect species since the same compound may be nontoxic to one species but toxic to the other species. Even after careful identification, methods are needed to produce a natural pesticide for commercial use and the most cost-effective means of production must be found. So far, only limited plant materials have been screened on limited pest species. Even the most effective compounds have not been tried on a commercial scale. If chemical synthesis of the compound is not feasible or viable, altering the biosynthetic pathways could be attempted. Genetically manipulating the plants by classical or biotechnological methods could also increase production of the terpenoids. Cell lines that produce higher levels of the compound can be identified and exploited. There are many thousands of terpenoid compounds produced by plants, but only a few of these have been tested for pesticidal activity. Only limited terpenoids, such as pyrethrins and azadirachtin, have been exploited on a commercial scale. Compounds like phytoecdysteroids, sesquiterpene lactones, limonene, and cineole could be further studied for use as natural pesticidal compounds.

Detailed mechanisms of action at the molecular level have to be studied for each group of terpenoids. More specific studies on structure–activity relationships and their mode of action should provide a better understanding of the bioactivity of terpenoids.

REFERENCES

1. Loomis, W.D.; Croteau, R. Biochemistry of terpenoids. In *Biochemistry of Plants, Lipids: Structures and Function*; Stumpf, P.K., Ed.; Academic Press: New York, 1980; Vol. 4, 363–418.
2. Ruzicka, L.; Eschenmoser, A.; Hausser, H. Isoprene rule and the biogenesis of terpenic compounds. Experientia. **1953**, *9*, 357–367.
3. Nes, W.R.; Mckean, M.L.; Biochemistry of Steroids and Other Isopentenoids. University Park Press: Baltimore, 1977.
4. Dev, S.; Nagasampagi, B.A. Terpenoids. In *Acyclic, Monocyclic, Bicyclic, Tricyclic and Tetracyclic Terpenoids*; Dev, S., Ed.; CRC Handbook of Terpenoids, CRC Press: Boca Raton, 1989; Vol. 1.
5. Dev, S.; Gupta, A.S.; Patwardhan, S.A. Triterpenoids. In *Pentacyclic and Hexacyclic Triterpenoids*; Dev, S., Ed.; CRC Handbook of Terpenoids, CRC Press: Boca Raton, 1989; Vol. 2.
6. Dev, S. Diterpenoids (4 volumes). CRC Press: Boca Raton, 1985–1986.
7. Connolly, J.D.; Hill, R.A. Terpenoids. In *Methods in Plant Biochemistry*; Banthorpe, D.V., Banthorpe, D.V., Eds.; Academic Press: London, 1991; Vol. 7, 361–368.
8. Hill, R.A.; Kirk, D.N.; Makin, H.L.J.; Murphy, G.M. Dictionary of Steroids. Chapman and Hall: London, 1991.
9. Bonner, J. Terpenoids. In *Plant Biochemistry*;. Bonner, J.; Varner, J.E., Eds.; Academic Press: New York, 1965; 665–692.
10. Devon, T.K.; Scott, A.I. Handbook of Naturally Occurring Compounds. Academic Press: New York, 1972; Vol. 2.
11. Nicholas, H.J. In *Phytochemistry*; Miller, L.P., Ed.; Van Nostrand–Reinhold: New York, 1973; Vol. 2, 254–309.
12. Robinson, T. The Organic Constituents of Higher Plants. 4th Ed.; Cordus Press: North Amberst, MA, 1980.
13. Guenther, E. The Essential Oils. Van Nostrand: New York, Vols. I–VI, 1976.
14. Banthorpe, D.V.; Charlwood, B.V. Terpenoids. In *Encyclopedia of Plant Physiology, New Series*; Pirson, A., Zimmermann, M.H., Eds.; Springer-Verlag: Berlin, 1980; Vol. 8, 185–220.
15. Dey, P.M.; Harborne, J.B. Isoprenoid metabolism. In *Plant Biochemistry*; Dey, P.M., Harborne, J.B., Eds.; Academic Press: New York, 1997; 417–438.
16. Hans-Walter Heldt. Plant Biochemistry and Molecular Biology. Oxford University Press: Oxford, 1997.
17. Torssell, K.B.G. Natural Product Chemistry. John Wiley and Sons: Chichester, 1983.
18. A Specialist Periodical Report, Terpenoids and Steroids. Chemical Society: London, 1977; Vol. 7.
19. Nes, W.D.; Fuller, G.; Tsai, L.S., Eds.; Isopentenoids in Plants: Biochemistry and Function, Marcel Dekker: New York, 1984.
20. Cordell, G.A. The occurrence, structure, elucidation and biosynthesis of sesterterpenes. Phytochemistry **1994**, *13*, 2343–2364.

21. Darnley, G.R. Chemotaxonomy of Flowering Plants. McGill Queons University Press: Montreal, 1994; Vols. 1–4.
22. Mahato, S.B.; Nandy, A.K.; Roy, G. Triterpenoids. Phytochemistry 1992, 31, 2199–2249.
23. Croteau, R. Monoterpenoids. Chem. Rev. 1987, 87, 929–954.
24. Adam, G.; Marquardt, V. Brassinosteroids. Phytochemistry 1986, 25, 1787–1799.
25. Luthra, R.; Luthra, P.M.; Kumar, S. Redefined role of mevalonate-isoprenoid pathway in terpenoid biosynthesis in higher plants. Curr. Sci. 1999, 76, 133–135.
26. Chappell, J. Biochemistry and molecular biology of the isoprenoid biosynthetic pathway in plants. Annu. Rev. Plant Physiol. Plant Mol. Biol. 1995, 46, 521–547.
27. Cane, D.E. Sesquiterpenoids. In Biosynthesis of Isoprenoid Compounds; Porter, J.W., Spurgeon, S.L., Eds.; John Wiley and Sons: New York, 1981; 283–374.
28. West, C.A. Diterpenes. In Biosynthesis of Isoprenoid Compounds; Porter, J.W.; Spurgeon, S.L.; Eds.; John Wiley and Sons: New York, 1981; 375–371.
29. Benveniste, P. Sterol biosynthesis. Annu. Rev. Plant Physiol. 1986, 37, 275–308.
30. Charlwood, B.V.; Charlwood, K.A. Monoterpenoids. In Methods in Plant Biochemistry: Terpenoids; Charlwood, B.V., Banthorpe, D.V., Eds.; Academic Press: London, 1991; Vol. 7, 43–98.
31. Klocke, J.A. Natural Plant compounds useful in insect control. In Allelochemicals: Role in Agriculture and Forestry. ACS. Symp. Ser. 1987, 330, 336–415.
32. Hwang, Y.S.; Wu, K.H.; Kumamoto, J.; Axelrod, H.; Mulla, M.S. Isolation and identification of mosquito repellents from Artemisia vulgaris. J. Chem. Ecol. 1985, 51, 1297–1306.
33. Obeng-Ofori, D.; Reichmuth, C.H.; Bexele, J.; Hassalali, A. Biological activity of 1,8-cineole, a major component of essential oil of Ocimum kenyense against stored product beetles. J. Appl. Entomol. 1997, 121, 237–244.
34. Desphande, R.S.; Adhikary, P.R.; Tipnis, H.P. Stored grain pest control agents from Nigella sativa and Pogostemon heyneanus. Bull. Grain Tech. 1974, 12, 232–234.
35. Desphande, R.S.; Tipnis, H.P. Insecticidal activity of Ocimum basilicum Linn. Pesticides 1977, 11, 11–12.
36. Mishra, A.K.; Dubey, N.K. Evaluation of some essential oils for their toxicity against fungi causing deterioration of stored food commodities. Appl. Environ. Microbiol. 1994, 60, 1101–1105.
37. Ellis, M.D.; Baxendale, F.P. Toxicity of seven monoterpenoids to trachaeal mites and their honey bee hosts when applied as fumigants. J. Econ. Entomol. 1997, 90, 1087–1091.
38. Selander, J.; Kalo, P.; Kangus, E.; Perttunen, V. Olfactory behaviour of Hylobius abietes L. 1. Response to several terpenoid fractions isolated from Scots pine phloem. Ann. Entomol. Fenn. 1974, 40, 109–115.
39. Brattsten, L.B. Cytochrome P450 involvement in the interactions between Plant terpenes and insect herbivores. ACS. Symp. Ser. 1983, 208, 173–195.

40. Hughes, P.R. Myrcene: a precursor of pheremones in Ips beetles. J. Insect Physiol. **1974**, *20*, 1271–1275.
41. Camacho, A.D.; Pierce, Jr, H.D.; Boron, J.H. Host compounds as kairomones for the western balsam bark beetle, *Dryocoetes confusus*. J. Appl. Entomol. **1998**, *122*, 289–293.
42. Jacobson, M. Glossary of Plant derived insect deterrents. CRC Press: Boca Raton, 1990.
43. Lindgren, B.S.; Nordiander, G.; Birgersson, G. Feeding deterrence of verbenone to the pine weevil, *Hyloblus abietis* (L.). J. Appl. Entomol. **1996**, *120*, 397–403.
44. Salom, S.M.; Gray, J.A.; Alford, A.R.; Mulesky, M.; Fettig, C.J.; Woods, S.A. Evaluation of natural products as antifeedants for pales weevil and as fungitoxins for *Leptographium procerum*. J. Entomol. Sci. **1996**, *31*, 453–465.
45. Bowers, M.D.; Puttick, G.M. Response of generalist and specialist insects to qualitative allelochemical variation. J. Chem. Ecol. **1988**, *14*, 319–334.
46. Henrick, C.A. Pyrethroids. In *Agrochemicals from Natural Products*; Godfrey, C.R.A., Ed.; Marcel Dekker: New York, 1995; 63–145.
47. Crombie, L. Chemistry and biosynthesis of natural pyrethrins. Pest. Sci. **1980**, *11*, 102–118.
48. Bowers, W.W. Phytochemical disruption of insect development and behavior. ACS. Symp. Ser. **1985**, *276*, 225–236.
49. Gershenzon, J.; Croteau, R. Terpenoids. In *Herbivores: Their Interactions with Secondary Metabolites, The Chemical Participants*; Rosenthal, G.A., Berenbaum, M.R., Eds.; Academic Press: New York, 1991; Vol. 1, 165–219.
50. Lee, S.; Tsao, R.; Coats, J.R. Influence of dietary applied monoterpenoids and derivatives on survival and growth of the European corn borer. J. Econ. Entomol. **1999**, *92*, 56–67.
51. Hassanali, A.; Lwande, W.; Antipest secondary metabolites from African plants. In Insecticides of Plant Origin. Arnason, J.T.; Philogene, B.J.R.; Morand, P. Eds. ACS. Symp. Ser. **1989**, *387*, 78–94.
52. Konstantopoulou, I.; Vassilopopoulou, L.; Tsipidou, P.M.; Scouras, Z.G. Insecticidal effects of essential oils. A study of the effects of essential oils extracted from eleven Greek aromatic plants on *Drosophila auraria*. Experientia. **1992**, *48*, 616–619.
53. Lee, S.; Tsao, R.; Peterson, C.; Coats, J.R. Insecticidal activity of monoterpenoids to western corn rootworm, two spotted spider mite and house fly. J. Econ. Entomol. **1997**, *90*, 883–892.
54. Rice, P.J.; Coats, J.R. Insecticidal properties of several monoterpenoids to the house fly, red flour beetle and southern corn root worm. J. Econ. Entomol. **1994**, *87*, 1172–1179.
55. Karr, L.L.; Coats, J.R. Effects of four monoterpenoids on growth and reproduction of the German cockroach. J. Econ. Entomol. **1992**, *85*, 424–429.
56. Harwood, S.H.; Moldenke, A.F.; Berry, R.E. Toxicity of peppermint monoterpenes to the variegated cutworm. J. Econ. Entomol. **1990**, *83*, 1761–1767.
57. Karr, L.L.; Coats, J.R. Insecticidal properties of *d*-limonene. J. Pest. Sci. **1988**, *13*, 287–290.

58. Rice, P.J.; Coats, R.R. Insecticidal properties of monoterpenoid derivatives to the house fly and red flour beetle. Pest. Sci. **1994**, *41*, 195–202.

59. Casida, J.E.; Lawrence, L.J. Structure-activity correlations for interactions of bicyclophosphorus esters and some polychlorocycloalkane and pyrethroid insecticides with the brain specific t-butylbicyclo-phosphorothionate receptor. J. Environ. Health Perspect. **1985**, *61*, 123–132.

60. Ahn, Y.J.; Lee, S.B.; Lee, H.S.; Kim, G.H. Insecticidal and acaricidal activity of carvacrol and β-thujaplicine derived from *Thujopsis dolabrata* var *hondai* sawdust. J. Chem. Ecol. **1998**, *24*, 81–90.

61. Picman, A.K. Effect of sesquiterpene lactones on insect feeding, development and survival. Biochem. Syst. Ecol. **1986**, *14*, 255–259.

62. Isman, M.B.; Rodriguez, E. Larval growth inhibitors from species of *Parthenium asteraceae*. Phytochemistry **1983**, *22*, 2709–2213.

63. Arnason, J.J.; Isman, M.B.; Philogene, B.J.R.; Woddell, T.G. Mode of action of the sesquiterpene lactone, tenulin from *Helinium amarum* against herbivorous insects. J. Nat. Prod. **1987**, *50*, 690–695.

64. Wang, M.T.; Quin, H.L.; Kong, M.; Li, Y.Z. Insecticidal sesquiterpene polyol ester from *Celastrus angulatus*. Phytochemistry **1991**, *30*, 3931–3934.

65. Wakabayashi, N.; Wu, W.J.; Waters, R.M.; Redfern, R.E.; Mills, G.M.; DeMilo, A.B.; Lusby, W.R.; Andregewski, D. Celangulin: a nonalkaloidal insect antifeedant from Chinese bittersweet, *Celastrus angulatus*. J. Nat. Prod. **1988**, *51*, 537–542.

66. McGovran, E.R.; Mayer, E.L. The toxicity of the natural bitter substances, quassin tenulin, helenalin and picrotoxin and some of their derivatives to certain insects. U.S. Dept. Agric. Bur. Entomol. Plant. Quarantine Entomol. Tech. E **1942**, 572.

67. Burnett, W.C.; Jones, S.B.; Mabry, T.J. Evolutionary implications of sesquiterpene lactones on veronia and mammalian herbivores. Taxon. **1977**, *26*, 203–207.

68. Ivie, G.W.; Witzel, D.A. Sesquiterpene lactones structural biological action and toxicological significance. In *Handbook of Natural Toxins*; Keeler, R.F., Tu, A.T., Eds.; Marcel Dekker: New York, 1983; Vol. 1, 543–584.

69. Jones, S.B.; Burnett, W.C.; Coile, N.C.; Mabry, T.J.; Betkouski, M.F. Sesquiterpene lactones of *Veronia*: influence of glaucolide A on the growth rate and survival of lepidopterous larvae. Oecologia. **1979**, *39*, 71–77.

70. Gegory, P.; Tingey, W.M.; Ave, D.A.; Boutheyette, PY. Potato glandular trichomes: a physicochemical defence mechanism against insects. ACS. Symp. Ser. **1986**, *296*, 160–167.

71. Stipanovic, R.D.; Williams, H.J.; Smith, L.A. Cotton terpenoid inhibition of *Heliothis virescens* development. ACS. Symp. Ser. **1986**, *296*, 79–94.

72. Yano, K. Minor components from growing buds of *Artemesia capillaris* that act as insect antifeedants. J. Agric. Food Chem. **1987**, *35*, 889–891.

73. Williams, L.A.D.; Gardner, M.T.; Singh, P.D.A.; The, T.L.; Fletcher, C.K.; Culed, W.L.; Kraus, W. Mode of action studies of the acaricidal agent epingalone. Invertebr. Reprod. Dev. **1997**, *31*, 231–236.

74. Gutierrez, C.; Fereres, A.; Reina, M.; Cabaer, R.; Gonzalez, CA. Behavioral and sublethal effects of structurally related lower terpenes on *Myzus persicae*. J. Chem. Ecol. **1997**, *23*, 1641–1650.
75. Altman, D.W.; Stipanovic, R.D.; Bell, A.A. Terpenoids in foliar pigment glands of A, D, and AD genome cottons: introgression potential for pest resistance. J. Hered. **1990**, *81*, 447–454.
76. Tsanuo, M.K.; Hassaneli, A.; Jondiko, I.J.O.; Torto, B. Mutangin, a dihydrogarofuranoid sesquiterpene insect antifeedant from *Elaedendron buchanani*. Phytochemistry **1993**, *34*, 65–667.
77. Asakawa, Y.; Dawson, G.W.; Griffiths, D.C.; Lallemand, J.Y.; Ley, S.V.; Mori, K.; Mudd, A.; Pezechik, L.M.; Pickett, J.; Watanabe, H.; Woodcock, C.M.; Zhang, Z.N. Activity of drimane antifeedant and related compounds against aphids and comparative biological effects and chemical reactivity of (+) and (−) polygodial. J. Chem. Ecol. **1988**, *14*, 1845–1855.
78. Zou, J.; Cates, R.G. Effects of terpenes and phenolic and flavonoid glycosides from Douglas fir on Western spruce budworm larval growth, pupal weight and adult weight. J. Chem. Ecol. **1997**, *23*, 2313–2326.
79. McDonough, L.M. Insect juvenile hormones. In *Isopentenoids in Plants*: *Biochemistry and Function*; Nes, W.D., Fuller, G., Tsai, L.S., Eds.; Marcel Dekker: New York, 1984; 81–102.
80. Judy, K.J.; Schooley, D.A.; Dunham, L.L.; Hall, M.S.; Bergot, B.J.; Siddall, J.B. Isolation, structure, and absolute configuration of a new natural insect juvenile hormone from *Manduca sexta*. Proc. Natl. Acad. Sci. USA **1973**, *70*, 1509–1513.
81. Henrick, C.A. Juvenoids In *Agrochemicals from Natural Products*; Godfrey, C.R.A., Ed.; Marcel Dekker: NewYork, 1995; 147–213.
82. Slama, K.; Williams, C.M. Juvenile hormone activity for the bug *Pyrrhocoris apterus*. Proc. Natl. Acad. Sci. USA **1965**, *54*, 411–414.
83. Bowers, W.S.; Fales, H.M.; Thompson, M.J.; Uebel, E.C. Identification of an active compound from balsam fir. Science. **1966**, *154*, 1020–1021.
84. Nishida, R.; Bowers, W.S.; Evans, P.H. Synthesis of highly active juvenile hormone analogs, Juvocimene I and II from the oil of sweet basil, *Ocimum basilicum*. J. Chem. Ecol. **1984**, *10*, 1435–1452.
85. Williams, C.M. The juvenile hormone of insects. Nature **1956**, *178*, 212–213.
86. Nishida, R.; Bowers, W.S.; Evans, P.H. Plant-derived sesquiterpenoids having insect sterilizing effects as juvenile hormone analogs. Arch. Insect Biochem. Physiol. **1983**, *1*, 17–21.
87. Bowers, W.S.; Ohta, T.; Cleere, J.S.; Marsella, P.A. Discovery of insect antijuvenile hormones in plants. Science **1976**, *193*, 542–547.
88. Marshall, G.T.; Klocke, J.A.; Lin, L.J.; Kinghorn, A.D. Effects of diterpene esters of tigliane, daprnane, ingenane, and lathyrane types on pink bollworm *Pectinophora gossypiella* (Lepidoptera: Gelechiidae). J. Chem. Ecol. **1985**, *11*, 191–206.
89. Jacobson, M.; Crosby, D.G. Naturally Occurring Insecticides, Marcel Dekker: New York, 1971.

90. Kloche, J.A.; Hu, M.Y.; Chiu, S.F.; Kubo, I. Grayanoid diterpene insect antifeedants and insecticides from *Rhododendron molle*. Phytochemistry **1991**, *30*, 1797–1800.

91. Xuan, L.J.; Xu, Y.M.; Fang, S.D. Three diterpene dilactone glycosides from *Podocarpus nagi*. Phytochemistry. **1995**, *39*, 1143–1145.

92. Cardellina, J.H. Natural products in the search for new agrochemicals. In *Biologically Active Natural Products, Potential Use in Agriculture*; Cutler, H.G., Ed.; ACS Sym. Ser. **1988**, *380*, 305–315.

93. Alexander, I.C.; Pascoe, K.O.; Manchand, P.; Williams, L.A.D. An insecticidal diterpene from *Croton linearis*. Phytochemistry **1991**, *30*, 1801–1803.

94. Elliger, C.A.; Zinkel, D.F.; Chan, G.B.; Waiss, A.C. Diterpene acids as larval growth inhibitors. Experientia. **1976**, *32*, 1364–1366.

95. Munakata, K. Nematocidal natural products. In *Natural Products for Innovative Pest Management*; Whitehead, D.L.; Bowers, W.S., Eds.; Pergamon Press: Oxford, 1983; 299–309.

96. Siddiqui, B.S.; Faizi, S.; Ghiasuddin, Siddiqui, S. Proc. World Neem Conference, Bangalore, India, 1993; 187–198.

97. Scheffrahn, R.H.; Hsu, R.C.; Su, N.Y.; Huffman, J.B.; Midland, S.L.; Sims, J.J. Allelochemical resistance of bald cypress *Taxodium distichum*, heartwood to the subterranean termite, *Coptotermes formaosanus*. J. Chem. Ecol. **1988**, *14*, 765–776.

98. Camps, F.; Coll, J. Insect allelochemicals from *Ajuga* plants. Phytochemistry **1993**, *32*, 1361–1370.

99. Champagne, D.E.; Koul, O.; Isman, M.B.; Scudder, G.G.E.; Towers, G.H.N. Biological activity of limonoids from the Rutales. Phytochemistry **1992**, *31*, 377–394.

100. Isogai, A.; Murakoshi, S.; Suzuki, A.; Tamura, S. Chemistry and biological activities of cinnzeylanine and cinnzeylanol, a insecticidal substances from *Cinnamonum zeylanium* Nees. Agric. Biol. Chem. **1979**, *41*, 1779–1784.

101. Bremmer, P.D.; Simmonds, M.S.J.; Blaney, W.M.; Vietch, N.C. Neo-clerodane diterpenoid insect antifeedants from *Ajuga reptans* cv catlins glant. Phytochemistry **1998**, *47*, 1227–1232.

102. Stipanovic, R.D.; Bell, A.A.; O'Brien, D.H.; Lukefahr, M.J. Heliocide H$_3$ an insecticidal terpenoid from *Gossypium hirsutum*. Phytochemistry **1978**, *17*, 151–152.

103. McAuslan, H.J.; Alborn, H.T.; Toth, J.P. Systemic induction of terpenoid aldehydes in cotton pigment glands by feeding of larval *Spodoptera exigua*. J. Chem. Ecol. **1997**, *23*, 2861–2879.

104. Panda, N.; Khush, G.S. Secondary Plant metabolites for insect resistance. In *Host Plant Resistance to Insects*. CAB International and International Rice Research Institute: Wallingford, Oxon, UK, 1995; 22–66.

105. Devakumar, C.; Dev, S. Chemistry of neem. In *Neem Research and Development*; Randhawa, N.S., Parmar, B.S., Eds.; Society of Pesticide Science: India. Publication No. 3, 1993; 63–96.

106. Schmutterer, H. Properties and potential of natural pesticides from the neem tree, *Azadirachta indica*. Annu. Rev. Entomol. **1990**, *35*, 271–297.
107. Lowery, D.T.; Isman, M.B. Inhibition of aphid reproduction by neem seed oil and azadirachtin. J. Econ. Entomol. **1996**, *89*, 602–607.
108. Aerts, R.J.; Mordue, A.J. Feeding deterrence and toxicity by neem triterpenoids. J. Chem. Ecol. **1997**, *23*, 2117–2132.
109. Schmutterer, H. The Neem Tree: Source of Unique Natural Products for Integrated Pest Management, Medicine, Industry and Other Purposes. VCH, New York, 1995.
110. Jimenez, A.R.; Mata, R.; Peredamiranda, R.; Calderon, J.; Esman, M.B.; Nicol, R.; Amason, J.T. Insecticidal limonoids from *Swietenia humilis* and *Cedrela salvadorensis*. J. Chem. Ecol. **1997**, *23*, 1225–1234.
111. Jayaprakasha, G.K.; Singh, R.P.; Pereira, J.; Sakaria, K.K. Limonoids from *Citrus reticulata* and their moult inhibiting activity in mosquito *Culex quinquesfasciatus* larvae. Phytochemistry **1997**, *44*, 843–846.
112. Yasui, H.; Kato, A.; Yazawa, M. Antifeedants to armyworms *Spodoptera litura* and *Pseudaletia separata* from bittergourd leaves. J. Chem. Ecol. **1998**, *24*, 803–813.
113. Tandon, M.; Shukla, Y.N.; Tripathi, A.K.; Singh, S.C. Insect antifeedant principles from *Veronia cinerea*. Phytother. Res. **1997**, *12*, 195–199.
114. Camps, F.; Coll, J. Insect allelochemicals from *Ajuga* plants. Phytochemistry **1993**, *32*, 1361–1370.
115. Lee, S.M.; Klocke, J.A.; Barnby, M.A.; Yamasaki, R.B.; Balandrin, M.F. Insecticidal constituents of *Azadirachta indica* and *Melia azedarach* (Meliaceae). In Naturally Occurring Pest Bioregulators; Hedin, P.H., Ed.; ACS Symp, Ser. **1991**, *449*, 293–304.
116. Richter, K.; Koolman, J. Antiecdysteroid effects of brassinosteroids in insects. In *Brassinosteroids: Chemistry, Bioactivity and Applications*; Cutler, H.G., Yokota, T., Adam, G., Eds.; American Chemical Society: Washington, DC, 1991; 265–278.
117. Dinan, L.; Whiting, P.; Girault, J.P.; Lafont, R.; Dhadialla, T.S.; Cress, D.E.; Mugat, B.; Antoniewski, C.; Lepesant, J.A. Cucurbitacins are insect steroid hormone antagonists acting at the ecdysteroid receptor. Biochem. J. **1997**, *327*, 643–650.
118. Bergamasco, R.; Horn, D.H.S. Distribution and role of insect hormones in plants. In *Endocrinology of Insects*; Downer, R.G.H., Laufer, H., Eds.; Alan R.Liss: New York, 1983; 627–654.
119. Slama, K. Ecdysteroids. Insect hormones, Plant defensive factors or human medicine? Phytoparasitica. **1993**, *21*, 3–8.
120. Wang, C.; Yang, Q.; Zhou, M. Effect of gossypol on growth of the cotton bollworm and development of its parasitoid *Campoletis chlorideae*. Entomologica Sinica. **1997**, *4*, 182–188.
121. Seiber, J.N.; Lee, S.M.; Benson, J.M. Characteristics and ecological significance of cardenolides in *Asclepias* (milkweed) species. In *Isopentenoids in Plants*. Nes, W.D., Fuller, G., Tsai, L.S., Eds.; Marcel Dekker: New York, 1984; 563–588.

122. Ishaaya, I.; Birk, Y.; Bondi, A.; Tencer, Y. Soyabean saponins. IX. Studies of their effect on birds, mammals and cold-blooded organisms. J. Sci. Food Agric. **1969**, *20*, 433–436.

123. Arnault, C.; Mauchamp, B. Ecdysis inhibition in *Acrolepiosis assectella* larvae by digitonin: antagonistic effects of cholesterol. Experientia **1985**, *41*, 1074–1077.

124. San Martin, A.; Negrete, A.; Rovirosa, J. Insecticide and acaricide activities of polyhalogenated monoterpenes from chilean *Plocamium cartilagineum*. Phytochemistry **1992**, *30*, 2165–2169.

125. Ma, W.C. Alteration of chemoreceptor function in armyworm larvae (*Spodoptera exempta*) by a Plant derived sesquiterpenoid and by sulfhydryl reagents. Physiol. Entomol. **1977**, *2*, 199–207.

126. Fritz, G.L.; Mills, G.D.; Warthen, J.D.; Waters, R.M. Reimer-Tiemann. adducts as potential insect antifeedant agents: reviewing the structure–activity relationship theory of the antifeedant warburganal. J. Chem. Ecol. **1989**, *15*, 2607–2623.

127. Isman, M.B. Toxicity and tolerance of sesquiterpene lactones in the migratory grasshopper, *Melanopus sanguinipes* (Acrididae). Pest. Biochem. Physiol. **1985**, *24*, 348–354.

128. Meisner, J.; Navon, A.; Zur, M.; Ascher, K.R.S. The response of *Spodoptera littoralis* larvae to gossypol incorporated in an artificial diet. Environ. Entomol. **1977**, *6*, 243–244.

129. Dhadialla, T.S.; Carlson, G.R.; Le, D.P. New insecticides with ecdysteroidal and juvenile hormone activity. Annu. Rev. Entomol. **1998**, *43*, 545–569.

130. Kleinin, H. The role of plastids in isoprenoid biosynthesis. Annu. Rev. Plant Physiol. Plant Mol. Biol. **1989**, *40*, 39–59.

131. Gray, J.C. Control of isoprenoid biosynthesis in higher plants. Adv. Bot. Res. **1987**, *14*, 25–91.

132. Bohlmann, J.; Meyer, G.G.; Croteau, R. Plant terpenoid synthases: molecular biology and phylogenetic analysis. Proc. Natl. Acad. Sci. USA **1998**, *95*, 4126–4133.

133. Choi, D.; Ward, B.L.; Bostock, R.M. Differential induction and suppression of potato 3-hydroxy-3-methylglutaryl CoA reductase genes in response to *Phytophthora infestans*. Plant Cell **1992**, *4*, 1333–1336.

134. Bartley, G.E.; Scolnik, P.A. cDNA cloning, expression during development and genome mapping of PSY 2, a second tomato gene encoding phytoene synthase. J. Biol. Chem. **1993**, *268*, 25718–25724.

135. Colby, S.M.; Alonso, W.R.; Katchira, E.J.; McGarvey, D.J.; Croteau, R. 4S-Limonene synthase from the oil glands of spearmint (*Mentha spicata*). cDNA isolation, characterization and bacterial expression of the catalytically active monoterpene cyclase. J. Biol. Chem. **1993**, *268*, 23016–23024.

136. Facchini, P.J.; Chappell, J. Gene family for an elicitor induced sesquiterpene cyclase in tobacco. Proc. Natl. Acad. Sci. USA **1992**, *89*, 11088–11092.

137. Mau, C.J.; West, C.A. Cloning of casbene synthase cDNA: evidence for conserved structural features among terpenoid cyclases in plants. Proc. Natl. Acad. Sci. USA. **1994**, *91*, 8479–8501.

138. Colby, S.M.; Crock, J.; Dowdle, R.B.; Lemaux, P.G.; Croteau, R. Germacrene C synthase from *Lycopersicon esculentum* cv VFNT cherry tomato: cDNA isolation, characterization and bacterial expression of the multiple product sesquiterpene cyclase. Proc. Natl. Acad. Sci. USA **1998**, *95*, 2216–2221.

139. Potrykus, I.; Armstrong, G.A.; Beyer, P.; Bieri, S.; Burkhardt, P.K.; Ding-Chen, H.; Ghosh Biswas, G.C.; Datta, S.K.; Futterer, J.; Kloti, A.; Spangenberg, G.; Terada, R.; Wunn, J.; Zhao, H. Transgenic indica rice for the benefit of less developed countries toward fungal, insect, and viral resistance and accumulation of beta carotene in the endosperm. In *Rice Genetics: Proceedings*, IRRI, Los Banos, Khush, G.S., Ed.; Laguna: Philippine, 1997; 179–187.

140. Champenoy, S.; Tourte, M. Expression of the yeast mevalonate kinase gene in transgenic tobacco. Mol. Breed. **1998**, *4*, 291–300.

141. Bouvier, F.; d'Harlingue, A.; Suire, C.; Backhaus, R.A.; Camara, B. Dedicated roles of plastid transketolases during the early onset of isoprenoid biogenesis in pepper fruits. Plant Physiol. **1998**, *117*, 1423–1431.

142. Steele, C.L.; Katoh, S.; Bohlmann, J.; Croteau, R. Regulation of oleoresinosis in grand fir (*Abies grandis*) differential transcriptional control of monoterpene, sesquiterpene and diterpene synthase genes in response to wounding. Plant Physiol. **1998**, *116*, 1497–1504.

143. Blanchand, L.; Karst, F. Characterization of lysine-to-glutamic acid mutation in a conservative sequence of farnesyl diphosphate synthase from *Saccharomyces cerevisiae*. Gene. **1993**, *125*, 185–189.

144. Cane, D.E.; Shim, J.H.; Xue, Q.; Fitzsimmons, B.C.; Hohn, T.M. Trichodiene synthase: identification of active residues by site directed mutagenesis. Biochemistry. **1995**, *34*, 2480–2488.

145. Croteau, R.; Karp, F.; Wagschal, K.C.; Satterwhite, D.M.; Hyatt, D.C.; Skotland, C.B. Biochemical characterization of a spearmint mutant that resembles peppermint in monoterpene content. Plant Physiol. **1991**, *96*, 744–752.

146. Yuba, A.; Yazaki, K.; Tabata, M.; Honda, G.; Croteau, R. c DNA ccloning, characterization and functional expression of 4S(−)limonene synthase from *Perilla frutescens*. Arch. Biochem. Biophys. **1996**, *332*, 280–287.

147. Barkovich, R.; Liao, J.C. Metabolic engineering of isoprenoids. Metab. Eng. **2001**, *3*, 27–39.

148. Broun, P.; Sommerville, C. Progress in Plant metabolic engineering. Proc. Natl. Acad. Sci. USA. **2001**, *98*, 8925–8927.

149. Mahmoud, S.S.; Croteau, R.B. Metabolic engineering of essential oil yield and composition in mint by altering expression of deoxyxylulose phosphate reductoisomerase and menthofuran synthase. Proc. Natl. Acad. Sci. USA **2001**, *98*, 8915–8920.

150. Lange, B.M.; Croteau, R.B. Genetic engineering of essential oil production in mint. Curr Opin Plant Biol. **1999**, *2*, 139–144.

151. Lange, B.M.; Rujan, T.; Martin, W.; Croteau, R.B. Isoprenoid biosynthesis: the evolution of two ancient and distinct pathways across genomes. Proc. Natl. Acad. Sci. USA. **2000**, *97*, 13172–13177.

152. Wang, E.; Wang, R.; DeParasis, J.; Loughrin, J.H.; Gan, S.; Wagner, G.J. Suppression of a P450 hydroxylase gene in Plant trichome glands enhances natural-product-based aphid resistance. Nat. Biotechnol. **2001**, *19*, 371–374.

153. Chen, D.H.; Ye, H.C.; Li, G.F. Expression of a chimeric farnesyl diphospahte synthase gene in *Artemisia annua* L. transgenic plants via *Agrobacterium tumefaciens*-mediated transformation. Plant Sci. **2000**, *155*, 179–185.

154. Lewinsohn, E.; Schalechet, F.; Wilkinson, J.; Matsui, K.; Tadmor, Y.; Nam, K.H.; Amar, O.; Lastochkin, E.; Ravid, U. Enhanced levels of the aroma and flavor compound S-linalool by metabolic engineering of the terpenoid pathway in tomato fruits. Plant Physiol. **2001**, *127*, 1256–1265.

11

Other Compounds: Epicuticular Lipids, Naphthoquinones, Acetogenins, Polyacetylenes, and Chromenes

11.1 INTRODUCTION

Plants protect themselves from insect pests with chemical and physical defenses that directly influence pest performance. In addition, plants, protect themselves indirectly through traits that attract natural enemies. The different groups of plant secondary metabolites involved in these processes have been discussed in the previous chapters. Screening, identification, and testing of the bioactivity of plant secondary metabolites active against pests is a continuous process; hence, newer compounds are added to the existing list. A detailed study on the biosynthetic pathway, bioactivities in various groups of insects, and mode of action are lacking for some of the compounds reported. The details available on these compounds are presented in this chapter. Among these, epicuticular lipids are discussed under different subheadings, whereas other compounds are presented without any subheadings for lack of information.

11.2 EPICUTICULAR LIPIDS

11.2.1 Introduction

The aerial parts of all plants are covered with a protective cuticle composed of a lipid polymer and a mixture of lipid compounds. These compounds are collectively called epicuticular lipids or surface waxes. These epicuticular lipids have diverse crystallization patterns (1,2), chemical compositions and relative abundance that change with plant age (3), development (4,5), and environment (6). The physical and chemical properties of these compounds play a key role in plant resistance to a variety of biotic and abiotic stresses, including those caused by fungal pathogens, phytophagous insects, drought, solar radiation, freezing temperatures, and mechanical abrasion. In addition, they play a protective role against acid rain and ozone. They also influence the uptake and efficiency of plant growth regulators, pesticides, and herbicides. Plant epicuticular lipids also have significant industrial value in such products as polishing agents, candles, cosmetics, protective coatings, lubricants, and medicines. They also contribute to the esthetic value of many ornamental plants. The composition of epicuticular lipids varies among species, among genotypes within species, and among parts within plants. The variability in chemical composition of epicuticular lipids suggests a variety of ecological functions. One of the important functions includes their interaction with insect herbivores. Chapman and Bernays (7) suggested that the compounds present on the plant surface inhibit biting or oviposition before any damage occurs. Cuticular lipids may also contribute to plant defense by adversely affecting insects through direct toxicity or by physical effects, such as interfering with movement. The mediator role of epicuticular lipids in insect–plant interactions has been emphasized. Many reviews (8–13) and books (14–18) on epicuticular lipids have been published.

11.2.2 Chemistry and Classification

The main components of plant epicuticular lipids can be classified into three major groups (19–23):

1. Hydrocarbons and their derivatives
2. Wax esters, free acids, free aldehydes, and free alcohols
3. Other compounds

Hydrocarbons are common components of epicuticular lipids. The proportion varies from less than 1% to more than 90%. They are mainly n-alkanes with an odd number of carbon atoms in the range C_{21}–C_{37} of which C_{29} and C_{31} are predominant. Branched as well as unsaturated hydrocarbons also occur as minor constituents. The chain lengths are similar to those of alkanes. Oxygenated derivatives of alkanes such as ketones and secondary alcohols occur as minor components. They are similar to alkanes in size and usually a carbonyl group or a hydroxyl group is found near the middle of the carbon chain. The bifunctional C_{29} α- and β- as well as γ-ketols have been identified as minor components along with β-diketones.

Wax esters are the most abundant component of epicuticular lipids. These are esters of n-alkanoic acids and n-alkanols (long chain). In the acyl portions, even numbered C_{20}–C_{24} carbon chains predominate whereas the alcoholic portion is principally composed of C_{24}–C_{28} carbon chains. Wax esters are accompanied by significant amounts of unesterified fatty alcohols and fatty acids. The chain length distribution of the free alcohols is similar to that of the alcohols in the wax esters. The chain length of the fatty acids is longer than that of wax ester acids. Fatty aldehydes with chain lengths similar to those of the alcohols are present in the wax of some plants. Other components, such as monoesters of phenolic acids and aliphatic alcohols, as well as mono-, sesqui-, di-, and triterpenoids, are also noted. The major classes of plant epicuticular lipids are given in Table 11.1.

11.2.3   Occurrence

The epicuticular lipids of a plant can vary with plant part, age, and environmental conditions (3–6,24–27). The abaxial leaf surface differs dramatically from that of the adaxial surface (28). There is also variation between younger and older leaves (29), and the composition of plants grown in greenhouses or growth chambers differs from those of field-grown plants (27).

Diketones are a major component of cuticular wax of barley spike, uppermost leaf sheath, and internode (30) and are found as minor components of other plants such as those of the family Brassicaceae and carnation (10). Different organs of the same plant often exhibit quite distinct surface wax characteristics. For example, leaf surfaces of the *Arabidopsis* lack wax crystals whereas stem surfaces have a white glaucous coating. The total amount of epicuticular lipids per area is 25-fold higher on *Arabidopsis* flowering stems than leaves. The leaves possess trace amount of secondary alcohols, ketones, and esters, which are major constituents of the stems. Major alkanes of *Arabidopsis* leaves are C_{31} homologues, whereas C_{29} homologues are the predominant alkanes in stem wax (31).

11.2.4   Biosynthesis

Several excellent reviews have covered cuticular lipid biosynthesis (32–35). The three important pathways for the biosynthesis of different cuticular lipid components include decarbonylation, acyl reduction and β-ketoacyl elongation. Fatty acid synthesis, catalyzed by a series of enzymatic components collectively referred to as fatty acid synthase (FAS), is initiated by the condensation of malonyl-ACP (acyl carrier protein) with acetyl-CoA.

Table 11.1 Major Classes of Plant Epicuticular Lipids of *Osimunda regalis* Fronds (20)

Lipids	Chemical structure	Chain length range
I. Hydrocarbons and their derivatives		
a. Alkanes	$CH_3(CH_2)_nCH_3$	C_{25}–C_{33}
b. Secondary alcohols	$CH_3(CH_2)CHOH(CH_2)_nCH_3$	C_{27}–C_{29}
c. Ketones	$CH_3(CH_2)_nCO(CH_2)_nCH_3$	C_{27}–C_{33}
d. Ketols	$CH_3(CH_2)_nCH_2CHOH(CH_2)_nCH_3$	C_{27}–C_{33}
e. Diketones	$CH_3(CH_2)_nCH_2OCH_2CO(CH_2)_nCH_3$	Not known
II. Wax esters and related compounds		
a. Wax esters (Alkyl esters)	$CH_3(CH_2)_nCOO(CH_2)_nCH_3$	C_{38}–C_{50}
b. Fatty acids	$CH_3(CH_2)_nCOOH$	C_{24}–C_{32}
c. Aldehydes	$CH_3(CH_2)_nCHO$	C_{24}–C_{34}
d. Fatty alcohols	$CH_3(CH_2)_nCH_2OH$	C_{22}–C_{32}
III. Other Compounds		
a. β-Sitosterol	Sterol	
b. Diterpene esters	Phytol + C_{16}–C_{24} fatty acids	
c. Triterpenoids	$C_{30}H_{50}O$ with ursane or oleane ring system	
d. Phenolic esters	Cinnamyl alcohol + C_{16}–C_{24} fatty acids	

Condensation reaction is followed by the sequential reduction of 3-ketoacyl-ACP, the dehydration of 3-hydroxyacyl-ACP, and the reduction of *trans*-Δ^2–enoyl-ACP. Further elongation is carried out by the addition of two carbons at a time in the form of malonyl-ACP following the sequential rounds of condensation, reduction, dehydration, and second-reduction steps (36–48). NAD(P)H serves as reducing equivalent for the two reductases. Fatty acid biosynthesis occurs ubiquitously, whereas cuticular lipid synthesis occurs almost exclusively in epidermal tissues (34). The majority of cuticular lipid components are derived from very long chain fatty acids that are 20–32 carbons in length. The very long chain fatty acids are produced from C_{16} or C_{18} fatty acid precursors that are elongated extraplastidially by microsomal enzymes, elongases. Because of their essential role in cuticular lipids production, elongases have been one of the most studied of the cuticular lipid biosynthetic steps (32,36). Ample evidence has been obtained for the operation of multiple elongation systems involved in cuticilar lipid biosynthesis, which are both sequential (generating a homologous series) and parallel reactions (generating different lipid classes) (35). Parallel elongases catalyze extensions leading to the production of different cuticular lipid classes as shown in Fig. 11.1. The involvement of decarbonylation, acyl reduction, and β-ketoacyl elongation are shown as parallel pathways. All three pathways are found in the epidermal tissue of most plants, but their relative contributions to the cuticilar lipid composition vary from organ to organ and species to species. Odd chain alkanes, secondary alcohols, ketones, and aldehydes are produced through a decarbonylation pathway (49,50). The acyl reduction pathway produces aldehydes, primary alcohols, and wax esters (34,35,51). The β-ketoacyl elongation pathway results in the production of β-diketones and their derivatives (52–54). Fatty acids produced in the plastid from *de novo* synthesis are utilized by at least three biosynthetic pathways that lead to the formation of glycerolipids, cuticular lipids, and cutin or suberin. In vegetative tissues, *de novo* synthesized C16:0 and C18:1 are the fatty acid precursors for glycerolipid biosynthesis, whereas the precursors for cuticular lipids biosynthesis are primarily saturated fatty acids, primarily derived from C18:0. Partitioning of precursors into these pathways may be accomplished by enzyme specificities or by substrate availabilities.

11.2.5 Bioactivity

Glossy and waxy plant genotypes are available in many plants. The glossy plants are also called bloomless or glazed. Glossy phenotypes have reduced amounts of epicuticular lipids relative to normal wax phenotypes. Glossy phenotypes are normally associated with a reduction in herbivore populations or susceptibility to insect damage. The cabbage aphid, (*Brevicoryne brassicae*) on glossy cruciferous vegetables (*Brassica oleracea*) is less than one-tenth the size of populations on plants with normal wax (55,56). Similarly, diamondback moth larvae (*Plutella xylostella*) on glossy *B. oleracea* are one-tenth the size of those on normal plants. On the contrary, the aphid population on glossy genotypes of barley, *Hordeum vulgare* vary from 2-fold to 10-fold greater than on normal plants (57). Thus, the reported effects of epicuticular lipids vary in magnitude depending on the plant species. Flea beetles (*Phyllotreta* spp). occur in much greater

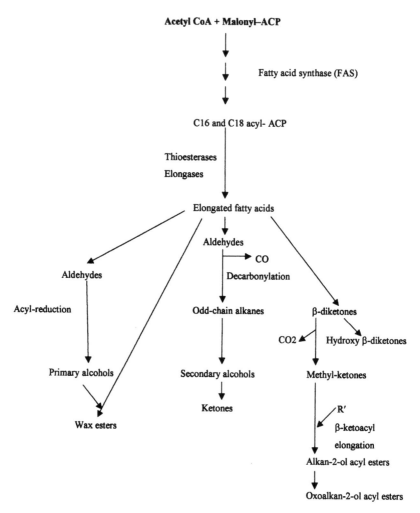

FIGURE 11.1 Biosythetic pathways for epicuticular lipids.

frequency on glossy *B. oleracea* than on normal plants (58). Glossy and normal genotypes of wheat and sorghum cultivars differ in susceptibility to key pests (59,60). Artificial production of the glossy phenotype either by mechanical polishing (61) or by systemic treatment with a thiocarbamate herbicide (62,63) strongly affects the rate of insect damage in *B. oleracea*. The cuticular compounds interfere with oviposition, feeding, and predator population.

11.2.5.1   Oviposition

Insects often oviposit more on glossy phenotypes than on normal plants, which is correlated with the composition of epicuticular lipids. Artificial removal or disruption of epicuticular lipids with mechanical polishing or detergents results in greater oviposition on *B. olearacea* leaves by *Plutella xylostella* (64) and *Delia radicum* (65). Similar results are reported on olive fruits oviposited by the olive fly, *Dacus oleae* (66). The Hessian fly, *Mayetiola destructor*, was shown to oviposit preferentially on epicuticular lipids of wheat applied to paper strips (67). The insect oviposited on chromatographic fractions of the lipids, but the whole extract elicited a response 2.5 times greater than the sum of the fractions. The cabbage rootfly, *D. radicum*, preferred to oviposit on leave surfaces treated with epicuticular lipid extract (68). Different fractions of this extract showed synergistic effect. But the specific lipid component responsible for the stimulation of oviposition was not identified. Plant epicuticular lipids may also deter oviposition in some insects. The diamondback moth oviposited on cabbage leaves more frequently when the surface lipid layer was removed (10). Prophenylbenzenes, coumarins, and a polyacetylene in leaf epicuticilar lipids of carrot, *Daucus carota* L., appeared to stimulate oviposition by carrot fly, *Psila rosae* Fabricius (69). Mixtures of wax constituents have been shown to act synergistically in affecting insect behavior (69,70).

The amount of olive surface waxes obtained from fruits belonging to five Greek cultivars ranged from 11.6 to 152.9 mg/100 g fruit (71). It varied significantly among varieties during fruit growth and maturation throughout the season, with two peaks around mid-August and the end of September up to mid-October. The two pentacyclic triterpene acids, oleanolic and maslinic, constituted the highest percentage of the total surface wax composition and were present in comparable amounts. The number of egg oviposited by *Bactrocera oleae* varied significantly under treatments with different wax solutions compared with controls, which showed the highest mean in egg number (6.4 eggs per fruit). Highly significant negative correlations were detected among the concentrations of the oleanene acids and the number of eggs oviposited; the correlation factor (R) was -0.76 for oleanolic acid and maslinic acid. Oleanolic acid had a dominant effect over maslinic acid. Choice experiments showed that the type of cultivar and the degree of olive fruit maturation affected the degree of infestation of olive fruits by *B. oleae*. The size of olive fruit was not significant in determining the preference of *B. oleae* for ovipositing. Neuenschwander et al. (72) noticed a strong deterrent effect of the waxy fruit covering of "Kalamon" olives on the oviposition behavior.

Three approaches were used to investigate effects of host plant epicuticular waxes on oviposition site selection by *Plutella xylostella* (L.) (73).

In the first approach, oviposition on canola [*Brassica napus* (L.)] that had epiculicular wax reduced by application of a carbamate herbicide (*S*-ethyl dipropylthiocarbamate) was compared with oviposition on untreated control plants. A second approach compared oviposition on sibling strains of *B. napus* with different wax blooms (glossy and waxy), and a third approach compared oviposition by *P. xylostella* on parafilm that had been applied to glossy and waxy *B. napus* strains for transfer of leaf components. Significantly more eggs were deposited on herbicide–treated plants (with reduced epicuticular wax) than on untreated controls. Similarly, more eggs were deposited on glossy than on waxy sibling strains of *B. napus*.

Grant et al. (74) investigated the effects of carboxylic acids on the oviposition behavior of the spruce budworm, a major defoliator of coniferous forests in North America. Carboxylic acids have been implicated as semiochemicals involved in lepidopteran host finding and oviposition, and they occur as free acids in the epicuticular wax of host (*Picea* and *Abies* spp) foliage where spruce budworm laid eggs. In a dual-choice laboratory bioassay, several straight chain and cyclic monocarboxylic acids, and two dicarboxylic acids, significantly enhanced oviposition. Peak activity was associated with saturated acids having 8–12 carbons. Unsaturated oleic and linoleic acids were also preferred. The lowest effective dosage occurred at $7.8 \, nmol/cm^2$ (1 mM solution). At higher dosages ($\geq 780 \, nmol/cm^2$), C_9–C_{10} acids became strongly deterrent and some shorter chain and longer chain acids became stimulating. Electroantennogram responses to C_6–C_{16} acids indicated that behaviorally active acids are detected by olfaction. The most active acids (C_8–C_{12} and oleic) have not been reported in the free fatty acid fraction of host cuticular waxes. However, long chain C_{14}–C_{28} acids are present as free acids, though they elicited significant oviposition responses only at doses that exceeded their levels in foliage waxes. Spruce budworm preference for carboxylic acids may represent a nonspecific response common to lepidopterans, which may have evolved because of the ubiquitous occurrence of carboxylic acids in plants.

11.2.5.2 Probing and Feeding

Initiation of stylet penetration by aphids depends on the sensory assessment of a number of plant surface features, including color, texture, and phytochemicals (volatile and nonvolatile) (75). Video recording behavior of the black bean aphid, *Aphis fabae*, showed that these insects rapidly inserted their stylets following contact with host plants (beans) but were reluctant to penetrate non hosts (oats). However, when epicuticular waxes were stripped from oats, using cellulose acetate, aphids penetrated the plant surface significantly earlier than on oats with the wax layer intact. Chloroform extraction of epicuticular lipids, followed by coupled gas chromatography–mass spectrometry (GC-MS), revealed a complex blend of wax components on beans, whereas one compound (1-hexacosanol) predominated on oats. Epicuticular lipids were applied to artificial (glass) substrates in order to investigate their behavioral activity. Initiation of a stylet penetration attempt by *A. fabae* was delayed when the oat extract or pure 1-hexacosanol was applied, but the bean exact had no behavioral effect. The results suggest that epicuticular lipids play an important role in early stages of host plant selection by *A. fabae*.

The effect of cereal leaf surface wax on *Diuraphis noxia* (Mordvilko), the Russian wheat aphid, probing behavior and nymphoposition was evaluated (76). Ultra structure of leaf epicuticular wax from wheat (*Triticum aestivum* L.) cv. 'Arapahoe' and 'Halt' was different from barley (*Hordeum vulgare* L.) cv. 'Morex', and oat (*Avena sativa* L.) cv. 'Border'. Both wheat cultivars had similar rod-shaped epicuticular wax, whereas barley and oat plants had flakes. The chemical composition comparison of gas chromatograms also indicated that the extract of the two wheat cultivars had similar pattern of peaks, whereas the barley and oat leaves had similar peaks. Cereal variety significantly affected aphid probing behavior ($P < 0.05$), but wax removal using ethyl ether swab did not ($P > 0.05$). Aphids initiated significantly more probes on Border oat leaves than on Morex barley irrespective of wax removal, although total probing duration per aphid was not significantly different among the four cereals examined. Nymphoposition of *D. noxia* on cereal leaves maintained on the benzimidazole-agar medium showed that aphids produced a greater number of nymphs on Morex barley and less on Border oat leaves, although wax removal did not affect aphid nymphoposition. Removal of leaf epicuticular waxes from the four cereal genotypes using ethyl ether swab indicated that the influence of wax on plant resistance to *D. noxia* probing and reproduction was limited.

Five genotypes of wheat—*Triticum aestivum* L. ('Arapahoe', 'Halt', PI 137739, PI 225245, and PI 262660)—were compared for their antixenotic effects on the Russian wheat aphid *D. noxia* (Mordvilko) (77). Comparisons were made by host plant choice tests, diurnal and nocturnal probing behavior of *D. noxia*, and examination of leaf trichome and epicuticular wax structure. The choice test showed that PI 137739 and PI 225245 were the least preferred hosts among the five genotypes evaluated. Electronic monitoring of *D. noxia* probing behavior showed that the aphids fed differently among the five wheat genotypes and aphid probing behavior was different between diurnal and nocturnal periods. Although the number of probes per aphid was not different among the five genotypes either diurnally or nocturnally, the accumulative probing duration per aphid during the 8-h recording period was different. The least probing duration per aphid was on

PI 137739 diurnally, but no differences were found among the five genotypes nocturnally. Furthermore, aphid piercing duration per probe was significantly longer on PI 137739 than on Arapahoe. *Diuraphis noxia* probing (i.e., salivation and ingestion) duration per probe was significantly different among the five wheat genotypes and between diurnal and nocturnal periods. The aphids had the shortest probing duration on PI 137739 among the five genotypes diurnally and nocturnally. Irrespective of genotype differences, total probing duration per aphid was significantly longer nocturnally than diurnally. Aphids made numerous short probes diurnally and fewer long probes nocturnally. Adaxial leaf surface examination showed that PI 137739 had the longest trichomes and medium density, whereas Arapahoe and Halt had the shortest trichomes and the highest density. Examination of adaxial leaf epicuticular wax showed that its ultrastructure was similar among the five wheat genotypes, although the density of wax flakes varied among the genotypes. It was concluded that Halt and PI 262660 were not antixenotic to *D. noxia*, but the PI 137739 and PI 225245 were antixenotic to *D. noxia* based on the host choice test, aphid probing behavior, and examination of leaf surface structure.

The chemical compositions of epicuticular lipids have been correlated with resistance of cultivated plants to insect herbivores. High levels of docosanol (C_{22} alcohol) in several tobaccos were reported to be associated with resistance to the tobacco budworm, *Heliothis virescens* (78). Similarly, alfalfa genotypes containing high levels of triacontanol (C_{30} alcohol) were associated with reduced feeding by the spotted alfalfa aphid, *Therioaphis maculata* (79). High levels of α- and β-amyrin in the epicuticular lipids of the azalea varieties was correlated to the resistance of azalea lace bug, *Stephanitis pyrioides* (80). The aphid-resistant sorghum had higher levels of triterpenols in the surface wax than did the susceptible plant (81). High levels of β-amyrin in several raspberry cultivars were correlated with resistance to the raspberry aphid, *Amphorophora idaei* (16). On the contrary, peanut species that are susceptible to insects had the highest levels of α- and β-amyrin (82). The triterpenols α- and β-amyrin inhibits feeding of *Locusta migratoria* when added to wheat flour discs (31). Cabbage lipid extracts containing α- and β-amyrin reduced acceptance behaviors by *Plutella xylostella* as compared with lipids of a susceptible line that did not contain amyrins.

Epicuticular components function as feeding deterrents or stimulants. Surface lipid extracts from rice varieties resistant to the brown planthopper, *Nilaparvata lugens*, deter feeding and increase restlessness of this insect when the extracts were applied to the surface of susceptible plants (83). The addition of extracts from nonhost plant to wheat flour discs reduced feeding by the grasshopper, *Chorthippus parallelus* (84). The tropical grasshopper *Microtylopteryx hebardi* preferred to feed on *Geonoma cuneata* (one of the palm species), and the insects preferentially bit filter paper discs containing the extracts of epicuticular lipids of this palm species in laboratory feeding by the boll weevil, *Anthonomus grandis* (85). Similarly, the surface extractants stimulated feeding of the tobacco hornworm, *Manduca sexta* (86), and neonate larvae of potato tuber moth, *Phthorinaea operculella* (87).

The hydrocarbon fractions containing C_{27}, C_{29}, C_{31}, and C_{33} *n*-alkanes from a broad bean, *Vicia faba*, stimulated the pea aphid, *Acyrthosiphon pisum*, to feed through an artificial membrane (88). Longer chains of *n*-alkanes ($> C_{23}$) from sorghum were not deterrent to *Locusta migratoria*, whereas the short chain *n*-alkanes (C_{19}, C_{21}, and C_{23}) deterred feeding (89). Several chrysomelid beetles were stimulated to feed by *n*-alkane fractions from the host plants. A fatty alcohol fraction also stimulated feeding by a chrysomelid beetle (90). The fatty alcohols hexacosanol (C_{26}) and ocatacosanol (C_{28}) present in surface lipids of mulberry leaves stimulate feeding in larvae of the silkworm *Bombyx mori* (91). Short chain free fatty acids (C_8–C_{13}) deter settling of the aphid *Myzus persicae* on both artificial and plant surfaces (92,93). Field experiments with application of dodecanoic acid (C_{12}) to crop plants showed reduced aphid settling and less insect damage (94,95).

11.2.5.3 Insect Behavior

The toxicity of epicuticular components was also exhibited to the insect herbivores after the ingestion of food materials. The growth of *Spodoptera frugiperda* was greater when larvae were fed diet containing corn foliage from which epicuticular lipids had been extracted than when they were fed diet containing unextracted foliage (96). Similar experiments revealed such growth retardation in *Helicoverpa zea* by corn silks (97). The *S. frugiperda* fed diet-containing foliage of wild and culivated peanut species from which the epicuticular lipids had been removed had increased larval weights and earlier pupation and adult emergence than insects reared on diet with untreated foliage (85,98). However, the lipid extracts had no effect on the insects when incorporated into artificial diets. The development of several lepidopteran species was interfered by the palmitate ester of α-amyrin isolated from the bark of the sandal tree, *Santalum album* (99). Leaf cutter bees (*Megachile* sp) made more cuts on the glaucous leaves of Mexican redbud, *Cercis canadensis* var *mexicana* L., than on the leaves of a glossy ecotype. In addition to lacking wax crystals on the adaxial leaf surface, the glossy ecotype had a sixfold reduction in the relative amount of triacontanol on its epicuticular lipids. Free and esterified triterpenols increased aphid resistance in sorghum when present at high levels (100).

Epicuticular waxes from the aphid-resistant red raspberry (*Rubus idaeus*) cultivar Autumn Bliss and the aphid-susceptible cultivar Malling Jewel were collected from the newly emerging crown leaves, and also from the group of four more mature leaves immediately below the crown (101). Resistance and susceptibility status of the leaves to infestation by the large raspberry aphid, *Amphorophora idaei*, were determined by bioassay with the insect just prior to collection of the wax. Analysis showed the waxes to consist of a complex mixture of free fatty acids; free primary alcohols and their acetates; secondary alcohols; ketones; terpenoids including squalene, phytosterols, tocopherol, and amyrins; alkanes and long chain alkyl and terpenyl esters. Compositional differences that may relate to *A. idaei* resistance status were noticeably higher levels of sterols, particularly cycloartenol, together with the presence of branched alkanes, and an absense of C_{29} ketones and the symmetrical C_{29} secondary alcohol in wax from the resistant cultivar Bliss. There were also differences between the cultivars in the distribution of individual amyrins and tocopherols and in the chain length distribution for homologues of fatty acids, primary alcohols, and alkanes, and these may also be related to resistance to *A. ideaei*. Emerging leaves had lower levels of primary alcohols and terpenes, but higher levels of long chain alkyl esters and, in general, more compounds of shorter chain length than the more mature leaves. During bioassay, *A. ideaei* displayed a preference to settle on the more mature leaves. This may be due to greater wax coverage and higher levels of the compounds of shorter chain length found in the newly emerged younger leaves.

The behavior of neonate *Plutella xylostella* was observed and quantified during the first 5 min of contact with cabbage surface waxes and surface wax components deposited as a film ($60\,\mu g/cm^2$) on glass (102). The time larvae spent biting was greater and the time walking was less on waxes extracted from the susceptible cabbage variety Round-up than on an insect-resistant glossy-wax breeding line, NY 9472. The waxes of both cabbage types were characterized, and some of the compounds present at higher concentrations in the glossy waxes were tested for their deterrent effects on larvae by adding them to the susceptible waxes. Adding a mixture of four *n*-alkane-l-ols or a mixture of α- and β-amyrins to wax from susceptible cabbage reduced the number of insects biting and, among those biting, reduced the time of biting and increased the time of walking in a dose-dependent manner. Among individual *n*-alkane-l-ols, adding C_{24} or C_{25} alcohols reduced the number of insects biting but only adding C_{25} alcohol reduced the time spent on biting among those insects that initiated biting. Adding a mixture of five *n*-alkanoic acids did not affect biting, but it increased the time spent palpating and decreased walking time. Among individual *n*-alkanoic acids, only adding C_{14} significantly increased the time of palpating. If the observed responses were gustatory, the results indicate that some primary wax components, including specific long chain alkyl components, have allelochemical activity influencing host acceptance behavior by a lepidopteran larva.

11.2.5.4 Predator Population

Seven accessions of varieties of cultivated pea *Pisum sativum* L. varying in surface wax bloom characteristics were grown in replicated small plots (1 m²) for two seasons to monitor natural infestations of insect herbivores and abundance of predatory insects (103). Wax bloom was quantified on the basis of the amount of wax extractable from leaf surfaces, densities of wax crystals visible with scanning electron microscopy, and visual appearance. During each season, pea aphid, *Acrythosiphon pisum* (Harris) (Homoptera: Aphidae), densities per plant were significantly lower on peas with reduced surface wax bloom as compared with peas with standard or normal surface wax bloom. This difference was greatest between two near isolines of peas differing in expression of a mutation that reduces surface waxes. Damage to leaves and stipules by the pea leaf weevil *Sitona lineatus* (L.) (Coleoptera: Curculionidae) were greater on reduced-wax bloom peas than normal wax bloom peas. Thus, as occurs in other crops, reduced wax bloom in peas was associated with lower natural infestations of an aphid, but also with increased susceptibility to a folivorous beetle. Populations of predatory coccinellids did not differ consistently between years on reduced wax bloom versus normal-wax bloom peas, failing to support a hypothesis that predator populations are higher on reduced wax bloom peas, contributing to the lower aphid populations on these plants.

Four mutations that reduce wax bloom in *Brassica oleracea* L. were examined for their effects on predation, mobility, and adhesion to the plant surface by the general predator *Hippodamia convergens* (Guerin-Meneville) (Coleoptera: Coccinellidae). The mutation reduces wax bloom to different degrees, but all produce a glossy phenotype (104). Plants tested were inbred lines, near isogenic lines, or segregating F(2) populations, depending on the mutation. In an experiment on caged leaves, predation of *Plutella xylostella* L. larvae by *H. convergens* adult females was significantly greater on glossy types as compared with normal wax or wild-type counterparts. Although the trend was the same for each mutation, individual comparisons between glossy and normal wax lines or segregants were only significant for two of them, those producing mutant alleles gl(a) and gl(d). Individual *H. convergens* were observed to spend more time walking on leaf edges and less time walking on leaf surfaces of normal-wax plants than glossy plants. *Hippodamia convergens* also obtained better adhesion to the surfaces of glossy plants than to normal-wax plants when tested using a centrifugal device. Two of the mutations produced similarly have strong effects on predation, behavior, and adhesion by *H. convergens*. These two are the same previously determined by Eigenbrode and Kabalo (104) to provide the strongest similar effects on another generalist predator, *Chrysoperla plorabunda* (Fitch). The results indicate that wax bloom variation in nature could affect herbivore populations through its effects on generalist predators.

11.2.6   Mechanism of Action

Epicuticular components stimulate or deter oviposition, movement, and feeding. The physical structure of plant surface lipids can affect insect herbivore attachment and locomotion. Epicuticular components also affect herbivores indirectly by influencing predatory and parasitic insects. Insect herbivores benefit from rapid assessment of host quality on the basis of cues (13) from surface lipids. They also adversely affect the insects through direct toxicity or by physical effects, such as interfering with movement. The cuticle provides a first line of defense between the plant and its environment. An in-depth study on the mode of action of epicuticular components is not available.

11.2.7   Molecular Biology

The structure and chemical characteristics of epicuticular lipids play a significant role in plant resistance to biotic and abiotic stresses. Therefore, the genetic modification of crops to alter their epicuticular lipid profiles has tremendous potential for improving crops' stress resistance. However, very little is known about how genes function in epicuticular lipid production. The molecular genetics of epicuticular lipids has been reviewed by many authors (12,105,106).

Mutagenesis of genes affecting cuticular lipids provides a means for identifying genes involved in biosynthesis of cuticular lipids. Mutagenesis has localized 85 unique loci in barley (12) and 24 loci in sorghum (107). The production of epicuticular lipids is a complex process that involves hundreds of individual genes, enzymes, and regulators.

Several genes involved in epicuticular lipids production have been cloned using insertion mutagenesis. For example, the *CER2* (108) and *CER3* genes from *Arabidopsis* (109) were cloned using T-DNA tagged alleles. *CER2* was also cloned using chromosome walking (110). Both *CER2* and *CER3* code for novel proteins, and hence their function is difficult to predict.

Interestingly, the *CER2* mRNA appears to be highly expressed in stems but not in leaves, and was dramatically altered in the *CER2* mutants. *CER2* has 63% sequence similarity over the entire protein with GL2 from maize. Like the *CER2* mutant in *Arabidopsis*, gl2 of maize has reduced chain length distribution for major wax constituents. This might suggest that both genes likely play a role in acyl-CoA elongation reactions, with *CER2* possibly being a stem-specific regulator. However, Xia et al. (111) found that the *CER2* protein was localized in the nucleus and thus does not catalyze wax elongation reactions. Jenks et al. (112) proposed that *CER3* may be involved in the hydrolysis of fatty acyl-CoA into free fatty acids and CoA, but Hannoufa et al. (109) found that the *CER3* gene lacked homology to members of the fatty acyl-CoA thioesterase gene family.

The *CER1* gene of *Arabidopsis* was cloned using the heterologous maize transposable element system Enhancer-Inhibitor (En/Spm) (113). Various transposon systems were also used to isolate the epicuticular wax genes, *GL1* (114), *GL8* (115), and *GL15* (116), from the seedlings' wax mutants in maize. The deduced amino acid sequence from GL1, although roughly twice as large, was similar to the proteins encoded by the *Arabidopsis CER1* and *Senecio odora* (*Def1*) *EPI23* genes. Moreover, *EPI23* was expressed only in the epidermis. Based on sequence analysis, Hansen et al. (114) proposed that the *GL1*, *CER1*, and *EPI23* belong to a family of membrane-bound receptors. If so, these gene products may be involved in wax secretion. Aarts et al. (113) contend that CER1 encodes an aldehyde decarboxylase based on regions of sequence homology to this group of enzymes. The *GL8* gene in maize has sequence homology to a gene coding the *E. coli* 3-oxoacyl-ACP reductase (115). Thus, *GL8* may play a role in reducing ketoacyl intermediates during the acyl-CoA elongation reactions. The *GL15* gene, by comparison, has high sequence homology to floral regulatory elements in *Arabidopsis*, suggesting a possible analogous regulatory role in wax biosynthesis (116). Further mutagenesis studies, using these exogenous and endogenous insertion elements, are needed to tag important epicuticular wax genes and facilitate their cloning. Identifying these genes will not only help to elucidate wax production processes but will also indicate candidate genes that might be used in crop improvement programs.

An *Arabidopsis* fatty acid elongase gene, *KCS1*, with a high degree of sequence identity to *FAE1*, encodes a 3-ketoacyl-CoA synthase that is involved in very long chain fatty acid synthesis in vegetative tissues and also plays a role in wax biosynthesis (117). Sequence analysis of *KCS1* predicted that this synthase was anchored to a membrane by two adjacent N-terminal, membrane-spanning domains. Analysis of a T-DNA-tagged kcs1-1 mutant demonstrated the involvement of *KCS1* in wax biosynthesis. Phenotypic changes in the *kcs1-1* mutant included thinner stems and less resistance to low-humidity stress at a young age. Complete loss of *KCS1* expression resulted in decreases of up to 80% in the levels of $C_{26}-C_{30}$ wax alcohols and aldehydes, but much smaller effects were observed on the major wax components, i.e., the C_{29} alkanes and C_{29} ketones on leaves, stems, and siliques. In no case did the loss of *KCS1* expression result in complete loss of any individual wax component or significantly decrease the total wax load. This indicated that there was redundancy in the elongase *KCS* activities involved in wax synthesis. Furthermore, since alcohol, aldehyde, alkane, and ketone levels were affected to varying degrees, involvement of the *KCS1* synthase in both the decarbonylation and acyl reduction in wax synthesis pathways were demonstrated.

Mutants of the *ECERIFERUM2* (*cer2*) gene of *Arabidopsis* condition bright green stems and siliques. This is indicative of the relatively low abundance of the cuticular wax crystals that compose the wax bloom on wild-type plants. Xia et al. (110) cloned the *CER2* gene via chromosome walking. Three lines of evidence establish that the cloned sequence represents the *CER2* gene: (a) this sequence is capable of complementing the cer2 mutant phenotype in transgenic plants; (b) the corresponding DNA sequence isolated from plants homozygous for the *cer2-2* mutant allele contains a sequence polymorphism that generates a premature stop codon; and (c) the deduced CER2 protein sequence exhibits sequence similarity to that of a maize gene (glossy2) also involved in cuticular wax accumulation. The *CER2* gene encodes a novel protein with a predicted mass of 47 kDa. The expression pattern of the *CER2* gene by *in situ* hybridization and analysis of transgenic *Arabidopsis* plants carrying a *CER2*-β-glucuronidase gene fusion that includes 1.0 kb immediately upstream of *CER2* and 0.2 kb of *CER2* coding sequences was studied. These studies demonstrate that the *CER2* gene is expressed in an organ- and tissue-specific manner; CER2 is expressed at high levels only in the epidermis of young siliques and stems. This finding is consistent with the visible phenotype associated with mutants of the *CER2* gene. Hence, the 1.2-kb fragment of the *CER2* gene used to construct the *CER2*-β-glucuronidase gene fusion includes all of the genetic information required for the epidermis-specific accumulation of CER2 mRNA.

The essential oils and leaf waxes of several putative somatic hybrid plants of *Lycopersicon esculentum* and *L. hirsutum* were compared with those of transgenic plants of *L. hirsutum* and nontransgenic plants of *L. esculentum*, and also of wild-type *L. hirsutum* (118). The latter contained β-myrcene, 3,7-dimethylocta-1,3,6-triene, and a high proportion of undecan-2-one, whereas the transgenic *L. hirsutum* and the somatic hybrids all contained a similar proportion of 2-carene and β-phellandrene, but only traces of alkan-2-ones.

Lycopersicon esculentum contained a very low proportion of essential oils and no alkan-2-ones. Hentriacontane (*n*-C31) and 3-methylhentriacontane (anteiso-C_{32}) were the predominant components in all the leaf waxes, and a comparison of the ratio between these compounds suggested that the somatic hybrids contained at least some contributions from both the parent species.

Molecular cloning of the *CER2* gene of *Arabidopsis* through the isolation of plant DNA flanking the site of T-DNA insertion as well as characterization of the two independent T-DNA insertion mutant alleles, *BRL5* and *BRL9*, of this gene were reported (108). In the mutant line BRL5, T-DNA was found to be inserted in the second exon of the *CER2* gene, whereas in *BRL9*, T-DNA is inserted in the only intron of this gene. Nucleotide sequence analysis suggests that the ORF encodes a 47.3-kDa polypeptide. High levels of *CER2* transcripts were detected in stems and flowers. The predicted amino acid sequence of the *CER2* gene product reveals little homology with known protein sequences. In accordance with structural characterization of the T-DNA insertion mutants, no evidence of transcripts derived from the *CER2* gene was found in either *BRL5* or *BRL9*.

The eceriferum3 (CER3) locus encodes one of 21 gene products known to be involved in wax biosynthesis in *Arabidopsis thaliana* (109). These loci are readily identified by their bright, dark green stems when compared with the more glaucous wild-type plant. Clones of a gene that encodes a 795-amino-acid open reading frame have been isolated by using plant DNA flanking the site of a T-DNA insertion in line BRL1. Molecular complementation of the cer3 mutant phenotype by clones of this gene establishes that it corresponds to the *CER3* gene. Although the 90-kDa predicted amino acid sequence of this gene shows no homology to any other known protein, the second exon of *CER3* encodes a RRX12KK nuclear localization sequence (NLS). Southern blotting and DNA sequence analysis revealed that the T-DNA is inserted 89 bp downstream of the translation termination codon of this gene. Northern blot hybridization of RNA isolated from the *BRL1* mutant with the *CER3* cDNA probe indicated that the transcript is absent in this mutant line. Unlike other *CER* genes that have been cloned to date, high levels of the *CER3* transcript were found in all tissues from wild-type plants, i.e., leaves, stems, roots, flowers, and apical meristems.

Xu et al. (119) have showed that the *gl8* gene is required for the normal accumulation of cuticular waxes on maize seedling leaves. Their findings indicated that the GL8 protein is a component of the acyl-CoA elongase and functions as a β-ketoacyl reductase during the elongation of very long chain fatty acids required for the production of cuticular waxes.

11.2.8 Perspectives

The biosynthesis of cuticular lipids has been investigated for more than three decades. Yet we know little about the factors that regulate the partitioning of fatty acid precursors. The cloning of biosynthetic genes has just begun, and it promises to open new paths and bring exciting discoveries (13).

Although epicuticular lipids can clearly act as semiochemicals, the mechanisms by which herbivores detect them remain largely unknown (7). The role of individual cuticular components and their synergistic effects on insects should be studied in detail. Elucidation of these chemosensory mechanisms will shed light on the means by which some insect predators select and identify hosts. The effects of lipid composition and microstructure on predators, parasitoids, and insect pathogens have to be investigated in detail before the mechanism of action can be understood.

11.3 NAPHTHOQUINONE DERIVATIVES

Quinones are ubiquitous in plant and animal cells, where they function in respiration, photosynthesis, and as defensive toxins (120). The quinones commonly occurring in nature are benzo-, naphtho-, and anthraquinones. Although they derive from different biosynthetic routes, their structures contain phenolic moieties. Many naphthoquinones occur in higher plants. Plumbagin, juglone, and lawsone are some of the well-known naphthoquinones from plants (121). The sporadic occurrence of naphthoquinones is known from about 20 families of higher plants, Ebenaceae, Droseraceae, Balsaminaceae, Plumbaginaceae, Bignoniaceae, Boraginaceae, and Juglandaceae.

Two naphthoquinone derivatives were isolated and identified from *Calceolaria andina* (122). These two compounds are 2-(1,1-dimethylprop-2-enyl)-3-hydroxy-1,4-naphthoquinone and the corresponding acetate. Two novel meroterpenoid naphthoquinones named cordiaquinone J and K have been isolated from the roots of *Cordia curassavica*, in addition to the known cordiaquinones A and B (123). Three new meroterpenoid naphthoquinones, the known cordiaquinone B and a new naphthoxirene have been isolated from the roots of *Cordia linnaei* (124). The structures of 2-(1,1-dimethylprop-2-enyl)-3-hydroxy-1,4-naphthoquinone, the corresponding acetate, plumbagin, and juglone are shown in Fig. 11.2.

The naphthoquinones are biosynthesized through succinylbenzoate pathway (125,126). This is an extension of the shikimate/arogenate pathway branching from chorismate. Isochorismate is formed from chorismate catalyzed by the enzymes isochorismate hydroxymutase. Isochorismate is in

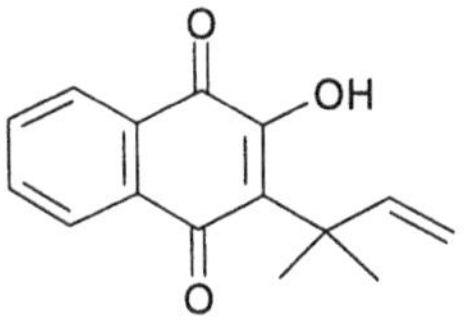

3-(1, 1-dimethylprop-2-enyl)-
2-hydroxy-1,4-naphthoquinone

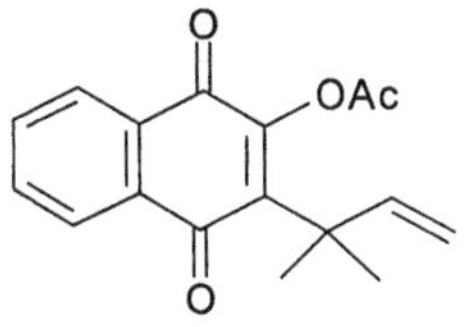

2-acetoxy-3-(1, 1-dimethylprop-2-enyl)-
1,4-naphthoquinone

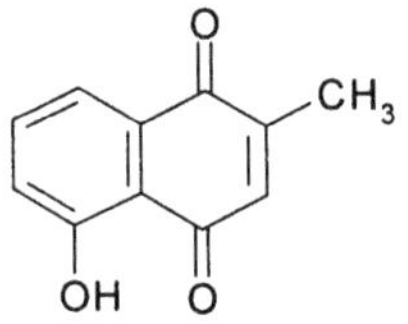

Plumbagin

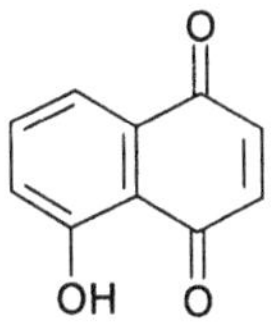

Juglone

FIGURE 11.2   Structures of 1,4-naphthoquinone derivatives.

turn converted 2-succinylbenzoate in the presence of 2-oxoglutarate and thiamine pyrophosphate. Succinylbenzoate is subsequently activated at the succinyl residue to give a mono-CoA ester in the presence of ATP. Ring closure of the CoA ester gives rise to 1,4-dihydroxy-2-naphthoate and then naphthoquinone (Fig. 11.3).

Several authors have observed altered development, molting, and ecdysis caused by plumbagin and other 1,4-naphthoquinones (127). Plumbagin caused molting failure to *Bombyx mori*. Mitchell and Smith (128) and Mitchell et al. (129) reported that both juglone and plumbagin inhibit ecdysone 20-monooxygenase activity in protein extracts from the larval *Aedes aegypti, Drosophila melanogaster*, and *Manduca sexta*. The performance of *Actias luna* and *Callosamia promethea* was compared when fourth-instar larvae of each were fed with birch foliage or supplemented with 0.05% juglone. *Actias luna* fed juglone exhibited no changes in developmental time or mortality compared to a diet without juglone. In contrast, juglone-supplemented diets, when fed to *C. promethea* caused negative growth rate and a 3.6-fold decrease in consumption rate. The performance of *A. luna* was also compared on birch and walnut; larvae grew more rapidly on an all-walnut diet than on all-birch diet. To examine the effect of 1,4-naphthoquinone structures on *A. luna* survival, first instars were fed on birch supplemented with varying concentrations of juglone, menadione, and plumbagin or lawsone. In diets supplemented at 0.05%, none of the

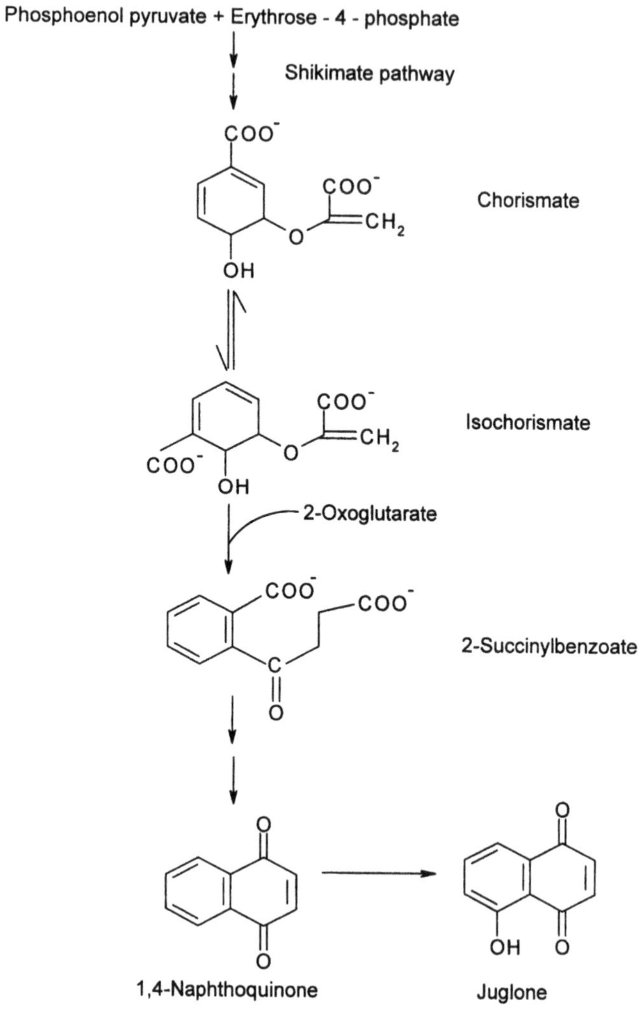

Phosphoenol pyruvate + Erythrose - 4 - phosphate

Shikimate pathway

Chorismate

Isochorismate

2-Oxoglutarate

2-Succinylbenzoate

1,4-Naphthoquinone

Juglone

FIGURE 11.3 Biosynthesis of 1,4-naphthoquinone and juglone.

compounds produced effect significantly different from that of controls. In diets supplemented at 0.05%, the treatments produced significant toxic effects in the order plumbagin > menadione = lawsone > juglone for mortality and plumbagin > lawsone > menadione = juglone for increased developmental time. A late-instar *A. luna* showed resistance to juglone compared to *C. promethea* and an early-instar *A. luna* showed resistance to several related

1,4-naphthoquinones. These results suggest a chemical basis for host choice among saturniids (121).

Khambay et al. (122) determined activities of 2-(1,1-dimethylprop-2-enyl)-3-hydroxy-1,4-naphthoquinone and the corresponding acetate against 29 pest species and 9 beneficial species of arthropod from a total of 11 orders. Activities against homopteran and acarine species were of the same order as those of established pesticides; significantly, cross-resistance is observed for strains resistant to established classes of insecticide. Mammalian toxicities are also low for these two compounds. The compounds effectively kill the notorious B type of the tobacco whitefly, *Bemisia tabaci*, and the peach potato aphid, *Myzus persicae*. These compounds are very easy to extract and account for 5% of the dry weight of the *Calceolaria andina*. The world's most insecticide-resistant whitefly, aphids, and mites died within minutes of exposure to either of these compounds. The scientists are also working with the structures, which are very simple chemically, to develop synthetic variants that are even more potent than these natural compounds.

A series of compounds with structures based on insecticidal/acaricidal naphthoquinones isolated from *Calceolaria andina* has been synthesized (130). A distinctive feature in the structure of the natural naphthoquinone derivatives is the quaternary (tetra-substituted) carbon at C-1 in the side chain at position 3. The importance of the gem dimethyl group at C-1 in the side chain of compound 1 for activity was confirmed by progressive loss of activity in compounds with only one or no methyl groups there. Insertion of a methyl group between the ring and the gem dimethyl group or after the latter also led to loss of activity. Replacing the vinyl group with methyl group produced a compound more active than the parent naphthoquine against *B. tabaci* but further extension to C_2H_5, C_3H_7, or C_4H_{10} led to progressive loss in activity against *B. tabaci* and *Tetranychus urticae*. The hydroxy analogue of compound 4 was particularly effective against *T. urticae*, but when a gem dimethyl group was introduced at C-1 substantial loss of activity against both species resulted.

Stepwise insertion of CH_2 groups at C-1 also led to progressive loss of activity against *B. tabaci*. In contrast, the activity against *T. urticae* peaked sharply at $n = 9$. In summary, synthesis and assay of a series of pesticidal naphthoquinones have clarified the significance of a quaternary carbon atom as a structural feature in the natural compounds, and led to the discovery of a range of synthetic compounds highly active against both susceptible and resistant strains.

Many larvicidal cordiaquinones (meroterpenoid naphthoquinones) were isolated from *Cordia curassavica* and *C. linnaei* and tested for the larvicidal effect against the larvae of the yellow fever transmitting mosquito

TABLE 11.2 Larvicidal Activities of Cordiaquinones and Plumbagin[123]

Cordiaquinones	Larvicidal activity against *Aedes aegypti* (µg/mL of compound required to kill the larvae after 24 h)
1	25
2	12.5
3	25
4	12.5
Plumbagin	6.25

Aedes aegypti (123,124). Dilution tests performed on *A. aegypti* showed larvicidal effects for all cordiaquinones. The activities were a little lower than those of the reference compound, plumbagin (Table 11.2).

Topical treatment of the plumbagin in doses ranging from 0.005 to 5 µg prevented oocyte development and affected fecundity and fertility in *Musca domestica* (131). The treatment to wandering larvae was less effective as the compound could only affect the fertility to a significant level whereas the fecundity was not significantly reduced. The compound also affected the oocyte maturation as it arrested the development of vitellogenic oocyte at stage six.

The toxicity of quinones is a consequence of their ability to interfere with the redox cycle. During redox cycling, quinones undergo a one-electron reduction to the corresponding semiquinone radical (120). The oxidative reversion of semiquinone to quinone correspondingly reduces oxygen to superoxide radical. Excess superoxide and its further metabolism to other reactive oxygen species such as H_2O_2 and OH results in oxidative cell damage. Cell toxicity from quinones may also result from covalent adduct formation between the semiquinone species and cell macromolecules and from the depletion of glutathione (132). As the juvenile hormone analogue methoprene and molting hormone 20-hydroxyecdysone or the mixture of these hormones could not restore the development of the oocyte in ovaries of plumbagin-treated *Musca domestica* L., it was concluded by Saxena et al. (133) that the plumbagin does not affect female houseflies through hormonal pathways, instead in all probability it acts like a cytotoxic compound. The chitin synthetase inhibitor plumbagin and its 2-demethyl derivative juglone were found to inhibit, in a dose–response fashion, the cytochrome P450-dependent ecdysone 20-monooxygenase activity associated with adult females of *A. aegypti*, wandering stage larvae of *D. melanogaster*, and fat body and midgut from last-instar larvae of *M. sexta* (128).

11.4 ACETOGENINS

Acetogenins, from trees and shrubs of the plant family Annonaceae, are C_{35}/ C_{37} natural fatty acid derivatives with long chain hydrocarbon portions connecting a variable number of tetrahydrafuran (THF) or tetrahydropyran (THP) rings and terminated with a 2,4-disubstituted γ-lactone or ketolactone (134–138). The three major classes of annonaceous acetogenins are mono-THF (e.g., gigantetrocin A, annomontacin), adjacent bis-THF (e.g., asimicin, parviflorin), and nonadjacent bis-THF (e.g., sylvaticin, bullatalicin). Gigantetrocin A, goniothalamicin, annonacin, annomontacin, goniotriocin, and xylomaticinones were isolated from the bark of *Goniothalamus gaganteus* Hook.f and Thomson (139); sylvaticin and bullatalicin were isolated from the leaves of *Rollinia sylvatica* (Jacq.) Baill (140); asimicin was isolated from the seeds of *Asimina triloba* (L.) Dunal (141–143); and parviflorin was isolated from the bark of *Annona squamosa* L. (144,145). A novel annonaceous acetogenin montanacin F, with a new type of terminal lactone unit, was isolated form the leaves of *Annona Montana* (146,147) Muricatenol, 2,4-*cis* gigantetrocinone, and 2,4-*trans*-gigantetrocinone have been isolated from the seeds of *Annona muricata* (148,149). Four C_{15} acetogenins were isolated from the red alga *Laurencia obtusa* (150) Annomolin and annocherimolin were isolated from *Annona Cherimolia* seeds (151). Acetogenins were also isolated from liverworts (152) and *Annona atermoya* (153).

The structures of a few important annonaceous acetogenins are shown in Fig. 11.4. The biogenetic pathway proposed for monotetrahydrofuran acetogenins is shown in Fig. 11.5.

Cohilins C and D have been show to be the two important metabolites in the biogenesis of acetogenins from *Annona muricata* and *A. nutans* (154). Acetogenins have been frequently described in the literature as being toxic to mosquito larvae, European corn borer, spider mites, melon aphids, Mexican bean beetles, bean leaf beetles, striped cucumber beetles, blowfly larvae, Colorado potato beetles, and free-living nematodes. The acetogenins typically occur as complex mixtures of more than 40 compounds within the plant extract; thus, application as a mixture has been suggested (155–158).

Most studies conducted previously with annonaceous acetogenins on other pests have used the complex plant extracts. Although development of resistance to complex plant extracts is less likely, standardization of activity is crucial. The potency of annonaceous plant extracts can vary significantly during the year and from tree to tree (159).

A concentration of 1000 ppm of acetogenins produced mortality and was associated with a delay in nymphal development when fed continuously to *Blatella germanica* (160). Production of acetogenins by synthesis (137) may be difficult, but crude extracts from the twigs or seeds of annonaceous trees

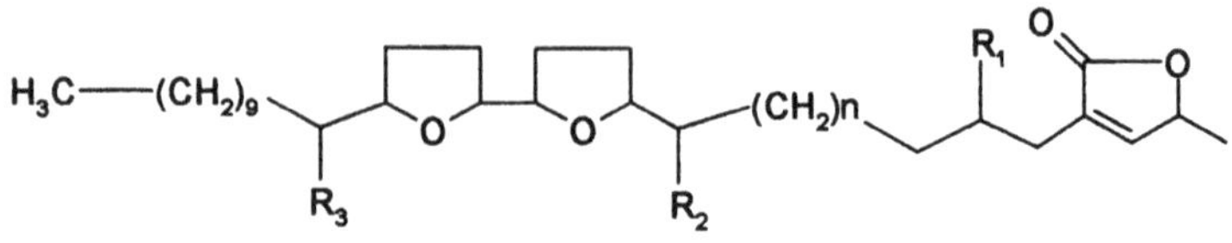

	R₁	R₂	R₃	n

	R_1	R_2	R_3	n
Annonin-1 (Squamosin)	H	OH	OH	9
Asimisin	OH	OH	OH	9
Parviflorin	OH	OH	OH	7
Neoannonin	H	OH	OH	7

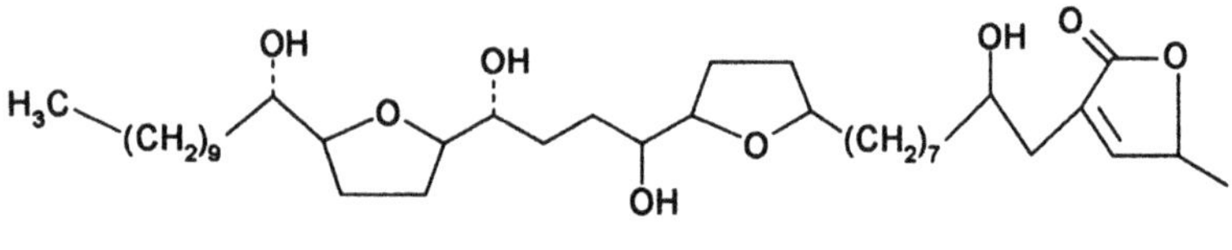

Bullatalicin

FIGURE 11.4 Structures of acetogenins.

would offer an economical source for the commercial production of acetogenin extracts as pesticides (156). Acetogenins are now considered to be the most potent inhibitors (nanomolar range) of complex I in the mitochondria. However, they are relatively safe compounds with low mammalian toxicity. Acetogenins were found to be only weak skin sensitizers.

Insecticides with delayed activity are important for the control of ants (161), termites (162), and cockroaches (163) because toxicant transfer among these insects is enhanced. Because of their slower mode of action, mainly against the resistant Muncie strain, bullatalicin and gigantetrocin A have potential for use against social insects, whereas faster-killing acetogenins, such as parviflorin, asimicin, annomontacin, and sylvaticin seem more suitable for nonsocial insects such as *B. germanica*. The pure annonaceous acetogenins can be effective against *B. germanica*, provided that their source of supply, effective application rates, and proper delivery system can be made economically feasible. The low resistance ratios, shown in this study, suggest an excellent potential for the use of annonaceous acetogenins in baits against cockroaches.

Six compounds, representing the mono-THF (gigantetrocin A, annomontacin), adjacent bis-THF (asimicin, parviflorin), and nonadjacent

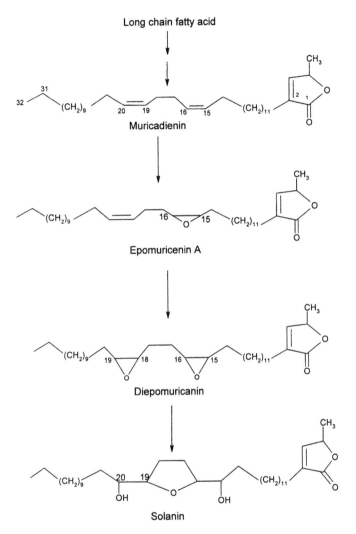

FIGURE 11.5 Bigenetic pathway for monotetrahydrofuran acetogenins.

bis–THF (sylvaticin, bullatalicin) classes of annonaceous acetogenins, were compared with technical grades of synthetic amidinohydrazone (hydramethylnon), carbamate (propoxur, bendiocarb), organophosphate (chlorpyrifos), and pyrethroid (cypermethrin) insecticides to determine their dietary toxicities to insecticide-resistant and insecticide-susceptible strains of the German cockroach, *Blattella germanica* (L.) (158). Differential susceptibility occurred among *B. germanica* nymphs of both strains to this variety of the acetogenins and the five conventional synthetic insecticides. The speed of kill (LT_{50}) values against insecticide-susceptible and insecticide-resistant second and fifth instars permitted ranking of all 11 compounds. The adjacent bis-THF acetogenins showed the highest potency among the 3-acetogenin classes. The acetogenins caused high percentages of mortality; delay in development was mainly affected by gigantetrocin A and annomontacin, whereas insecticide-resistant nymphal development was mainly affected by gigantetrocin A and bullatalicin. Most tested acetogenins performed better than the conventional insecticides against both stages of both strains. No growth regulation effects were caused by any of the compounds tested. Low resistance ratios were obtained for most compounds (except chlorpyrifos). Low resistance ratios values for second instars ranged from 0.9 to 2.2 with the natural acetogenins and from 1.0 to 3.8 with the synthetic compounds; the fifth instars ranged from 0.2 to 3.9 with the natural acetogenins and from 0.6 to 0.8 with the synthetic compounds.

Gaudano et al. (157) have further evaluated the antifeedant and insecticidal effects of squamocin and annonacin, two annonaceous acetogenins, on *Spodoptera littoralis*, *Leptinotarsa decemlineata*, and *Myzus persicae*. In addition, to partially assess the environmental risk of these substances, they also tested their mutagenicity in *Salmonella typhimurium* strains TA98, TA100, and TA102 in the presence and absence of a metabolic activation system. Among the test compounds, annonacin showed antifeedant effects on *L. decemlineata*, while squamocin was toxic to *L. decemlineata* and *M. persicae*. Neither acetogenin was mutagenic although both were toxic in the absence of a metabolic activation system. They have compared these results with those obtained with rotenone, a well-known respiratory inhibitor that was highly toxic to *L. decemlineata* and *M. persicae* and showed no mutagenicity/toxicity in the *S. typhimurium* strains tested up to a concentration of 1000 mg per plate.

Two new acetogenins, 2,4-*cis* and 2,4-*trans*-gigantecinone, isolated as admixture, and 4-deoxygigantecin, a known acetogenin whose absolute stereochemistry has not been determined previously, were isolated using activity-directed fractionation from the bark of *Goniothalamus giganteus* (164). A key step in solving their absolute stereochemistries was preparation of 1,4-diol formaldehyde acetal derivatives. Using the advanced Mosher ester method and circular dichroism, the absolute stereochemistries were found to be the same as that of gigantecin, which supports a common biogenetic origin. Both of them showed potent and selective cytotoxicities against the PC-3 human prostate adenocarcinoma cell line. Against yellow fever mosquito larvae, both compounds were more potent than rotenone in pesticidal activity.

These acetogenins are among the most potent of the known inhibitors of NADH: ubiquinone oxidoreductase (165), an essential enzyme in complex I of the electron transport system that eventually leads to oxidative phosphorylation in mitochondria (137,166). Thus, the end result of this mode of action is adenosine triphosphate (ATP) deprivation.

11.5 POLYACETYLENES AND CHROMENES (LIGHT-ACTIVATED INSECTICIDES)

An increasing number of phytochemicals, including phenols, terpenoids, polyketides, and alkaloids, are being recognized as photochemically active substances or light-activated insecticides (167). These compounds, unlike the photosynthetic pigments of the phytochromes, do not have any known functions in the plant species in which they occur. However, when introduced into the biological systems, e.g., cells or complex organisms, many of them are extremely toxic in light. The cellular targets and the photochemical processes for some of them have been defined. In one type of process there is cyclo-addition of the plant sensitizer with a nucleic acid base, e.g., the formation of a photoadduct of a furanocoumarin such as 8-methoxypsoralen with thymine in a nucleic acid. Certain alkaloids, including furanoquinolines, β-carbolines, canthinones, and certain furanochromones and furanochromenes, appear to be of this type; they are photogenotoxic, giving rise to gross chromosomal aberration in light. A second type of photoreaction, inherently bimolecular, often leads to oxidations. A very large group of photochemicals, particularly the polyacetylenes and their thiophene derivatives, belong here. These compounds characteristics of the Compositae and about 20 other families of flowering plants, as well as basidiomycetous fungi, are powerful photochemicals whose main targets in the phototoxicity process are cell membranes.

Plant-derived phototoxins with insecticidal properties are (a) polyacetylenes, (b) benzopyrans and benzofurans, (c) β-carboline alkaloids, (d) extended quinones, (e) furanocoumarins, (f) furanochromenes, (g) furoquinoline alkaloids, (h) benzylisoquinoline alkaloids, and (i) thiophenes (168). The structures of a few important polyacetylenic compounds and chromenes are shown in Fig. 11.6. Major work demonstrating the importance of light in promoting insecticidal activity for a variety of plant–derived compounds began in 1970s. The toxicity of furanocoumarin, a xanthotoxin, to armyworm *Spodoptera eridania* was markedly increased in the presence of light (169,170). Camm et al. (171) established that plant-derived polyacetylenes have photosensitizing properties. Since then a variety of other biologically active phototoxic compounds of differing chemical types have been derived

Polyacetylenes

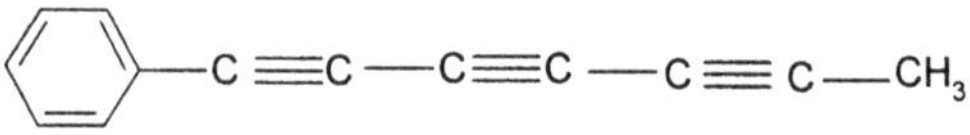

Phenylheptatriyne (PHT)

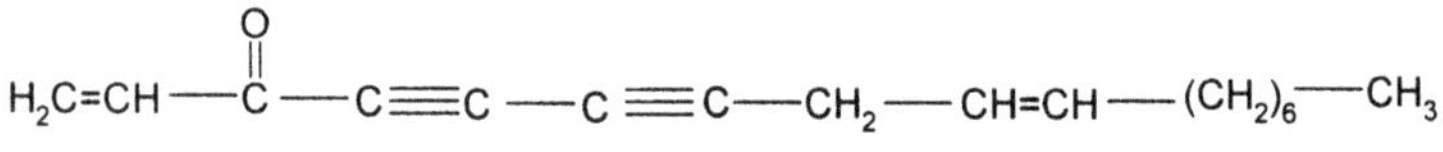

Falcarinone

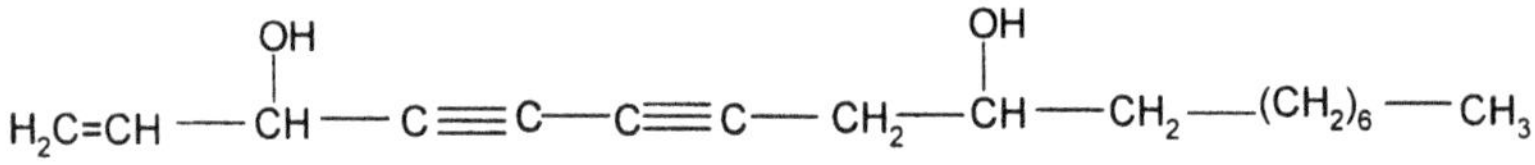

Falcarindiol

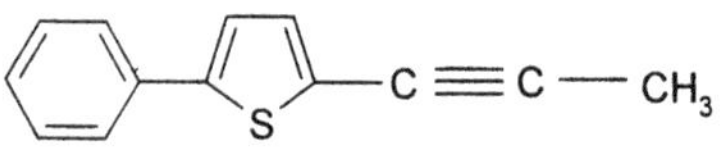

2-Phenyl-5- (1' propynl)-thiophene

Chromenes

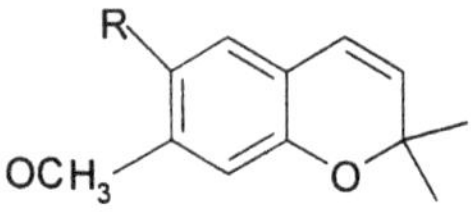

Precocene-1; R = H
Precocene- 2; R = OCH₃

Encecalin;   R = CH₂-C-CH₃

FIGURE 11.6 Structures of polyacetylenes and chromenes.

from plant sources (169). Most prominent among them are polyacetylenic compounds. Most photoactivated polyacetylenes appear to come from the Asteraceae. More than two dozen polyacetylenes from Asteraceae have been extensively tested for photoactivity against a variety of organisms. Furano-acetylenes isolated from *Chrysanthemum leucanthemum* L. were tested against mosquito, *Aedes atropalpus*, whose activity was equal to that of α-terthienyl (172). Experiments with microorganisms show that at least two conjugated acetylenic bonds are needed to cause phototoxic activity (173), phenylhepta-tryine, which is widely distributed in Asteraceae species (174), occur in the stems and leaves of *Bidens pilosa* (175). Phenylheptatryine reduced feeding and weight gain in larvae of the polyphagous insect *Euxoa messoria* when incorporated into artificial diet at 10–300 ppm (174). It was also identified from *Coleopsis lanceolata* (176) and shown to possess ovicidal activity against *Drosophila melanogaster*. The ovicidal activity was increased 37-fold on light activation of phenylheptatryine when compared to the dark (177). The ovicidal activity of α-terthienyl was increased 4333-fold on light activation when tested against *D. melanogaster*.

Another group of acetylenic compounds in Compositae are thiarubrines such as 1-(2-methylethynyl)-4-(hexa-1,3-diyn-5-enyl)-2,3-dithiacyclohexa-4,6-diene, thiarubrine-A). It exhibited higher activity against the root knot nematode *Meloidogyne incognita*, but in this case light was not a factor (178). The family Compositae contains α-terthienyl and 5-(3-buten-1-yl)-2,2′-bithienyl. A number of herbivorous insects are sensitive to topically applied α-terthienyl in the presence of near-UV irradiation or sunlight including the larvae of *Manduca sexta* and *Pieris rapae* (179). On the other hand, *Heliothis virescens* and some other species are relatively insensitive. High potency for α-terthienyl was reported for light-exposed water-living larvae of the mosquito *Aedes aegypti* and the black fly *Simulium verecundum* (180). A field trial with the synthesized α-terthienyl in eastern Canada at deciduous forest sites resulted in reliable control of *Aedes* spp larvae at an application rate of 50 or 100 g of α-terthienyl per hectare (180).

Polyacetylene derivatives (PADs) occurring in *Rudbeckia hirta* (Asteraceae) were isolated or chemically synthesized and examined for their insecticidal properties against mosquito larvae under different light regimes: dark (D), visible (VIS), and visible + near-UV (VIS+UV) (181). A straight chain polyine, 1-tridecene-3,5,7,9,11-pentayne, was highly toxic under all light regimes tested although a thiarubrine, 3-(1-propynyl)-6(3,5-hexadien-1-y1)-1,2 dithiacyclohexa-3,5-diene, and a thiophene, 2-(1-propyny-1)-6-3,5-hexadien-1-yl)-thiophene showed a toxicity that significantly varied between the light regimes. The thiarubrine was more toxic against mosquito larvae under D or VIS+UV conditions, whereas the thiophene had a more pronounced toxicity only in presence of VIS+UV irradiation. The distinctive insecticidal properties in darkness of the thiarubrine compared with thiophene were also confirmed in trials with larvae of a herbivorous insect, *M. sexta*. Such variability in the light-modulated toxicity to insects for different biosynthetically related PADs emphasizes a diversity in the insecticidal mechanisms of action.

Undamaged leaves of 12 host plant species differing widely in acceptability to ovipositing carrot flies were extracted with a microwave-assisted method with hexane as solvent (182). The highly stimulatory diethyl ether fraction obtained by separation on a silica gel column was semi-quantitatively analyzed by GC-MS for previously identified oviposition stimulants of the carrot fly (phenylpropenes, furanocoumarins, polyacetylenes). Various plant species exhibited widely differing profiles of these compounds. In choice assays, moderate numbers of eggs were deposited underneath surrogate leaves sprayed with fractions that contained high amounts of just one type of compound and low amounts of the other two types. Only fractions with medium to high levels of at least two compound classes elicited strong ovipositional responses (e.g., phenylpropenes and polyacetylenes in *Daucus carota*, furanocoumarins and polyacetylenes in *Heracleum sphondylium* and *Conium maculatum*). None of the examined plants contained high quantities of all three compound classes. The content of the stimulants seemed to account in a synergistic manner for the variation in activity of the diethyl ether fraction. However, they could not adequately explain the observed preference hierarchy of the carrot fly for the host plant species.

Two acetylenic compounds were isolated form *Cryptotaenia canadensis*, a native North American umbellifer frequently encountered in moist woodlands (183). Fresh foliage, roots, and fruits were extracted with a hot ethanol and water mixture and then dried. The extract was partitioned into chloroform and water, and both phases were bioassayed against fourth instars of *Culex pipiens* at concentrations between 5 and 50 ppm. Only the organic phase was active. Gas–liquid chromatography revealed that the two main components of the organic phase were polyacetylenes. The compounds were isolated by vacuum chromatography on silica gel and identified as falcarinol and falcarindiol based on GC-MS-EI and GC-MS-CI fragmentation patterns, high-resolution MS (167), and H and C NMR spectrometric data (179). The dose–response curves with mosquito larvae were determined by probit analysis. The LC_{50} values were 3.5 ppm for falcarinol and 6.5 ppm for falcarindiol. The distribution of polyacetylenes varied among plant parts. Fruits contained an unknown compound not found in either foliage or roots.

Benzopyrans and benzofurans of diverse structures are known from many species of higher plants, but the majority occur in the Asteraceae. The structures and the bioactivities of 167 isolated compounds have been reviewed (184). The compound 6,7-dimethoxy-2,2-dimethylchromene and 7-methoxy-2,2-dimethylchromene induce precocious metamorphosis in some insects by destroying the gland that secretes juvenile hormone. Several other chromenes have attracted attention as insecticides. Acetylchromenes, i.e., 6-acetyl-7-methoxy-2,2-dimethylchromene, 6-acetyl-7-hydroxy-2,2-dimethylchromene, and 6-acetyl-2,2-dimethylchromene— have been isolated from some species of desert sunflowers of the genus *Encelia* (Asteraceae). The compound encecalin when incorporated in artificial diet reduced the feeding of larvae of bollworms, (*Helicoverpa zea*), loopers (*Plusia gamma*), and cutworms (*Peridroma saucia*) (185). The residual contact test with encecalin (coated on the inner surfaces of

TABLE 11.3 Newer Compounds with Bioactivities Against Insect Pests

S. No	Name of the compound	Source	Bioactivity tested against	Mechanism	Ref.
1.	Dearomatized isoprenylated phloroglucinol (DIPs) (Hypercalin A)	*Hypericum calycinum*	Deterrent and toxic to *Uteheisa ornatrix*	Visible and attractive UV pigments to insects	191
2.	Rocaglamide derivatives	*Aglaia spectabilis* (Meliaceae)	*Spodoptera littoralis* (neonate larvae)	Not known	192
3.	Withanolides	*Salpichroa origanifolia*	*Musca domestica*	Antifeedant	193
4.	Tricin 5-*O*-glucoside	*Oryza sativa*	White-back planthopper, *Sogoitella furcifera*	Probing stimulant	194
5.	Cinnamomin	Camphor tree (*Cinnamomum camphora*)	*Helicoverpa armigera* and mosquito (*Culex pipiens*)	Ribosome inactivating protein	195
6.	Acylphloroglucinols	*Humulus lupulus* (Cannabaceae)	Cereal aphids	Antifeedant	196
7.	Alkamides: pellitorine, neopellitorine A&B	*Artemisia dracuculus*	*Sitophilus oryzae* *Rhyzopertha dominica*	Insecticidal	197
8.	Cytochrome P450 enzymes: hydroperoxide lyase, peroxygenase	Alfalfa	—	Release of volatiles such as hexenal	198
9.	Chitinase	Beans, tomato	Homoptera, Lepidoptera	Act on insect peritrophic membrane	199, 200
10.	Anionic peroxidase	Tobacco	Lepidoptera, Coleoptera, Homoptera	—	201

(continued)

TABLE 11.3 Continued

S. No	Name of the compound	Source	Bioactivity tested against	Mechanism	Ref.
11.	Tryptophan decarboxylase	*Catharanthus roseus*	Homoptera	Alkaloid synthesis	202
12.	Eremophilanolidae	*Senecio miser*	—	Antifeedant	203
13.	Sandoricum	*Sandorium koetjape*	*Spodoptera frugiperda, Ostrina nubilalis*	Antifeedant	204
14.	Osmundalin	*Osmunda japonica*	*Eurema hecabe mandarina*	Antifeedant	205
15.	3-Hexenal, 3-hexenyl acetate, α-bergamotene	Lima bean, *Nicotiana attenuata*	*Tetranychus urticae, Phytosciulus persimilis*	Activation of defense genes	206, 207
16.	10-Deacetylbaccatin III & V	*Taxus baccata* (yew tree)	*Tribolium confusum Sitophilus granarius Trigoderma granarium*	Antifeedant	208
17.	Anacardic acids	Pelargonium	Spider mites	Inhibition of prostaglandin endoperoxide synthase and lipoxygenase	209

glass vials) afforded LC_{50} values of 20, 3, and 12 μg for neonate milkweed bugs (*Oncopeltus fasicatus*), cutworms (*Peridroma sancia*), and glasshoppers (*Melanoplus sanguinipes*), respectively (186). Demethoxyencecalin was equivalent in topical test to that of encecalin, whereas demethylencecalin was less active by topical application. None of the acetylchromenes showed antijuvenile hormone effects characteristic of the precocenes (187). Chromene and 6-vinyl-7-methoxy–2,2*p*-dimethylchromene isolated from tarweed, *Hemizonia fitchii* A. Grey (Asteraceae), showed a LC_{50} value of 3.0 and 1.8 ppm against the first-instar *Culex pipiens* (188). There was no effect on development noted for survivors through subsequent larval, pupal, and adult stages. These two compounds when tested topically showed less active or no effect in the second-and third-instar nymph of milkweed bugs. Cytotoxic polyacetylenes were also isolated from marine sponge *Petrosia* species (189) and from the twigs of *Ochanostachys amentacea* (190).

11.6 NEWER COMPOUNDS

The emerging newer compounds, for which details are not available, are presented in Table 11.3.

REFERENCES

1. Eigenbrode, S.D.; Stoner, K.A.; Shelton, A.M.; Kain, W.C. Characteristics of glossy leaf waxes associated with resistance to diamondback moth (Lepidoptera: Plutellidae) in *Brassica oleracea*. J. Econ. Entomol. **1991**, *84*, 1609–1618.
2. Dorset, D.L.; Ghiradella, H. Insect wax secretion: the growth of tubular crystals. Biochem. Biophys. Acta. **1983**, *760*, 136–142.
3. Jenks, M.A.; Tuttle, H.A.; Feldmann, K.A. Changes in epicuticular waxes on wildtype and *Eceriferum* mutants in *Arabidopsis* during development. Phytochemistry **1995**, *42*, 29–34.
4. Atkin, D.S.J.; Hamilton, R.J. The changes with age in the epicuticular wax of *Sorghum bicolor*. J. Nat. Prod. **1982**, *45*, 697–703.
5. Baker, E.A.; Hunt, G.M. Developmental changes in leaf epicuticular waxes in relation to foliar penetration. New Phytol. **1981**, *88*, 731–747.
6. Baker, E.A. The influence of environment on leaf wax development in *Brassica oleracea* var. *gemmifera*. New Phytol. **1974**, *73*, 955–966.
7. Chapman, R.F.; Bernays, E.A. Insect behavior at the leaf surface and learning aspects of host plant selection. Experientia **1989**, *45*, 215–222.
8. Eigenbrode, S.D. Influence of plant surface waxes on insect behavior. In *Plant Cuticles: An Integrated Functional Approach*; Kerstiens, G., Ed.; Bios Press: Oxford, 1966; 201–222.

9.   Eigenbrode, S.D.; Espelie, K.E. Effects of plant epicuticular lipids on insect herbivores. Annu. Rev. Entomol. **1995**, *40*, 171–194.
10. Walton, T.J. Waxes, cutin, and suberin. In *Methods in Plant Biochemistry*; Hardwood, J.L., Bowyer, J.R., Eds.; Academic Press: San Diego, 1990; Vol. 4, 105–158 pp.
11. Woodhead, S.; Chapman, R.F. Insect behavior and the chemistry of plant surface waxes. In *Insects and the Plant Surface*; Juniper, B., Southwood, T.R.E., Eds.; Edward Arnold: London, 1986; 123–135 pp.
12. von Wettstein-Knowles, P.M. Biosynthesis and genetics of waxes. In *Waxes: Chemistry, Molecular Biology and Functions*; Hamilton, R.J., Ed.; Oily Press Ltd: Dundee, Scotland, 1995; 91–130 pp.
13. Jenks, M.A.; Ashworth, E.N. Plant epicuticular waxes. Function production and genetics. Hort. Rev. **1999**, *23*, 1–68.
14. Cutler, D.F.; Alvin, K.L.; Price, C.E. *The Plant Cuticle*; Academic Press: New York, 1982.
15. Hamilton, R.J.;. Waxes: *Chemistry, Molecular Biology and Functions*; Oily Press Ltd: Dundee, Scotland, 1995.
16. Kerstiens, G. *Plant Cuticles: An Integrated Functional Approach*; BIOS: Oxford, 1996.
17. Hadley, N.F. *The Adaptive Role of Lipids in Biological Systems*; Wiley-Interscience: New York, 1985.
18. Juniper, B.E.; Southwood, T.R.E., Eds.; *Insect and the Plant Surface*; Edward Arnold: London, 1986.
19. Jeffree, C.E. The cuticle, epicuticular waxes and trichomes of plants, with reference to their structure, functions and evolution. In *Insects and the Plant Surface*; Juniper, B.E., Southwood, T.R.E., Eds.; Edward Arnold: London, 1986; 23–64 pp.
20. Jetter, R.; Riederer, M. Composition of cuticular waxes on *Osmunda regalis* fronds. J. Chem. Ecol. **2000**, *26*, 399–412.
21. Jeffree, C.E.; Baker, E.A.; Holloway, P.J. Ultrastructure and recrystallization of plant epicuticular waxes. New Phytol. **1975**, *75*, 539–549.
22. Yang, G.; Wiseman, B.R.; Isenhour, D.J.; Espelie, K.E. Chemical and ultrastructural analysis of corn cuticular lipids and their effect on feeding by fall armyworm larvae. J. Chem. Ecol. **1993**, *19*, 2055–2074.
23. Gulz, P.G.; Muller, E.; Schmitz, K.; Marner, F.J.; Guth, S. Chemical composition and surface structures of epicuticular leaf waxes of *Ginkgo biloba*, *Magnolia grandiflora* and *Liriodendron tulipifera*. Z Naturforsch. **1992**, *47C*, 516–526.
24. Baker, E.A.; Bukovac, M.J.; Flore, J.A. Ontogenetic variations in the composition of peach leaf wax. Phytochemistry **1979**, *18*, 781–784.
25. Bukovac, M.J.;.Flore, J.A.; Baker, E.A. Peach leaf surfaces: changes in wettability, retention, cuticular permeability, and epicuticular wax chemistry during expansion with special reference to spray application. J. Am. Soc. Hort. Sci. **1979**, *104*, 611–617.
26. Riederer, M.; Schneider, G. The effect of the environment on the permeability and composition of *Citrus* leaf cuticles. Planta **1990**, *180*, 154–165.

27. Woodhead, S. Environmental and biotic factors affecting the phenolic content of different cultivars of *Sorghum bicolor*. J. Chem. Ecol. **1981**, *7*, 1035–1047.
28. Premachandra, G.S.; Hahn, D.T.; Joly, R.J. A simple method for determination of abaxial and adaxial epicuticular wax loads in intact leaves of *Sorghum bicolor* L. Can. J. Plant Sci. **1993**, *73*, 521–524.
29. Bergman, D.K.; Dillwith, J.W.; Zarrabi, A.A.; Berberet, R.C. Epicuticular lipids of alfalfa leaves relative to position on the stem and their correlation with aphid (Homoptera: Aphididae) distributions. Environ. Entomol. **1991**, *20*, 470–476.
30. von Wettstein–Knowles, P.M.; Sogaard, B. *Genetic Evidence that cer-cqu is a Cluster Gene. Barley Genetics* IV, Proc. 4th Int. Barley Genet Symp. Edinburgh University Press: Edinburgh, 1981; 625–630 pp.
31. Jenks, M.A.; Tuttle, H.A.; Eigenbrode, S.D.; Feldmann, K.A. Leaf epicuticular waxes of the *Eceriferum* mutants in *Arabidopsis*. Plant Physiol. **1995**, *108*, 369–377.
32. Assagne, C.C.; Lessire, R.; Bessoule, J.J.; Moreau, P. Plant elongases. In *The Metabolism, Structure and Function of Plant Lipids*; Stumpf, P.K., Mudd, J.B., Nes, W.D., Eds.; Plenum Press: New York, 1987; 481–488 pp.
33. Ohlrogge, J.B.; Jawirski, J.G.; Post-Beittenmiller, D. De novo fatty acid biosynthesis. In *Lipid Metabolism in Plants*; Moore, T.S., Ed.; CRC Press: Boca Raton, 1993; 3–32 pp.
34. Pollard, M.; McKeon, T.; Gupta, L.; Stumpf, P. Studies on biosynthesis of waxes by developing jojoba seed. II. The demonstration of wax biosynthesis by cell-free homogenates. Lipids **1979**, *14*, 651–662.
35. von Wettstein-Knowles, P.M. Biosynthesis and genetics of waxes. In *Waxes: Chemistry, Molecular Biology and Functions*; Hamilton, R.J., Ed.; Oily Press: Dundee, Scotland, 1995; Vol. 6, 91–130 pp.
36. von Wettstein-Knowles, P.M. Elongases and epicuticular wax biosynthesis. Physiol. Veg. **1982**, *20*, 797–809.
37. Liu, D.; Post-Beittenmiller, D. Discovery of an epidermal stearoyl-acyl carrier protein thioesterase: its potential role in wax biosynthesis. J. Biol. Chem. **1995**, *270*, 16962–16969.
38. Cassagne, C.; Lessire, R.; Bessolue, J.J.; Moreau, P.; Creach, A. Biosynthesis of very long chain fatty acids in higher plants. Prog. Lipid Res. **1994**, *33*, 55–69.
39. Agrawal, V.P.; Lessire, R.; Stumpf, P.K. Biosynthesis of very long chain fatty acids in microsomes from epidermal cells of *Allium porrum* L. Arch. Biochem. Biophys. **1984**, *230*, 580–589.
40. Cassagne, C.; Lessire, R. Biosynthesis of saturated very long chain fatty acids by purified membrane fractions from leek epidermal cells. Arch. Biochem. Biophys. **1978**, *191*, 146–152.
41. Ichihara, K.; Nakagawa, M.; Tanaka, K. Acyl-CoA synthetase in maturing safflower seeds. Plant Cell Physiol. **1993**, *34*, 557–566.
42. Jaworski, J.G.; Post-Beittenmiller, D.; Ohlrogge, J.B. Acetyl-acyl carrier protein is not a major intermediate in fatty acid biosynthesis in spinach. Eur. J. Biochem. **1993**, *213*, 981–987.

43. Lessire, R.; Hartmann-Bouillon, M.A.; Cassagne, C. Very long chain fatty acids: occurrence and biosynthesis in membrane fractions from etiolated maize coleoptiles. Phytochemistry **1982**, *21*, 55–59.

44. Ohlrogge, J.B.; Shine, W.E.; Stumpf, P.K. Fat metabolism in higher plants: characterization of plant acyl-ACP and acyl-CoA hydrolases. Arch. Biochem. Biophys. **1978**, *189*, 382–391.

45. Lessire, R.; Bessoule, J.J.; Cassagne, C. Involement of a α-ketoacyl-CoA intermediate in acyl-CoA elongation by an acyl-CoA elongase purified from leek epidermal cells. Biochim. Biophys. Acta. **1989**, *1006*, 35–40.

46. Lessire, R.; Cassagne, C. Long chain fatty acid CoA-activation by microsomes from *Allium porrum* epidermal cells. Plant Sci. Lett. **1979**, *16*, 31–39.

47. Whitfield, H.V.; Murphy, D.J.; Hills, M.J. Sub-cellular localization of fatty-acid elongase in developing seeds of *Lunaria annua* and *Brassica napus*. Phytochemistry **1993**, *32*, 255–258.

48. Agarwal, V.P.; Stumpf, P.K. Characterization and solubilization of an acyl chain elongation system in microsomes of leek epidermal cells. Arch. Biochem. Biophys. **1985**, *240*, 154–165.

49. Cheesbrough, T.M.; Kolattukudy, P.E. Alkane biosynthesis by decarbonylation of aldehydes catalyzed by a particulate preparation from *Pisum sativum*. Proc. Natl. Acad. Sci. USA **1984**, *81*, 6613–6617.

50. Dennis, M.W.; Kolattukudy, P.E. Alkane biosynthesis by decarbonylation of aldehyde catalyzed by a microsomal preparation from *Botryococcus braunii*. Arch. Biochem. Biophys. **1991**, *287*, 268–275.

51. Kolattukudy, P.E. Enzymatic synthesis of fatty alcohols in *Brassica oleracea*. Arch. Biochem. Biophys. **1971**, *142*, 701–709.

52. Mikkelsen, J.D. Structure and biosynthesis of α-diketones in barley spike epicuticular wax. Carlsberg Res. Commun. **1979**, *44*, 133–147.

53. Mikkelsen, J.D.; von Wettstein-Knowles, P.M. Biosynthesis of β-diketones and hydrocarbons in barley spike epicuticular wax. Arch. Biochem. Biophys. **1978**, *1288*, 172–181.

54. Shimakata, T.; Stumpf, P.K. Isoaltion and function of spinach leaf β-ketoacyl (acyl-carrier-protein) synthases. Proc. Natl. Acad. Sci. USA **1982**, *79*, 5808–5812.

55. Stoner, K.A. Glossy leaf wax and plant resistance to insects in *Brassica oleracea* under natural infestation. Environ. Entomol. **1990**, *19*, 730–739.

56. Way, M.J.; Murdie, G. An example of varietal variations in resistance of Brussels sprouts. Ann. Appl. Biol. **1965**, *56*, 326–328.

57. Tsumuki, H.; Kanehisa, K.; Kawada, K. Leaf surface wax as a possible resistance factor of barley to cereal aphids. Appl. Entomol. Zool. **1989**, *24*, 295–301.

58. Bodnaryk, R.P. Leaf epicuticular wax, an antixenotic factor in Brassicaceae that affects the rate and pattern of feeding of flea beetles *Phyllotreta cruciferae* (Goeze). Can. J. Plant Sci. **1992**, *72*, 1295–1303.

59. Lowe, H.J.B.; Murphy, G.J.P.; Parker, M.L. Non-glaucousness, a probable aphid-resistance character of wheat. Ann. Appl. Biol. **1985**, *106*, 555–560.

60. Schwager, B.; Pitre, H.; Gourley, L. Field evaluation of sorghum characteristics for resistance to fall armyworm. J. GA Entomol. Soc. **1984**, *19*, 333–339.

61. Weibel, D.E.; Starks, K.J. Greenbug nonpreference for bloomless sorghum. Crop Sci. **1980**, *26*, 1151–1153.

62. Eigenbrode, S.D.; Shelton, A.M. Survival and behavior of *Plutella xylostella* larvae on cabbages with leaf waxes altered by treatment with S-ethyl dipropylthiocarbamate. Entomol. Exp. Appl. **1992**, *62*, 139–145.

63. Eigenbrode, S.D.; Shelton, A.M.; Kain, W.C.; Leichtweis, H.; Spittler, T.D. Managing lepidopteran pests in cabbage with herbicide-induced resistance, in combination with a pyrethroid insecticide. Entomol. Exp. Appl. **1993**, *69*, 41–50.

64. Uematsu, H.; Sakanoshita, A. Possible role of cabbage leaf wax bloom in suppressing diamondback moth *Plutella xylostella* (Lepidoptera: Yponomeutidae) oviposition. Appl. Entomol. Zool. **1989**, *24*, 253–327.

65. Prokopy, R.J.; Collier, R.H.; Finch, S. Leaf color used by cabbage root flies to distinguish among host plants. Science **1983**, *221*, 190–192.

66. Neuenschwander, P.; Michelakis, S.; Holloway, P.; Berchtold, W. Factors affecting the susceptibility of fruits of different olive varieties to attack by *Dacus oleae* (Gmel.) (Dipt., Tephritidae). Z. Angew Entomol. **1985**, *100*, 174–188.

67. Foster, S.P.; Harris, M.O. Foliar chemicals of wheat and related grasses influencing oviposition by Hessian fly, *Mayetiola destructor* (Say) (Diptera: Cecidomyiidae).J. Chem. Ecol. **1992**, *18*, 1965–1980.

68. Stadler, E.; Schoni, R. Oviposition behavior of the cabbage root fly, *Delia radicum* (L.), influenced by host plant extracts. J. Insect Behav. **1990**, *3*, 195–209.

69. Stadler, E.; Buser, H.R. Defense chemicals in leaf surface wax synergistically stimulate oviposition by a phytophagous insect. Experientia **1984**, *40*, 1157–1159.

70. Spencer, J.L. Waxes enhance Plutella xylostella oviposition in response to sinigrin and cabbage homogenates. Entomol. Exp. Appl. **1996**, *81*, 165–173.

71. Kombargi, W.S.; Michelakis, S.E.; Petrakis, C.A. Effect of olive surface waxes on oviposition by *Bactrocera oleae* (Diptera: Tephritidae). J. Econ. Entomol. **1998**, *91*, 993–998.

72. Neuenschwander, P.; Michelakis, S.; Holloway, P.; Berchtold, W. Factors affecting the susceptibility of fruits of different olive varieties to attack by *Dacus oleae* (Gmel.) (Dipt., Tephritidae). Z. Angew Entomol. **1985**, *100*, 174–188.

73. Justus, K.A.; Dosdall, L.M.; Mitchell, B.K. Ovisposition by *Plutella xylostella* (Lepidoptera: Plutellidae) and effects of phylloplane waxiness. J. Econ. Entomol. **2000**, *93*, 1153–1159.

74. Grant, G.G.; Zhao, B.; Langevin, D. Oviposition response of spruce budworm (Lepidoptera: Tortricidae) to aliphatic acids. Environ. Entomol. **2000**, *29*, 164–170.

75. Powell, G.; Maniar, S.P.; Pickett, J.A.; Hardie, J. Aphid responses to non-host epicuticular lipids. Entomol. Exp. Appl. **1999**, *91*, 115–123.

76. Ni, X.Z.; Quisenberry, S.S.; Siegfried, B.D.; Lee, K.W. Influence of cereal leaf epicuticular wax on *Diuraphis noxia* probing behavior and nymphoposition. Entomol. Exp. Appl. **1998**, *89*, 111–118.

77. Ni, X.Z.; Quisenberry, S.S. Effect of wheat leaf epicuticular structure on host selection and porbing rhythm of Russian wheat aphid (Homoptera: Aphididae). J. Econ. Entomol. **1997**, *90*, 1400–1407.

78. Johnson, A.W.; Severson, R.F. Leaf surface chemistry of tabacco budworm resistant tobacco. J. Agric. Entomol. **1984**, *1*, 23–32.

79. Bergman, D.K.; Dillwith, J.W.; Zarrabi, A.A.; Caddel, J.L.; Berberet, R.C. Epicuticular lipids of alfalfa relative to its susceptibility to spotted alfalfa aphids (Homoptera: Aphididae). Environ. Entomol. **1991**, *20*, 781–785.

80. Balsdon, J.A.; Espelie, K.E.; Braman, S.K. Cuticular lipids from azalea (*Rhododendron* spp.) and their potential role in host plant acceptance by azalea lace bug, *Stephanitis pyrioides* (Scott). Biochem. Syst. Ecol. **1995**, *23*, 477–485.

81. Heupel, R.C. Varietal similarities and differences in the polycyclic isopentenoid composition of sorghum. Phytochemistry **1985**, *24*, 2929–2937.

82. Yang, G.; Espelie, K.E.; Toad, J.W.; Culbreath, A.K.; Pittman, R.N.; Demski, J.W. Cuticular lipids from wild and cultivated peanuts and the relative resistance of these peanut species to fall armyworm and thrips. J. Agric. Food Chem. **1993**, *41*, 814–818.

83. Woodhead, S.; Padgham, D E. The effect of plant surface characteristics on resistance of rice to the brown planthopper, *Nilaparvata lugens*. Entomol. Exp. Appl. **1988**, *47*, 15–22.

84. Bernays, E.A.; Chapman, R.F. The importance of chemical inhibition of feeding in host-plant selection by *Chorthippus parallelus* (Zetterstedt). Acrida. **1975**, *4*, 83–93.

85. Braker, E.; Chazdon, R.L. Ecological behavioral and nutritional factors influencing use of palms as host plants by a neotropical forest grasshopper. J. Trop. Ecol. **1993**, *9*, 183–197.

86. de Boer, G.; Hanson, F.E. The role of leaf lipids in food selection by larvae of the tobacco hornworm, *Manduca sexta*. J. Chem. Ecol. **1988**, *14*, 669–679.

87. Varela, L.; Bernays, E.A. Behavior of newly hatched potato tuber moth larvae, *Phthorimaea operculella* Zell. (Lepidoptera: Gelechiidae) in relation to their host plants. J. Insect Behav. **1988**, *1*, 261–275.

88. Klingauf, F.; Nocker–Wenzel, K.; Klein, W. Einfluss einiger Wachskomponenten von Vicia faba L. auf das Wirtswahlverhalten von Acyrthosiphon pisum (Harris) (Homoptera: Aphididae). Z Pflanzenkr Pflanzensch **1971**, *78*, 641–648.

89. Woodhead, S. Surface chemistry of *Sorghum bicolor* and the importance in feeding by *Locusta migratoria*. Physiol. Entomol. **1983**, *8*, 345–352.

90. Adati, T.; Matsuda, K. Feeding stimulants for various leaf beetles (Coleoptera: Chrysomelidae) in the leaf surface wax of their host plants. Appl. Entomol. Zool. **1993**, *28*, 319–324.

91. Mori, M. n-Hexacosanol and n-octacosanol: feeding stimulants for larvae of the silkworm, *Bombyx mori*. J. Insect Physiol. **1982**, *11*, 969–973.

92. Greenway, A.R.; Griffiths, D.C.; Lloyd, S.L. Response of *Myzus persicae* to components of aphid extracts and to carboxylic acids. Entomol. Exp. Appl. **1978**, *24*, 369–374.

93. Sherwood, M.H.; Greenway, A.R.; Griffiths, D.C. Responses of *Myzus persicae* (Sulzer) (Hemiptera: Aphididae) to plants treated with fatty acids. Bull. Entomol. Res. **1981**, *71*, 133–136.

94. Herbach, E. Effect of dodecanoic acid on the colonisation of sugar beet by aphids and the secondary spread of virus yellows. Ann. Appl. Biol. **1987**, *111*, 477–482.

95. Phelan, P.L.; Miller, J.R. Post-landing behavior of alate *Myzus persicae* as altered by (E)-β-farnesene and three carboxylic acids. Entomol. Exp. Appl. **1982**, *32*, 46–53.

96. Yang, G.; Isenhour, D.J.; Espelie, K.E. Activity of maize leaf cuticular lipids in resistance to leaf feeding by the fall armyworm. Fla. Entomol. **1991**, *74*, 229–236.

97. Yang, G.; Wiseman, B.R.; Espelie, K.E. Cuticular lipids from silks of seven corn genotypes and their effect on development of corn earworm larvae [*Helicoverpa zea* (Boddie)]. J. Agric. Food Chem. **1992**, *40*, 1058–1061.

98. Yang, G.; Espelie, K.E.; Todd, J.W.; Culbreath, A.K.; Pittman, R.N.; Demski, J.W. Characterization of cuticular lipids from cultivated and wild peanut species and their effect on feeding by fall armyworm (Lepidoptera: Noctuide). Peanut Sci. **1995**, *22*, 49–54.

99. Shankaranarayana, K.H.; Ayyar, K.S.; Krishna Rao, G.S. Insect growth inhibitor from the bark of *Santalum album*. Phytochemistry **1980**, *19*, 1239–1240.

100. Heupel, R.C. Varietal similarities and differences in the polycyclic isopentenoid composition of sorghum. Phytochemistry **1985**, *24*, 2929–2937.

101. Sheperd, T.; Robertson, G.W.; Griffiths, D.W.; Birch, A.N.E. Epicuticular wax composition in relation to aphid infestation and resistance in red raspberry (*Rubus idaeus* L.). Phytochemistry **1999**, *52*, 1239–1254.

102. Eigenbrode, S.D.; Pillai, S.K. Neonate *Plutella xylostella* responses to surface wax components of resistant cabbage (*Brassica oleracea*). J. Chem. Ecol. **1998**, *24*, 1611–1627.

103. White, C.; Eigenbrode, S.D. Effects of surface wax variation in *Pisum sativum* on herbivorous and entomphagous insects in the field. Environ. Entomol. **2000**, *29*, 773–780.

104. Eigenbrode, S.D.; Kabalo, N.N. Effects of *Brassica oleracea* waxblooms on predation and attachment by *Hippodamia convergens*. Entomol. Exp. Appl. **1999**, *91*, 125–130.

105. Schnable, P.S.; Stinard, P.S.; Wen, T.J.; Heinen, S.; Weber, D.; Schneerman, M.; Zhang, L.; Hansen, J.D.; Nikolau, B.J. The genetics of cuticular wax biosynthesis. Maydica **1994**, *39*, 279–287.

106. Post–Beittenmiller, D. Biochemistry and molecular biology of wax production in plants. Annu. Rev. Plant Physiol. Plant Mol. Biol. **1996**, *47*, 405–430.

107. Peters, P.J. Genetics and drought reaction of epicuticular wax mutants in *sorghum bicolor*. Ph.D dissertation, Purdue University, West Lafayette, IN, 1996.

108. Negruk, V.; Yang, P.; Subramanian, M.; McNevin, J.P.; Lemieux, B. Molecular cloning and characterization of the CER2 gene of *Arabidopsis thaliana*. Plant J. **1996**, *9*, 137–145.

109. Hannoufa, A.; Negruk, V.; Eisner, G.; Lemieux, B. The CER3 gene of *Arabidopsis thaliana* is expressed in leaves, stems, roots, flowers and apical meristems. Plant J. **1996**, *10*, 459–467.

110. Xia, Y.; Nikolau, B.J.; Schnable, P.S. Cloning and characteization of CER2, an *Arabidopsis* gene that affects cuticular wax accumulation. Plant Cell **1996**, *8*, 1291–1304.

111. Xia, Y.; Nikolau, B.J.; Schnable, P.S. Developmental and hormonal regulation of the *Arabidopsis* CER2 gene that codes for a nuclear-localized protein required for the normal accumulation of cuticular waxes. Plant Physiol. **1997**, *115*, 925–937.

112. Jenks, M.A.; Tuttle, H.A.; Eigenbrode, S.D.; Feldmann, K.A. Leaf cuticular waxes of the eceriferum mutants in Arabidopsis. Plant Physiol. **1995**, *108*, 369–377.

113. Aarts, M.G.M.; Keijzer, C.J.; Stiekema, W.J.; Pereira, A. Molecular characterization of the CER1 gene of *Arabidopsis* involved in epicuticular wax biosynthesis and pollen fertility. Plant Cell **1995**, *7*, 2115–2127.

114. Hansen, J.D.; Pyee, J.; Xia, Y.; Wen, T.J.; Robertson, D.S.; Kolattukudy, P.E.; Nikolau, B.J.; Schnable, P.S. The glossy1 locus of maize and an epidermis-specific cDNA from *Kleinia odora* define a class of receptor-like proteins required for the normal accumulation of cuticular waxes. Plant Physiol. **1997**, *113*, 1091–1100.

115. Xu, X.; Heinan, S.; Wen, T.J.; Delledonne, M.; Nikolau, B.J.; Schnable, P.S. Cloning and sequence analysis of a maize acetyl CoA carboxylase gene. In *Cloning Plant Genes Known Only by Phenotype*; Plant Molecular Genetic Institute: Saint Paul: MN, 1994; 34 p.

116. Moose, S.P.; Sisco, P.H. The maize homeotic gene CLOSSY15 is a member of the APETALA 2 gene family. J. Cell. Biochem. **1995**, *21A*, 458.

117. Todd, J.; Post-Beittenmiller, D.; Jaworski, J.G. KCS1 encodes a fatty acid elongase 3-ketoacyl CoA synthase affecting wax biosynthesis in *Arabidopsis thaliana*. Plant J. **1999**, *17*, 119–130.

118. Smith, R.M.; Marshall, J.A.; Davey, M.R.; Lowe, K.C.; Power, J.B. Composition of volatiles and waxes in leaves of genetically engineered tomatoes. Phytochemistry **1996**, *43*, 753–758.

119. Xu, X.; Dietrich, C.R.; Lessire, R.; Nikolau, B.J.; Schnable, P.S. The endoplasmic reticulum-associated maize GL8 protein is a component of the acyl coenzyme A elongase involved in the production of cuticular waxes. Plant Physiol. **2002**, *128*, 924–934.

120. O'Brien, P.J. Molecular mechanisms of quinone cytotoxicity. Chem. Biol. Interact. **1991**, *80*, 1–41.

121. Thiboldeaux, R.L.; Lindroth, R.L.; Tracy, J.W. Differential toxicity of juglone (5-hydroxy 1,4-naphthoquinone) and related naphthoquinones to saturniid moths. J. Chem. Ecol. **1994**, *20*, 1631–1641.

122. Khambay, B.P.S.; Batty, D.; Cahill, M.; Denholm, I.; Mead-Briggs, M.; Vinall, S.; Niemeyer, H.M.; Simmonds, M.S. Isolation, characterization and biological activity of naphthoquinones from *Calceolaria andina* L. J. Agric. Food Chem. **1999**, *47*, 770–775.

123. Ioset, J.R.; Marston, A.; Gupta, M.P.; Hostettmann, K. Antifungal and larvicidal cordiaquinones from the roots of *Cordia curassavica.* Phytochemistry **2000**, *53*, 613–617.

124. Ioset, J.R.; Marston, A.; Gupta, M.P.; Hostettmann, K. Antifungal and larvicidal meroterpenoid naphthoquinones and a naphthoxirene from the roots of *Cordia linnaei.* Phytochemistry **1998**, *47*, 729–734.

125. Stack, D. Phenolic metabolism. In *Plant Biochemistry*; Dey, P.M., Horborne, J.B., Eds.; Academic Press: New York, 1997; 387–416 pp.

126. Muller, W.U.; Liestner, E. 1,4-Naphthoquinone, an intermediate in juglone (5-hydroxy-1,4-naphthoquinone) biosynthesis. Phytochemistry **1976**, *15*, 407–410.

127. Thompson, R.H. *Naturally Occurring Quinones*; 2nd Ed.; Academic Press: London, 1971.

128. Mitchell, M.J.; Smith, S.L. Effects of the chitin synthetase inhibitor plumbagin and its 2-demethyl derivative juglone on insect ecdysone 20-monooxygenase activity. Experientia **1988**, *44*, 990–991.

129. Mitchell, M.M.; Keogh, D.P.; Crooks, J.R.; Smith, S.L. Effects of plant flavonoids and other allelochemicals on insect cytochrome P-450 dependent steroid hydroxylase activity. J. Biochem. Mol. Biol. **1993**, *23*, 65–71.

130. Khambay, B.P.S.; Batty, D.; Beddie, D.G.; Denholm, I.; Cahill, M.R. A new group of plant–derived naphthoquinone pesticides. Pest. Sci. **1997**, *50*, 291–296.

131. Jacobsen, N.; Pedersen, L.E.K. Activity of 2-(1-alkenyl)–3-hydroxy 1,4-naphthoquinones and related compounds against *Musca domestica.* Pest. Sci. **1986**, *17*, 511–516.

132. Thiboldeaux, R.L.; Lindroth, R.L.; Tracy, J.W. Effects of juglone (5-hydroxy-1,4-naphthoquinone) on mudgut morphology and glutathione status on Saturniid moth larvae. Com. Biochem. Physiol. C. Pharmacol. Toxicol. Endocrinol. **1998**, *120*, 481–487.

133. Saxena, B.P.; Thappa, R.K.; Tikku, K.; Sharma, A.; Suri, O.P. Effect of plumbagin on gonodotrophic cycle of the housefly, *Musca domestica* L. Indian J. Exp. Biol. **1996**, *34*, 739–744.

134. Fang, X.P.; Rieser, M.J.; Gu, Z.M.; Zhao, G.X.; McLaughlin, J.I. Annonaceous acetogenins: an updated review. Phytochem. Anal. **1993**, *4*, 27–67.

135. Rupprecht, J.K.; Hui, Y.H.; McLaughlin, J.L. Annonaceous acetogenin: a review. J. Nat. Prod. **1990**, *53*, 237–278.

136. Gu, Z.M.; Zhao, G.X.; Oberlies, N.H.; Zeng, L.; McLaughlin, J.L. Annonaceous acetogenins: potent mitochondrial inhibitors with diverse applications. In *Recent Advances in phytochemsitry*; Arnason, J.T., Mata, R., Romeo, J.T., Eds.; Plenum Press: New York, 1995; 249–310 pp.

137. Cave, A.; Figadere, B.; Laurens, A.; Cortes, D. Acetogenins from Annonaceae. In *Progress in the Chemistry of Natural Products*; Herz, W., Kirby, G.W., Moore, R.E., Steglich, W., Tamm, C., Eds.; Springer-verlag: New York, 1997; 81–288 pp.

138. Zeng, L.; Ye, Q.; Oberlies, N.H.; Shi, G.; Gu, Z.M.; He, K.; McLaughlin, J.L. Recent advances in annonaceous acetogenins, Nat. Prod. Rep. **1996**, *13*, 275–306.

139. Alali, F.Q.; Rogers, L.; Zhang, Y.; McLaughlin, J.L. Goniotriocin and (2,4-cis-and trans)-xylomaticinones, bioactive annonaceous acetogenins from *Goniothalamus giganteus*. J. Nat. Prod. **1999**, *62*, 31–34.

140. Mikolajezak, K.J.; Madrigal, R.V.; Rupprecht, J.K.; Hui, Y.H.; Liu, Y.M.; Smith, DL.; McLaughlin, J.L. Sylvaticin: a new cytotoxic and insecticidal acetogenin from *Rollina sylvatica* (Annonaceae). Experientia **1990**, *46*, 324–327.

141. Rupprecht, J.K.; Chang, C.J.; Cassady, J.M.; McLaughlin, J.L. Asimicin, a new cytotoxic and pesticidal acetogenin from the Paw Paw, *Asimina triloba* (Annonaceae.) Heterocycles (Tokyo) **1986**, *24*, 1197–1201.

142. Lewis, M.A.; Arnason, J.T.; Philogene, B.J.; Rupprecht, J.K.; McLaughlin, J.L. Inhibition of respiration at site I by asimicin, an insecticidal acetogenin of the Paw Paw *Asimina triloba* (Annonaceae). Pest. Biochem. Physiol. **1993**, *45*, 15–23.

143. Ratnayake, S.; Ruppreeht, J.K.; Potter, W.M.; McLaughlin, J.L. Evaluation of various parts of the Paw Paw tree, *Asimina triloba* (Annonaceae), as commercial sources of the pesticidal annonaceous acetogenins. J. Econ. Entomol. **1992**, *85*, 2353–2356.

144. Singh, A.; Singh, D.K. Molluscidal activity of the custard apple (*Annona squamosa* L.) alone and in combination with plant derived molluscides. J. Herbs Spices Med. Plants **2001**, *8*, 23–29.

145. Kawazu, K.; Alcantara, J.P.; Kobayashi, A. Isolation and structure of neononin, a novel insecticidal compound from the seeds of *Annona squamosa*. Agric. Biol. Chem. **1989**, *53*, 2719–2722.

146. Wang, L.Q.; Min, B.S.; Li, Y.; Nakamura, N.; Qin, G.W.; Li, C.J.; Hattori, M. Annonaceous acetogenins from the leaves of *Annona montana*. Bioorg. Med. Chem. **2002**, *10*, 561–565.

147. Wang, L.Q.; Li, Y.; Min, B.S.; Nakamura, N.; Qin, G.W.; Li, C.J.; Hattori, M. Cytotoxic mono-tetrahydrofuran ring acetogenins from leaves of *Annona montana*. Planta Med. **2001**, *67*, 847–852.

148. Li, D.Y.; Yu, J.G.; Zhu, J.X.; Yu, D.L.; Luo, X.Z.; Sun, L.; Yang, S.L. Annonaceous acetogenins of the seeds from *Annona muricata*. J. Asian Nat. Prod. Res. **2001**, *3*, 267–276.

149. Roblot, F.; Laugel, T.; Lebceuf, M.; Cave, A.; Laprevote, O. Two acetogenins from *Annona muricata* seeds. Phytochemistry **1993**, *34*, 281–285.

150. Iliopoulou, D.; Vigias, C.; Harvala, C.; Roussis, V. C (15) Acetogenins from the red alga *Laurencia obtusa*. Phytochemistry **2002**, *59*, 111–116.

151. Kim, D.H.; Ma, E.S.; Suk, K.D.; Son, J.K.; Lee, J.S.; Woo, M.H. Annomolin and annocherimolins new cytotoxic annonaceous acetogenins from *Annona cherimolia* seeds. J. Nat. Prod. **2001**, *64*, 502–506.

152. Asakawa, Y. Recent advances in phytochemistry of bryophytes–acetogenins, terpenoids, and bis(bibenzyl) from selected Japanese, Taiwanese, New Zealand, Argentinean, and European liverworths. Phytochemistry **2001**, *56*, 297–312.

153. Chang, F.R.; Chen, J.L.; Lin, C.Y.; Chiu, H.F.; Wu, M.J.; Wu, Y.C. Bioactive acetogenins from the seeds of *Annona atermoya*. Phytochemistry **1999**, *51*, 883–889.

154. Gleye, C.; Raynaud, S.; Fourneau, C.; Laurens, A.; Laprevote, O.; Serani, L.; Fournet, A.; Hocqueoniller, R. Cohibins C and D two important metabolites in the biogenesis of acetogenins from *Annona muricata* and *A. nutans*. J. Nat. Prod. **2000**, *63*, 1192–1196.

155. He, K.; Zeng, L.; Ye, Q.; Shi, G.; Oberlies, N.H.; Zhao, G.X.; Njoku, C.J.; McLaughlin, J.L. Comparative SAR evaluation of annonaceous acetogenins for pesticidal activity. Pest. Sci. **1997**, *49*, 327–378.

156. McLaughlin, J.I.; Zeng, L.; Oberlies, N.H.; Alfonso, D.; Johnson, H.A.; Cummings, B.A. Annonaceous acetogenins as new natural pesticides: recent progress. In *Phytochemical Pest Control Agents*; Hedin, P., Hollingworth, R., Mujamoto, J., Masler, E., Thompson, D., Eds.; Am. Chem. Soc: Washington, DC, 1997; 117–133 pp.

157. Guadano, A.; Gutierrez, C.; de La Pena, E.; Cortes, D.; Gonzalez-Coloma, A. Insecticidal and mutagenic evaluation of two annonaceous acetogenins. J. Nat. Prod. **2000**, *63*, 773–776.

158. Alali, F.Q.; Kaakeh, W.; Bennett, G.W.; McLaughlin, J.L. Annonaceous acetogenins as natural pesticides: potent toxicity against insecticide-susceptible and resistant German cockroaches (Dictyoptera: Blattellidae). J. Econ. Entomol. **1998**, *91*, 641–649.

159. Alali, F.Q.; Liu, X.X.; McLaughlin, J.L. Annonaceous acetogenins: recent progress. J. Nat. Prod. **1999**, *62*, 504–540.

160. Alkofahi, A.; Rupprecht, J.K.; Anderson, J.E.; McLaughlin, J.L.; Mikolajezak, K.L.; Scott, B.A. Search for new pesticides from higher plants. In Insecticides of plant origin. ACS Symp Seri. Arnason, J.T., Philogene, B.J.R., Morand, P., Eds.; 1989; Vol. 387, 24–43.

161. Su, T.H.; Beardsley, J.W.; McEwen, F.L. AC 217, 300, a promising new insecticide for use in baits for control of the bigheaded ant in pineapple. J. Econ. Entomol. **1982**, *73*, 755–756.

162. Klotz, J.H.; Reid, B.L. Oral toxicity of chlordane hydramethylnon, and imidacloprid to free-foraging workers of *Camponotus pennsylvanicus* (Hymenoptera: Formicidae). J. Econ. Entomol. **1993**, *86*, 1730–1737.

163. Kaakeh, W.; Reid, B.L.; Bennett, G.W. Horizontal transmission of the entomopathogenic fungus, *Metarhizium anisopliae* (imperfect fungi: hypho-mycetes) and hydramethylnon among German cockroaches (Dictyoptera: Blattellidae). J. Entomol. Sci. **1996**, *31*, 378–390.

164. Alali, F.Q.; Zhang, Y.; Rogers, L.; McLaughlin, J.L. (2,4-cis and trans)–gigantecinone and 4-deoxy gigantecin, bioactive nonadjacent bis- tetra-hydrofuran annonaceous acetogenins from *Goniothalamus giganteus*. J. Nat. Prod. **1997**, *60*, 929–933.

165. Londershausen, M.; Leicht, W.; Lieb, F.; Moeschler, H.; Weiss, H. Mode of action of annonins. Pest. Sci. **1991**, *33*, 427–438.
166. Tormo, J.R.; Zafra-Polo, M.C.; Serrano, A.; Estornell, E.; Cortes, D. Epoxy-acetogenins and other polyketide epoxyderivatives as inhibitors of the mitochondrial respiratory chain complex I. Planta Medica. **2000**, *66*, 318–323.
167. Towers, G.H.N. Interactions of light with phytochemicals in some natural and novel systems. Can. J. Bot. **1984**, *62*, 2900–2911.
168. Gommers, F.J.; Bakker, J. In *Chemistry and Biology of Naturally Occurring Acetylenes and Related Compounds: Bioactive Molecules*; Lam, J., Breteler, H., Arnason, T., Hansen, L., Eds.; Elsevier: New York, 1998; Vol. 7, 61 p.
169. Berenbaum, M. Toxicity of furanocoumarin to armyworms: a case of biosynthetic escape from insect herbivores. Science **1978**, *201*, 532–534.
170. Berenbaum, M.R., Charge of the light brigade. Phototoxicity as defence against insects. In Light–activated pesticides. ACS Symp. Ser. Heitz, J.R., Downum, K.R., Eds.; 1987; Vol. 339, 206–216.
171. Camm, E.L.; Towers, G.H.N.; Mitchell, J.C. UV-mediated antibiotic activity of some compositae species. Phytochemistry **1975**, *14*, 2007–2011.
172. Arnason, J.T.; Philogene, B.J.R.; Berg, C.; MacEachern, A.; Kaminski, J.; Leitch, L.C.; Morand, P.; Lam, J. Phototoxicity of naturally occurring and synthetic thiophene and acetylene analogues to mosquito larvae. Phytochemsitry **1981**, *25*, 1609–1611.
173. Ivie, G.W.; Bull, D.L.; Beier, R.C.; Waller, N. Allelochemicals: Role in Agriculture and Forestry. In ACS Symp. Ser. Waller, G.R., Ed.; 1987; Vol. 330, 455 p.
174. McLachlan, D.; Arnason, J.T.; Philogene, B.J.R.; Champagne, D. Antifeedant activity of the polyacetylene, phenylheptatriyne (PHT), from the Asteraceae to *Euxoa messoria* (Lepidoptera: Noctuidae). Experientia **1982**, *38*, 1061–1062.
175. Wat, C.K.; Biswas, R.K.; Graham, E.A.; Bohm, L.; Towers, G.H.N.; Waygood, E.R. Ultraviolet–mediated cytotoxic activity of phenylheptatriyne from *Bidens pilosa* L. J. Nat. Prod. **1979**, *42*, 103–111.
176. Nakajima, S.; Kawazu, K. Insect development inhibitors from *Coleopsis lanceolata* L. Agric. Biol. Chem. **1980**, *44*, 1529–1533.
177. Kagan, J.; Chan, G. The phototoxicidal activity of plant components towards *Drosophila melanogaster*. Experientia **1983**, *39*, 402–403.
178. Robrigues, E. Dithiopolyacetylenes as potential pesticides. In Biologically active natural products: potential use in agriculture; ACS Symp. Ser. Cutler, H.G., Ed.; 1988; Vol. 380, 432 p.
179. Arnason, J.T.; Philogene, B.J.R.; Morand, P.; Scaiano, J.C.; Werstiuk, N.; Lam, J. Thiophenes and acetylenes as phototoxic agents to herbivorous and blood feeding insects. In Light–Activated Pesticides. ACS Symp. Ser. 339, Heitz, J.R., Downum, K.R., Eds.; 1987; 255–264.
180. Arnason, J.T.; Philogene, B.J.R.; Morand, P.; Imrie, K.; Iyengar, S.; Duval, F.; Breau, C.S.; Scaiano, J.C.; Werstiuk, N.H.; Hasspieler, B.; Downe, A.E.R. Naturally occurring and synthetic thiophenes as photoactivated insecticides. In Insecticides of Plant Origin. ACS Symp. Ser. Arnason, J.A., Philogene, B.J.R., Morand, P., Eds.; 1989; 387, 166–172.

181. Guillet, G.; O'Meara, J.; Philogene, B.J.R.; Arnason, J.T. Multiple modes of insecticidal action of three classes of polyacetylene derivatives form *Rudbeckia hirta*. Phytochemistry **1997**, *46*, 495–498.

182. Degen, T.; Stadler, E. Patterns of oviposition stimulants for carrot fly in leaves of various host plants. J. Chem. Ecol. **1999**, *25*, 67–87.

183. Eckenbach, U.; Lampman, R.L.; Seigler, D.S.; Novak, R.J.; Ebinger, J. Mosquitocidal activity of acetylenic compounds from *Cryptotaenia canadensis*. J. Chem. Ecol. **1999**, *25*, 1885–1893.

184. Proksch, P.; Rodriguez, E. Chromenes and benzofurans of the Asteraceae, their chemistry and biological significance. Phytochemistry **1983**, *22*, 2335–2348.

185. Wisdom, C.S.; Smiley, J.T.; Rodriguez, E. Toxicity and deterrency of sesquiterpene lactones and chromenes to the corn earworm (Lepidoptera: Noctuidae). J. Econ. Entomol. **1983**, *76*, 993–998.

186. Isman, M.B.; Proksch, P. Deterrent and insecticidal chromenes and benzofurans from *Encelia* (Asteraceae). Phytochemistry **1985**, *24*, 1949–1951.

187. Proksch, P.; Proksch, M.; Towers, G.H.N.; Rodriguez, E. Phototoxic and insecticidal activities of chromenes and benzofurans from *Encelia*. J. Nat. Prod. **1983**, *46*, 331–334.

188. Bohlmann, F.; Jakupovic, J.; Ahmed, M.; Wallmeyer, M.; Robinson, H.; King, R.M. Labdane derivatives from *Hemizonia* species. Phytochemistry **1981**, *20*, 2383–2387.

189. Lim, Y.J.; Park, H.S.; Im, K.S.; Lee, C.O.; Hong, J.; Lee, M.Y.; Kim, D.K.; Jung, J.H. Additional cytotoxic polyacetylenes from the marine sponge *Petrosia* species. J. Nat. Prod. **2001**, *64*, 46–53.

190. Ito, A.; Cui, B.; Chavez, D.; Chai, H.B.; Shin, Y.G.; Kawanishi, K.; Kardono, L.B.S.; Riswan, S.; Farnworth, N.R.; Cordell, G.A. Cytotoxic polyacetylenes from the twigs of *Ochanostachys amentacea*. J. Nat. Prod. **2001**, *64*, 246–248.

191. Gronquist, M.; Bezzerides, A.; Attygalle, A.; Meinwald, J.; Eisner, M.; Eisner, T. Attractive and defensive functions of the ultraviolet pigments of a flower (*Hypericum calycinum*). Proc. Natl. Acad. Sci. USA **2001**, *98*, 13745–13750.

192. Schneider, C.; Bohnenstengel, F.I.; Nugroho, B.W.; Wray, V.; Witte, L.; Hung, P.D.; Kiet, L.C.; Proksch, P. Insecticidal rocaglamide derivatives from *Aglaia spectabilis* (Meliaceae). Phytochemistry **2000**, *54*, 731–736.

193. Mareggiani, G.; Picollo, M.I.; Veleiro, A.S.; Tettamanzi, M.C.; Benedetti-Doctorovich, M.O.; Burton, G.; Zerba, E. Response of *Tribolium castaneum* (Coleoptera, Tenebrionidae) to *Salipichroa origanifolia* Withanolides. J. Agric. Food Chem. **2002**, *50*, 104–107.

194. Afriye, F.A.; Chulsa, K.; Takemura, M.; Ishikawa, M.; Horike, M. Isoaltion and identification of the probing stimulants in the rice plant for the whiteback planthopper *Sogoitella furcifera* (Homoptera: Delphacidae). Biosci. Biotechnol. Biochem. **2000**, *64*, 443–446.

195. Zhou, X.; Dong, L.X.; Zchong, Y.J.; Hua, T.Z.; Wangy, L. Toxicity of cinnamomin. A new type II ribosome–inactivating protein to bollworm and mosquito. Insect Biochem. Mol. Biol. **2000**, *30*, 259–264.

196. Powell, G.; Hardie, J.; Pickett, J.A. Laboratory evaluation of antifeedant compounds for inhibiting settling by cereal aphids. Entomol. Exp. Appl. **1997**, *84*, 189–193.
197. Saadali, B.; Boriky, D.; Blaghen, M.; Vanhaelen, M.; Talbi, M. Alkamides from *Artemisia dracunculus*. Phytochemistry **2001**, *58*, 1083–1086.
198. Noordermeer, M.A.; Veldink, G.A.; Vliegenthart, J.F.G. Fatty acid hydroperoxide lyase: a plant cytochrome P450 enzyme involved in wound healing and pest resistance. Chembiochem. **2001**, *2*, 494–504.
199. Picard–Nizou, A.L.; Pham-Deleue, M.H.; Kergurlen, V.; Douault, P.; Marilleau, R.; Olsen, L.; Grison, R.; Toppan, A.; Masson, C. Foraging behavior of honey bees (*Apis mellifera* L.) on transgenic oilseed rape (*Brassica napus* L. var. *oleifera*). Transgenic Res. **1995**, *4*, 270–276.
200. Gatehouse, A.M.R.; Down, R.E.; Powell, K.S.; Sauvion, N.; Rahbe, Y.; Newell, C.A.; Merryweather, A.; Hamilton, W.D.O.; Gatehouse, J.A. Transgenic potato plants with enhanced resistance to the peach-potato aphid *Mycus persicae*. Entomol. Exp. Appl. **1996**, *79*, 295–307.
201. Down, P.F.; Lagrimini, L.M. In *Advances in Insect Control: The Role of Transgenic Plants*; Carozzi, N., Koziel, M., Eds.; Taylor and Francis: New York, 1997; 195–223 pp.
202. Thomas, J.C.; Adams, D.G.; Nessler, C.L.; Brown, J.K.; Bohnert, H.J. Tryptophan decarboxylase, tryptamine and reproduction of the whitefly. Plant Physiol. **1995**, *109*, 717–720.
203. Reina, M.; Gonzalez-Coloma, A.; Gutierrez, C.; Cabrera, R.; Rodriguez, M.L.; Fajardo, V.; Villarroel, L. Defensive chemistry of *Senecio miser*. J. Nat. Prod. **2001**, *64*, 6–11.
204. Powell, R.G.; Mikolajczak, K.L.; Zilkowski, B.W.; Mantus, E.K.; Cherry, D.; Clardy, J. Limonoid antifeedants from seed of *Sandoricum koetjape*. J. Nat. Prod. **1991**, *54*, 241–246.
205. Numata, A.; Takahashi, C.; Fujiki, R.; Kitano, E.; Kitajima, A.; Takemura, T. Plant constituents biologically active to insects. VI. Antifeedants for larvae of the yellow butterfly, *Eurema hecabe mandarina*, in *Osmunda japonica*. Chem. Pharm. Bull. (Tokyo) **1990**, *38*, 2862–2865.
206. Arimura, G.I.; Ozawa, R.; Shimoda, T.; Nishioka, T.; Boland, W.; Takabayashi, J. Herbivory–induced volatiles elicit defense genes in lima bean leaves. Nature **2000**, *406*, 512–515.
207. Kessler, A.; Baldwin, I.T. Defensive function of herbivore-induced plant volatile emissions in nature. Science **2001**, *291*, 2141–2144.
208. Daniewski, W.M.; Gumulka, M.; Anczewski, W.; Masnyk, M.; Bloszyk, E.; Gupta, K.K. Why the yew tree (*Taxus baccata*) is not attacked by insects. Phytochemistry **1998**, *49*, 1279–1282.
209. Schultz, D.J.; Medford, J.I.; Cox-Foster, D.; Grazzini, R.A.; Craig, R.; Mumma, R.O. Anacardic acids in trichomes of *Pelargonium*: biosynthesis, molecular biology and ecological effects. Adv. Bot. Res. **2000**, *31*, 175–192.

12

Signaling and Insect-Inducible Compounds in Plants

12.1 INTRODUCTION

Plants are sessile organisms, anchored to the ground through the root system for acquisition of nutrients and water, and thus are devoid of any possible escaping mechanism to prevent injuries caused by insects and other herbivores. However, plants are endowed with preexisting physical barriers that limit damage, such as the cuticle and hardened woody covers that may successfully withstand the aggression of small herbivores, or else have trichomes, thorns, and other specialized organs that may further restrict pest access to the more nutritious parts of the plant. Once an injury occurs there is no possibility of mobilizing specialized cells devoted to wound healing such as in mammals, as plant cells are encapsulated inside rigid walls. However, plants have the capacity of making each cell competent for the activation of defense responses, which largely depend on the transcriptional activation of specific genes. These wound-activated responses are directed to heal the damaged tissues and to the activation of defense mechanisms that prevent further damage. Much research has been focused in recent years on the interactions that occur among insect herbivores, plant pathogens, and their hosts. This research has revealed many interesting and valuable responses that are used by plant to limit damage that can occur from insect feeding and pathogen infection (1–6). In addition, tritropic interactions show the ability of many plants to bring on additional forces to combat herbivore attack, in the form of herbivore parasitoids (7). Most of the induced responses occur in a time window between a few minutes to several hours after wounding, and include the generation, perception, and transduction of specific signals for the subsequent activation of wound-related defense genes. Proteins encoded by those wound–inducible genes may have one of the following functions: (a) repair of damaged plant tissue; (b) production of substances that inhibit growth and development of the insect pest, i.e., those lowering the digestibility of the plant tissue or producing a toxin; (c) participation in the activation of wound defense signaling pathways; or (d) adjustment of plant metabolism to the imposed nutritional demands (4).

It has been shown that chemical pathways are induced in plants in response to attack by insect herbivores and pathogens. Attack by insect herbivores activates the octadecanoid pathway *via* lipoxygenation of linolenic acid, which in turn produces jasmonic acid (6). Jasmonic acid serves as a signal for expression of a number of compounds, such as proteinase inhibitor, polyphenol oxidase, and steroid glycoalkaloids, which appear to contribute to plant resistance against many insect pests and some pathogens. Fungal, bacterial, and viral pathogen infections induce a different biochemical pathway that produces salicylic acid, which in turn signals the formation of pathogenesis-related proteins, and phytoalexins (8). These compounds have been shown to confer resistance to many pathogens and some insects. These pathways lead to systemic responses by the plant called induced resistance (IR) and systemic acquired resistance (SAR). These induced plant defenses can cause reciprocal-induced resistance as well as signaling conflicts, which may compromise resistance to other pests (9–11).

12.2 CHEMISTRY OF WOUND-INDUCIBLE COMPOUNDS

Induced plant responses against insect attack include the release of volatile chemicals. These compounds are classified as follows:

1. *Green leaf volatiles.* Green leaf volatiles derived from octadecanoid pathway include *cis*-3-hexen-1-ol, *cis*-2-hexenylacetate, and *cis*-3-hexenylbutyrate (12).
2. *Terpenoid volatiles.* Terpenoid volatiles include *trans*-β-ocimene, *cis*-α-bergamotene, *trans*-β-farnesene, and linalool. The two acyclic homoterpenes are 4,8-dimethyl-1,3*E*,7-dimethylnonatriene

(DMNT or homoterpene I) and 4,8,12-trimethyl-1,3E,7E,11-tridecatetraene (TMNT or homoterpene II), which are of sesquiterpenoid and diterpenoid origin, respectively (13).

3. *Shikimate-derived compounds.* The shikimate-derived compounds include methylsalicylate, methylbenzoate, and indole (14).

4. *Signal compounds.* Many structurally different molecules play regulatory roles in wound signaling, including oligopeptide systemin, oligosaccharides released from the damaged cell wall, and molecules with hormonal activity such as jasmonates and ethylene (15,16).

5. *Insect-derived compounds.* A plant's response to herbivory can be influenced by factors present in insect oral secretions. Volicitin isolated from the regurgitant of *Spodoptera exigua* is N-(17-hydroxylinolenoyl)-L-glutamine (17–20). It is an elicitor of maize volatiles that attract parasitoids. Bruchins occur in pea and cowpea weevils. Bruchins are mono- and diesters of C_{22-24} α,ω-diols with 3-hydroxypropanoic acid (21). The structures of important volatile and signal molecules are shown in Fig. 12.1. The β-glucosidase of the herbivores is considered as the true elicitor and enhanced the emission of volatiles to the environment (22). These volatiles are used as host location signals by foraging parasitoids, which are natural enemies of insect herbivores.

12.3 OCCURRENCE

Induced resistance is ubiquitous in plants and reported in more than 100 plant species from approximately 30 families. Cotton plants released volatiles from undamaged leaves after 2–3 days of continuous feeding by beet armyworm (23). All systemically released compounds are induced by caterpillar damage and are not released in significant amounts by undamaged plants. Tomato plants *in vivo* treated with jasmonic acid also produced volatile compounds (24). Release of volatiles by insect-damaged plants was shown in corn (25), potato (20), lima bean (26,27), tobacco (28,29), *Arabidopsis* (16), *Bursera* which comprises of about 100 species (Burseraceae), gymnosperms such as *Abies, Larix, Picea, Pinus, Thuja, Tsuga*, and angiosperms such as *Salix, Acer, Populus, Fraxinus, Ulmus, Castanea, Prunus, Alnus, Betula*, and *Tilia* (30).

Induction of volatiles for defense has also been reported in *Ulmus minor* (31), *Melaleuca quinquenna* (32), peanuts (*Arachis hypogeae*) (33), strawberry (34), *Pinus sylvestris* (35), *Brassica oleracea* (36), *Thuja occidentalis* (37), barley (*Hordeum vulgare*) (38), and willow (*Salix*) (39).

12.4   BIOSYNTHESIS

12.4.1   Jasmonates

The biosynthesis of jasmonic acid occurs *via* the octadecanoid pathway, starting with linolenic acid. In *Arabidopsis*, there is a hexadecanoid pathway in parallel with the octadecanoid pathway. The first cyclic compounds in jasmonate biosynthesis are the cyclopentenones 12-oxophytodienoic acid (OPDA) and di-*nor*-OPDA (dnOPDA), which are subsequently reduced to cyclopentanones and β-oxidized to jasmonic acid (1,6).

i. Jasmonates

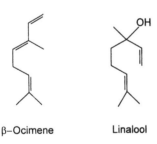

Methyl Jasmonate   Jasmone

ii. Terpenes

β–Ocimene   Linalool

iii. Benzenoids

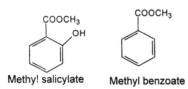

Methyl salicylate   Methyl benzoate

FIGURE 12.1   Structure of some important plant- and insect-derived compounds.

iv. Green leaf volatiles

3-Hexenal

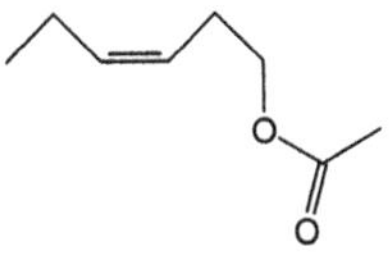

3-Hexenyl acetate

v. Insect-derived fatty acid derivatives

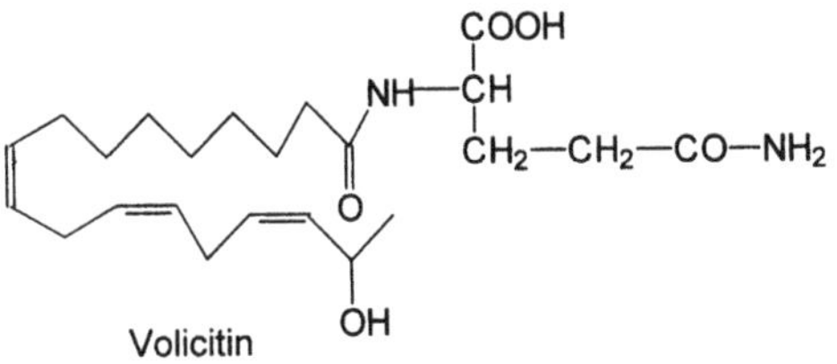

Volicitin

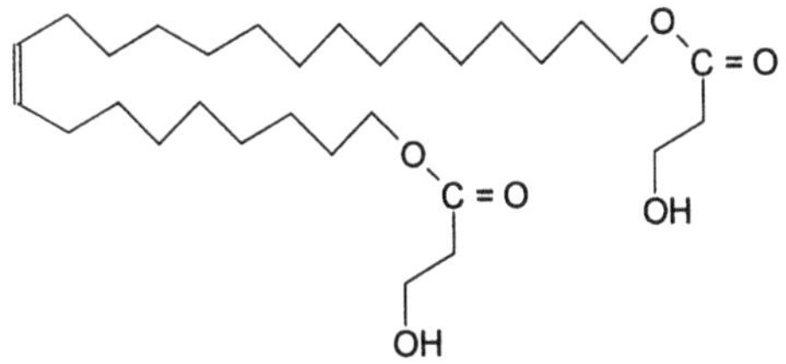

Bruchin C

FIGURE 12.1 Continued.

Biosynthesis of 12-oxophytodienoic acid from α-linolenic acid occurs in plastids, mainly in chloroplasts, and is thought to start with free linolenic acid liberated from membrane lipids by lipase action (40). In *Arabidopsis thaliana*, the glycerolipid fraction contains esterified 12-oxophytodienoic acid, which can be released enzymatically by sn1-specific, but not by sn2-specific, lipases. The 12-oxophytodienoyl glycerolipid fraction was isolated, purified, and characterized. Enzymatic, mass spectrometric, and nuclear magnetic resonance (NMR) spectroscopic data were used to establish the structure of the novel oxylipin as sn1-*O*-(12-oxophytodienoyl)-sn2-*O*-(hexadecatrienoyl)-monogalactosyldiglyceride (14). The novel class of lipids is localized in plastids. Purified monogalactosyldiglyceride was not converted to the sn1-(12-oxophytodienoyl) derivative by the combined action of (soybean) lipoxygenase and (*A. thaliana*) allene oxide synthase, an enzyme that converts free α-linolenic acid to free 12-oxophytodienoic acid. When leaves were wounded, a significant and transient increase in the level of (12-oxophytodienoyl)-monogalactosyldiglyceride was observed. There might be continuous exchange of free and esterified OPDA in the chloroplast. The cyclopentenone jasmonates have to leave the chloroplast either to act as signals or to be further metabolized in the peroxisome, where reduction of the cyclopentenone ring and β-oxidation are thought to take place.

The conversion of OPDA to the end product of the pathway, jasmonic acid, a C_{12} compound, first proceeds through reduction to 3-oxo-2-(pent-2′-enyl)-cyclopentane-1-octanoic acid, which is then converted to jasmonic acid by three cycles of β-oxidation (Scheme 12.1) (41–43). The first of these conversions is a decisive point in the biosynthetic sequence

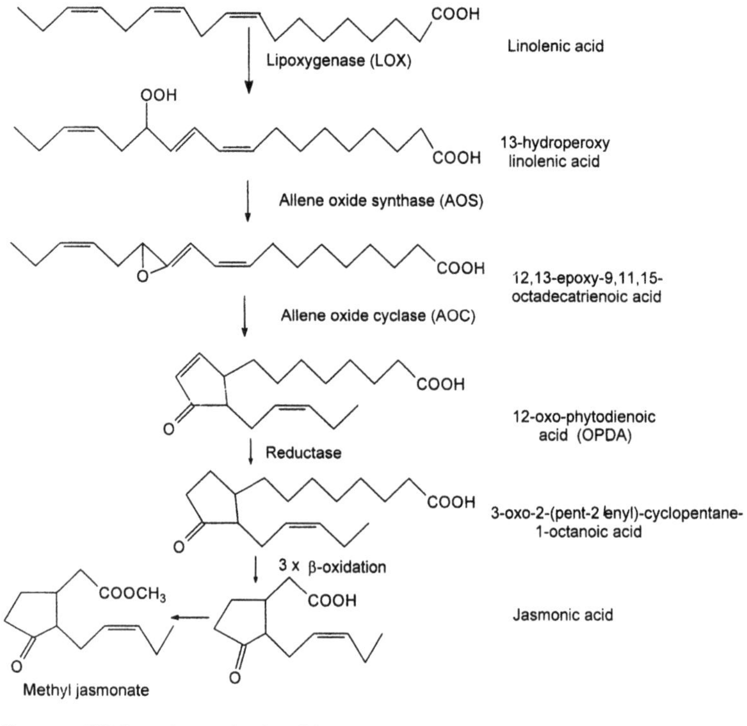

SCHEME 12.1   Biosynthesis of jasmonate and methyl jasmonate.

in that it channels the octadecanoid into the pathway of β-oxidation. 12-Oxophytodienoate reductase is soluble and a monomer of apparent molecular mass 41 kDa that prefers NADPH over NADH to reduce the 10,11 double bond of 12-Oxophytodienoic acid (41). The structure of the reaction product was proved by derivatization, gas chromatography–mass spectroscopy (GC/MS), and NMR analysis. The enzyme accepts both the cis and the trans isomers of 12-oxophytodienoic acid, with a preference for the cis isomer (6:1). 12-Oxophytodienoate reductase will also convert the synthetic substrate 2-cyclohexenone to cyclohexanone, but the enzyme did not reduce some other cyclic αβ-unsaturated ketones tested (the plant hormone abscisic acid or the steroids testosterone and progesterone). Jasmonic acid and perhaps oxopentenylcyclopentanes (OPCs) leave the peroxisome to act as signals and jasmonic acid can be methylated in the cytosol to its volatile counterpart methyl jasmonate. This model of jasmonate biosynthesis and intracellular fluxes in *Arabidopsis* is based on many experimental findings, especially enzyme location studies (1,6).

12.4.2   Leaf Volatiles

While the chemistry of plant volatiles is well understood, less is known about the biosynthesis of this diverse group of volatile compounds. This is particularly the case for nonterpenoid components such as volatile acyclic alcohols and their esters. Kandra and Wagner (44) have studied metabolic pathways leading to the formation of the anteiso-branched alcohol, 4-methyl-1-hexanol volatilized by petal tissue of *Nicotiana sylvestris*. Evidence presented supports the involvement of steps in the pathways of both biosynthesis and degradation of isoleucine to form 2-oxo-3-methyl-valeric acid and 2-methylbutyryl-CoA. Results indicate that 2-methylbu-tyryl CoA is then elongated by addition of one acetate molecule *via* fatty acid synthesis system followed by reduction to yield 4-methyl-1-hexanol. This pathway is in contrast to elongation of 2-oxo-3-methylvaleric acid *via* α-ketoacid elongation leading to the formation of 4-methylhexanoylacyl groups of tobacco leaf trichome–secreted sugar esters.

The 12-hydroperoxylinolenic acid formed from linolenic acid by the action of lipoxygenase. Fatty acid hydroperoxides are reduced to the corresponding hydroxy and epoxy alcohols by the action of peroxygenases (45). Alternatively, one group of hydroperoxide lyase–mediated cleavage of fatty acid hydroperoxides leads to the production of 12-carbon-oxoacids and 6-carbon aldehydes. Another group of hydroperoxide lyase cleaves the 9-hydroperoxyoctadecatrienoic acid (9-HPOT) to 9-oxononanoic acid and nonadienal (46) (Scheme 12.2). The 13-hydroperoxylinolenic is also converted

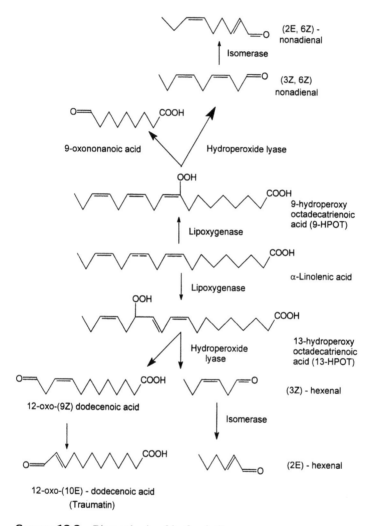

Scheme 12.2 Biosynthesis of leaf volatiles.

to 12,13-epoxyoctodecatrienoic acid by allene oxide synthase, which is then hydrolyzed to release volatile α- and γ-ketols (2,46).

Volatile terpenoids (mono- and diterpenoids) are synthesized *de novo* along the novel deoxy-D-xylulose (DOX) pathway, while the biosynthesis of sesquiterpenes may be fueled from both the DOX and the mevalonate pathway. Minor changes in the amino acid moiety may result in different volatile profiles (sesqui- and diterpenoids) (47).

12.4.3   Insect–Derived Compounds

The fatty acid portion of volicitin is plant derived, whereas the 17-hydroxylation reaction and the conjugation with glutamine are carried out by the caterpillar by using glutamine of insect origin (18).

Frey et al. (48) reported on a previously unidentified enzyme, indole-3-glycerol phosphate lyase (IGL), which catalyzes the formation of free indole and is selectively activated by volicitin.

12.4.4   Systemin

Prosystemin is the 200-amino-acid precursor of the 18-amino-acid polypeptide defense hormone systemin (15,49). Prosystemin was found to be as biologically active as systemin when assayed for proteinase inhibitor induction in young tomato plants and nearly as active in the alkalinization response in *Lycopersicon esculentum* suspension-cultured cells. Similar to many animal prohormones that harbor multiple signals, the systemin precursor contains five imperfect repetitive domains. N-terminal deletions of prosystemin had little effect on its activity in tomato plants or suspension–cultured cells. Deletion of the C-terminal region or prosystemin containing the 18-amino-acid systemin domain completely abolished its proteinase inhibitor induction and alkalinization activities. The apoplastic fluid from tomato leaves and the medium of cultured cells were analyzed for proteolytic activity that could process prosystemin to systemin. These experiments showed that proteolytic enzymes present in the apoplasm and medium could cleave prosystemin into large fragments. But these enzymes did not produce detectable levels of systemin. In addition, inhibitors of these proteolytic enzymes did not affect the biological activity of prosystemin. The cumulative data indicated that prosystemin and/or large fragments of prosystemin can be active inducers of defense responses in both tomato leaves and suspension-cultured cells.

12.5   BIOACTIVITY

Jasmonic acid and its methyl ester are known to induce various pathways associated with the biosynthesis of secondary metabolites in plants (50). Methyl jasmonate being a volatile compound can act as an airborne signal for such processes (51). Other products from the lipoxygenase pathway of plants, from which jasmonic acid is derived, e.g., (*E*)-2-hexenal and (*Z*)-jasmone, are potent semiochemicals for herbivorous insects. Release of volatile organic compounds (VOCs) after herbivory attack is known to attract parasitoids and predators to actively feeding larvae in the laboratory (52). Evidence from agricultural systems suggests a role for herbivore-induced VOCs in increasing predation pressure (53). Herbivore-induced plant VOCs may also influence herbivore host location behavior. By releasing VOCs after herbivore attack, a plant can profoundly influence both oviposition and predation rates in nature and thereby influence both "bottom-up" and "top-down" control over its herbivore population (14). The effect of herbivore-induced compounds and VOCs on host location, egg parasitism, predation, and control of mites, nematodes, and insects is presented.

Constitutively produced plant volatiles play a role in attracting pollinators and seed-dispersing animals; in addition, they repel a wide range of potential herbivores and attract a smaller number of pest species that have evolved to take advantage of these chemicals in finding food. Plant volatiles that are induced on damage to repel insect attack also may act as an indirect plant defense mechanism by attracting other insects that prey on or parasitize the herbivores. Such compounds also may act as signals between plants, whereby defense mechanisms are induced in undamaged plants in response to volatiles produced by neighboring infested plants, and specific volatiles, methyl salicylate and methyl jasmonate, have been implicated. Compounds containing six carbon atoms, such as (E)-2-hexenal, which are rapidly emitted from damaged or wounded plant tissue, also have recently been shown to induce the expression of defense-related genes in intact plants.

12.5.1 Host Location

The volatiles are used as host location signals by foraging parasitoids, which are natural enemies of insect herbivores. A plant's response to herbivory can be influenced by factors present in insect oral secretions. Volicitin (N-(17-hydroxylinolenoyl)-L-glutamine), identified in beet armyworm (*Spodoptera exigua*) oral secretions, stimulates volatile release in corn (*Zea mays* L.) seedlings in a manner similar to that of beet armyworm herbivory. Volicitin is hypothesized to trigger release of induced volatiles, at least in part, by modulating levels of the wound hormone, jasmonic acid (JA). Schmelz et al. (54) compared the sesquiterpene volatile release of damaged leaves treated with aqueous buffer alone, or with the same buffer containing volicitin or JA. Leaves were damaged by scratching with a razor and test solutions were applied to the scratched area. The leaves were either excised from the plant or left intact shortly after this treatment. Plants were treated at three different times (designated as evening, midnight, and morning) and volatiles were collected in the subsequent photoperiod. Jasmonic acid and volicitin treatments stimulated the release of volatile sesquiterpenes, namely, β-caryophyllene, (*E*)-α-bergamotene, and (*E*)-β-farnesene. In all cases, JA stimulated significant sesquiterpene release on mechanical damage alone. Volicitin induced an increase in sesquiterpene volatiles in all excised-leaf bioassays and the midnight intact plants. Volicitin treatments in the evening and morning intact plants produced more sesquiterpenes than the untreated controls, whereas mechanical damage alone produced an intermediate response that did not differ from either treatment group. Excised leaves produced a 2.5- to 8.0-fold greater volatile response than similarly treated intact plants. Excision also altered the ratio of JA- and volicitin-induced sesquiterpene release by preferentially increasing (*E*)-β-farnesene levels relative to β-caryophyllene. The inducibility of volatile release varied with time of treatment. Sesquiterpene release was highest in the midnight excised leaves and lowest in the morning intact plants. The duration of induced volatile release also differed between treatments. JA produced a sustained release of sesquiterpenes over time, with more than 20% of the combined sesquiterpenes released in the third and final volatile collection period. In contrast, less than 8% of the combined sesquiterpenes induced by volicitin were emitted during this period. The large quantitative differences between intact plants and detached leaves suggest that the results of assays using excised tissues should be cautiously interpreted when considering intact plant models.

Learning of host-induced plant volatiles by *Cotesia kariyai* females was examined with synthetic chemicals in a wind tunnel. Wasps were preconditioned by exposure to volatiles and feces simultaneously (55). A blend of four chemicals—geranyl acetate, β-caryophyllene, (*E*)-β-farnesene, and indole—were released from plants infested by host larvae *Mythimna separata*. These host-induced blends elicited a response in naive *C. kariyai* but did not enhance the response after conditioning. A blend of five chemicals—(*E*)-2-hexenal, (*Z*)-3-hexen-1-ol, (*Z*)-3-hexen-1-yl acetate, β-myrcene, and linalool—which are known to be released not only from plants infested by the host larvae but also from artificially damaged plants or undamaged ones (unspecific blend), elicited little response in naive wasps, but significantly enhanced the wasps' response after conditioning. With a blend of the above nine chemicals, wasps could learn the blend at lower concentrations than they did in the nonspecific blend. Hence, both the host-induced and nonspecific volatile compounds appear to be important for *C. kariyai* females to learn the chemical cues in host location.

If predators lack information on the prey's position, prey have more chance to escape predation and will therefore reach higher population densities. One of the many possible cues that predators may use to find their prey is herbivore-induced plant volatiles. Although their effects on the behavior of foraging predators have been well studied, little is known about how these prey-related odors affect predator–prey dynamics on a plant. Pel and Sabelis (56) hypothesize that herbivore-induced plant volatiles provide the major cue eliciting predator arrestment on prey-infested leaves and that the response to these volatiles ultimately leads to lower prey densities. To test this hypothesis experimentally, Pel and Sabelis created two types of odor-saturated environments: one with herbivore-induced plant volatiles (treatment), and one with green leaf volatiles (control). An odor-free environment could not be tested because herbivores require plants for population growth. They measured the rate at which predatory mites (*Phytoseiulus persimilis*) immigrate, emigrate, and exploit a single leaf infested by two-spotted spider mites (*Tetranychus urticae*). The experiments did not show a significant difference between treatment and control. At best, there was a somewhat higher rate of predator (and possibly also prey) emigration in the treatment. The lack of a pronounced difference between treatment and control indicates that at the spatial scale of the experiments random searching for prey was as effective as directional searching. Alternatively, predators were arrested in the prey patch by responding not merely to herbivore-induced plant volatiles, but also to other prey-related cues, such as web and feces.

     The role of a general green leaf volatile (glv) in host finding by larvae of the oligophagous chrysomelid *Cassida denticollis* was investigated using a new bioassay that takes into account the need for neonate larvae of this species to climb fresh host plants from the ground (57). A "stem arena" was designed in which plant stems of the host, tansy (*Tanacetum vulgare*), and stem dummies (tooth picks), both wrapped in perforated filter paper, were offered to neonate larvae. The wrapping allowed olfactory responses to be tested by preventing access to contact stimuli of stems and dummies. Larvae significantly preferred to climb the wrapped tansy stems over dummies after a period of 15 min. The test glv, (Z)-3-hexen-1-ol, was not attractive when applied to dummies. However, when the glv was applied to the bottom of the arena, the ability of larvae to discriminate between host stems and untreated dummies was significantly enhanced. More larvae climbed wrapped host stems than dummies even within 5 min. While numerous other herbivorous insects are known to be directly attracted by glv, this study shows that a singly offered glv on its own is unattractive to an herbivore but enhances the herbivore's ability to differentiate between host and nonhost plants.

12.5.2   Egg Parasitization and Predation

Plant resistance to herbivores mediated by the octadecanoid pathway has both positive and negative influence on parasitoids. The net effect of this interaction on herbivore mortality was a doubling of the number of naturally occurring parasitized caterpillars on induced plants compared with controls. Production of volatiles that influence attraction of natural enemies may be vital to the plant's defense system, overcoming the negative influence of induction on wasp performance (58). There are specific interactions between herbivores and plants that can result in a varied blend of volatiles being released from plants, which selectively attract parasitoids. The parasitoids could have utilized either herbivore-specific or nonspecific plant cues. Because JA was used to induce plants and attract parasitoids of caterpillars, activation of the octadecanoid pathway in tomato plants with either specifically stimulates the same plant responses as *S. exigua* feeding or stimulates nonherbivore-specific cues.

The octadecanoid pathway contributes to plant resistance both by directly killing herbivores and by enhancing the action of natural enemies of herbivores. In tomato plants, jasmonate-induced resistance resulted in reduced abundance of herbivores in three feeding guilds, including leaf chewers, phloem feeders, and cell content feeders. The finding that jasmonate-induced resistance results in increased mortality of herbivores due to parasitism is potentially of great importance in pest control. These data point to conditions where induced plant resistance and biological control may provide compatible and potentially synergistic control of herbivores (59).

Plants synthesize and emit a large variety of VOCs with terpenoids and fatty acid derivatives as the dominant classes. Whereas some volatiles are probably common to most plants, others are specific to only one or a few related taxa. Floral volatiles serve as attractants for species-specific pollinators whereas the volatiles emitted from vegetative parts, especially those released after herbivory, appear to protect plants by deterring herbivores and by attracting the enemies of herbivores (12).

Oviposition of the elm leaf beetle *Xanthogaleruca luteola* causes the emission of volatiles from its food plant, *Ulmus minor*. These volatiles are exploited by the egg parasitoid, *Oosmyzus gallerucae*, to locate its host (31). In contrast to other tritrophic systems, the release of volatiles is not induced by feeding but by oviposition. Previous investigations showed that the release is systemic and can be triggered by JA. Comparison of headspace analysis revealed similarities in the blend of volatiles emitted following oviposition and feeding. The mixture consists of more than 40 compounds; most of the substances are terpenoids. Leaves next to those carrying eggs emit fewer compounds. When treated with JA, leaves emit a blend that consists almost exclusively of terpenoids. Dichloromethane extracts of leaves treated with JA were also investigated. After separation of extracts of jasmonate induced elm leaves on silica, Wegener et al. (31) obtained a fraction of terpenoid hydrocarbons that was attractive to the parasitoids. This indicates that JA stimulates the production of terpenoid hydrocarbons that convey information of oviposition to the parasitoid.

Plants under attack by a herbivore emit characteristic volatiles that are implicated in the attraction of the natural enemies of the herbivore. The signal cascade between leaf damage and the volatile production is stimulated by high or low molecular weight elicitors from the secretions of the herbivore (60). Besides compounds from the octadecanoid signaling pathway, several structurally nonrelated amino acid conjugates, such as the bacterial phytotoxin coronatine, the synthetic indanoyl isoleucine, or amino acid conjugates of linolenic acid, likewise induce volatile biosynthesis. Minor changes in the amino acid moiety may result in different volatile profiles (sesqui- and diterpenoids), attributed to the amino acid substructure having a specific role for recognition and selective induction.

Plants can respond to arthropod herbivory with the induction of a blend of volatiles that attracts predators and/or parasitoids of herbivores. Carnivorous arthropods can discriminate between infested plants and mechanically wounded plants, and between plants infested by different herbivore species (61). The volatile blends emitted by different plant species infested by the same herbivore species show large qualitative differences, whereas blends emitted by plants of the same species but infested by different herbivore species are mostly qualitatively similar with quantitative variation. Carnivores can discriminate between blends that differ qualitatively and/or quantitatively. However, it remains unknown what differences in blends are used by carnivorous arthropods in this discrimination. Signal transduction pathways involved in the induction of direct and indirect defense seem to overlap. Direct and indirect defense may interfere with each other's effectiveness. For application of direct and indirect defense in agriculture, it is important to compare the relative importance of these two defense types in the same plant species.

Oviposition by an herbivorous insect was shown to induce a gymnosperm plant to emit volatiles that attract egg parasitoids (35). Odor from twigs of *Pinus sylvestris* laden with egg masses of the pine sawfly *Diprion pini* attracts the eulophid egg parasitoid *Chrysonotomyia ruforum*. Volatiles released from pine twigs without *Diprion* eggs are not attractive. Oviposition by the sawfly onto pine needles induces not only a local response in pine needles laden with eggs but also a systemic reaction. Needles without eggs but adjacent to those bearing *Diprion* eggs also release the volatiles that attract the egg parasitoid. The elicitor of the attractive volatiles was shown to be present in the oviduct secretion coating the eggs of *D. pini*. When pine twigs are treated with jasmonic acid, a well-known plant wound signal, they emit volatiles that attract the egg parasitoid. These results show for the first time that a gymnosperm plant is able to attract parasitoids as soon as a herbivore has deposited eggs on it. Thus, the plant appears to defend itself against herbivores prior to being damaged by feeding larvae.

Corn seedlings respond to insect herbivore-inflicted injury by releasing relatively large amounts of several characteristic terpenoids and as a result become highly attractive to parasitic wasps that attack the herbivores (25). Chemical evidence showed that the induced emission of volatiles is not limited to the sites of damage but occurs throughout the plant. This evidence was obtained by comparing the release of volatiles from undamaged leaves of seedlings with two injured leaves treated with caterpillar regurgitant. Immediately after injury no differences were measured in the released volatiles, but several hours later the undamaged leaves of injured plants released the terpenoids linalool, (3E)-4,8-dimethyl-1,3,7-nonatriene, and (3E,7E)-4,8,12-trimethyl-1,3,7,11-tridecatetraene in significantly larger amounts than leaves of unharmed plants. Other volatiles that are released by herbivore-injured leaves were detected occasionally only in trace amounts from the undamaged leaves of a damaged seedling. The systemic release of volatiles by injured corn coincided with attractiveness to the parasitoid *Cotesia marginiventris*; undamaged leaves of injured plants became significantly more attractive than leaves from control seedlings. These findings show conclusively that when a plant is injured by an insect herbivore the whole plant emits chemical signals.

Herbivore attack is known to increase the emission of volatiles, which attract predators to herbivore-damaged plants in the laboratory and agricultural systems. Kessler and Baldwin (14) quantified volatile emissions from *Nicotiana attenuata* plants growing in natural populations during attack by three species of leaf-feeding herbivores and mimicked the release of five commonly emitted volatiles individually. Three compounds (*cis*-3-hexen-1-ol, linalool, and *cis*-α-bergamotene) increased egg predation rates by a generalist predator; linalool and the complete blend decreased lepidopteran oviposition rates. As a consequence, a plant could reduce the number of herbivores by more than 90% by releasing volatiles. These results confirm that indirect defenses can operate in nature.

12.5.3   Control of Mites, Nematodes, and Insects

Jasmonic acid applications to gerbera and lima bean plants also resulted in induced volatile production and the attraction of natural enemies of herbivorous mites (62). The profiles of the volatiles emitted after application of JA were similar to those induced by natural herbivory, although not identical. In contrast to mite-induced plants, methyl salicylate, a well-known attractant of predatory mites, was conspicuously missing from JA-induced plants, which were consequently less preferred by predators than were mite-induced plants. The dose of JA and method of its delivery is likely to be an important determinant of some of the differences between the attraction of predators to JA-induced and mite-induced plants. A novel class of elicitors of volatiles, fungal peptaibols, induces both jasmonate products and methyl salicylate. Fungal peptaibols were previously shown to induce phytoalexin and are particularly intriguing because they appear to circumvent the typical antagonistic interaction between jasmonate and salicylate signaling in plants (63).

Methyl salicylate, which was found to repel aphids such as the black bean aphid (*Aphis fabae*) and cereal aphids including the grain aphid (*Sitobion avenae*), and also to inhibit attraction to their host plants (64), originally was discovered to be an aphid signal by single-cell recording (SCR) on the antenna of the bird–cherry–oat aphid, *Rhopalosiphum padi*. More than 30 species of insects, both plant feeders and their natural enemies, from five orders subsequently have been found, by SCR and by recording from the whole antenna (electroantennography, or EAG), to possess olfactory receptors for this compound. Subsequently, Raskin's group (65) showed that methyl salicylate might act as an airborne plant signal mediating plant pathogen resistance.

Plant roots in the soil are under attack from many soil organisms. Although many ecologists are aware of the presence and importance of natural enemies in the soil that protect the plants from herbivores, the existence and nature of tritrophic interactions are poorly understood. So far, attention has focused on how plants protect their above-ground parts against herbivorous arthropods, either directly or indirectly (i.e., by getting help from the herbivore's enemies). Van Tol et al. (37) showed that indirect plant defenses also operate underground. The roots of a coniferous plant (*Thuja occidentalis*) release chemicals upon attack by weevil grub (*Otiorhynchus sulcatus*) and these chemicals in turn attract parasitic nematodes (*Heterorhabditis megidis*).

More candidate airborne plant-to-plant signals have been reported for bean plants (*Phaseolus lunatus*) infested with herbivorous spider mites (*Tetranychus urticae*). In response to attack, infested leaves release volatiles that can increase the resistance of uninfested leaves to attack by spider mites, as well as inducing the expression of several defense-related genes in neighboring uninfected lima bean leaves (66). β-Ocimene and two other related terpenoids are thought to be responsible for this effect. These compounds were shown recently to activate the expression of a number of defense genes in detached bean leaves, but it will be important to test whether the compounds up-regulate gene expression in intact bean plants.

Potentially related to this work is the interesting biology of (Z)-jasmone, a product of JA metabolism. (Z)-Jasmone released by plants was found to be electrophysiologically active in insects, both herbivores and their predators. Treatment of healthy bean (*Vicia faba*) plants with (Z)-jasmone induced both (E)-β-ocimene release and also α-tubulin gene expression (67). One could speculate that, in bean, (Z)-jasmone treatment causes release of β-ocimene, which itself activates gene expression, but at present the evidence is lacking.

Using spider mites (*Tetranychus urticae*) and predatory mites (*Phytoseiulus persimilis*), it has been shown that not only the attacked plant but also neighboring plants are affected, becoming more attractive to predatory mites and less susceptible to spider mites. However, the mechanism involved in such interactions, remains elusive. Arimura et al. (13) show that uninfested lima bean leaves activate five separate defense genes when exposed to volatiles from conspecific leaves infested with *T. urticae*, but not when exposed to volatiles from artificially wounded leaves. The expression pattern of these genes is similar to that produced by exposure to JA. At least three terpenoids in the volatiles are responsible for this gene activation; they are released in response to herbivory but not artificial wounding. Expression of these genes requires calcium influx and protein phosphorylation/dephosphorylation.

Peanut plants, *Arachis hypogaea*, infected with white mold, *Sclerotium rolfsii*, emit a blend of organic compounds that differs both quantitatively and qualitatively from the blend emitted from plants damaged by beet armyworm (BAW; *Spodoptera exigua*) larvae or from uninfected, undamaged plants. Attack by BAW induced release of lipoxygenase products (hexenols, hexenals, and hexenyl esters), terpenoids, and indole. The plant-derived compound methyl salicylate and the fungal-derived compound 3-octanone were found only in headspace samples from white mold–infected plants (33). White mold–infected plants exposed to BAW damage released all the volatiles emitted by healthy plants fed on BAW in addition to those emitted in response to white mold–infection alone. When BAW larvae were given a choice of feeding on leaves from healthy or white mold–infected plants, they consumed larger quantities of the leaves from infected plants. Exposure to commercially available (Z)-3-hexenyl acetate, linalool, and methyl salicylate, compounds emitted by white mold–infected plants, significantly reduced the growth of the white mold in solid media cultures. Thus, emission of these compounds by infected plants may constitute a direct defense against this pathogen.

Plant responses to herbivores are complex. Genes activated on herbivore attack are strongly correlated with the mode of herbivore feeding and the degree of tissue damage at the feeding site. Phloem-feeding whiteflies and aphids that produce little injury to plant foliage are perceived as pathogens and activate the salicylic acid–dependent and jasmonic acid/ ethylene–dependent signaling pathways (68). Differential expression of plant genes in response to closely related insect species suggests that some elicitors generated by phloem-feeding insects are species specific and are dependent on the herbivore's developmental stage. Other elicitors for defense gene activation are likely to be more ubiquitous. Analogies to the pathogen-incompatible reactions are found. Chewing insects such as caterpillars and beetles and cell-content feeders such as mites and thrips cause more extensive tissue damage and activate wound signaling pathways. Herbivore feeding is not equivalent to mechanical wounding. Wound responses are a part of the induced responses that accompany herbivore feeding. Herbivores induce direct defenses that interfere with herbivore feeding, growth and development, fecundity, and fertility. In addition, herbivores induce an array of volatiles that creates an indirect mechanism of defense. Volatile blends provide specific cues to attract herbivore parasitoids and predators to infested plants.

The volatile compounds emitted by leaves of 10 willow varieties that differ in their susceptibility to damage by blue (*Phratora vulgatissima*), brassy (*P. vitellinae*), and brown (*Galerucella lineola*) willow beetles were examined both before and after mechanical damage and correlated with feeding preferences of these beetles determined under laboratory conditions (39). Three compounds were identified from intact undamaged leaves of six willow varieties: *cis*-3-hexenyl acetate, *cis*-3-hexenol, and benzaldehyde. After mechanical damage, the yield and number of volatile compounds increased for all varieties. There were significant differences among willow varieties for both the concentration of *cis*-3-hexenyl acetate and the relative proportion of this compound to *cis*-3-hexenol (green leaf volatile ratio). The 10 varieties collectively showed a significant negative correlation between the relative resistance of each variety to blue and brown willow beetles and the yield of *cis*-3-hexenyl acetate from damaged plants. The green leaf volatile ratio of damaged plants was also negatively correlated with the relative resistance of willow variety to these two beetle species.

Feeding by *Pieris brassicae* caterpillars on the lower leaves of Brussels sprouts (*Brassica oleracea* var. *gemmifera*) plants triggers the release of volatiles from upper leaves (36). The volatiles are attractive to its parasitoid *Cotesia glomerata*. Parasitoids are attracted only if additional damage is inflicted on the systemically induced upper leaves and only after at least 3 days of herbivore feeding on the lower leaves. Upon termination of caterpillar feeding, the systemic signal is emitted for a maximum of one more day. Systemic induction did not occur at low levels of herbivore infestation. Systemically induced leaves emitted green leaf volatiles, cyclic monoterpenoids, and sesquiterpenes. GC-MS profiles of systemically induced and herbivore-infested leaves did not differ for most compounds, although herbivore infested plants emitted higher amounts of green leaf volatiles. Emission of systemically induced volatiles in Brussels sprouts might function as an induced defense that is activated only when needed, i.e., at the time of caterpillar attack. In this way plants may adopt a flexible management of inducible defensive resources to minimize costs of defense and to maximize fitness in response to unpredictable herbivore attack.

Induction of octadecanoid pathway increases resistance against herbivorous insect in tomato plants, in part by causing production of toxic and antinutritive proteinase inhibitors and oxidative enzymes. Herbivore-infested tomato plants release increased amounts of volatiles and attract natural enemies of the herbivores, as do other plants. The octadecanoid pathway may regulate production of these volatiles, which attract host-seeking parasitic wasps. However, plant resistance compounds can adversely affect parasitoids as well as herbivores. It is unclear whether the combination of increased retention and/or attractiveness of parasitic wasps to induced plants and the adverse effects of plant defense compounds on both caterpillars and parasitoids results in a net increase in parasitization of herbivores feeding on induced plants (69).

Until recently, little was known about how phloem-feeding aphids and their natural enemies respond to induced plant responses. Although induced direct defense to aphids is still poorly documented, a body of literature on the indirect attraction of natural enemies of aphids has emerged (70). As demonstrated in studies of plant responses to lepidopteran and mite herbivores, volatiles induced in response to aphid attack are systemically released by the plant and appear to be positively dependent on the level of infestation. It was found that volatile extracts from host-infested plants are more attractive to parasitoids than extracts from plants infested with nonhost aphids. This result exemplifies specificity in the response of plants to different aphid herbivores. Volatile isothiocyanates (produced from preformed glucosinolates) in the Brassicaceae family attract aphid parasitoids. Bradburne and Mithen (71) have shown that breeding for the release of specific volatiles from *Brassica napus* and *B. oleracea* results in the attraction of aphid parasitoids in the laboratory and field.

Research on the model plant *Arabidopsis thaliana* (L.) Heynh. has contributed considerably to the unraveling of signal transduction pathways involved in direct plant defense mechanisms against pathogens. Van Poccke et al. (72) demonstrated that *Arabidopsis* is also a good candidate for studying signal transduction pathways involved in indirect defense mechanisms. They showed that the adult females of *Cotesia rubecula*, a specialist parasitic wasp of *Pieris rapae* caterpillars, were attracted to *P. rapae*-infested *Arabidopsis* plants and *Arabidopsis* infested by *P. rapae* emited volatiles from several major biosynthetic pathways, including terpenoids and green leaf volatiles. The blends from herbivore-infested and artificially damaged plants are similar. However, differences can be found with respect to a few components of the blend, such as two nitriles and the monoterpene myrcene, which were produced exclusively by caterpillar-infested plants, and methyl salicylate, which was produced in larger amounts by caterpillar-infested plants.

cis-Jasmone, (Z)-jasmone, is well known as a component of plant volatiles, and its release can be induced by damage, e.g., during insect herbivory. (Z)-Jasmone was found to be electrophysiologically active and also to be repellent in laboratory choice tests. In field studies, repellency was demonstrated for the damson-hop aphid, and with cereal aphids numbers were reduced in plots of winter wheat treated with (Z)-jasmone (73). In contrast, attractant activity was found in laboratory and wind tunnel tests for insects act antagonistically to aphids, namely, the seven-spotted ladybird and an aphid parasitoid. When applied in the vapor phase to intact bean plants, (Z)-jasmone induced the production of volatile compounds, including the monoterpene (E)-β-ocimene, which increases plant defense, such as by stimulating the activity of parasitic insects. These plants were more attractive to the aphid parasitoid in the wind tunnel when tested 48 h after exposure to (Z)-jasmone had ceased. This possible signaling role of (Z)-jasmone is qualitatively different from that of the biosynthetically related methyl jasmonate and produces a long-lasting effect after removal of the stimulus.

12.5.4 Exogenous Application

Exogenous applications of jasmonates increase the production of volatile compounds. In tomato plants, spider mite damage and jasmonate spray result in similar induction of volatile compounds. Both results in a large increase in the proportion of methyl salicylate and terpenoids found in the volatile blend emitted; minor inducible constituents include (Z)-3-hexen-1-ol, a C6 leafy green volatile (74). The localized production of C6 leafy green volatiles may be due to activation of the octadecanoid cascade through the action of lipoxygenase, an enzyme involved in the induction of both foliar and volatile phytochemicals. In addition, it has been proposed that volicitin, a compound found in insect oral secretions, might interact with the octadecanoid pathway and result in increased production of volatile compounds in corn plants that are attractive to natural enemies. The multiplicity of effects provided by the octadecanoid pathway indicates that plants coordinate an interrelated, complex system of defense.

Herbivore attack is widely known to reduce food quality and to increase chemical defenses and other traits responsible for herbivore resistance. Inducible defenses are commonly thought to allow plants to forego the costs of defense when not needed; however, neither their defensive function (increasing a plant's fitness) nor their cost-saving functions have been demonstrated in nature. Thaler (75) demonstrated that exogenous applications of JA to field-grown tomato plants cause an increase in the parasitism of naturally occurring beet armyworm larvae by a solitary ichneumonid parasitoid. This result was a direct effect of changes in plant quality, not an effect of changes in herbivore quantity or quality. The JA-mediated attraction of parasitoids was also shown to be associated with foliar defenses that negatively affected several herbivores by increasing their development time. However, the pupal weights of the parasitoids of these herbivores were also decreased. Thus, there were positive and negative effects associated with the JA induction, with the net benefits apparently being more important than the costs. Similar positive (i.e., parasitoid attraction) and negative effects (i.e., 40% fewer emerging parasitoid progeny) have been reported for a gregarious braconid parasitoid attacking gypsy moth larvae on induced poplar trees (76).

Rodriguez-Saona et al. (77) investigated the effect of exogenous methyl jasmonate (MeJA) on the emission of herbivore-induced volatiles; these volatile chemicals can signal natural enemies of the herbivore to the damaged plant. Exogenous treatment of cotton cv. Deltapine 5415 plants with MeJA induced the emission of the same volatile compounds as observed for herbivore-damaged plants. Cotton plants treated with MeJA emitted elevated levels of the terpenes (E)-β-ocimene, linalool, $(3E)$-4,8-dimethyl-1,3,7-nonatriene, (E,E)-α-farnesene, (E)-β-farnesene, and (E,E)-4,8,12-trimethyl-1,3,7,11-tridecatetraene compared with untreated controls. Other induced components included (Z)-3-hexenyl acetate, methyl salicylate, and indole. Methyl jasmonate treatment did not cause the release of any of the stored terpenes such as α-pinene, β-pinene, α-humulene, and (E)-β-caryophyllene. In contrast, these compounds were emitted in relatively large amounts from cotton due to physical disruption of glands by the herbivores. The timing of volatile release from plants treated with MeJA or herbivores followed a diurnal pattern, with maximal volatile release during the middle of the photoperiod. Similar to herbivore-treated plants, MeJA treatment led to the systemic induction of (Z)-3-hexenyl acetate, (E)-β-ocimene, linalool, $(3E)$-4,8-dimethyl-1,3,7-nonatriene, (E,E)-α-farnesene, (E)-β-farnesene, and (E,E)-4,8,12-trimethyl-1,3,7,11-tridecatetraene. The results indicated that treatment of cotton with MeJA could directly and systemically induce the emission of volatiles that may serve as odor cues in the host-search behavior of natural enemies.

The weevil *Oxyops vitiosa* is an Australian species imported to Florida, USA, for the biological control of the invasive weed species *Melaleuca quinquenervia*. Grubs of this species feed on leaves of their host and produce a shiny orange secretion that covers the integument. When this secretion was applied at physiological concentrations to dog food bait, fire ant consumption and visitation were significantly reduced. Gas chromatographic analysis indicated that the secretion qualitatively and quantitatively resembled the terpenoid composition of the host foliage. When the combination of 10 major terpenoids from the *O. vitiosa* secretion was applied to dog food bait, fire ant consumption and visitation were reduced. When these 10 terpenoids were tested individually, the sesquiterpene viridiflorol was the most active component in decreasing fire ant consumption (32). Fire ant visitation was initially (15 min after initiation of the study) decreased for dog food bait treated with viridiflorol and the monoterpenes 1,8-cineole and α-terpineol. Fire ants continued to avoid the bait treated with viridiflorol at 18 µg/mg dog food for up to 6 h after the initiation of the experiment. Moreover, ants avoided bait treated with 1.8 µg/mg for up to 3 h. The concentrations of viridiflorol, 1,8-cineole, and α-terpineol in larval washes were about twice that of the host foliage, suggesting that the larvae sequester these plant-derived compounds for defense against generalist predators.

Herbivory induces both direct and indirect defenses in plants; however, some combinations of these defenses may not be compatible. The jasmonate signal cascade activated both direct (nicotine accumulations) and indirect (mono- and sesquiterpene emissions) whole–plant defense responses in the native tobacco *Nicotiana attenuata* Torr. ex Wats. Nicotine accumulations were proportional to the amount of leaf wounding and the resulting increases in JA concentrations (78). However, when larvae of the nicotine-tolerant herbivore, *Manduca sexta*, fed on plants or their oral secretions were applied on leaf punctures, the normal wound response was dramatically altered, as evidenced by large (4- to 10-fold) increases in the release of (a) volatile terpenoids and (b) ethylene, (c) increased (4- to 30-fold) accumulations of endogenous JA pools, but (d) decreased or unchanged nicotine accumulations. The ethylene release, which was insensitive to inhibitors of induced JA accumulation, was sufficient to account for the attenuated nicotine response. Applications of ethylene and ethephon suppressed the induced nicotine response, and pretreatment of plants with a competitive inhibitor of ethylene receptors, 1-Methylcyclopropene, restored the full nicotine response. However, the ethylene burst did not inhibit the release of volatile terpenoids. Because parasitoids of *Manduca* larvae are sensitive to the dietary intake of nicotine by their hosts, this ethylene-mediated switching from direct to a putative indirect defense may represent an adaptive tailoring of a plant's defense response.

12.5.5   Diurnal Variation

Cotton plants attacked by herbivorous insect pests emit relatively large amounts of characteristic volatile terpenoids that have been implicated in the attraction of natural enemies of the herbivores (23). However, the composition of the blend of volatile terpenes released by the plants varies remarkably throughout the photoperiod. Some components are emitted in at least 10-fold greater quantities during the photophase than during the scotophase, whereas others are released continuously, without conforming to a pattern, during the entire time that the plants are under herbivore attack. The diurnal pattern of emission of volatile terpenoids was determined by collecting and analyzing the volatile compounds emitted by cotton plants subjected to feeding damage by beet armyworm larvae *in situ*. The damage was allowed to proceed for 3 days, and volatile emission was monitored continuously. During early stages of damage high levels of lipoxygenase-derived volatile compounds [e.g., (Z)-3-hexenal, (Z)-3-hexenyl acetate] and several terpene hydrocarbons [e.g., α-pinene, caryophyllene] were emitted. As damage progressed, high levels of other terpenes, all acyclic [e.g., (E)-β-ocimene, (E)-α-farnesene], were emitted in a pronounced diurnal fashion; maximal emissions occurred in the afternoon. These acyclic terpenes followed this diurnal pattern of emission, even after removal of the caterpillars, although emission was in somewhat smaller amounts. In contrast, the emission of cyclic terpenes almost ceased after the caterpillars were removed.

The effects of volatile blends on herbivore behavior have been investigated to only a limited extent, in part because of the assumption that herbivore-induced volatile emissions occur mainly during the light phase of the photoperiod. Because many moths whose larvae are some of the most important insect herbivores are nocturnal, herbivore-induced plant volatiles have not hitherto been considered to be temporally available as host location cues for ovipositing females. De Moraes et al. (79) present chemical and behavioral assays showing that tobacco plants (*Nicotiana tabacum*) release herbivore-induced volatiles during both night and day. Moreover, several volatile compounds are released exclusively at night and are highly repellent to female moths (*Heliothis virescens*). The demonstration that tobacco plants release temporally different volatile blends and that lepidopteran herbivores use induced plant signals released during the dark phase to choose sites for oviposition adds a new dimension to our understanding of the role of chemical cues in mediating tritrophic interactions.

12.6   MECHANISM OF ACTION

Jasmonic acid signaling is now known to be crucial for plant stress responses following wounding, interaction with various pathogens, UV light, and insect attack. Jasmonic acid is also indispensable for anther dehiscence and pollen development. Both the biosynthesis and the action of JA seem to be regulated by the following factors: (a) substrate generation as a result of external stimuli; (b) intracellular sequestration; (c) tissue-specific generation and accumulation of JA, e.g., in vascular bundles; (d) temporally different patterns of JA accumulation; and, finally, (e) metabolic transformation of JA into its methyl ester or into the volatile degradation product *cis*-jasmone, leading to spatially and temporally variable removal of the JA signal.

12.6.1   Signal Transduction

Plant defense responses to wounding and herbivore attack are regulated by signal transduction pathways that operate both at the site of wounding and in undamaged distal leaves. Genetic analysis in tomato indicates that systemin and its precursor protein, prosystemin, are upstream components of a wound-induced, intercellular signaling pathway that involves both the biosynthesis and action of JA. To examine the role of JA in systemic signaling, reciprocal grafting experiments were used to analyze wound-induced expression of the proteinase inhibitor II gene in a JA biosynthetic mutant (spr-2) and a JA response mutant (jai-1) (24). The results showed that spr-2 plants are defective in the production, but not recognition, of a graft-transmissible wound signal. Conversely, jai-1 plants are compromised in the recognition of this signal but not its production. It was also determined that a graft-transmissible signal produced in response to ectopic expression of prosystemin in root stocks was recognized by spr-2 but not by jai-1 scions. Taken together, the results show that activation of the jasmonate biosynthetic pathway in response to wounding or (pro)systemin is required for the production of a long-distance signal whose recognition in distal leaves depends on jasmonate signaling. These findings suggest that JA or a related compound derived from the octadecanoid pathway may act as a transmissible wound signal

The induction of plant defenses by insect feeding is regulated via multiple signaling cascades. One of them, ethylene signaling, increases susceptibility of *Arabidopsis* to the generalist herbivore Egyptian cotton worm (*Spodoptera littoralis*; Lepidoptera: Noctuidae). The hookless1 mutation, which affects a downstream component of ethylene signaling, conferred resistance to Egyptian cotton worm as compared with wild-type plants. Likewise, ein2, a mutant in a central component of the ethylene signaling pathway, caused enhanced resistance to Egyptian cotton worm that was similar in magnitude to hookless1 (80). Moreover, pretreatment of plants with ethephon (2-chloroethanephosphonic acid), a chemical that releases ethylene, elevated plant susceptibility to Egyptian cotton worm. By contrast, these mutations in the ethylene signaling pathway had no detectable effects on diamondback moth (*Plutella xylostella*) feeding. It is surprising that this is not due to nonactivation of defense signaling because the diamondback moth does induce genes that relate to wound-response pathways. Of these wound-related genes, jasmonic acid regulates a novel β-glucosidase 1 (BGL1), whereas ethylene controls a putative calcium-binding elongation factor hand protein. These results suggest that a specialist insect herbivore triggers general wound-response pathways in *Arabidopsis* but, unlike a generalist herbivore, does not react to ethylene-mediated physiological changes.

Signaling cross-talk between wound- and pathogen-response pathways influences resistance of plants to insects and disease. To elucidate potential interactions between salicylic acid (SA) and jasmonic acid (JA) defense pathways, Stotz et al. (81) exploited the availability of characterized mutants of *Arabidopsis thaliana* (L.) Heynh. and monitored resistance to Egyptian cotton worm. This generalist herbivore is sensitive to induced plant defense pathways and is thus a useful model for a mechanistic analysis of insect resistance. As expected, treatment of wild-type *Arabidopsis* with JA enhanced resistance to Egyptian cotton worm. Conversely, the coil mutant, with a deficiency in the JA response pathway, was more susceptible to Egyptian cotton worm than wild-type *Arabidopsis*. By contrast, the npr1 mutant, with defects in systemic disease resistance, exhibited enhanced resistance to Egyptian cotton worm. Pretreatment with SA significantly reduced this enhanced resistance of npr1 plants but had no influence on the resistance of wild-type plants. However, exogenous SA reduced the amount of JA that Egyptian cotton worm induced in both npr1 mutant and wild-type plants. Thus, this generalist herbivore engages two different induced defense pathways that interact to mediate resistance in *Arabidopsis*.

Ozawa et al. (27) compared volatiles from lima bean leaves (*Phaseolus lunatus*) infested by beet armyworm (*Spodoptera exigua*), common army-worm (*Mythimna separata*), or two-spotted spider mite (*Tetranychus urticae*). They have also analyzed volatiles from the leaves treated with JA and/or methyl salicylate (MeSA). The volatiles induced by aqueous JA treatment were qualitatively and quantitatively similar to those induced by *S. exigua* or *M. separata* damage. Furthermore, both *S. exigua* and aqueous JA treatment induced the expression of the same basic PR (pathogenesis-related) genes. In contrast, gaseous MeSA treatment, and aqueous JA treatment followed by gaseous MeSA treatment, induced volatiles that were qualitatively and quantitatively more similar to the *T. urticae*-induced volatiles than those induced by aqueous JA treatment. In addition, *T. urticae* damage resulted in the expression of the acidic and basic PR genes that were induced by gaseous MeSA treatment and by aqueous JA treatment, respectively. Based on these data, Ozawa et al. (27) suggested that in lima bean leaves, the JA-related signaling pathway is involved in the production of caterpillar-induced volatiles, while both the SA-related signaling pathway and the JA-related signaling pathway are involved in the production of *T. urticae*-induced volatiles.

Little is known about molecular responses in plants to phloem feeding by insects. The induction of genes associated with wound and pathogen response pathways was investigated following green peach aphid (*Myzus persicae*) feeding on *Arabidopsis* (82). Aphid feeding on rosette leaves induced transcription of two genes associated with SA-dependent responses to pathogens (PR-1 and BGL2) 10- and 23-fold, respectively. Induction of PR-1 and BGL2 mRNA was reduced in npr1 mutant plants, which are deficient in SA signaling. Application of the SA analogue benzothiadiazole led to decreases in aphid reproduction on leaves of both wild-type and mutant plants deficient in responsiveness to SA, suggesting that wild-type SA-dependent responses do not influence resistance to aphids. Twofold increases occurred in mRNA levels of PDF1.2, which encodes defensin, a peptide involved in the JA-ethylene-dependent response pathway. Transcripts encoding JA-inducible lipoxygenase (LOX2) and SA/JA-inducible phenylalanine ammonia lyase (PAL) increased 1.5- to 2-fold. PDF1.2 and LOX2 induction by aphids did not occur in infested leaves of the JA-resistant coil-1 mutant. Aphid feeding induced 10-fold increases in mRNA levels of a stress-related monosaccharide symporter gene, *STP4*. Phloem feeding on *Arabidopsis* leads to stimulation of response pathways associated with pathogen infection and wounding.

Felton et al. (11) reported that reducing phenylpropanoid biosynthesis by silencing the expression of phenylalanine PAL reduces systemic acquired resistance (SAR) to tobacco mosaic virus (TMV), whereas overexpression of PAL enhances SAR. Tobacco plants with reduced SAR exhibited more effective grazing-induced systemic resistance to larvae of *Heliothis virescens*, but larval resistance was reduced in plants with elevated phenylpropanoid levels. Furthermore, genetic modification of components involved in phenylpropanoid synthesis revealed an inverse relationship between SA and JA levels. These results demonstrate phenylpropanoid-mediated cross-talk *in vivo* between microbially induced and herbivore-induced pathways of systemic resistance.

At least two different signaling pathways have been identified that result in clearly distinguishable volatile profiles' response to pathogens and herbivores in the lima bean *Phaseolus lunatus* (83). Alamethicin, a voltage-gated ion channel–forming peptide from *Trichoderma viride*, is a potent inducer of volatile biosynthesis in the lima bean. Unlike elicitation with cellulysin or herbivore damage, which act through the JA pathway and result in a complex pattern of volatile compounds, the emitted blend comprises only the two homoterpenes 4,11-dimethylnona-1,3,7-triene and 4,8,12-trimethyl-trideca-1,3,7,11-tetrene, and MeSA. Both pathways, represented by JA and alamethicin, depend on lipid-derived signaling compounds, set off by the activation of a phospholipase A and further processing by lipoxygenase activity. The alamethicin-induced signal transduction pathway interferes with the octadecanoid cascade, probably due to increased SA levels, resulting in an inhibition of the typical JA-induced volatile profile.

12.6.2   Precursor Deficient Mutants

The signaling pathways that allow plants to mount defenses against chewing insects are known to be complex. To investigate the role of jasmonate in wound signaling in *Arabidopsis* and to test whether parallel or redundant pathways exist for insect defense, McConn et al. (16) studied a mutant (fad3-2 fad7-2 fad8) that is deficient in the jasmonate precursor linolenic acid. Mutant plants contained negligible levels of jasmonate and showed extremely high mortality (~80%) from attack by larvae of a common saprophagous fungal gnat, *Bradysia impatiens* (Diptera: Sciaridae), even though neighboring wild-type plants were largely unaffected. Application of exogenous methyl jasmonate (MeJA) substantially protected the mutant plants and reduced mortality to about 12%. These experiments precisely define the role of jasmonate as being essential for the induction of biologically effective defense in this plant–insect interaction. The transcripts of three wound-responsive genes were shown not to be induced by wounding of mutuant plants but the same transcripts could be induced by application of MeJA. By contrast, measurements of transcript levels for a gene encoding glutathione *S*-transferase demonstrated that wound induction of this gene is independent of jasmonate synthesis. These results indicate that the mutant will be a good genetic model for testing the practical effectiveness of candidate defense genes. The fad3-2 fad7-2 fad8 mutant, but not neighboring wild-type plants, was devastated by larval attack but could be protected by applications of jasmonate. Although parallel mechanisms may operate at other stages in the pathway, McConn et al. (16) have shown that, for the *Arabidopsis–Bradysia* interaction, the synthesis and action of jasmonate is a nonredundant step if the plant is to mount a biologically effective response. Analogous results have recently been obtained for a tomato mutant with reduced jasmonate synthesis that is susceptible to damage from *Manduca sexta* larvae.

The activation of defense genes in tomato plants has been shown to be mediated by an octadecanoic acid–based signaling pathway in response to herbivore attack or other mechanical wounding. Howe et al. (80) have reported that a tomato mutant (JL5) deficient in the activation of would-inducible defense genes is also compromised in resistance toward the lepidopteran pest *Manduca sexta* (tobacco hornworm). Thus, they propose the name defenseless1 (def1) for the mutation in the JL5 line that mediates this altered defense response. In experiments designed to define the normal function of def1, they found that def1 plants are defective in defense gene signaling initiated by prosystemin overexpression in transgenic plants as well as by oligosaccharide (chitosan and polygalacturonide) and polypeptide (systemin) elicitors. Supplementation of plants through their cut stems with intermediates of the octadecanoid pathway indicates that def1 plants are affected in octadecanoid metabolism between the synthesis of hydroperoxylinolenic acid and 12-oxophytodienoic acid. Consistent with this defect, def1 plants are also compromised in their ability to accumulate JA, the end product of the pathway, in response to wounding and the aforementioned elicitors. Taken together, these results show that octadecanoid metabolism plays an essential role in the transduction of upstream wound signals to the activation of antiherbivore plant defenses.

12.6.3 Receptors

A wound-inducible systemin cell surface receptor with a molecular weight of 160 kDa has recently been identified (84). The receptor regulates an intracellular cascade including depolarization of the plasma membrane, the opening of ion channels, an increase in intracellular Ca^{2+}, activation of a MAP kinase activity and a phospholipase A (2) activity. These rapid changes appear to play important roles leading to the intracellular release of linolenic acid from membranes and its subsequent conversion to JA, a potent activator of defense gene transcription. Although the mechanisms for systemin processing, release, and transport are still unclear, studies of the timing of the synthesis and of the intracellular localization of wound- and systemin-inducible mRNAs and proteins indicates that differential synthesis of signal pathway genes and defensive genes occurs in different cell types. This signaling cascade in plants exhibits extraordinary analogies with the signaling cascade for the inflammatory response in animals.

12.7 MOLECULAR BIOLOGY

Events that lead to the synthesis of oxylipins from fatty acids through the octadecanoid pathway are relatively well characterized, and most of the genes encoding enzymes of the biosynthetic pathway have already been cloned (85). Wounding activates the expression of most of the genes encoding enzymes of the JA biosynthetic pathway. The genes involved in regulation of JA pathway as well as prosystemin and nicotine have also been investigated.

12.7.1   JA Biosynthetic Enzymes

Genes from major biosynthetic pathways involved in volatile production are induced by caterpillar feeding (72). These include *AtTPS10*, encoding a terpene synthase involved in myrcene production, *AtPAL1*, encoding PAL involved in MeSA production, and *AtLOX2* and *AtHPL*, encoding lipoxygenase and hydroperoxide lyase, respectively, both involved in the production of green leaf volatiles. *AtAOS*, encoding allene oxide synthase, involved in the production of JA, was also induced by herbivory (72).

Plant–plant interactions *via* herbivory-induced leaf volatiles could result in the induction of defense responses against aggressive biotic agents in plants (86). cDNA microarray technology showed comprehensive gene activation in lima bean leaves that were exposed to volatiles released from the neighboring leaves infested with spider mites. The infestation with spider mites and the herbivory-induced volatiles enhanced 97 and 227 gene spots, respectively, on the microarray tip printed with 2032 lima bean cDNA. These genes are related to such broad functions as responses to pathogenesis, wounding, hormones, ethylene biosynthesis, flavonoid biosynthesis, (post)transcriptional modifications, translations, chaperones, secondary signaling messengers, membrane transports, protein/peptide degradations, and photosynthesis.

Exogenous jasmonate treatment of *Nicotiana attenuata* Torr. ex Wats. plants elicits durable resistance against herbivores and attack from its specialist herbivore, *M. sexta*, results in an amplification of the transient wound-induced increase in endogenous JA levels. To understand whether this "JA burst" is under transcriptional control, Zeigler et al. (87) cloned allene oxide synthase (AOS), the enzyme that catalyzes the dehydration of 13(*S*)-hydroperoxyoctadecatrienoic acid to an allene oxide, the first specific reaction in JA biosynthesis. An AOS cDNA coding for a 520 amino acid protein (58.6 kDa) with an isoelectric point of 8.74 was overexpressed in bacteria and determined to be a functional AOS. Southern blot analysis indicated the presence of more than one gene and AOS transcripts were detected in all organs, with the highest levels in stems, stem leaves, and flowers. Attack by *M. sexta* larvae resulted in a sustained JA burst producing an endogenous JA amount ninefold above control levels and threefold above maximal wound-induced levels. Such responses could be mimicked by the addition of *Manduca* oral secretion and regurgitant to puncture wounds. *Manduca sexta* attack, wounding, and regurgitant treatment transiently increased AOS transcript in the wounded leaf, but increases were not proportional to the JA response. Moreover, transcript accumulation lagged behind JA accumulation. Systemic wound-induced increases in AOS transcript, as well as AOS activity or JA accumulation, could not be detected. It was concluded that increase in AOS transcript does not contribute to the initial increase in endogenous JA but may contribute to sustaining the JA burst.

One of the exciting features of JA signaling is that plants can transform JA by methylation to its volatile counterpart MeJA, which can act as a signal in interplant communication and might also move in the intercellular spaces within the plant. Hence, one could argue that the only role of MeJA is to mobilize JA. However, a surprising finding came from a study in which the S-adenosyl-L-methionine–jasmonic acid carboxyl methyltransferase (JMT) (the JA-methylating enzyme in *Arabidopsis*) was overproduced (88). Transgenic *Arabidopsis* plants overproducing JMT had constantly elevated levels (threefold more than the control) of MeJA but normal levels of JA. The plants also showed a high constitutive expression of JA-responsive genes. This is remarkable because overproduction of another jasmonate biosynthetic enzyme, allene oxide synthase, either did not result in changes in the amounts of jasmonates in unwounded plants (89) or, in the case of a study performed in potato, resulted in enhanced JA levels without having an effect on gene expression (90). Therefore, the hypothesis that changes in concentration of JA rather than the absolute levels are responsible for a biological effect does not apply to MeJA. One might speculate that MeJA is perceived differently than other jasmonates by the plant. JMT knockout plants that are unable to synthesize MeJA might provide an answer to this question.

12.7.2 Regulatory Genes

Sporamin, a tuberous storage protein of sweet potato, was systemically expressed in leaves and stems by wound stimulation (91). In an effort to demonstrate the regulatory mechanism of wound response on the sporamin gene, a 1.25-kb sporamin promoter was isolated for studying the wound-induced signal transduction. Two wound response–like elements, a G-box-like element and a GCC core-like sequence, were found in this promoter. A construct containing the sporamin promoter fused to a β-glucuronidase (GUS) gene was transferred into tobacco plants by *Agrobacterium*-mediated transformation. The wound-induced high level of GUS activity was observed in stems and leaves of transgenic tobacco, but not in roots. This expression pattern was similar to that of the sporamin gene in sweet potato.

Exogenous application of MeJA activated the sporamin promoter in leaves and stems of sweet potato and transgenic tobacco plants. A competitive inhibitor of ethylene (2,5-norbornadiene) down-regulated the effect of MeJA on sporamin gene expression. In contrast, salicylic acid, an inhibitor of the octadecanoid pathway, strongly suppressed the sporamin promoter function that was stimulated by wound and MeJA treatments. In conclusion, wound-response expression of the sporamin gene in aerial parts of plants is regulated by the octadecanoid signal pathway.

Several mRNAs coding for enzymes of JA biosynthesis are up-regulated upon JA treatment or endogenous increase of the JA level. Miersch and Wasternack (92) investigated the positive feedback of endogenous JA on JA formation, as well as its β-oxidation steps. JA-responsive gene expression was recorded in terms of proteinase inhibitor 2 (pin2) mRNA accumulation. JA formed upon treatment of tomato (*Lycopersicon esculentum* cv. Moneymaker) leaves with JA derivatives carrying different lengths of the carboxylic acid side chain was quantified by GC-MS. The data revealed that β oxidation of the side chain occurs up to a butyric acid moiety. The amount of JA formed from side chain–modified JA derivatives correlated with pin2-mRNA accumulation. JA derivatives with a carboxylic side chain of 3.5 or 7 carbon atoms were unable to form JA and express on pin2, whereas even-numbered derivatives were active.

12.7.3 Prosystemin

Constitutive overexpression of the prosystemin gene in transgenic tomato plants resulted in the overproduction of prosystemin and the abnormal release of systemin, conferring a constitutive overproduction of several systemic wound-response proteins (SWRPs). The data indicate that systemin is a master signal for defense against attacking herbivores (93). The same defensive proteins induced by wounding are synthesized in response to oligosaccharide elicitors that are generated in leaf cells in response to pathogen attacks. Inhibitors of the octadecanoid pathway, and a mutation that interrupts this pathway, block the induction of SWRPs by wounding, systemin, and oligosaccharide elicitors, indicating that the octadecanoid pathway is essential for the activation of defense genes by all of these signal. The tomato mutant line that is functionally deficient in the octadecanoid pathway is highly susceptible to attack by *M. sexta* larvae. The similarities between the defense signaling pathway in tomato leaves and those of the defense signaling pathways of macrophages and mast cells of animals suggest that both the plant and animal pathways may have evolved from a common ancestral origin.

12.7.4 Nicotine

To understand whether this herbivore-induced signal cross-talk occurs at the level of transcript accumulation, Winz and Baldwin (28) cloned the putrescine N-methyltransferase genes (*NaPMT1* and *NaPMT2*) of *N. attenuata*, which are thought to represent the rate-limiting step in nicotine biosynthesis, and measured transcript accumulations by Northern analysis after various jasmonate, 1-MCP, ethephon, and herbivory treatments. Transcripts of both root putrescine N-methyltransferase genes and nicotine accumulation increased dramatically within 10 h of shoot MeJA treatment and immediately after root treatments. Root ethephon treatments suppressed this response, which could be reversed by 1-MCP pretreatment. Moreover, 1-MCP pretreatment dramatically amplified the transcript accumulation resulting from both wounding and *M. sexta* herbivory. They concluded that attack from this nicotine-tolerant specialist insect causes *N. attenuata* to produce ethylene, which directly suppresses the nitrogen-intensive biosynthesis of nicotine.

12.8   PERSPECTIVE

The jasmonate perception pathway is currently being delineated through the use of gain-of-function and loss-of-function mutants. Protein kinases are implicated in early events of jasmonate signaling (94). More detailed knowledge about how the jasmonate pathway contributes to the coordination of direct defense responses and indirect defense responses will permit an understanding of how plants extend their defense umbrella. Advances in the manipulation of the pathway hold promise for future strategies in agriculture.

Receptors are known for the volatile intraplant signal ethylene but not for the larger molecular weight herbivore-induced volatile organic compounds. Identifying and characterizing receptors for wound signals is likely to be a main focus of future research in the field of plant responses to wounding (84). Elucidating the compartmentalization and cell type specificity of the different wound signaling pathways is likely to follow suit. Promoter elements and the corresponding transcription factors directing wound-activated gene expression will provide useful tools for genetic manipulation of plants toward enhanced resistance to stress. Finally, it will also be of major interest to unravel the regulatory connections between wound-activated signaling pathways and signal transduction pathways triggered by other stress factors. Receptors for these putative volatile signals are to be discovered to understand how the signals enter a plant. If the ongoing transcriptional analyses identify genes that are strongly regulated by herbivore-induced VOCs, these are likely to provide a source of promoters that could be fused to easily characterized reporters, e.g., GUS or green fluorescent protein. Plants that have been transformed with such reporter genes could provide much needed information about how plants respond to volatile signals under natural conditions.

Extreme caution will be needed when deploying genetically modifying plants; natural enemies of herbivores are extremely good learners, and the lack of prey at sites emitting volatiles causes natural enemies to negatively associate the odors with prey.

REFERENCES

1. Weber, H. Fatty acid–derived signals in plants. Trends Plant Sci. **2002**, *7*, 217–223.
2. Creelman, R.A.; Mullet, J.E. Biosynthesis and action of jasmonates in plants. Annu. Rev. Plant Physiol. Plant Mol. Biol. **1997**, *48*, 355–381.
3. Baldwin, I.T.; Kessler, A.; Halitschke, R. Volatile signaling in plant–plant–herbivore interactions. Curr. Opin. Plant Biol. **2002**, *5*, 1–4.
4. Leon, J.; Rojo, E.; Sanchez-Sarrano, J.J. Wound signaling in plants. J. Exp. Bot. **2001**, *52*, 1–9.
5. Farmer, E.E. Surface-to-air signals. Nature **2001**, *411*, 854–856.
6. Feussner, I.; Wasternack, C. The lipoxygenase pathway. Annu. Rev. Plant Physiol. Plant Mol. Biol. **2002**, *53*, 275–297.
7. Agrawal, A.A. Mechanisms, ecological consequences and agricultural implications of tri-trophic interactions. Curr. Opin. Plant Biol. **2000**, *3*, 329–335.
8. Fidantsef, A.L.; Stout, M.J.; Thaler, J.S.; Duffey, S.S.; Bostock, R.M. Signal interactions in pathogen and insect attack: expression of lipoxygenase, proteinase inhibitor II, and pathogenesis-related protein P4 in the tomato, *Lycopersicon esculentum*. Physiol. Mol. Plant Pathol. **1999**, *54*, 97–114.
9. Ryals, J.; Uknes, S.; Ward, E. Systemic acquired resistance. Plant Physiol. **1994**, *104*, 1109–1112.
10. Stout, M.J.; Fidantsef, A.L.; Duffey, S.S.; Bostock, R.M. Signal interactions in pathogen and insect attack: systemic plant–mediated interactions between pathogens and herbivores of the tomato, *Lycopersicon esculentum*. Physiol. Mol. Plant Pathol. **1999**, *54*, 115–130.
11. Felton, G.W.; Korth, K.L.; Bi, J.L.; Wesley, S.V.; Huhman, D.V.; Mathews, M.C.; Murphy, J.B.; Lamb, C.; Dixon, R.A. Inverse relationship between systemic resistance of plants to microorganisms and to insect herbivory. Curr. Biol. **1999**, *9*, 317–320.
12. Pickersky, E.; Gershenzon, J. The formation and function of plant volatiles: perfumes for pollinator attraction and defense. Curr. Opin. Plant Biol. **2002**, *5*, 237–243.
13. Arimura, G.; Ozawa, R.; Shimoda, T.; Nishioka, T.; Boland, W.; Takabayashi, J. Herbivory-induced volatiles elicit defense genes in lima bean leaves. Nature **2000**, *406*, 512–515.

14. Kessler, A.; Baldwin, I.T. Defensive function of herbivore-induced plant volatile emissions in nature. Science **2001**, *291*, 2141–2144.
15. Pearce, G.; Moura, D.S.; Stratmann, J.; Ryan, C.A. Production of multiple plant hormones from a single polyprotein precursor. Nature **2001**, *411*, 817–820.
16. McConn, M.; Creelman, R.A.; Bell, E.; Mullet, J.E.; Browse, J. Jasmonate is essential for insect defense in *Arabidopsis*. Proc. Natl. Acad. Sci. USA **1997**, *94*, 5473–5477.
17. Musser, R.O.; Hum-Musser, S.M.; Eichenseer, H.; Peiffer, M.; Erwin, G.; Murray, J.B.; Felton, G.W. Herbivory: caterpillar saliva beats plant defences. Nature **2002**, *416*, 599–600.
18. Pare, P.W.; Alborn, H.T.; Tumlinson, J.H. Concerted biosynthesis of an insect elicitor of plant volatiles. Proc. Natl. Acad. Sci. USA **1998**, *95*, 13971–13975.
19. Turlings, T.C.J.; Alborn, H.T.; Loughrin, J.H.; Tumlinson, J.H. Volicitin, an elicitor of maize volatiles in oral secretion of *Spodoptera exigua*: its isolation and bioactivity. J. Chem. Ecol. **2000**, *26*, 189–202.
20. Landolt, P.J.; Tumilson, J.H.; Alborn, D.H. Attraction of Colorodo potato beetle (Coleoptera: Chrysomelidae) to damaged and chemically induced potato plants. Environ. Entomol. **1999**, *28*, 973–978.
21. Doss, R.P.; Oliver, J.E.; Proebsting, W.M.; Potter, S.W.; Kuy, S.; Clement, S.L.; Williamson, R.T.; Carney, J.R.; De Vilbiss, E.D. Bruchins: insect-derived plant regulators that stimulate neoplasm formation. Proc. Natl. Acad. Sci. USA **2000**, *97*, 6218–6223.
22. Mattiacci, L.; Dicke, M.; Posthumus, M.A. Beta-glucosidase: an elicitor of herbivore-induced plant odor that attracts host-searching parasitic wasps. Proc. Natl. Acad. Sci. USA **1995**, *92*, 2036–2040.
23. Loughrin, J.H.; Manukian, A.; Heath, R.R.; Turlings, T.C.; Tumlinson, J.H. Diurnal cycle of emission of induced volatile terpenoids by herbivore–injured cotton plant. Proc. Natl. Acad. Sci. USA **1994**, *91*, 11836–11840.
24. Li, L.; Li, C.; Lee, G.I.; Howe, G.A. Distinct roles for jasmonate synthesis and action in the systemic wound response of tomato. Proc. Natl. Acad. Sci. USA **2002**, *99*, 6416–6421.
25. Turlings, T.C.; Tumlinson, J.H. Systemic release of chemical signals by herbivore-injured corn. Proc. Natl. Acad. Sci. USA **1992**, *89*, 8399–8402.
26. Arimura, G.; Ozawa, R.; Nishioka, T.; Boland, W.; Koch, T.; Kuhnemann, F.; Takabayashi, J. Herbivore-induced volatiles induce the emission of ethylene in neighbouring lima bean plants. Plant J. **2002**, *29*, 87–98.
27. Ozawa, R.; Arimura, G.; Takabayashi, J.; Shimoda, T.; Nishioka, T. Involvement of jasmonate- and salicylate-related signaling pathways for the production of specific herbivore-induced volatiles in plants. Plant Cell Physiol. **2000**, *41*, 391–398.
28. Winz, R.A.; Baldwin, I.T. Molecular interactions between the specialist herbivore *Manduca sexta* (Lepidoptera. Sphingidae) and its natural host *Nicotiana attenuata*. IV. Insect-induced ethylene reduces jasmonate-induced nicotine accumulation by regulating putrescene N-methyltransferase transcripts. Plant Physiol. **2001**, *125*, 2189–2202.

29. Karban, R. Communication between sagebrush and wild tobacco in the field. Biochem. Syst. Ecol. **2001**, *29*, 995–1005.
30. Lill, J.T.; Marquis, R.J.; Ricklefs, R.E. Host plants influence parasitism of forest caterpillars. Nature **2002**, *417*, 170–173.
31. Wegener, R.; Schulz, S.; Meiners, T.; Hadwich, K.; Hilker, M. Analysis of volatiles induced by oviposition of elm leaf beetle *Xanthogaleruca luteola* on *Ulmus minor*. J. Chem. Ecol. **2001**, *27*, 499–515.
32. Wheeler, G.S.; Massey, L.M.; Southwell, I.A. Antipredator defense of biological control agent *Oxyops vitiosa* is mediated by plant volatiles sequestered from the host plant *Melaleuca quinquenervia*. J. Chem. Ecol. **2002**, *28*, 297–315
33. Cardoza, Y.J.; Alborn, H.T.; Tumlinson, J.H. In vivo volatile emissions from peanut plants induced by simultaneous fungal infection and insect damage. J. Chem. Ecol. **2002**, *28*, 161–174.
34. Hakala, M.A.; Lapvetelainen, A.T.; Kallio, H.P. Volatile compounds of selected strawberry varieties analyzed by purge-and-trap headspace GC-MS. J. Agric. Food Chem. **2002**, *50*, 1133–1142.
35. Hilker, M.; Kobs, C.; Varma, M.; Schrank, K. Insect egg deposition induces *Pinus sylvestris* to attract egg parasitoids. J. Exp. Biol. **2002**, *205*, 455–461.
36. Mattiacci, L.; Rocca, B.A.; Scascighini, N.; D'Alessandro, M.; Hern, A.; Dorn, S. Systemically induced plant volatiles emitted at the time of danger. J. Chem. Ecol. **2001**, *27*, 2233–2352.
37. van Tol, R.W.; van Der Sommen, A.T.; Boff, M.I.; van Bezooijen, J.; Sabelis, M.W.; Smits, P.H. Plants protect their roots by alerting the enemies of grubs. Ecol. Lett. **2001**, *4*, 292–294.
38. Kramell, R.; Miersch, O.; Atzorn, R.; Parthier, B.; Wasternack, C. Octadecanoid derived alteration of gene expression and the "oxylipin signature" in stressed barley leaves. Implications for different signaling pathways. Plant Physiol. **2000**, *123*, 177–188.
39. Peacock, L.; Lewis, M.; Powers, S. Volatile compounds from *Salix* spp. Varieties differing in susceptibility to three willow beetle species. J. Chem. Ecol. **2001**, *27*, 1943–1951.
40. Gundlach, H.; Zenk, M.H. Biological activity and biosynthesis of pentacyclic oxylipins: the linoleic acid pathway. Phytochemistry **1998**, *47*, 527–537.
41. Schaller, F.; Weiler, E.W. Enzymes of octadecanoid biosynthesis in plants— 12-oxophytodienoate 10,11-reductase. Eur. J. Biochem. **1997**, *245*, 294–299.
42. Schaller, F. Enzymes of the biosynthesis of octadecanoid-derived signaling molecules. J. Exp. Bot. **2001**, *52*, 11–23.
43. Liechti, R.; Farmer, E. The jasmonate pathway. Science **2002**, *296*, 1649–1650.
44. Kandra, L.; Wagner, G.J. Pathway for the biosynthesis of 4-methyl-1-hexanol volatilized from petal tissue of *Nicotiana sylvestris*. Phytochemistry **1998**, *49*, 1599–1604.
45. Blee, E. Phytooxylipins and plant defense reactions. Progr. Lipid. Res. **1998**, *37*, 33–72.

46. Noordermeer, M.A.; Veldink, G.A.; Vliegenthart, J.F.G. Fatty acid hydroperoxide lyase: a plant cytochrome P450 enzyme involved in wound healing and pest resistance. Chembiochem. **2001**, *2*, 494–504.
47. Boland, W.; Koch, T.; Krumm, T.; Piel, J.; Jux, A. Induced biosynthesis of insect semiochemicals in plants. Novartis Found Symp. **1999**, *223*, 110–126.
48. Frey, M.; Stettner, C.; Pare, P.W.; Schmelz, E.A.; Tumlinson, J.H.; Gierl, A. An herbivore elicitor activates the gene for indole emission in maize. Proc. Natl. Acad. Sci. USA **2000**, *97*, 14801–14806.
49. Lindsey, K.; Casson, S.; Chilley, P. Peptides: new signaling molecules in plants. Trends Plant Sci. **2002**, *7*, 78–83.
50. Blechert, S.; Brodschelm, W.; Holder, S.; Kammerer, L.; Kutchan, T.M.; Mueller, M.J.; Xia, Z.Q.; Zenk, M.H. The octadecanoid pathway: signal molecules for the regulation of secondary pathways. Proc. Natl. Acad. Sci. USA **1995**, *92*, 4099–4105.
51. Farmer, E.E. New fatty acid–based signals: a lesson from the plant world. Science **1997**, *276*, 912–913.
52. Turlings, T.C.J.; Benrey, B. Effects of plant metabolites on the behavior and development of parasitic wasps. Ecoscience **1998**, *5*, 321–333.
53. Scutareanu, P.; Drukker, B.; Bruin, J.; Posthumus, M.A.; Sabelis, M.W. Volatiles from psylla-infested pear trees and their possible involvement in attraction of anthocorid predators. J. Chem. Ecol. **1997**, *23*, 2241–2260.
54. Schmelz, E.A.; Alborn, H.T.; Tumlinson, J.H. The influence of intact-plant and excised-leaf bioassay designs on volicitin- and jasmonic acid-induced sesquiterpene volatiles release in *Zea mays*. Planta **2001**, *214*, 171–179.
55. Fukushima, J.; Kainoh, R.; Honda, H.; Takabayashi, J. Learning of herbivore-induced and nonspecific plant volatiles by a parasitoid, *Cotesia kariyai*. J. Chem. Ecol. **2002**, *28*, 579–586.
56. Pels, B.; Sabelis, M.W. Do herbivore-induced plant volatiles influence predator migration and local dynamics of herbivorous and predatory mites? Exp. Appl. Acarol. **2000**, *24*, 427–440.
57. Muller, C.; Hilker, M. The effect of a green leaf volatile on host plant finding by larvae of a herbivorous insect. Naturwissenschaften **2000**, *87*, 216–219.
58. Dicke, M.; Takabayashi, J.; Posthumus, M.A.; Schutte, C.; Krips, O.E. Plant–phytoseiid interactions mediated by herbivore-induced plant volatiles: variation in production of cues and in responses of predatory mites. Exp. Appl. Acarol. **1998**, *2*, 311–333.
59. Dicke, M.; Sabelis, M.; Takabayashi, W.; Bruin, J.; Posthumus, M.A. Plant strategies of manipulating predator–prey interactions through allelochemicals: prospects for application to pest control. J. Chem. Ecol. **1990**, *16*, 3091–3118.
60. Tumlinson, J.H.; Pare, P.W.; Lewis, W.J. Plant production of volatile semiochemicals in response to insect derived elicitors. Novartis Found Symp. **1999**, *223*, 95–105.
61. Dicke, M. Specificity of herbivore-induced plant defenses. Novartis Found Symp. **1999**, *223*, 43–54.

62. Dicke, M.; Gols, R.; Ludeking, D.; Posthumus, M.A. Jasmonic acid and herbivory differentially induce carnivore-attracting plant volatiles in lima bean plants. J. Chem. Ecol. **1999**, *25*, 1907–1922.
63. Engelberth, J.; Koch, T.; Kuhnemann, F.; Boland, W. Channel-forming paptaibols are a novel class of potent elicitors of plant secondary metabolism and tendril coiling. Angew. Chem. **2000**, *39*, 1860–1862.
64. Pettersson, J.; Pickett, J.S.; Pye, B.J.; Quiroz, A.; Smart, L.E.; Wadhams, L.J.; Woodcock, C.M. Winter host component reduces colonization of summer hosts by the bird cherry-oat aphid *Rhopalosiphum padi* and other aphids in cereal fields. J. Chem. Ecol. **1994**, *20*, 2565–2574.
65. Shulaev, V.; Silverman, P.; Raskin, I. Airborne signaling by methyl salicylate in plant pathogen resistance. Nature **1997**, *385*, 718–721.
66. Pickett, J.A.; Poppy, G.M. Switching on plant genes by external chemical signals. Trends Plant Sci. **2001**, *6*, 137–139.
67. Pare, P.W.; Tumlinson, J.H. De novo biosynthesis of volatiles induced by insect herbivory in cotton plants. Plant Physiol. **1997**, *114*, 1161–1167.
68. Walling, L.L. The myriad plant responses to herbivores. J. Plant Growth Regul. **2000**, *19*, 195–216.
69. Karban, R.; Baldwin, I.T. Induced Responses to Herbivory. University of Chicago Press, 1997.
70. Guerrieri, E.; Poppy, G.M.; Powell, W.; Tremblay, E.; Pennacchio, F. Induction and systemic release of herbivore-induced plant volatiles mediating in-flight orientation of *Aphidius ervi*. J. Chem. Ecol. **1999**, *25*, 1247–1262.
71. Bradburne, R.; Mithen, R. Glucosinolate genetics and the attraction of the aphid parasitoid *Diaretiella rapae* to *Brassica*. Proc. Royal Soc. Lond. Ser. B **2000**, *267*, 89–95.
72. Van Poecke, R.M.; Posthumus, M.A.; Dicke, M. Herbivore-induced volatile production by *Arabidopsis thaliana* leads to attraction of the parasitoid *Cotesia rubecula*: chemical, behavioral and gene-expression analysis. J. Chem. Ecol. **2001**, *27*, 1911–1928.
73. Birkett, M.A.; Campbell, C.A.M.; Chamberlain, K.; Guerrier, E.; Hick, A.J.; Martin, J.L.; Matthes, M.; Napier, J.A.; Patersson, J.; Pickett, J.A.; Poppy, G.M.; Pow, E.M.; Pye, B.J.; Smart, L.E.; Wadhams, G.H.; Wadhams, L.J.; Woodcock, C.M. New roles for cis-jasmone as an insect semiochemical and in plant defence. Proc. Natl. Acad. Sci. USA **2000**, *97*, 9329–9334.
74. Hopke, J.; Donath, J.; Blechert, S.; Boland, W. Herbivore-induced volatiles: the emission of acyclic homoterpenes from leaves of *Phaseolus lunatus* and *Zea mays* can be triggered by a β-glucosidase and jasmonic acid. FEBS Lett. **1994**, *352*, 146–150.
75. Thaler, J.S. Jasmonate-inducible plant defences cause increased parasitism of herbivores. Nature **1999**, *399*, 686–688.
76. Baldwin, I.T. Jasmonate-induced responses are costly but benefit plants under attack in native populations. Proc. Natl. Acad. Sci. USA **1998**, *95*, 8113–8118.

77. Rodriguez-Saona, C.; Crafts-Brandner, S.J.; Pare, P.W.; Henneberry, T.J. Exogenous methyl jasmonate induces volatile emissions in cotton plants. J. Chem. Ecol. **2001**, *27*, 679–695.
78. Kahl, J.; Seiemens, D.H.; Aerts, R.J.; Gabler, R.; Kuhnemann, F.; Preston, C.A.; Baldwin, I.T. Herbivore-induced ethylene suppresses a direct defense but not a putative indirect defense against an adapted herbivore. Planta **2000**, *210*, 336–342.
79. De Moraes, C.M.; Mescher, M.C.; Tumlinson, J.H. Caterpillar-induced nocturnal plant volatiles repel conspecific females. Nature **2001**, *410*, 577–580.
80. Howe, H.A.; Lightner, J.; Browse, J.; Ryan, C.A. An octadecanoid pathway mutant (JL5) of tomato is compromised in signaling for defense against insect. Plant Cell **1996**, *8*, 2067–2077.
81. Stotz, H.U.; Koch, T.; Biedermann, A.; Weniger, K.; Boland, W.; Olds, T.M. Evidence for regulation of resistance in *Arabidopsis* to Egyptian cottonworm by salicylic and jasmonic acid signaling pathways. Planta **2002**, *214*, 648–652.
82. Moran, P.J.; Thompson, G.A. Molecular responses to aphid feeding in *Arabidopsis* in relation to plant defense pathways. Plant Physiol. **2001**, *125*, 1074–1085.
83. Engelberth, J. Differential signaling and plant-volatile biosynthesis. Biochem. Soc. Trans. **2000**, *28*, 871–872.
84. Meindl, T.; Boller, T.; Felix, G. The plant wound hormone systemin binds with the N-terminal part to its receptor but needs the C-terminal part to activate it. Plant Cell **1998**, *10*, 1561–1570.
85. Leon, J.; Sanchez-Serrano, J.J. Molecular biology of JA biosynthesis in plants. Plant Physiol. Biochem. **1999**, *37*, 373–380.
86. Arimura, G.; Tashiro, K.; Kuhara, S.; Nishioka, T.; Ozawa, R.; Takabayashi, J. Gene responses in bean leaves induced by herbivory and by herbivore-induced volatiles. Biochem. Biophys. Res. Commun. **2000**, *277*, 305–310.
87. Ziegler, J.; Keinanen, M.; Baldwin, I.T. Herbivore-induced allene oxide synthase transcripts and jasmonic acid in *Nicotiana attenuata*. Phytochemistry **2001**, *58*, 729–738.
88. Seo, H.S.; Song, J.T.; Cheong, J.J.; Lee, Y.H.; Lee, Y.W.; Hwang, I.; Lee, J.S.; Choi, Y.D. Jasmonic acid carboxyl methyltransferase: a key enzyme for jasmonate regulated plant responses. Proc. Natl. Acad. Sci. USA **2001**, *98*, 4788–4793.
89. Laudert, D.; Schaller, F.; Weiler, E.W. Transgenic *Nicotiana tabacum* and *Arabidopsis thaliana* plants overexpressing allene oxide synthase. Planta **2000**, *211*, 163–165.
90. Harms, K.; Atzorn, R.; Brash, A.; Kuhn, H.; Wasternack, C.; Willmitzer, L.; Pena-Cortes, H. Expression of a flax allene oxide synthase cDNA leads to increased endogenous jasmonic acid (JA) levels in transgenic potato plant but not to a corresponding activation of JA-responding genes. Plant Cell **1995**, *7*, 1645–1654.
91. Wang, S.J.; Lan, Y.C.; Chen, S.F.; Chen, Y.M.; Yeh, K.W. Wound-response regulation of the sweet potato sporamin gene promoter region. Plant Mol. Biol. **2002**, *48*, 223–231.

92. Miersch, O.; Wasternack, C. Octadecanoid and jasmonate signaling in tomato (*Lycopersicon esculentum* Mill) leaves: endogenous jasmonates do not induce jasmonate biosynthesis. Biol. Chem. **2000**, *381*, 715–722.

93. Bergy, D.R.; Howe, G.A.; Ryan, C.A. Polypeptide signaling for plant defensive genes exhibits analogies to defense signaling in animals. Proc. Natl. Acad. Sci. USA **1996**, *93*, 12053–12058.

94. Petersen, M.; Brodersen, P.; Naested, H.; Andreasson, E.; Lindhart, U.; Johansen, B.; Nielsen, H.B.; Lacy, M.; Austin, M.J.; Parker, J.E.; Sharma, S.B.; Klessig, D.F.; Martienssen, R.; Mattsson, O.; Jensen, A.B.; Mundy, J. *Arabidopsis* map kinase-4 negatively regulates systemic acquired resistance. Cell **2000**, *103*, 1111–1120.

13

Plant Resistance from a Sustainable Agricultural and Environmental Perspective

13.1 INTRODUCTION

The world's population is expected to reach 10.9 billion by 2050, and a majority of people will live in developing countries, intensifying the struggle against poverty and malnutrition (1,2). Food production has to be doubled without relying on specialized fertilizers and pesticides, which would further disturb the ecological balance. Factors such as population growth; increasing affluence with rising consumption, pollution, and waste; and persistent poverty are putting increased pressure on the environment. Rain forests are being destroyed at the highest rate in history, taking with them the crucial sources of biodiversity and contributing to global warming, thereby boosting the already rising sea levels (3). Improved food production has been achieved through technological advances such as chemical fertilizers, pesticides, and farm machinery, sometimes with undesirable consequences for the underlying natural resources of the production system. Even the green revolution, which provided a major boost for food production (particularly important for third-world countries), has resulted in large-scale damage to the agricultural ecosystem. Stockpiles of food grain have decreased each year since 1986 and a further decline is predicted (4,5). In spite of this we could still feed the entire world's population today with an adequate, though not generous, diet. The International Food Policy

Research Institute (IFPRI) projects that global demand for cereals may increase by 41% between 1993 and 2020 to reach 2490 million metric tons, and for roots and tubers demand may increase by 40% to 855 million tons. With an expected 40% population increase and an average annual income growth rate of 4.3%, developing countries are projected to account for most of the increase in global demand for cereals between 1993 and 2020. Demand for cereals for feeding livestock will increase considerably in coming decades, especially in developing countries, in response to strong demand for livestock products. Between 1993 and 2020, developing countries' demand for cereals for animal feed is projected to double, while demand for cereals for food for direct human consumption is projected to increase by 47% (6,7). Hence, rejuvenated drive is necessary in developing countries to meet the food demand.

Food grains under storage must be protected from insect and rodent infestations. Pest attack on stored grain causes serious deterioration in quality and value through spoilage, weight loss, and reduction in nutritional content. Heavy infestations lead to an increase in grain temperature and moisture, encouraging mold growth with further loss in food value and tainting of the grain. The Food and Agricultural Organization (FAO) has estimated that the average loss of cereal during storage owing to insect attack is 10% of the harvest. In certain tropical or subtropical areas this loss can reach up to 30% (8,9).

13.2 PESTICIDES IN AGRICULTURE

Pesticides pose severe health hazards and have other environmental, ecological, economic, and social implications. Several hundred pesticides (active ingredients) are currently used worldwide. Of the 700-plus chemical pesticides considered to be in current use, 33 have been classified as extremely hazardous to human health by the World Health Organization (class Ia) (10), 48 as highly hazardous (class Ib), 118 as moderately hazardous (class II), and 239 as slightly hazardous (class III). Another 149 pesticides were considered as unlikely to cause acute hazard in normal use (class IV) and WHO has not yet classified 164 of the *Pesticide Manual's* chemical entries according to acute health hazard parameters.

In 1995, world pesticide consumption reached 2.6 million metric tons of active ingredients with a market value of U.S.$ 38,000 million. Roughly 85% of this consumption was used in agriculture (11). About three fourths of pesticide use occurs in developed countries, mostly in North America, western Europe, and Japan. Although the volume of pesticides used in developing countries is small relative to that in many developed countries, it is nonetheless substantial and increasing rapidly. In addition, insecticides dominate the pesticide market in developing countries with a higher acute toxicity than herbicides, which dominate in the developed countries.

13.3 SUSTAINABLE AGRICULTURE AND PEST MANAGEMENT

Sustainable agriculture has been often equated with low input or subsistence agriculture. The goals of a sustainable agriculture are the same as for the long-term success of any other enterprise, oriented ultimately to societal well-being, as well as economic and environmental sustainability. These three are interdependent, and economic well-being is integral to underpinning environmental stability and sustainability (12,13). To create more sustainable agricultural ecosystems, we need to change their design to prevent pest problems rather than relying so heavily on pesticides (14). One fundamental element of the redesigned agricultural system must be an understanding of how crop plants resist pest attack. The interactions between plants and pests are very complex. Characteristics of the plant that influence attractiveness, suitability, and tolerance to pests are affected by many environmental and plant factors (15–18), including plant nutrition, secondary plant chemicals, moisture status, light, temperature, exposure to pesticides, growth regulators, pollutants, and injury by mechanical damage, insects, and plant disease.

On the other side of the interaction, the ability of pests to attack plants is also affected by biological as well as physical factors in their environment, including surrounding plants (19–21), biological control agents such as parasites, predators, and pathogens (22,23), and competitors. Pests can respond to resistance factors in plants as individuals by making adjustments in their own biochemistry, physiology, and behavior (24), as well as through evolution of the populations (25). In breeding plants for resistance to insects, entomologists and plant breeders have focused on the goal of developing crop cultivars with genes for permanent, rather than inducible resistance to insects. This approach has been successful in developing cultivars used as the primary control method for some major pests. As it has become possible to transfer genes directly into plants through genetic engineering rather than relying only on transfer through pollination, new possibilities and approaches to plant resistance to insects has developed. New sources and mechanisms of resistance beyond the gene pool of the crop plant have become available. The third Consultative Group of International Agricultural Research (CGIAR) system review has proposed that CGIAR centers should promote a global initiative for integrated gene management. Agricultural biotechnologies have major potential for facilitating and promoting sustainable agriculture and rural development. They could also generate environmental benefits, especially where renewable genetic inputs can be effectively used to substitute for dependency on externally provided agrochemical inputs. The fact that genes or genotypes (e.g., varieties, breeds) can constitute locally renewable resources has profound significance to the further development of sustainable agriculture and rural development. The specific products engineered for sustainable agriculture would be different from those that are being developed to fit into industrial agricultural systems and their development should probably await the wider adoption of such systems (12,13).

13.4 ROLE OF HOST PLANT RESISTANCE IN SUSTAINABLE AGRICULTURE

Traits conferring host plant resistance (HPR) to insects and pathogens are among the most important for crop improvement. Insect resistance has also been introduced into several hundred crop varieties during the last 20 years (26), and its importance is increasing as insecticides lose efficacy due to pest adaptation or are removed from use to protect the environment and human health (27). Multiple genes are required for sustained resistance to counter pest adaptation. Thus, maintaining agricultural productivity to meet world food needs depends on access by agricultural scientists to plentiful sources of HPR genes. Existing *ex situ* collections may not be adequate for HPR breeding in the future. Pest resistance genes are rare and predominantly found in unimproved varieties or wild accessions. For example, resistance to insects is found in only 0.01–2% of rice accessions (28), and much of this occurs in exotic land races. In potato, high levels of resistance to the green peach aphid (*Myzus persicae*) has been identified in about 6% of examined accessions of wild *Solanum* species, but in none of 360 accessions of *S. tuberosum* and other cultivated *Solanum* species (29). The pattern is similar for resistance to other insect pests of potatoes. In cultivated tomato (*Lycopersicon esculentum*) insect resistance is rare, but it is more prevalent in wild accessions of *L. esculentum* var. *cerasiforme* (27) and common in more distantly related *Lycopersicon hirsutum* (30). Many wheat breeding lines in the United States carry exotic genes for resistance to insects Hessian fly (*Mayetiola destructor*) and greenbug (*Schizaphis graminum*). Resistance to an important new pest, the Russian wheat aphid (*Diuraphis noxia*), has been found in about 50 accessions of wheat and barley (0.5% of the accessions screened) and most of these are landraces and are from Afghanistan and the former Soviet Union. But wild accessions and landraces that provide the most host plant resistance genes are under represented in many germplasm collections. Only about 8% of accessions in the world crop germplasm collections are wild crop relatives, and some collections have little or no wild germplasm. In most crops, excepting wheat, oats, potato, and tomato, it is estimated that less than 20% of the genetic diversity represented in wild species and unimproved cultivars is conserved *ex situ* (31). Time and resource pressures to obtain material have resulted in relative under sampling of geographically remote areas (32). But more germplasm collection is needed continuously. Resources are already strained to maintain collections at their present size. Losses and erosion in extensive collections inevitably occur due to contamination, drift, errors in record keeping, physical facilities failures, inadvertent selection, and other causes. Collections require backups at one or more locations to protect against these kinds of losses, but accessions in some crops do not have sufficient backups. In other crops, germplasm duplication is excessive among international collections, but coordinating optimal reductions is logistically difficult. Curators and managers of these collections are working hard to minimize these problems. Another difficulty with extensive *ex situ* collections is that many accessions are inadequately characterized or have undesirable traits that discourage breeders from using them. Evaluation and enhancement (improving agronomic suitability by breeding) within the context of curation requires considerable resources. To address these problems with available resources, many collections are now managed using the *ëcoreí* concept (33). A subset of accessions is chosen to maximize representation of the alleles in the entire collection.

13.5 INTEGRATION OF HOST PLANT RESISTANCE WITH BIOLOGICAL CONTROL

The integration of host plant resistance with biological control tactics may be additive, or even synergistic, in their effect on decreasing pest populations (34). Complexes of biological control agents have been identified for most major rice insect pests, and studies indicate that such agents have the potential to regulate rice insect populations with host plant resistance (35,36). This provides long-term significance of their interaction in delaying or preventing the development of biotypes capable of overcoming previously resistant rice cultivars (37). On the other hand, insect-resistant cultivars may have an adverse effect on natural enemies by reducing prey density but generally have a positive effect on natural enemies, especially parasites and predators, by minimizing the need for the application of toxic insecticides. The green leafhopper is an important rice pest in Asia because it is a vector of the disease causing rice tungro virus (RTV). Combinations of green leafhopper-resistant cultivars and predation by the mirid bug, *Cyrtorhinus lividipennis* Reuter, have a cumulative effect on reducing green leafhopper population. Insect resistant cultivars can enhance the activity of predators resulting in synergistic effect of two control tactics. For example, the *Cyrtorhinus lividipennis* predation rate increased when the prey, brown planthopper nymphs, fed on the resistant rice cultivars, "IR36". Increased movement of planthopper nymphs on resistant plants facilitates detection of prey by the mirid bug (38).

13.6 INTEGRATION OF HOST PLANT RESISTANCE WITH CHEMICAL CONTROL

Control of rice insects with insecticides has been shown to be more effective when the host plant is an insect-resistant cultivar as compared with a susceptible cultivar. Depending on the insect–pest complex, the level of insect pressure, and the degree of plant resistance, insecticides in combination with an insect-resistant cultivar can increase grain yields above that of a resistant cultivar alone. A cumulative effect was observed when plant resistance and insecticides were integrated in the control of the green leafhopper in field studies in the Philippines (39). The cumulative effect was based on the percentage of RTV-infected plants in the various cultivars tested and was dependent on the level of green leafhopper resistance in the cultivars and insecticide rate. Cultivar "IR28", with a high level of green leafhopper resistance, had a low pest population and a low incidence of RTV at all insecticide rates including the untreated control. Conversely, the cultivar with high susceptibility to green leafhopper (IR22) had a high level of RTV regardless of the rate of insecticide applied. However, the moderately resistant cultivar "IR36" showed a cumulative effect. Plant resistance is important in minimizing the extent of insecticide-induced brown planthopper resurgence. In field studies at International Rice Research Institute, Philippines, brown planthopper populations on rice plants treated with a resurgence-inducing insecticide only reached 10 insects per hill on a resistant rice cultivar, whereas populations on a susceptible cultivar reached 1100 per hill (40). Thus, when insecticide is needed to control a defoliator, the level of insecticide-induced brown planthopper resurgence can be reduced, or even eliminated, by the planting of a hopper-resistant cultivar.

13.7 INTEGRATION OF HOST PLANT RESISTANCE WITH OTHER METHODS

The integration of insect-resistant cultivars with cultural management can be a powerful tool in managing pests (41). This pest control involves two basic approaches; one is to make the environment less favorable to pests and the other is to favor the pests natural enemies. Early maturing cultivars and planting date are the usual cultural practices employed to evade rice insect attack. In field studies in the Philippines, brown planthopper predator ratios on early-maturing rice cultivars were significantly lower than those on later maturing cultivars because the insect was not able to complete as many generations on the early-maturing cultivars as on the long-duration cultivars (42). Similarly, the population of boll weevil and pink bollworm in cotton during winter season can be effectively reduced by adjusting synchrony of plant and herbivore phenologies through uniform planting, early maturing varieties, defoliation, and stalk destruction before the larvae are forced into diapauses by short days and cool nights (43). The vegetational diversity also plays a key role in reducing the pest damage from specialist insect because of the presence of confusing or masking chemical stimuli, physical barriers to movement, and other adverse environmental factors. Moreover, these factors reduce the insect survival (44). The vegetational diversity can also be increased by the way of polyculture, growing more than one crop in the same area. These areas are ecologically complex because interspecific and intraspecific plant competition occurs simultaneously with herbivores, insect predators, and insect parasitoids (45). The varietal mixture cultivation of different proportion of resistant and susceptible plants will also reduce virulent nature of insect and development of new biotypes. Thus, the combined effect of varietal mixture and natural enemies are likely to be effective in suppressing the pest populations. The trap crops reduce the pest damage on target crops by attracting the insects on it. Trap crops are offered at critical times of pest phenology. Such tactics allow the concentration of pest at the desired site, i.e., trap crop. In cotton/sesame intercrop trials 5% of sesame crop, was used as a trap crop to attract *Helicoverpa* spp from the main crop of cotton. Sesame is highly attractive to *Helicoverpa* spp and its parasitoid *Campoletis sonorensis* from seedling stage to senescence which ultimately parasitised large numbers of *Helicoverpa* larvae (46).

Nitrogenous fertilizer is a major component for production of high yields of modern rice cultivars. High plant nitrogen generally favors insect pest populations and is manifested in greater pest survival, increased feeding rate, increased fecundity, and faster growth (47). In a greenhouse study, nitrogenous fertilizer was shown to favor population growth of brown planthoppers on rice, even on a resistant cultivar. Thus, when high nitrogen levels are needed to maximize rice production, planting of resistant cultivars will minimize brown planthopper populations.

13.8 IMPACT OF TRADITIONAL PLANT BREEDING IN DEVELOPING RESISTANT VARIETIES

Conventional breeding programs for insect resistance often require 12 years or more of sustained, cooperative effort between entomologists and plant breeders to begin producing resistant varieties (25,48,49). The close relative sources of resistance are always easier and transfer of genes for resistance is genetically through conventional plant breeding to a locally adapted, commercially acceptable type (50,51). It is often difficult to determine which chemical or physical characteristics confer resistance to an insect, especially when comparisons of resistance and susceptibility are made between distantly related plants (52,53). Many characteristics, especially qualitative and quantitative differences in chemistry, may vary between two populations. When individual chemical factors are tested in isolation, their effects may be changed when separated from interactions in the plant or with the environment (54). Many teams have successfully bred insect-resistant varieties without identifying specific chemical or physical factors involved (55). Once the adapted plant is produced with useful resistance to insects combined with all the necessary traits of disease resistance, growth rate, yield quality, and adaptation to requirements of the crop production system, the material is released to commercial breeders. If the crop is ready to be multiplied and then used by farmers, the insect-resistant material is released as a variety. If the resistant material would be used as one parent in a hybrid, or if further breeding may be needed, the material is released as a parental line or as germplasm for its resistant source.

13.9   PLANT RESISTANCE AND ENVIRONMENTAL PERSPECTIVES

Plants may produce compounds that directly or indirectly impair their biological environment (56). These compounds fall within a broad category of compounds called allelochemicals, and are those exclusive of food that influence the growth, health, or behavior of other organisms, whether plant or animal (57). One reason for the interest in allelochemicals is their potential for use in alternative pest management system. Using plant-produced allelochemicals in agricultural and horticultural practices could minimize synthetic pesticide use, reduce associated potential for environmental contamination, and contribute to sustainable agricultural systems.

Plants have many ways of coping with their insect herbivores. Research has elucidated the mechanisms of direct and indirect plant defenses, and has provided the first proof of a protective function for indirect defenses in nature (58). Plants can respond actively to damage by herbivores and also its mode of defense that is directly aimed at the herbivore itself. Plants can emit volatiles that attract carnivores, i.e., the enemies of their enemies. Knowledge of the mechanisms underlying the induction of these herbivore-induced plant volatiles and of the responses of the carnivores is progressing rapidly. Inferences on the initial causes of evolution of herbivore-induced plant volatiles remain conjectural. However, once plant–carnivore interactions have evolved to the net benefit of both participants. This mutualism is expected to have evolutionary and ecological consequences for the three trophic levels: plant, herbivore, and natural enemy. When plant selection and foraging behavior of natural enemies is linked to plant fitness, this can influence different aspects of the plant defense strategy. The way carnivores perceive and process plant information may influence the evolution of the plant signal. Large transcriptional reorganization caused by herbivore differs from that elicited by mechanical wounding. Elicitors in herbivore oral secretions can account for herbivore-specific responses. Patterns of transcriptional changes point to the existence of central herbivore-activated regulators of metabolism. High or low molecular weight elicitors from the secretions of the herbivore stimulate the signal cascade between leaf damage and the volatile production. Besides compounds from the octadecanoid signaling pathway, several structurally non related amino acid conjugates such as the bacterial phytotoxin coronatine, the synthetic indanoyl isoleucine, or amino acid conjugates of linolenic acid likewise induce volatile biosynthesis. Minor changes in the amino acid moiety may result in different volatile profiles (sesqui- and diterpenoids), attributing to the amino acid substructure a specific role for the recognition and the selective induction (58). Signaling pathway will be important for the plant defense in the case of introduction of inhibitors together with the salivary secretion of herbivores into the leaf tissue. Recent studies shows partially overlapping signal transduction pathways controlling responses to wounding, insects, and pathogens. Chemical and behavioral assays show that plants release herbivore-specific volatiles and that parasitic wasps can distinguish between these emission patterns.

For example, *Nicotiana attenuata* has both direct (induced nicotine production) and indirect (induced release of mono- and sesquiterpenes) defenses induced by herbivore attack; the jasmonate cascade activates both, albeit in different tissues (roots and shoots, respectively) (59). The fact that both types of defenses are induced suggests that their benefits are conditional. Because inducing nicotine production can make 6% of a plant's nitrogen budget unavailable for seed production, it can cause a resource-based cost. Volatile production is likely to be less costly but could make plants more "apparent" to herbivores and thereby exert an ecological cost. Direct defenses could also have ecological costs if they are sequestered by specialist herbivores and used against their enemies. Herbivory by the nicotine-tolerant herbivore *Manduca sexta* dramatically amplifies the increase in jasmonates and the quantity of volatiles released, but decreases the nicotine response in comparison to mechanical simulations of the wounding that larval feeding causes. The apparent switching from nicotine production to the release of volatiles may reflect incompatibilities in the use of direct and indirect defenses with specialist herbivores (59).

Host plant quality is a key determinant of the fecundity of herbivorous insects. Components of host plant quality directly affect herbivore fecundity. The responses of insect herbivores to changes in host plant quality vary within and between feeding guilds. Host plant quality also affects insect reproductive strategies; egg size and quality, the allocation of resources to eggs, and the choice of oviposition sites may all be influenced by plant quality, as may egg or embryo resorption on poor-quality hosts. Many insect herbivores change the quality of their host plants, affecting both inter- and intraspecific interactions. Higher tropic level interactions, such as the performance of predators and parasitoids, may also be affected by host plant quality.

Habitat management, a form of conservation biological control, is an ecologically based approach aimed at favoring natural enemies and enhancing biological control in agricultural systems. The goal of habitat management is to create a suitable ecological infrastructure within the agricultural landscape to provide resources such as food for adult natural enemies, alternative prey or hosts, and shelter from adverse conditions (60). These resources must be integrated into the landscape in a way that is spatially and temporally favorable to natural enemies and practical for producers to implement (61).

13.10 VERTICAL AND HORIZONTAL RESISTANCE

Most resistance breeding to date has focused on methods that result in vertical resistance wherein resistance is based on a single gene. It has gene-for-gene relationship whereby each gene of resistance in the host has a matching gene of parasitic ability in the parasite. Qualitatively, the resistance is completely present or absent against the specific pest. This is contrary to horizontal resistance breeding, whereby resistance is based on several polygenes or minor genes. Quantitatively, horizontal resistance is exhibited in varying degrees, from minimal to maximal each with a small contribution to the resistance trait and shows no differential interaction when infested with different biotypes of insect. Horizontal resistance is moderate, does not exert a high selection pressure on the insect, and thus is more stable or durable. Vertical resistance is convenient because high levels of resistance can be achieved and the method is compatible with breeding schemes used for enhancing crop performance through control of major genes. However, its gene-for-gene nature can sometimes lead to its breakdown through the evolution of resistance breaking pest genotypes, as in the case of brown planthopper on rice. In general context, the single-technology solution promised by a high level of vertical resistance is not necessarily desirable if it brings the risk of resistance by the pest. This means that partial resistance, or other forms of resistance such as horizontal or polygenic resistance which is built on the quantitative effect of many genes, can be effective for sustainable agriculture. Unfortunately, the tradition of plant breeding and now biotechnology for resistance to pests favors vertical resistance, with its inherent risks (62–64). Suggested solutions to resistance problems involve more complex strategies of gene deployment. This includes mixed or intercropped populations of resistant and susceptible plants, or genetic methods to restrict expression of genes to certain parts of plant or certain times.

Horizontal or polygenic resistance is synonymous with tolerence. Parents with polygenic resistance are generally landraces with poor agronomic trait. In breeding program, such parents are crossed with agronomically superior parents. But not all the polygenes or quantitative trait loci are transferred and so the resistance is diluted. Conventionally, recurrent selection method is obviously used (65,66).

Hybrid is usually meant for higher yield or performance of F_1 hybrid in the commercial production. This method exploits the phenomenon of heterosis or hybrid vigor and involves the use of genetically diverse parental lines. The insect-resistant inbred lines are developed by wide hybridization and upon subsequent selfing. These lines are used as parents in hybrid breeding. If the insect-resistant gene is dominant, the hybrid variety is resistant. Many insect-resistant rice hybrids to green leafhopper and brown planthopper have been developed in China. Even though male sterile female parents are susceptible, the restorer parents developed at the IRRI (IR26, IR54 and IR9761-19-1) had dominant genes for resistance and so F_1 hybrids were resistant (67,68). Moreover, two dominant genes for insect resistance can be deployed in F_1 hybrid simultaneously by incorporating one resistant gene to female and the other to male parent. The F_1 hybrid has higher level of resistance due to combined effect of two dominant genes. Resistance to more than one biotype of the insect can also be built in the hybrid breeding. In addition, hybrid breeding also allows combination of genes for horizontal resistance from one parent and vertical resistance from the another parent. This could serve for long-term resistance for sustainability in interaction with the host pests.

The breeding strategies for durable resistance include the sequential release of insect-resistant crop varieties with major genes, e.g., IR36, IR38, IR42 for BPH (69), rotation of insect resistant crop varieties season to season with major genes (70), and development of multiline varieties. These strategies will provide long life to resistance of crop plants in interaction with insect pests as well as slow down their development of new biotypes.

One way of developing a higher level of resistance is gene pyramiding. Gene pyramiding is a powerful technique aimed at combining two or more major genes into the same variety that will provide a long term or a durability of resistance. Moreover, the development of new biotypes will be at a slower rate. The introduction of both cowpea trypsin inhibitor CpTI and pea lectin (P-Lec) genes into tobacco was achieved successfully. These plants were obtained by cross-breeding plants derived from the two primary transformed lines. The insecticidal effects of the two genes were additive, not synergistic, with insect biomass on the double expressors reduced by nearly 90% compared with those from control plants and 50% compared with those from plants expressing either CpTI or P-Lec alone. Also, leaf damage was the least on the double expressing plants (71). More recently, potato plants expressing both snowdrop lectin (GNA) and bean chitinase (BCH) have been obtained. When tested for enhanced aphid resistance, fecundity was reduced by approximately 95% compared with control plants (72).

13.11 MARKER AIDED SELECTION FOR INSECT-RESISTANT CROPS

Wild species and/or nondomesticated crop relatives possess many valuable genes for resistance to insects. Molecular markers are used to identify and track genes of interest (73). The most common molecular markers currently in use are restriction fragment length polymorphisms (RFLPs), random amplified polymorphic DNA (RAPD), amplified fragment length polymorphisms (AFLPs), and single sequence repeats (SSRs) or microsatellites (74). They can be used to select genes of interest while simultaneously exercising selection against unwanted genomic segments, thus reducing linkage drag and speedy cultivar development. Expressed sequence tags (ESTs) are used to rapidly identify expressed genes for insect resistance using near-isolines or insect-resistant and susceptible sibs (75). Thirty genes for insect resistance have been mapped in six crop species conferring resistance to species from Homoptera, Hemiptera, Diptera, Lepidoptera, and Coleoptera (74). The *Mi* gene from tomato was identified as a dominant gene for resistance to a root-knot nematode, *Meloidogyne incognita*. The *Mi* gene is the first gene cloned for insect resistance from tomato (76). Once an insect resistance gene is cloned, a large number of accessions can be screened to search for additional alleles at that locus, some of which may confer a greater level of resistance than the initial cloned version. In cases in which resistance is attributed to a known metabolic pathway, another strategy would be to search for variation in genes of that pathway to identify loci and alleles with the best prospects for maximizing resistance levels while minimizing negative consequences.

13.12   GENETIC ENGINEERING

The important benefits derived by growing transgenic crops included significant agronomic, environmental, health, and economic advantages. More than 3.5 million farmers grew transgenic crops in 2000, and the number is expected to grow to 5 million farmers or more. The total area covered by transgenic crops (45 million hectares) is also expected to show phenomenal growth. The two important traits used in first-generation transgenic crops until 2000 are herbicide tolerance and insect resistance, independently or as stacked genes on more than 99% of the 45 million hectares of transgenic crops (Table 13.1). There are eight industrial and five

TABLE 13.1   Global Area of Transgenic Crops in 2000

Category	Area (million hectares)	%
By country		
United States	30.3	68
Argentina	10.0	23
Canada	3.0	7
China	0.5	1
South Africa	0.2	< 1
Australia	0.2	< 1
Romania	< 0.1	< 1
Mexico	< 0.1	< 1
Bulgaria	< 0.1	< 1
Spain	< 0.1	< 1
Germany	< 0.1	< 1
France	< 0.1	< 1
Uruguay	< 0.1	< 1
By Crops		
Soybean	25.8	58
Maize	10.3	23
Cotton	5.3	12
Canola	2.8	7
Potato	< 0.1	< 1
Squash	< 0.1	< 1
Papaya	< 0.1	< 1
By traits		
Herbicide tolerance	32.7	74
Insect resistance (Bt)	8.3	19
Bt/herbicide tolerance	3.2	7
Virus resistance/others	< 0.1	< 1

Source: C James. Global review of commercialized transgenic crops: 2000. ISAAA Briefs No. 21: Preview ISAAA, Ithaca, NY, p 8.

developing countries that grew transgenic crops in 2000 (Table 13.1). Of the top four countries that grew 99% of the global transgenic crop area the United States 68%, Argentina 23%, Canada 7%, and China 1%. The other nine countries grew the remaining 1%. The transgenic soybean occupied 58% of the global area of transgenic crops (herbicide tolerant) followed by transgenic corn, cotton, and canola. The insect-resistant maize and cotton occupied 26% of the cultivated area in 2000 (77). Transgenic crops offered a simplified pest management system, which resulted in the increasing overall efficiency of crop production.

The primary benefit of the Bt and other transgenic crops is that they provide producers with improved methods of pest control. Bt toxins have been transferred and expressed in at least 26 plant species. However, the level of resistance they confer will depend on whether native bacterial or truncated, codon-optimized genes have been used. The transformed plants express Bt toxin at levels sufficient to cause high mortality of target pests in the field. An alternative strategy attempted was to integrate the native Cry 1Ac gene into the chloroplast genome of tobacco plants resulting in high expression levels (78). Twenty-eight transgenic maize hybrids containing Bt endotoxin and several of their non-Bt isolines were evaluated for controlling the southwestern corn borer, *Diatraea grandiosella* Dyar, the European corn borer, *Ostrinia nubilalis* (Hubner), and the corn earworm, *Helicoverpa zea* (Boddie) in the shank and ear (79). Hybrids provided good control of corn borer in shanks and ears and consistent control of corn earworm larvae on kernels.

Farmers are better able to control the insect pest with Bt crops, under some circumstances, than with conventional insecticide sprays, and this effective control results in higher yields (80). An 80% reduction in insecticide use on Bt cotton was also reported in China (81). Thus, by using transgenic crops, the cost of cultivation could be reduced and the yield could be increased.

Maintaining the biodiversity of insect species is an important component in IPM. It has been reported that populations of predatory bugs (*Orius* and *Geocoris* species), spiders, and ants are all significantly higher in Bt cotton fields compared with conventional cotton fields treated with insecticides (82). Pray et al. (83) investigated the enhancement of insects biodiversity by the adoption of Bt cotton in China. Moreover, lower insecticide use in Bt cotton fields is associated with predator populations that are 24% higher than in fields of conventional cotton (84). The report by Losey et al. (85) led to premature speculation and extrapolations that caterpillars of monarch butterflies were being poisoned and killed by pollen from commercial Bt maize. On the contrary, the recent publications (86,87) confirmed that the Bt maize planted in the United States is not a threat to monarch butterflies feeding on milkweed on which corn pollen is deposited.

A group of 22 corn entomologists and ecologists reported that the evidence to date supported the appropriate use of Bt corn as one component in the economically and ecologically sound management of lepidopteran pests. The impact of new technologies at the field level will be premature if based on extrapolation from laboratory experiments that was evident from the monarch butterfly experience (88). The cultivation of Bt corn offered significant real benefits to ecosystems and human health, including those from a reduction in use of more broad-spectrum foliar insecticides.

Transgenic technology offers the potential to reduce the occurrence of mycotoxins in food crops. This could be achieved by using transgenic crops to control the broad spectrum of insect pests that cause damage and thus prelude toxigenic fungi from invading the plants (89). There is a reduction in use of pesticides and pesticide poisonings by the introduction of transgenic crops (83). Several studies have confirmed that farmers planting transgenic crops are more efficient in managing their weed and insect pests. The use of transgenic crops result in more sustainable and efficient crop management practices; conservation of natural resources; more effective control of insect pests and weeds; lesser use of conventional pesticides; improved ground and surface water quality with less pesticide residues; improved pest control combined with improved yields all contribute to a greater economic advantage to farmers for a more sustainable farming systems (77).

13.13 PERSPECTIVES

Millions of farmers around the world, both industrial and developing countries, continue to increase their cultivation of transgenic crops due to the significant multiple benefits they offer. Governments, supported by the global scientific and international development community, must ensure continued safe and effective testing and introduction of transgenic crops and implement regulatory programs that inspire public confidence. The public should be well informed about the impact of the technology and global food security. Biotechnology and transgenic crops are the potential areas that will contribute to global food security and the alleviation of hunger in the third world. The cost-effective, environmentally safe, and sustainable way to ensure global food security have to be achieved by the plant breeders using both conventional and biotechnology tools.

Biotechnological innovations that facilitate broad-based quantitative crop resistance to pests will be valuable to the sustainable agriculture in future. Biotechnology has considerable potential applications in improving mass production technologies for natural enemies of pests, and for improving diagnostic systems that allow scientists to recognize desirable plant genes and natural enemies, and that allow farmers to recognize potential pest problems before damage occurs. In its next generation, biotechnology stands to contribute greatly to sustainable pest management.

Public concerns have been voiced about the safety of genetically engineered food and the possible adverse impact of transgenic crops on the environment. An example of the former concern is that the use of antibiotic marker genes used in the genetic engineering technology might lead to antibiotic resistance in human pathogens. The current trend is either not to use antibiotic resistance marker genes or to deliver the marker gene in a different locus so that it can later be bred away. The genes coding for 5-enolpyruvylshikimate-3-phosphate synthase (EPSPS) and phosphinothricin-N-acetyltransferase (PAT) have been used as selectable markers in maize (90). Syngenta announced the development of Positech, a new marker system that provides a selection system based on a marker gene that enables the plant cells to use mannose to grow and form new plants (91). It is an alternative to antibiotic resistance genes as markers in future transgenic crops. Positech has been used successfully in cassava.

A further concern is the possibility of unexpected, harmful effects from introduced genes due to the random nature of the transformation process. There is a possibility of gene flow to weedy or wild relatives of crop plants. Transgenes can reach weed populations in viable pollen and, since most crops have weedy/wild relatives, the escape of transgenes cannot be ruled out. However, risk assessment is available. The industry is aware of this, and good agricultural procedures have been designed accordingly to reduce the risks (92).

Concerns have been raised about the potential effects of transgenic introductions on the genetic diversity of crop landraces and wild relatives in areas of crop origin and diversification, as this diversity is considered essential for global food security. Direct effects on nontarget species and the possibility of unintentionally transferring traits of ecological relevance onto landraces and wild relatives have also been sources of concern. Quist and Chapela (93) reported the presence of introgressed transgenic DNA constructs in native maize landraces grown in remote mountains in Oaxaco, Mexico, part of the Meso-American center of origin and diversification of this crop. However, serious concerns about the finding were raised almost immediately when a Mexican group found it could not repeat the experiment, and several researchers questioned the techniques used and the interpretation of the results. Christou (94) critically analyzed the data and commented that the data presented in the published article are more artifacts resulting from poor experimental design and practices. The journal *Nature* admitted that it was wrong in publishing the flawed research and withdrew its approval from the study, a move thought to be the first in the magazine's 133-year history.

The recent controversy in the United Kingdom over the genetically manipulated, lectin-containing potatoes illustrates the need for new processing technologies to be rigorously tested for potential allergenic, toxic, and antimetabolic effects (95). The biosafety assessment committees have to look into these aspects and utmost care must be taken before releasing any transgenic crops for commercial use.

The promoter CaMV 35S has been used in the majority of insect-resistant transgenic plants. It produces continuous gene expression in most tissues of the plant. As a result of continuous expression there is a possibility of yield loss due to energy costs of transgenic crops, and there is also an increasing risk of the pests' developing resistance (96). Considerable research effort is now directed to concentrating expression in those parts of the plant attacked by insects. The use of phloem-specific or seed-specific promoters has been attempted. The new plant vector system MAT used it as a selectable marker gene, a gene that has the capacity to eliminate itself from the plant genome overtime (90). Documenting the benefits, as well as the constraints, associated with transgenic crops is an important step in the sharing of information with the society.

In addition to Bt toxin, lectins, proteinase inhibitors, amylase inhibitors (97), and other natural plant defenses can be utilized and manipulated to impart insect resistance upon transgenic crops. To produce transgenic crops into an effective weapon in pest control it is important to deploy genes with different modes of action in the same plant. Serine proteinase inhibitors or tannic acid can enhance the activity of Bt in transgenic plants. Cornu et al. (98) reported that transgenic poplars expressing proteinase inhibitor and Cry IIIA genes exhibited reduced larval growth, altered development, and increased mortality as compared to the control.

For a better understanding of the plant defenses against pest, a thorough knowledge of plant secondary metabolic pathway and their regulation is essential. Although a few of the secondary metabolites and the secondary metabolic pathway have been exploited for genetic engineering towards enhancing insect resistance, there are many more metabolites/pathways to be attempted. Plants defense responses often are customerized for certain, interactive, multitrophic situations. Genetic engineering is a powerful tool and if its deployment is to be sustainable must be used in conjunction with a solid appreciation of multitrophic interactions in ways that anticipate countermoves within the systems (99). The next generations of transgenic insect-resistant crops to sustain agricultural production will need to employ gene pyramiding strategies based on a thorough examination of the interactions that occur between insecticidal proteins/metabolites, the plant and its products, and the characteristics and habits of insect pests.

A total system approach is essential as the guiding premise of pest management.

REFERENCES

1. United Nations Population Division. World Population Prospects: The 1996 Revisions. 1997, UN, New York.
2. United Nations Population Division. World Population Prospects: The 1998 Revisions. 1998, Electronic version, UN, New York.
3. World Bank. World Development Indicators. Washington, DC, 1997.
4. Pinstrup-Andersen, P.; Pandya-Lorch, R.; Rosegrant, M.W. The World Food Situation: Recent Developments, Emerging Issues, and Long-Term Prospects. 2020 Vision Food Policy Report. International Food Policy Research Institute, Washington, DC, 1997.
5. Rosegrant, M.W.; Agcaoili-Sombilla, M.; Perez, N.D. Global Food Projections to 2020: Implications for Investment. 2020 Vision for Food, Agriculture, and the Environment Discussion Paper No. 5. International Food Policy Research Institute, Washington, DC, 1995.
6. Yudelman, M.; Ratta, A.; Nygaard, D. Pest Management and Food Production: Looking to the Future. 2020 Vision for Food, Agriculture, and the Environment Discussion Paper No. 25. International Food Policy Research Institute, Washington, DC, 1998.
7. Pinstrup-Andersen, P.; Cohen, M.J. The world food outlook and the role of crop protection. Prepared for presentation at the 65th Annual Meeting of the American Crop Protection Association, White Sulphur Springs, West Virginia, 27–29 September 1998.
8. FAO (Food and Agriculture Organization of the United Nations). Data for 1961–1996: FAOSTAT database, http://apps.fao.org, 1997.
9. Sharma, H.C.; Sharma, K.K.; Seetharama, N.; Ortiz, R. Prospects for using transgenic resistance to insects in crop improvement. Elec. J. Biotechnol. May 24, **2000**, *3*(2).
10. The WHO Recommended Classification of Pesticides by Hazard, and Guidelines to Classification 1998–1999, WHO/PCS/98.21, International Programme on Chemical Safety, Geneva, 1998.
11. Current pesticide spectrum, and major concerns. AGROW World Crop Protection News, No 301, 1998, p 14.
12. Pinstrup-Andersen, P.; Pandya-Lorch, R.; Rosegrant, M.W. World Food Prospects: Critical Issues for the Early Twenty-first Century, 2020 Vision Food Policy Report, International Food Policy Research Institute (IFPRI), Washington, DC, 1999.
13. Lyson, T.A. Advanced agricultural biotechnologies and sustainable agriculture. Trends Biotechnol. **2002**, *20*, 193–196.
14. Hill, S.B. Pest control in sustainable agriculture. Proceedings of the Entomological Society of Ontario. **1990**, *121*, 5–12.

15. Tingey, W.M.; Singh, S.R. Environmental factors influencing the magnitude and expression of resistance. In *Breeding Plant Resistance to Insects*; Maxwell, F.G., Jennings, P.R., Eds.; John Wiley and Sons: New York, 1980; 87–113.

16. Hare, J.D. Manipulation of host suitability for herbivore pest management. In *Variable Plants and Herbivores in Natural and Managed Systems*; Denno, R.F., McClure, M.S., Eds.; Academic Press: New York, 1983; 655–680.

17. Hilder, V.A.; Gatehouse, A.M.R.; Sheerman, S.E.; Barker, R.F.; Boulter, D. A novel mechanism of insect resistance engineered into tobacco. Nature **1987**, *330*, 160–163.

18. Hedin, P.A. Bioregulator-induced changes in allelochemicals and their effects on plant resistance to pests. CRC Crit. Rev. Plant Sci. **1990**, *9*, 371–379.

19. Wink, M. Functions of plant secondary metabolites and their exploitation in biotechnology. Annu. Plant Rev. 3: CRC Press, Boca Raton, 2000.

20. Risch, S.J.; Andow, D.A.; Altieri, M.A. Agroecosystem diversity and pest control: data, tentative conclusions, and new research directions. Environ. Entomol. **1983**, *12*, 625–629.

21. Baliddawa, C.W. Plant species diversity and crop pest control: an analytical review. Insect Sci. Appl. **1985**, *16*, 479–488.

22. Price, P.W.; Bouton, C.E.; Gross, P.; McPheron, B.A.; Tompson, J.N.; Weis, A.E. Interactions among three tropic levels: influence of plants on interactions between insect herbivores and natural enemies. Annu. Rev. Ecol. Syst. **1980**, *11*, 41–65.

23. Boethel, D.J.; Eikenbary, R.D. Interactions of Plant Resistance and Parasitoids and Predators of Insects. Ellis Harwood, Chichester, 1986.

24. Slansky, F. Insect nutritional ecology as a basis for studying host plant resistance. Flo. Entomol. **1990**, *73*, 359–378.

25. Gould, F. Genetics of plant-herbivore system: interactions between applied and basic study. In *Variable Plants and Herbivores in Natural and Managed Systems*; Denno, R.F.; McClure, M.S., Eds.; Academic Press: New York, 1983; 599–654.

26. Smith, C.M. Plant Resistance to Insects: A Fundamental Approach. John Wiley and Sons: New York, 1989.

27. Eigenbrode, S.D.; Trumble, J.T. Plant resistance to insects in integrated pest management in vegetables. J. Agrl. Entomol. **1994**, *11*, 201–224.

28. Heinrichs, E.A. Perspectives and directions for the continued development of insect-resistant rice varieties. Agriculture Ecosyst. Environ. **1986**, *18*, 9–36.

29. Flanders, K.L.; Hawkes, J.G.; Radcliffe, E.B.; Lauer, F.I. Insect resistance in potatoes: sources, evolutionary relationships, morphological and chemical defenses, and ecogeographical associations. Euphytica **1992**, *61*, 83–111.

30. Farrar, R.R.; Kennedy, G.G. Sources of insect and mite resistance in tomato in *Lycopersicon*. In *Genetic Improvement of Tomato*; Kalloo, G., Ed.; Springer-Verlag: Berlin, 1992; 121–142.

31. Reid, W.V.; Miller, K.R. Keeping Options Alive: The Scientific Basis for Conserving Biodiversity. World Resources Institute, Washington, DC, 1989.

32. Fowler, C.; Mooney, P. *Shattering Food Politics and the Loss of Genetic Diversity*. University of Arizona Press: Tucson, 1990; 1–278.
33. Frankel, O.H.; Brown, A.H.D. Plant genetic resources today: a critical appraisal. In *Crop Genetic Resources: Conservation and Evaluation*. Holden, J.H.W., Williams, J.T., Eds.; Allen and Unwin: London, 1984; 249–257.
34. Smith, C.M. Integration of rice insect control strategies and tactics, In *Biology and Management of Rice Insects*; Heinrichs, E.A., Ed.; Wiley Eastern: New Delhi, 1994; 517–548.
35. Ooi, P.A.C.; Shepard, B.M. Predators and parasitoids of rice insect pests. In *Biology and Management of Rice Insects*; Heinrichs, E.A., Ed.; Wiley Eastern: New Delhi, 1994.
36. Mc Rombach, Roberts, D.W.; Aguda, R.M. Biological control of the brown planthopper, *Nilaparvata lugens* (Homoptera, Delphacidae) with dry mycelium applications of *Metarhizium anisopliae* (Deuteromycotina: Hyphrmycetes). Phillipp. Entomol. **1986**, *6*, 613–619.
37. Way, M.J.; Heong, K.L. The role of biodiversity in the dynamics and management of insect pests of tropical irrigated rice—a review. Bull. Entomol. Res. **1994**, *84*, 567–587.
38. Kartohardjono, A.; Heinrichs, E.A. Populations of the brown planthopper, *Nilaparvata lugens* (Homoptera: Delphacidae), and its predators on rice cultivars with different levels of resistance. Environ. Entomol. **1984**, *13*, 359–365.
39. Heinrichs, E.A.; Rapusas, H.R.; Aquino, G.B.; Palis, F. Integration of host plant resistance and insecticides in the control of *Nephotettix virescens* (Homoptera: Cicadellidae), a vector of rice tungro virus. J. Econ. Entomol. **1986**, *79*, 437–443.
40. Aquino, G.B.; Heinrichs, E.A. Brown planthopper populations on resistant cultivar treated with resurgence causing insecticide. Int. Rice Res. Newslett. **1979**, *4*(5), 12.
41. Maxwell, F.G. Use of insect resistant plants in integrated pest management programmes. FAO Plant Prot. Bull. **1991**, *39*, 139–146.
42. Heinrichs, E.A.; Adesina, A.A. The contribution of multiple pest resistance to tropical crop production. In *Proc. Nelson Memorial Symposium on Host Plant Resistance*; Wiseman, B.R., Webster, J.A., Eds.; Thomas Say Pubs, Entomological Society of America, Lanham, 1994.
43. Adkisson, P.L.; Gaines, J.C. Pink bollworm control as related to the total cotton insect control program of Central Texas. Tex Agric. Exp. Stn. Misc. Publ. 1960.
44. Baliddawa, C.W. Plant species diversity and crop pest control. Insect. Sci. Appl. **1985**, *6*, 479–487.
45. Van Emden, H.F. The role of uncultivated land in the biology of crop pests and beneficial insects. Sci. Hort. **1965**, *17*, 121–136.
46. Pair, S.D.; Laster, M.L.; Martin, D.F. Parasitoids of *Heliothis* spp (Lepidopterea: Noctuidae) larvae in Mississippi associated with sesame interplanting in cotton, 1971–1974: implications of host–habitat interaction. Environ. Entomol. **1982**, *11*, 509–512.

47. Heinrichs, E.A.; Medrano, F.G. Influence of nitrogen fertilizer on the population development of brown planthopper. Int. Rice Res. Newslett. **1985**, *10*, 20–21.

48. Roberts, J.J.; Foster, J.E.; Patterson, F.L. The Purdue-USDA small grain improvement programs: a model of research productivity. J. Prod. Agric. **1988**, *1*, 239–241.

49. Stoner, K.A. Bibliography of plant resistance to arthropods in vegetables, 1977–1991. Phytoparasitica **1992**, *20*, 125–180.

50. Stoner, A.K. Breeding for insect resistance in vegetables. Hort. Sci. **1970**, *5*, 76–79.

51. Harian, J.R.; Starks, K.J. Germplasm resources and needs. In *Breeding Plants Resistance to Insects*; Maxwell, E.G., Jennings, P.R., Eds.; John Wiley and Sons: New York, 1980; 254–273.

52. Kennedy, G.G.; Nienhuis, J.; Helentjaris, T. Mechanisms of arthropod resistance in tomatoes. In *Tomato Biotechnology*; Nevin, D.J., Jones, R.A., Eds.; Alan R. Liss: New York, 1987; 145–154.

53. Flanders, K.L.; Hawkes, J.G.; Radeliffe, E.B.; Lauer, F.I. Insect resistance in potatoes: sources, evolutionary relationships, morphological and chemical defenses and ecogeographical associations. Euphytica **1992**, *61*, 83–111.

54. Wolfson, J.L.; Murdock, L.L. Growth of *Manduca sexta* on wounded tomato plants: role of induced proteinase inhibitors. Entomol. Exp. Appl. **1990**, *54*, 257–264.

55. Heinrichs, E.A. Perspectives and directions for the continued development of insect-resistant rice varieties. Agriculture Ecosyst. Environ. **1986**, *18*, 9–36.

56. Brown, D.; Morra, M.J. Control of soil-borne plant pests using glucosinolate containing plants. Adv. Agron. **1997**, *61*, 167–231.

57. Whittaker, R.H.; Feeney, P.P. Alleochemics: chemical interaction between species. Science **1971**, *171*, 757–770.

58. Baldwin, I.T.; Halitschke, R.; Kessler, U.; Schittko, A. Merging molecular and ecological approaches in plant–insect interactions. Curr. Opin. Plant Biol. **2001**, *4*, 351–358.

59. Schittko, U.; Hermsmeier, D.; Baldwin, I.T. Molecular interactions between the specialist herbivore *Manduca sexta* (Lepidoptera, Sphingidae) and its natural host *Nicotiana attenuata*. II. Accumulation of plant mRNAs in response to insect-derived cues. Plant Physiol. **2001**, *125*, 683–700.

60. Landis, D.A.; Wratten, S.D.; Gurr, G.M. Habitat management to conserve natural enemies of arthropod pests in agriculture. Annu. Rev. Entomol. **2000**, *45*, 175–201.

61. Landis, D.; Marino, P. Landscape structure and extra field process: impact on management of pest and beneficials. In *Handbook of Pest Management*; Rubeson, J., Ed.; Marcel Dekker: New York, 1999; 79–106.

62. Waage, J.K. Integrated pest management and biotechnology: an analysis of their potential for integration. In *Biotechnology and Integrated Pest Management*; Persley, G.J., Ed.; CAB International, Oxon: UK, 1996.

63. Waage, J.K. Biopesticides at the crossroads: IPM products or chemical clones? In *Microbial Insecticides: Novelty or Necessity?* BCPC Symposium Proceedings No. 68, Bracknell, 1997; 41–50 pp.
64. Thomas, M.B.; Waage, J.K. Integrating Biological Control and Host Plant Resistance Breeding: A Scientific and Literature Review. Technical Centre for Agricultural and Rural Cooperation of the European Union (CTA), Waeningen, 1996.
65. Allard, R.W. Principles of Plant Breeding. John Wiley and Sons: New York, 1960.
66. Khush, G.S.; Brar, D.S. Genetics of resistance to insects in crop plants. Adv. Agron. **1991**, *45*, 223–274.
67. Lin, S.C.; Yuan, L.P. Hybrid rice breeding in China. In *Innovative Approaches to Rice Breeding*. International Rice Research Institute, Manila, Philippines, 1980; 35–51.
68. Yuan, L.P.; Virmani, S.S.; Khush, G.S. Wei You 64: an early duration hybrid for China. Int. Rice Res. Newslett. **1985**, *10*, 11–12.
69. Gallun, R.L.; Khush, G.S. Genetic factors affecting expression and stability of resistance. In *Breeding Plants Resistance to Insects*; Maxwell, F.G., Jennings, P.R., Eds.; John Wiley and Sons: New York, 1980; 63–85.
70. Manwan, S.; Sama, I.; Rizvi, S.A. Management strategy to control Tungro virus in Indonesia. In *Proceedings of the Workshop on rice Tungro Virus.* Ministry of Agriculture, Jakarta, Indonesia, 1977.
71. Boulter, D.; Edwards, G.A.; Gatehouse, A.M.R.; Gatehouse, J.A.; Hilder, V.A. Additive protective effectives of incorporating two different higher plants derived insect resistance genes in transgenic tobacco plants. Crop Prot. **1990**, *9*, 351–354.
72. Down, R.E.; Gatehouse, A.M.R.; Hamilton, W.D.O.; Gatehouse, J.A. Snowdrop lectin inhibits development and decreases fecundity of the glasshouse potato aphid *Audacorthum solani* when administered in vitro and via transgenic plants both in laboratory and glass house trials. J. Insect. Physiol. **1996**, *42*, 1035–1045.
73. Joshi, S.P.; Ranjekar, P.K.; Gupta, V.S. Molecular markers in plant genome analysis. Curr. Sci. **1999**, *77*, 230–240.
74. Yenco, G.C.; Cohen, M.B.; Byrne, P.F. Applications of tagging and mapping insect resistance loci in plants. Annu. Rev. Entomol. **2000**, *45*, 393–422.
75. Hannappel, U.; Balzer, H.J.; Ganal, M.W. Direct isolation of cDNA sequences form specific chromosomal regions of the tomato genome by the differential display technique. Mol. Gen. Genet. **1995**, *249*, 19–24.
76. Kaloshian, I.; Yaghoobi, J.; Liharska, T.; Hontelez, J.; Hanson, D. Genetic and physical localization of the root-knot nematode resistance locus Mi in tomato. Mol. Gen. Genet. **1998**, *257*, 376–385.
77. James, C. Global review of commercialized transgenic crops: 2000. ISAAA Briefs No. 23, 2001.
78. McBride, K.E.; Svav, Z.; Schaaf, D.J.; Hogan, P.S.; Stalker, D.M.; Maliga, P. Application of a chimere of an insecticidal protein in tobacco. Biotechnology **1995**, *13*, 362–365.

79. Archer, T.L.; Patrick, C.; Schuster.; Cronholm, G.; Bynum, E.D.; Morrison W.P. Jr. Ear and shank damage by corn borers and corn earworms to four events of *Bacillus thuringiensis* transgenic maize. Crop Prot. **2001**, *20*, 139–144.
80. Carpenter, J.E.; Gianessi, L.P. Agricultural Biotechnology Updated Benefit Estimates. National Center for Food and Agricultural Policy, Washington, DC, 2001.
81. Pray, C.E.; Huang, J.; Ma, D.; Quiao, F. Impact of Bt cotton in China. Conference on the Economics of Agricultural Biotechnology, Ravello, Italy, 2000.
82. Head, G.; Freeman, B.; Moar, W.; Ruberson, J. Turnispeed, S. Natural enemy abundance in commercial bollgard and conventional cotton fields. In Proceedings of the Beltwide Cotton Conference. National Cotton Council, Memphis, TN, 2001, Jan. 9–13.
83. Pray, C.E.; Huang, J.; Ma, D.; Qiao, F. Impact of Bt cotton in China. World Dev. **2001**, *29*, 813–825.
84. Xia, J.Y.; Cui-Jin, J.; Ma, L.H.; Dong, S.L.; Cui, X.F. The role of transgenic Bt cotton in integrated insect pest management. Acta. Gossipii Sinica **1999**, *11*, 57–64.
85. Losey, J.E.; Rayor, L.S.; Carter, M.E. Transgenic pollen harms monarch larvae. Nature **1999**, *399*, 214.
86. Hellmich, R.L.; Siegried, B.D.; Sears, M.K.; Stanley-Horn, D.E.; Daniels, M.J.; Mattila, H.R.; Spencer, T.; Bidne, K.G.; Lewis, L.C. Monarch larvae sensitivity to *Bacillus thuringiensis* purified proteins and pollen. Proc. Natl. Acad. Sci. USA. **2001**, *98*, 11925–11930.
87. Sears, M.K.; Hellmich, R.L.; Stanley-Horn, D.E.; Oberhauser, K.S.; Pleasants, J.M.; Mattila, H.R.; Siegfried, B.D.; Dively, G.P. Impact of Bt corn pollen on monarch butterfly populations: a risk assessment. Proc. Natl. Acad. Sci. USA. 2001, 98, 11937–19942.
88. Shelton, A.M.; Seats, M.K. The monarch butterfly controversy: scientific interpretations of a phenomenon. Plant J. **2001**, *27*, 483–488.
89. Pietri, A.; Piva, G. Occurrence and control of mycotoxins in maize grown in Italy. In *Proceedings of the 6th International Feed Conference*; Piva, G., Masoero, F., Eds.; Food Safety: Current Situation and Perspectives in the European Community. Piacenza: Italy, 2000; 226–236.
90. Ebinuma, H.; Sugita, K.; Matsunago, E.; Yamakado, M. Selection of marker-free transgenic plants using the isopentenyltransferase gene. Proc. Natl. Acad. Sci. USA **1997**, *94*, 2117–2121.
91. Syngenta Foundation. Launching of the foundation. WWW.syngenta.com, 2001.
92. Shelton, A.M.; Zhao, J.Z.; Roush, R.T. Economic, ecological, food safety and social consequences of deployment of Bt transgenic plants. Annu. Rev. Entomol. **2002**, *47*, 845–881.
93. Quist, D.; Chapela, I.H. Transgenic DNA introgressed into traditional maize landraces in Oaxaca, Mexico. Nature **2001**, *414*, 541–543.

94. Christou, P. No credible scientific evidence is presented to support claims that transgenic DNA was introgressed into traditional maize landraces in Oaxaco, Mexico. Transgenic Res. 11: iii–v, 2002.

95. Gillard, M.S.; Flynn, L.; Rowell, A. Food scandal exposed. The Guardian, 12/2/99:1, 1999.

96. Schuler, T.H.; Poppy, G.M.; Kerry, B.R.; Denholm, I. Insect-resistant transgenic plants. Trends Biotechnol. **1998**, *16*, 168–174.

97. Franco, O.L.; Rigden, D.J.; Melo, F.R.; Grossi-De-Sa, M.F. Plant α-amylase inhibitors and their interaction with insect α-amylases. Eur. J. Biochem. **2002**, *269*, 397–412.

98. Cornu, D.; Leple, J.C.; Bonade-Bottino, M.; Ross, A.; Augustin, S.; Delplanque, A.; Jouanin, L.; Pllate, G.; Ahuja, M.R. Expression of a proteinase inhibitor and a *Bacillus thuringiensis* delta endotoxin in transgenic poplars. In *Somatic Cell Genetics and Molecular Genetics of Trees*; Boerjan, W., Neale, D.B. Eds., Kluwer Academic: Dordrecht, 1996; 131–136.

99. Lewis, W.J.; Van Lenteren, J.C.; Phatak, S.C.; Tumlinson, J.H. A total system approach to sustainable pest management. Proc. Natl. Acad. Sci. USA **1997**, *94*, 12243–12248.

Index